Lecture Notes in Computer Science 5124

Lecture Notes in Computer Science 5324

Commenced Publication in 1973
Founding and Former Series Editors:
Gerhard Goos, Juris Hartmanis, and Jan van Leeuwen

Joachim Gudmundsson (Ed.)

Algorithm Theory – SWAT 2008

11th Scandinavian Workshop on Algorithm Theory
Gothenburg, Sweden, July 2-4, 2008
Proceedings

Volume Editor

Joachim Gudmundsson
NICTA
Locked Bag 9013
Alexandria NSW 1435, Australia
E-mail: joachim.gudmundsson@nicta.com.au

Library of Congress Control Number: Applied for

CR Subject Classification (1998): F.2, E.1, G.2, I.3.5, C.2

LNCS Sublibrary: SL 1 – Theoretical Computer Science and General Issues

ISSN 0302-9743
ISBN-10 3-540-69900-7 Springer Berlin Heidelberg New York
ISBN-13 978-3-540-69900-2 Springer Berlin Heidelberg New York

Springer is a part of Springer Science+Business Media

springer.com

Printed in Germany

Typesetting: Camera-ready by author, data conversion by Scientific Publishing Services, Chennai, India
Printed on acid-free paper SPIN: 12321810 06/3180 5 4 3 2 1 0

Preface

The Scandinavian Workshop on Algorithm Theory (SWAT) is a biennial international conference intended as a forum for researchers in the area of design and analysis of algorithms and data structures. The first SWAT workshop was held in Halmstad, Sweden, in 1988. Since then it has been held biennially, rotating between the five Nordic countries – Denmark, Finland, Iceland, Norway and Sweden, with the exception of 2006 when it was in Riga. Earlier SWATs were held in Humlebæk, Denmark (2004), Turku, Finland (2002), Bergen, Norway (2000), Stockholm, Sweden (1998), Reykjavik, Iceland (1996), Århus, Denmark (1994), Helsinki, Finland (1992), Bergen, Norway (1990) and Halmstad, Sweden (1988).

This volume contains the contributed papers presented at the 11th Scandinavian Workshop on Algorithm Theory (SWAT 2008), held in Gothenburg, Sweden, July 2–4, 2008. In addition, the volume also includes abstracts of an invited talk by Michael Mitzenmacher on "A Survey of Results for Deletion Channels and Related Synchronization Channels" and by Vijay V. Vazirani on "Nash Bargaining via Flexible Budget Markets."

Papers were solicited for original research on algorithms and data structures in all areas, including but not limited to: approximation algorithms, computational biology, computational geometry, distributed algorithms, external-memory algorithms, graph algorithms, online algorithms, optimization algorithms, parallel algorithms, randomized algorithms, string algorithms and algorithmic game theory. The 36 contributed papers were chosen from 111 submissions. Revised and expanded versions of selected papers will be considered for publication in a special issue of *Algorithmica.* The best paper award was given to Yossi Azar, Uriel Feige and Daniel Glasner for their paper "A Preemptive Algorithm for Maximizing Disjoint Paths on Trees."

Each paper was reviewed by at least three referees, and evaluated on the quality, originality and relevance to the symposium. The challenging task of selecting the papers for presentation was performed by the members of our Program Committee and external reviewers. All of them deserve special thanks for their hard work.

We thank all of those who submitted papers for contributing to an interesting and diverse conference.

April 2008 Joachim Gudmundsson

Preface

Organization

Program Chair

Joachim Gudmundsson, NICTA, Australia

Program Committee

Mark de Berg, TU Eindhoven, The Netherlands
Michael Elkin, Ben-Gurion University, Israel
Leah Epstein, University of Haifa, Israel
Fedor Fomin, University of Bergen, Norway
Leszek Gąsieniec, University of Liverpool, UK
Anupam Gupta, Carnegie Mellon University, USA
Thore Husfeldt, Lund University, Sweden
Johan Håstad, Royal Institute of Technology, Sweden
Nicole Immorlica, Centrum voor Wiskunde en Informatica, The Netherlands
Mikko Koivisto, HIIT, University of Helsinki, Finland
Peter Bro Miltersen, University of Aarhus, Denmark
Pat Morin, Carleton University, Canada
Rasmus Pagh, IT University of Copenhagen, Denmark
Marina Papatriantafilou, Chalmers University of Technology, Sweden
Mihai Pătraşcu, MIT, USA
Kunihiko Sadakane, Kyushu University, Japan
Martin Skutella, TU Berlin, Germany
Christian Sohler, University of Paderborn, Germany

Local Organization

Ewa Cederheim-Wäingelin
Rebecca Cyrén
Georgios Georgiadis
Catharina Jerkbrant
Marina Papatriantafilou
Tiina Rankanen
Philippas Tsigas

Steering Committee

Lars Arge, University of Aarhus, Denmark
Magnús M. Halldórsson, University of Iceland, Iceland
Rolf Karlsson, Lund University, Sweden
Andrzej Lingas, Lund University, Sweden
Jan Arne Telle, University of Bergen, Norway
Esko Ukkonen, University of Helsinki, Finland

Sponsoring Institutions

LINDHOLMEN
SCIENCE PARK

Security Arena
Lindholmen Science Park

ERICSSON

External Reviewers

Marcel R. Ackermann
Dror Aiger
Mohammad Ali Abam
Nina Amenta
Daniel Andersson
Sunil Arya
Hideo Bannai
Amitabh Basu
Boaz Ben-Moshe
Michael Bender
Marcin Bienkowski
Philip Bille
Vincenzo Bonifaci
Peter Braß
Kevin Buchin
Maike Buchin
Jaroslaw Byrka
Daniel Cederman
Janka Chlebíková
Marek Chrobak
Sébastien Collette
Derek Corneil
Annalisa DeBonis
Brian Dean
Bastian Degener
Kedar Dhamdhere
Karim Douïeb
Vida Dujmović
Lene M. Favrholdt
Fernando Mario Oliveira Filho
Sorelle Friedler
Zhang Fu
Naveen Garg
Serge Gaspers
Giorgos Georgiadis
Anders Gidenstam
Inge Li Gørtz
Peter Golovach
Lee-Ad Gottlieb
Phuong Ha
Mikael Hammar
Sariel Har-Peled
Refael Hassin
Herman Haverkort
Meng He
Stefan Hougardy
John Iacono
Csanád Imreh
Klaus Jansen
Jesper Jansson
Juha Karkkainen
Petteri Kaski
Hyosil Kim

Ralf Klasing
Rolf Klein
Boris Koldehofe
Guy Kortsarz
Mirosław Korzeniowski
Marc van Kreveld
Danny Krizanc
Alexander Kröller
Sven O. Krumke
Marcin Kubica
Jochen Könemann
Christiane Lammersen
Stefan Langerman
Erik Jan van Leeuwen
Asaf Levin
Christian Liebchen
Andrzej Lingas
Tamás Lukovszki
Anil Maheshwari
Hamid Mahini
Azarakhsh Malekian
Nunkesser Marc
Dániel Marx
Daniel Meister
George Mertzios
Julian Mestre
Shuichi Miyazaki
Morteza Monemizadeh
Haiko Müller
Viswanath Nagarajan
Alfredo Navarra
Bengt Nilsson
Zeev Nutov
Joseph O'Rourke
Hirotaka Ono
Aris Pagourtzis
Seth Pettie
Igor Potapov
David Pritchard
Artem Pyatkin
Vijaya Ramachandran
R. Ravi
Dror Rawitz
Daniel Reichman
Romeo Rizzi
Geneviève Roberge
Liam Roditty
Aaron Roth
Milan Ružić
Harald Räcke
Peter Sanders
Saket Saurabh
Elad Michael Schiller
Jiří Sgall
Anastasios Sidiropoulos
Shakhar Smorodinsky
Bettina Speckmann
Rob van Stee
Jukka Suomela
Maxim Sviridenko
Xuehou Tan
Orestis Telelis
Peter Tiedemann
Laura Toma
Ryuhei Uehara
Taso Viglas
Antoine Vigneron
Yngve Villanger
Berthold Vöcking
Renato Werneck
Alexander Wolff
Thomas Wolle
Maverick Woo
Koichi Yamazaki
Norbert Zeh
Paweł Żyliński

Table of Contents

Invited Lectures

Contributed Papers

A Survey of Results for Deletion Channels and Related Synchronization Channels

Michael Mitzenmacher*

Harvard University
School of Engineering and Applied Sciences
michaelm@eecs.harvard.edu

The binary symmetric channel, where each bit is independently received in error with probability p, and the binary erasure channel, where each bit is erased with probability p, enjoy a long and rich history. Shannon developed the fundamental results on the capacity of such channels in the 1940's [19], and in recent years, through the development and analysis of low-density parity-check codes and related families of codes, we understand how to achieve near-capacity performance for such channels extremely efficiently [2,13,17].

Now consider the following channel: n bits are sent, but each bit is independently deleted with fixed probability p. This is the *binary i.i.d. deletion channel*, which we may refer to more succinctly as the *binary deletion channel* or just the *deletion channel*. A deletion channel should not be confused with an erasure channel. With an erasure channel, when n bits are sent, n symbols are received; a third symbol, often denoted by '?', is obtained at the receiver to denote an erasure. In contrast, with a deletion channel, there is no sign that a bit is deleted. For example, if 10101010 was sent, the receiver would obtain 10011 if the third, sixth, and eighth bits were deleted, and would obtain 10?01?1? if the bits were erased.

What is the capacity of this channel? Surprisingly, we do not know. Currently, we have no closed-form expression for the capacity, nor do we have an efficient algorithmic means to numerically compute this capacity. Not surprisingly, this lack of understanding of channel capacity goes hand in hand with a lack of good codes for the deletion channel.

More generally, channels with synchronization errors, including both insertions and deletions and more general timing errors, are simply not adequately understood by current theory. Given the near-complete knowledge we have channels with erasures and errors, in terms of both the capacity and codes that can nearly achieve capacity, our lack of understanding about channels with synchronization errors is truly remarkable.

On the other hand, substantial progress has been made in just the last few years. A recent result that we will highlight is that the capacity for the binary deletion channel is at least $(1-p)/9$ for *every* value of p [8,16]. Another way of thinking about this result is that the capacity of the deletion channel is always within a (relatively small) constant factor of the corresponding erasure channel, even as the deletion probability p

* Supported in part by NSF grant CCF-0634923.

J. Gudmundsson (Ed.): SWAT 2008, LNCS 5124, pp. 1–3, 2008.

goes to 1! As the erasure channel gives a clear upper bound on the capacity, this result represents a significant step forward; while in retrospect the fact that these capacities are always within a constant factor seems obvious, the fact that this factor is so small still appears somewhat surprising.

The purpose of this talk is to present recent progress along with a clear description of open problems, with the hope of spurring further research in this area. The presentation is necessarily somewhat biased, focusing on my own recent research in the area. Background information can be found in, for example, [1,3,4,10,11,18,20]. Results to be presented will include capacity lower bound arguments [6,7,8,16], capacity upper bound arguments [5], codes and capacity bounds alternative channel models [12,14,15], and related problems such as trace reconstruction [9].

Before beginning, it is worth asking why this class of problems is important. From a strictly practical perspective, such channels are arguably harder to justify than channels with errors or erasures. While codes for synchronization have been suggested for disk drives, watermarking, or general channels where timing errors may occur, immediate applications are much less clear than for advances in erasure-correcting and error-correcting codes. However, this may be changing. In the past, symbol synchronization has been handled separately from coding, using timing recovery techniques that were expensive but reasonable given overall system performance. With improvements in coding theory, specifically in iteratively decodable codes, it may become increasingly common that synchronization errors will prove a bottleneck for practical channels. Moreover, because we are currently so far away from having good coding schemes for even the most basic synchronization channels, in practice coding is rarely if ever considered as a viable solution to synchronization. If efficient codes for synchronization problems can be found, it is likely that applications will follow. If such codes are even a fraction as useful as codes for erasures or errors have been, they will have a significant impact. The work on capacity lower bounds demonstrates that there is more potential here than has perhaps been realized.

Of course, coding theory often has applications outside of engineering, and channels with deletions and insertions prove no exception, appearing naturally in biology. Symbols from DNA and RNA are deleted and inserted (and transposed, and otherwise changed) as errors in genetic processes. Understanding deletion channels and related problems may therefore give us important insight into genetic processes.

But regardless of possible applications, scientific interest alone provides compelling reasons to tackle these channels. As stated initially, while the deletion channel appears almost as natural and simple as the binary erasure and error channels, it has eluded similar understanding for decades, and appears to hide a great deal more complexity. The fact that we know so little about something so apparently basic is quite simply disturbing. Besides the standard questions of capacity and coding schemes for this and related channels, there appear to be many further easily stated and natural related variations worthy of study. Finally, the combinatorial nature of these channels brings together information theory and computer science in ways that should lead to interesting cross-fertilization of techniques and ideas between the two fields.

References

1. Batu, T., Kannan, S., Khanna, S., McGregor, A.: Reconstructing strings from random traces. In: Proceedings of the Fifteenth Annual ACM-SIAM Symposium on Discrete Algorithms, pp. 910–918 (2004)
2. Chung, S.Y., Forney Jr., G.D., Richardson, T.J., Urbanke, R.: On the design of low-density parity-check codes within 0.0045 dB of the Shannon limit. IEEE Communications Letters 5(2), 58–60 (2001)
3. Davey, M.C., Mackay, D.J.C.: Reliable communication over channels with insertions, deletions, and substitutions. IEEE Transactions on Information Theory 47(2), 687–698 (2001)
4. Diggavi, S., Grossglauser, M.: On information transmission over a finite buffer channel. IEEE Transactions on Information Theory 52(3), 1226–1237 (2006)
5. Diggavi, S., Mitzenmacher, M., Pfister, H.: Capacity upper bounds for deletion channels. In: Proceedings of the 2007 IEEE International Symposium on Information Theory (ISIT), pp. 1716–1720 (2007)
6. Drinea, E., Kirsch, A.: Directly lower bounding the information capacity for channels with i.i.d. deletions and duplications. In: Proceedings of the 2007 IEEE International Symposium on Information Theory (ISIT), pp. 1731–1735 (2007)
7. Drinea, E., Mitzenmacher, M.: On lower bounds for the capacity of deletion channels. IEEE Transactions on Information Theory 52(10), 4648–4657 (2006)
8. Drinea, E., Mitzenmacher, M.: Improved lower bounds for the capacity of i.i.d. deletion and duplication channels. IEEE Transactions on Information Theory 53(8), 2693–2714 (2007)
9. Holenstein, T., Mitzenmacher, M., Panigrahy, R., Wieder, U.: Trace reconstruction with constant deletion probability and related results. In: Proceedings of the Nineteenth Annual ACM-SIAM Symposium on Discrete Algorithms, pp. 389–398 (2008)
10. Levenshtein, V.I.: Binary codes capable of correcting deletions, insertions, and reversals. Soviet Phys. -Dokl 10(8), 707–710 (1966), (English translation)
11. Levenshtein, V.I.: Efficient reconstruction of sequences. IEEE Transactions on Information Theory 47(1), 2–22 (2001)
12. Liu, Z., Mitzenmacher, M.: Codes for deletion and insertion channels with segmented errors. In: Proceedings of the 2007 IEEE International Symposium on Information Theory (ISIT), pp. 846–850 (2007)
13. Luby, M.G., Mitzenmacher, M., Shokrollahi, M.A., Spielman, D.A.: Efficient erasure correcting codes. IEEE Transactions on Information Theory 47(2), 569–584 (2001)
14. Mitzenmacher, M.: Polynomial time low-density parity-check codes with rates very close to the capacity of the q-ary random deletion channel for large q. IEEE Transactions on Information Theory 52(12), 5496–5501 (2006)
15. Mitzenmacher, M.: Capacity bounds for sticky channels. IEEE Transactions on Information Theory 54(1), 72 (2008)
16. Mitzenmacher, M., Drinea, E.: A simple lower bound for the capacity of the deletion channel. IEEE Transactions on Information Theory 52(10), 4657–4660 (2006)
17. Richardson, T.J., Shokrollahi, M.A., Urbanke, R.L.: Design of capacity-approaching irregular low-density parity-check codes. IEEE Transactions on Information Theory 47(2), 619–637 (2001)
18. Schulman, L.J., Zuckerman, D.: Asymptotically good codes correcting insertions, deletions, and transpositions. IEEE Transactions on Information Theory 45(7), 2552–2557 (1999)
19. Shannon, C.E.: A mathematical theory of communication. Bell Systems Technical Journal 27(3), 379–423 (1948)
20. Sloane, N.J.A.: On single-deletion-correcting codes. Codes and Designs. In: Proceedings of a Conference Honoring Professor Dijen K. Ray-Chaudhuri on the Occasion of His 65th Birthday, Ohio State University, May 18-21, 2000 (2002)

Nash Bargaining Via Flexible Budget Markets

Vijay V. Vazirani

College of Computing, Georgia Institute of Technology,
Atlanta, GA 30332–0280
vazirani@cc.gatech.edu

Abstract. In his seminal 1950 paper, John Nash defined the bargaining problem; the ensuing theory of bargaining lies today at the heart of game theory. In this work, we initiate an algorithmic study of Nash bargaining problems.

We consider a class of Nash bargaining problems whose solution can be stated as a convex program. For these problems, we show that there corresponds a market whose equilibrium allocations yield the solution to the convex program and hence the bargaining problem. For several of these markets, we give combinatorial, polynomial time algorithms, using the primal-dual paradigm.

Unlike the traditional Fisher market model, in which buyers spend a fixed amount of money, in these markets, each buyer declares a lower bound on the amount of utility she wishes to derive. The amount of money she actually spends is a specific function of this bound and the announced prices of goods.

Over the years, a fascinating theory has started forming around a convex program given by Eisenberg and Gale in 1959. Besides market equilibria, this theory touches on such disparate topics as TCP congestion control and efficient solvability of nonlinear programs by combinatorial means. Our work shows that the Nash bargaining problem fits harmoniously in this collage of ideas.

J. Gudmundsson (Ed.): SWAT 2008, LNCS 5124, p. 4, 2008.

Simplified Planar Coresets for Data Streams

John Hershberger[1] and Subhash Suri[2,⋆]

[1] Mentor Graphics Corp.
8005 SW Boeckman Road, Wilsonville, OR 97070, USA
john_hershberger@mentor.com
[2] Computer Science Department
University of California, Santa Barbara, CA 93106, USA
suri@cs.ucsb.edu

Abstract. We propose a simple, optimal-space algorithm for maintaining an extent coreset for a data stream of points in the plane. For a given parameter r, the coreset consists of a sample of $\Theta(r)$ points from the stream seen so far. The coreset property is that for any slab (defined by two parallel lines) that encloses the sample, a $(1+\epsilon)$ expansion of the slab encloses all the stream points seen so far, for $\epsilon = O(1/r^2)$. The coreset can be maintained in $O(\log r)$ amortized time per point of the stream.

1 Introduction

The data stream model of computation has received considerable attention recently in the research communities of algorithms, databases, and networking. This model assumes that the input to an algorithm is presented in a fixed (possibly adversarial) order $x_1, x_2, \ldots, x_n, \ldots$, where x_1 is the first (oldest) element and x_n is the most recent element seen so far. The algorithm has a limited amount of memory r, which is considerably smaller than the stream size. Any input value not explicitly stored by the algorithm is essentially lost, and the algorithm uses its memory budget of r to construct a careful summary data structure, the so-called "synopsis," that attempts to capture application-dependent important characteristics of the whole set. The algorithm uses this summary structure to produce approximate query answers, and good summaries are those for which the approximation quality improves with r. Well-known examples of such summary structures include sketches [6,7], histograms [12], order statistics [8], and several others [10].

In computational geometry, *coresets* have been proposed as a versatile summary structure for low-dimensional queries that depend on the "shape" of the underlying point stream. For instance, it has been shown that coresets provide guaranteed-quality approximations for such spatial measures as diameter, minimum enclosing ball, smallest enclosing box, width, and directional extent, among others [1,5].

⋆ The second author gratefully acknowledges the support of the National Science Foundation through grants CCF-0514738 and CCF-0702798.

J. Gudmundsson (Ed.): SWAT 2008, LNCS 5124, pp. 5–16, 2008.

The main contribution of our paper is a new, space-optimal streaming algorithm for constructing an extent coreset of a two-dimensional set of points. Specifically, for a given parameter r, our coreset (also called an ϵ-kernel in [3]) consists of a sample of $\Theta(r)$ points from the stream seen so far with the property that for any slab (defined by two parallel lines) that encloses the sample, a $(1+\epsilon)$ expansion of the slab encloses all the stream points seen so far, for $\epsilon = O(1/r^2)$. This is known to be optimal in space, and we show that our algorithm can maintain the coreset in $O(\log r)$ amortized time per point of the stream.

Our algorithm combines and builds on ideas from several previous papers, including our own adaptive convex hull [9], Agarwal, Har-Peled, and Varadarajan's affine scaling [2], and the algorithmic structure proposed by Chan [5]. The original scheme proposed by Agarwal, Har-Peled and Varadarajan [2] requires $O(r \log^2 n)$ space for two-dimensional extent coresets, with $O(r^3)$ per-point processing. This was subsequently improved to $O(r \log^2 r)$ space and $O(r)$ per-point processing by Chan [5]. Independently, Hershberger and Suri [9] obtained an $O(r)$-space streaming scheme for approximate convex hulls, but it lacks the strong properties of an extent coreset—the Hausdorff distance between the approximate convex hull and the true convex hull is $O(D/r^2)$, where D is the diameter of the set. The existence of space-optimal coresets for planar streams was only recently settled by Agarwal and Yu [3], who achieved an $O(r)$ size structure with $O(1/r^2)$ relative extent error and $O(\log r)$ per-point processing.

The Agarwal-Yu scheme is an adaptation of Chan's original scheme. However, it requires several clever ideas and fairly complex data structures to avoid the logarithmic factors inherent in Chan's construction. A possible downside of the Agarwal-Yu scheme is that in its quest for space optimality, it loses the simplicity (both conceptual and implementation) of Chan's scheme. The main virtue of our new algorithm is its simple structure, which lends itself to both an easier implementation and simpler proofs.

The original adaptive convex hull [9] works by finding extreme vertices of the stream in $O(r)$ *sample directions*. These sample directions hierarchically refine an initial set of r uniformly spaced directions. Our new algorithm chooses sample directions hierarchically, as in [9], but with the uniformly spaced directions of that paper replaced by a variable distribution guided by the smallest enclosing box of the stream. Our algorithm also simplifies the algorithm and data structures of the original adaptive convex hull.

2 Preliminaries

Our coreset algorithm combines our adaptive convex hull [9] with the affine scaling idea of Agarwal, Har-Peled, and Varadarajan [2] and the algorithmic structure proposed by Chan [5]. In this section we review the last two of these; Section 3 discusses the adaptive convex hull.

2.1 Extent Definitions

The extent of a point set in a direction θ is defined in terms of the set's extrema in directions θ and $-\theta$. We use the notation $\max_\theta(S)$ to denote an extremum of a set S in direction θ. The inner product between two vectors u and v is denoted $\langle u, v\rangle$. Within an inner product, we often use the notation for a direction θ as shorthand for the unit vector pointing in direction θ. Thus for two directions θ and ϕ, $\langle\theta, \phi\rangle$ is the cosine of the angle between the directions. With this notation, we can express the *extent* of a set S in direction θ as

$$E(S, \theta) = \langle \max_\theta(S) - \max_{-\theta}(S), \theta\rangle.$$

When the set S is clear from context, we use the simpler notation $E(\theta)$ instead of $E(S, \theta)$. Note that $E(\theta) = E(-\theta)$. The minimum and maximum extents of a set are called the set's *width* and *diameter*.

Our algorithm maintains a sample P drawn from a larger set S. The convex hull of P is used as an approximation for that of S. The absolute error incurred by this approximation for a particular direction θ is

$$err(P, S, \theta) = \langle \max_\theta(S) - \max_\theta(P), \theta\rangle,$$

that is, the distance in direction θ between the extrema of S and P in that direction. When P and S are clear from context, we write $err(\theta)$ in place of $err(P, S, \theta)$.

The extent error of a sample P is $E(S, \theta) - E(P, \theta)$, which is the same as $err(P, S, \theta) + err(P, S, -\theta)$. The relative extent error of a sample, which is an important measure of coreset quality, is $E(S, \theta)/E(P, \theta) - 1$.

2.2 Bounding Boxes

A key insight of Agarwal, Har-Peled, and Varadarajan [2] is that an approximation scheme whose extent error for any direction θ is at most ϵ times the diameter of the point set can be used to produce an approximation with relative extent error $O(\epsilon)$ for every direction; that is, $E(S, \theta)/E(P, \theta) = 1 + O(\epsilon)$ for every θ. The idea is to apply an affine transform to the point set such that the resulting set has width and diameter within a constant factor of each other.

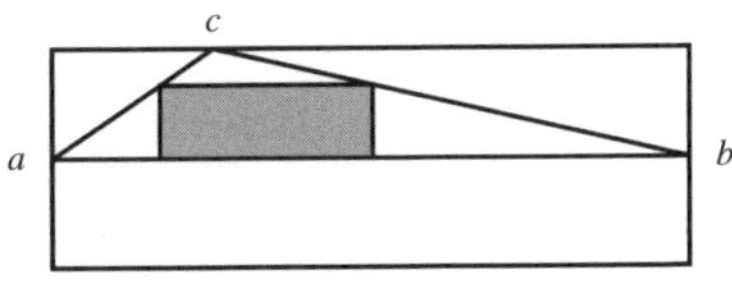

Fig. 1. The bbox B contains a similar rectangle of one-third size

We describe the transform for a static point set S first. Let $\overline{ab}$ be a diametral pair chosen from S. Draw a minimal rectangle B that encloses S and whose sides are parallel and perpendicular to $\overline{ab}$. We call B the *bbox*, short for *bounding box*. Let $c \in S$ be a point on the side of B parallel to $\overline{ab}$ and farther from it. Observe that a similar rectangle to B with side length at least one-third that of B can be nested inside $\triangle abc$, and hence inside the convex hull of S. See Figure 1.

Define an affine transform T that stretches B perpendicular to $\overline{ab}$ until B becomes a square. Apply T to S, mapping each point from the *bbox frame of reference* (untransformed space) into the *square frame of reference*. In the square frame of reference, the convex hull of $T(S)$ contains a square with one-third the side length of $T(B)$. Therefore the extent of $T(S)$ in any direction is between $\frac{1}{3}|ab|$ and $\sqrt{2}|ab|$.

Let $P \subseteq S$ be a sample such that $E(T(S),\phi)/E(T(P),\phi)$ is at most $1+\epsilon$ for any direction ϕ in the square frame of reference. To see that P has good relative extent error, consider the supporting lines $\ell_1, \ell_2, \ell_3, \ell_4$ that determine $E(S,\theta)$ and $E(P,\theta)$ for some direction θ. See Figure 2. $E(S,\theta)$ is the distance between ℓ_1 and ℓ_4, and $E(P,\theta)$ is the distance between ℓ_2 and ℓ_3. The transform of the supporting lines from bbox space to square space changes their slope, but it does not change their relative spacing. If $d(\cdot,\cdot)$ denotes distance, then $d(\ell_1,\ell_4)/d(\ell_2,\ell_3) = d(T(\ell_1),T(\ell_4))/d(T(\ell_2),T(\ell_3))$. The first quantity is $E(S,\theta)/E(P,\theta)$, and the second is $E(T(S),\phi)/E(T(P),\phi)$ for some ϕ. By the choice of P, the latter quantity is at most $1+\epsilon$, and hence so is the former.

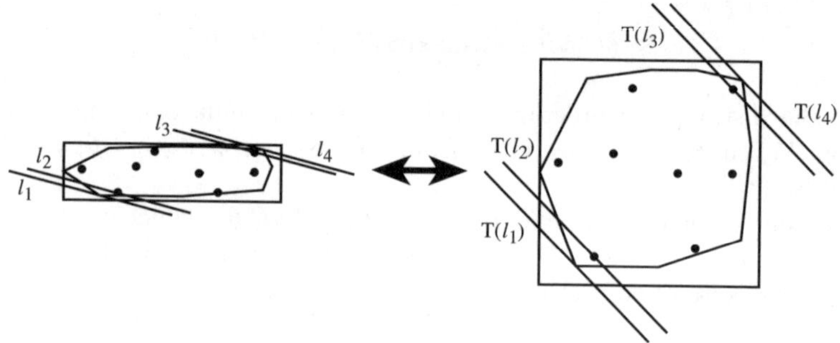

Fig. 2. The relative separation of parallel lines is preserved by the bbox transform. The lines support S and P; S is represented by a convex polygon, and P by dots within it.

2.3 Streaming, Epochs, and Subepochs

When the points of S are not known *a priori*, but arrive one at a time in a data stream, the bbox B described above is not fixed, but depends on the prefix of the stream seen so far. Chan [5] proposed a scheme that adapts the static bbox transform to a data stream setting; our formulation below follows that of Agarwal and Yu [3].

Let the *origin* be the first point of the stream. Let the *baseline* be the line segment from the origin to the farthest point seen so far within a constant factor (more on this below). The *bounding box B* (bbox) is a rectangle aligned with the baseline, centered on the origin, with length twice the length of the baseline and height twice the distance from the origin to the point perpendicularly farthest from the baseline. (We think of the baseline as horizontally oriented.) We define the length of the bbox to be D, the height to be W, and the aspect ratio to be $A = D/W$. Note that if the baseline connects the origin to the point farthest from it, then S is contained in B. Once a bbox B is defined, we use it until a point arrives that lies outside the box B' twice as big as B and concentric with it.

When a point arrives outside the doubled bbox B', we update the bbox. If the distance from the point to the origin is more than twice the length of the current baseline, we reset the baseline to use the new point and define the new bbox from scratch, based on the points seen so far. This starts a new *epoch*. Otherwise, we increase the height of the bbox to contain the new point, but keep its length and orientation unchanged. This starts a new *subepoch*. (Note that the bbox always remains centered on the origin, so the increase in height is symmetric about the origin.) At every instant, the current set S is contained in B', and the convex hull of S contains a rectangle similar to B and with at least one-sixth the side length. See Figure 3. Thus the affine transform described above can be applied to S to get a good extent approximation.

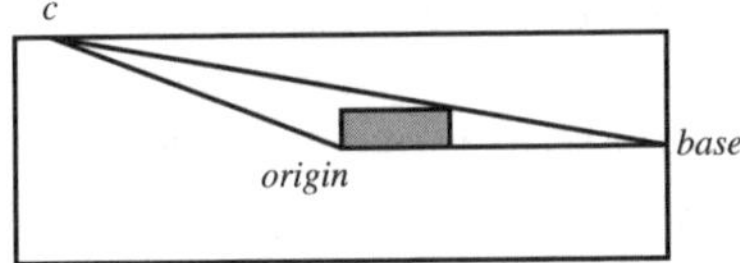

Fig. 3. The stream bbox B contains a similar rectangle of one-sixth size

Although the preceding description talks about updating a bbox by selecting farthest vertices from all the points seen so far, the same scheme works, with only a trivial error in the bbox parameters, if the farthest vertices are selected from an extent coreset that approximates the stream seen so far.

3 The Adaptive Convex Hull: One Subepoch

The adaptive convex hull [9] is a convex hull approximation for streaming points that has $err(\theta) = O(D/r^2)$, where D is the diameter and r is the approximation size, for any θ. As such, it could be plugged directly into the bbox transform scheme described in Section 2.2. However, because we want to be able to update the approximation when the bbox changes, we examine and modify some of the details of the adaptive convex hull in this section.

The original adaptive convex hull selects sample points by maintaining extrema of S in $O(r)$ directions. There are r directions spaced uniformly $\frac{2\pi}{r}$ apart in angle space, and up to r additional directions chosen adaptively to reduce the approximation error. Each sample point is extreme in at least one sample direction. If H is the convex hull of the sample points, then every edge e of H has an *uncertainty triangle* above it. The uncertainty triangle of an edge e of H is the triangle formed by e and the supporting lines (normal to sample directions) at the left/right ends of e whose directions are closest to each other. Any point of S outside H must lie in some uncertainty triangle.[1] The height of an uncertainty triangle over its base e is a measure of the adaptive convex hull error. We define $\alpha(e)$ as the angle associated with an edge of H; it is the smallest difference between the directions associated with the two ends of e. The uncertainty triangle angle opposite e has measure $\pi - \alpha(e)$. We define $len(e)$ as

[1] When S arrives as a stream, some points may lie outside uncertainty triangles, but they always lie close to H.

the total length of the two uncertainty triangle edges opposite e. This is closely related to the length of e itself, but a bit longer. (The closer $\alpha(e)$ is to zero, the closer $len(e)$ is to the length of e. Specifically, $len(e) = |e|(1 + O(\alpha(e))^2)$.)

Whereas the original adaptive convex hull starts with r uniformly spaced directions, in the present case we must use additional directions dependent on the bbox. We want to achieve the effect of mapping uniformly-spaced directions in the square frame of reference into normal space, but in a way that does not change sample directions unnecessarily when the bbox changes. We use a *bbox distribution function*, defined as

$$\Phi(\theta) = \arctan\left(\frac{\tan\theta}{A}\right),$$

where A is the aspect ratio of the bbox. This function maps directions in normal space into the square frame of reference. Here we choose direction $\theta = 0$ to be aligned with the bbox baseline. To obtain this function, we consider directions ϕ in the square frame of reference, with direction $\phi = 0$ also aligned with the bbox baseline. We associate each direction ϕ with its perpendicular $\phi + \pi/2$, map the perpendiculars from square space back to the original bbox space, and then take the normals to those directions to get directions θ in the original space. (The mapping from directions to their perpendiculars and back is necessary because the affine transformation does not preserve right angles, and what we really care about in the adaptive convex hull is the supporting lines, not the directions in which points are extreme.) For any two directions θ_1 and θ_2 in normal space, $|\Phi(\theta_1) - \Phi(\theta_2)| \times (r/2\pi)$ is approximately how many uniformly spaced directions in square space lie between the images of θ_1 and θ_2.

In the bbox distribution, directions are bunched up on the top and bottom of the bbox, centered on the direction of minimum extent, and spread out on the left and right sides. Intuitively, in a distribution with large aspect ratio we need more samples on the top and bottom to ensure that the width is approximated well, and fewer at the left and right, because the diameter is easy to approximate. See Figure 4.

We obtain a set of *base sample directions* as follows. First choose a set of r uniformly spaced directions $\{\theta_1, \ldots, \theta_r\}$—these directions will be base sample directions for the lifetime of the algorithm; they do not change when the bbox changes. Second choose additional directions based on the bbox: so long as some base direction interval (θ_i, θ_{i+1}) has $|\Phi(\theta_i) - \Phi(\theta_{i+1})| > \frac{2\pi}{r}$, choose a new base direction that bisects the interval. (Here we assume wraparound of indices and appropriate branch choices for the tangent and arctangent.)

Lemma 1. *The number of base sample directions is $O(r + \log A)$.*

Proof. (Sketch) We argue that if a base direction interval is not one of $O(1)$ intervals near $\theta = \pi/2$ or $3\pi/2$, then bisection splits the Φ "mass" in the interval into two pieces differing by at most a constant factor. This means that bisections happen $O(r)$ times except near $\pi/2$ or $3\pi/2$.

If an interval includes $\pi/2$ or $3\pi/2$ (the worst cases), some algebra reveals that the interval will be bisected unless $\frac{2\pi}{rA} \geq |\theta_1 - \theta_2|$. Therefore the number

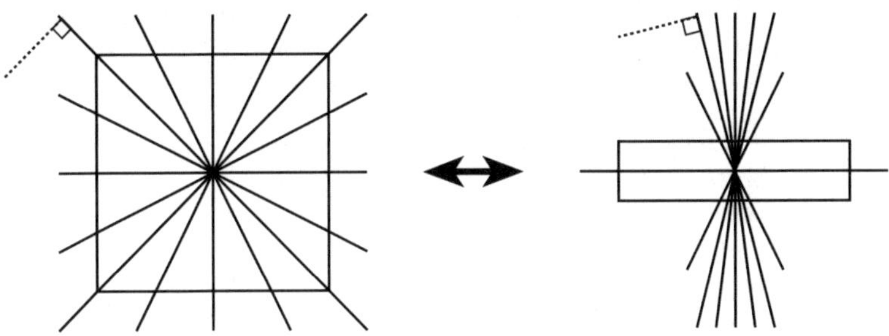

Fig. 4. The bbox distribution. Note that the unit-slope normal in the square is mapped to a normal with slope $1/A$ in the bbox.

of bisections to be applied to reach this width, starting from an initial interval width of $2\pi/r$, is $O(\log A)$. □

If $A > 2^r$, simple bisection gives too many base sample directions. There are $O(\log A)$ intervals in which bisection is *ineffective*—it does not split the Φ mass in a balanced way. In this case we use a more complicated scheme to find and represent the ineffective bisections in $O(r)$ time and space. A path of ineffective bisections in the hierarchy of base direction interval splits is represented as a single entity. The details of this scheme appear in the full paper.

Lemma 2. *The number of base sample directions can be reduced to $O(r)$.*

In the remainder of this abstract, operations whose details are affected by the implicit bisection described in the preceding lemma are flagged by a note [Lemma 2].

The choice of base sample directions ensures that uniform direction sampling in the square frame of reference is simulated in normal space. If θ_1 and θ_2 are directions corresponding to neighboring uniformly-spaced directions in the square frame of reference, then those transformed directions are $\phi_1 = \Phi(\theta_1)$ and $\phi_2 = \Phi(\theta_2)$, and $|\phi_1 - \phi_2| = \frac{2\pi}{r}$. The bisection rule for base sample directions ensures that θ_1 and θ_2 do not lie inside the same base direction interval.

Let us define a conversion factor $scale(\theta) = \sqrt{\sin^2\theta + A^2\cos^2\theta}$. The mapping between the square and bbox frames of reference means that an edge e in bbox space whose normal has direction θ (where the bbox baseline has direction $\theta = 0$) maps to an edge in the square frame of reference with length $|e| \cdot scale(\theta)$. Mapping the property of uniform direction sampling in the square frame of reference into the bbox frame of reference gives the following lemma.

Lemma 3. *Let θ_1 and θ_2 be a pair of adjacent base sample directions, let v_1 and v_2 be extrema in those directions, and let $\theta \in [\theta_1, \theta_2]$ be the normal to $\overline{v_1 v_2}$. Then there is a constant c such that $E(\theta) \cdot scale(\theta) > c \cdot r \cdot D \cdot |\theta_1 - \theta_2|$.*

Although the lemma's statement looks rather mysterious, it follows easily from the transform between the square and bbox frames of reference. This lemma lets us bound uncertainty triangle heights in terms of edge lengths. Consider a base edge e with normal θ that is associated with a base sample interval (θ_1, θ_2). The lemma says that if e has length proportional to $D/(r \cdot \mathit{scale}(\theta))$, then the height of the uncertainty triangle of e is at most $c \cdot E(\theta)/r^2$ for some constant c.

Lemma 4. *Let θ_1, θ_2, v_1, v_2, and θ be as in Lemma 3. Let p be the third vertex of the uncertainty triangle of $\overline{v_1 v_2}$. Define $d(\alpha)$ to be $\langle u - \max_\alpha(v_1, v_2), \alpha\rangle$, the directional distance from $\overline{v_1 v_2}$ to p. Then for every $\alpha \in [\theta_1, \theta_2]$, $d(\alpha)/E(\alpha) = O(d(\theta)/E(\theta))$.*

Proof. This is easy to see by mapping to the square frame of reference. The analogue of $d(\alpha)$ is maximized for $\Phi(\theta)$—it is the height of the uncertainty triangle, and the extent is proportional to D for any ϕ, including $\Phi(\theta)$ and $\Phi(\alpha)$. The ratios of directional distances are preserved by the bbox transform, and this completes the proof. □

Given a set of base sample directions and the convex hull of the extrema in those directions, the algorithm for computing an adaptive convex hull is essentially the same as in our earlier paper [9]. Each convex hull edge e has an angle $\alpha(e)$ and a length $\mathit{len}(e)$, and the product of those two approximates the maximum height of the edge's uncertainty triangle. If the height is too great, we *refine* the direction interval associated with e. That is, we add a new sample direction θ midway between the sample directions bounding e and update the convex hull by adding the extremum in direction θ, which may just be an endpoint of e (but cf. Lemma 2). To express the refinement rule we define the *depth* of a sample direction θ, $\mathit{depth}(\theta)$, as the number of refinements of a base direction interval needed to produce θ. If an edge e has its uncertainty triangle defined by directions θ_1 and θ_2, then $\mathit{depth}(e) \equiv \max(\mathit{depth}(\theta_1), \mathit{depth}(\theta_2))$. Instead of $\mathit{len}(e)$ we use the scaled length $\mathit{scale\_len}(e)$, equal to the total scaled length of the two uncertainty triangle edges opposite e, where *scaled length* means the true length of an edge times $\mathit{scale}(\theta)$ for its normal direction θ. The rule for adaptive refinement is that we refine an angle (either a base angle or one produced by prior refinement) when

$$\left(\frac{r}{D}\right) \mathit{scale\_len}(e) - \mathit{depth}(e) > 1.$$

To produce an adaptive convex hull from the base convex hull, we refine edges until the refinement rule no longer applies to any edge. Note that as in [9], every sample direction is either a uniformly spaced direction or obtained by a power-of-two refinement of a $(\frac{2\pi}{r})$-width uniform interval.

Adaptive refinement does not asymptotically increase the size of the sample:

Lemma 5. *The adaptive convex hull of a set of points has $O(r)$ vertices.*

Proof. The proof is essentially the same as in the original paper [9]. If we define $w(e) = (r/D)\mathit{scale\_len}(e) - \mathit{depth}(e)$, the sum of $w(e)$ over all e such that $w(e) > 1$ is at most $8r$ (because the total edge length of a convex polygon inside a square with side length $2D$ is at most $8D$). Each refinement reduces this sum by at least 1, so the total number of directions (and extrema) added by refinement is $O(r)$. □

By mapping the original proofs for the adaptive convex hull from the square frame of reference into the bbox frame of reference as in Lemma 3, we obtain the following two theorems. (Lemma 2 affects the proofs slightly.)

Theorem 1. *The adaptive convex hull of a static set of points has directional error* $err(\theta) = O(E(\theta)/r^2)$.

Theorem 2. *Let* Θ *be the set of base directions associated with the bbox of a single subepoch in a stream, and suppose that the extrema of the stream in directions* Θ *are known exactly at the start of the subepoch. Then during the subepoch the adaptive convex hull has directional error* $err(\theta) = O(E(\theta)/r^2)$.

4 The Adaptive Convex Hull: Multiple Epochs and Subepochs

The algorithm for applying the adaptive convex hull to a stream of points, including updates to the bbox when required, is quite straightforward. We process the stream in batches of r points in order to simplify maintenance of the adaptive convex hull. At any instant our coreset consists of an adaptive convex hull plus the partial batch of points awaiting the next round of processing. This means that our algorithm's per-point time bounds are amortized, but they can be converted to worst-case if desired, as discussed below. Here is the algorithm:

1. (Initialization)
   - Collect the first r points of the stream.
   - Define the bbox B, the bbox distribution Φ, and the base sample directions Θ.
   - Build an adaptive convex hull for the r points using the base sample directions Θ and the refinement rule given in Section 3.
2. (Maintenance)
   - Collect the next r points of the stream.
   - If none of the new points lies outside the doubled bbox B' concentric with B:
     - Merge the new points into the current adaptive convex hull.
   - Else (there is a new epoch or subepoch):
     - Define the bbox, the bbox distribution, and the base sample directions based on the origin, the current hull vertices, and the k new points. (The new bbox may be aligned with the old one, if this is a new subepoch, or not, if this is a new epoch.)
     - Build an adaptive convex hull for all the points using the new base sample directions.

Given a set of $O(r)$ points, with one designated as the origin, we can find the bbox B with two linear scans over the data. The bbox distribution function $\Phi()$ can then be defined in constant time. To compute the base sample directions, we build a list of r uniformly spaced directions, then scan through the list evaluating

$|\Phi(\theta_1) - \Phi(\theta_2)|$ on each interval between consecutive directions θ_1 and θ_2. For each interval where this quantity is greater than $2\pi/r$, we insert $(\theta_1 + \theta_2)/2$ between θ_1 and θ_2, then back up the scan to consider the first of the subintervals produced by bisection. This produces a set of base sample directions in $O(r + \log A)$ time, by Lemma 1. If $\log A > r$ and we want to reduce the time complexity and the size of Θ to $O(r)$ as in Lemma 2, the processing is slightly more complicated.

Once the base sample directions are known, we can build an adaptive convex hull for $O(r)$ points in $O(r \log r)$ time. This holds whether the points come from an unprocessed batch of r points, as in the initialization step, or from an existing adaptive convex hull and a new batch of r points, as in the maintenance step. In either case we build the new adaptive convex hull from scratch, rather than trying to update an existing structure.

We can use any standard convex hull algorithm for the batch of r new points [4]. If there is an existing adaptive convex hull (we are in the maintenance step), we merge the adaptive hull and the new hull with a linear scan over the two hulls in slope order of their edges.

To compute the adaptive convex hull, we put all the convex hull vertices in a circular list. We scan through the convex hull and the list of base sample directions in tandem, marking each convex hull vertex that is extreme in a base sample direction as belonging to the adaptive convex hull and storing its directions of extremity with it. For every pair of consecutive marked vertices (defining an edge of the base convex hull), we push the pair onto a stack of edges to process. Then we repeatedly remove an edge e from the stack and check whether the refinement rule of Section 3 applies; if so, we bisect e's angle (or do the more complicated partitioning of Lemma 2), find the extremum v in the refining direction, mark v as a sample point, and push onto the stack each nontrivial edge defined by v and the endpoints of e. The extremum v can be found by a dovetailed pair of linear scans on the convex hull list, starting at the ends of e and walking over the interval between them. We find the extremum in time proportional to its distance in the list from the nearer endpoint, which leads to an overall $O(r \log r)$ cost for extremum-finding. Once the stack of edges to process is empty, we discard the unmarked vertices and retain the marked ones as the new adaptive convex hull.

This completes the description of our coreset algorithm. It is clear that each batch of r points takes $O(r \log r)$ time to process, so the amortized time per point is $O(\log r)$. This bound can be made worst-case if necessary, at the cost of a little more bookkeeping. Details appear in the full paper.

Error Analysis

We now show that the multi-bbox version of the adaptive convex hull achieves the same error bounds as the version for a single bbox (Theorem 2). Our approach extends the technique used in the original adaptive convex hull proof of error bounds for data streams [9, Section 5.3]. For each active sample direction θ we maintain a half-plane bounded by a line perpendicular to θ. The intersection of all the half-planes is guaranteed to contain all points of S, and simultaneously is not far from the convex hull of the sample points. Note that these half-planes do not appear in the adaptive convex hull itself; they are used only in the analysis.

For every active sample direction θ we maintain two lines $L_{\text{curr}}(\theta)$ and $L_{\text{hist}}(\theta)$, both perpendicular to θ, and a third line $L(\theta) = \max_\theta(L_{\text{curr}}(\theta), L_{\text{hist}}(\theta))$. Each line has a corresponding half-plane $H_{\text{curr}}(\theta)/H_{\text{hist}}(\theta)/H(\theta)$ that lies on the side opposite direction θ. For each set of half-planes there is a corresponding intersection region $I_{\text{curr}}/I_{\text{hist}}/I$, defined, for example, as

$$I = \bigcap_{\substack{\theta \text{ is an active} \\ \text{sample direction}}} H(\theta).$$

For any active θ, the current line $L_{\text{curr}}(\theta)$ is the line used in the proof of Theorem 2, which maps the bounding lines used in Section 5.3 of [9] from the square frame of reference to normal (bbox) space. A sufficient characterization for our purposes is that $L_{\text{curr}}(\theta)$ is at most $O(E(\theta)/r^2)$ in direction θ beyond the current sample point (extremum) in direction θ. If P_{start} is the set of sample points active when the current bbox B was updated, and S_{curr} is the set of stream points seen since that time, then each $H_{\text{curr}}(\theta)$ contains $P_{\text{start}} \cup S_{\text{curr}}$, and so $(P_{\text{start}} \cup S_{\text{curr}}) \subset I_{\text{curr}}$, as guaranteed by Theorem 2.

For any active θ, the historical line $L_{\text{hist}}(\theta)$ is chosen such that $H_{\text{hist}}(\theta)$ contains all points of S that arrived before the current bbox was updated, a set denoted S_{hist}. Let us define $I_{\text{curr}}(t)$ to be the current approximation boundary at some time t, and define t_i to be the time of the i'th bbox update (the end of a subepoch). Then every point that arrived during the i'th subepoch is contained in $I_{\text{curr}}(t_i)$. We choose $L_{\text{hist}}(\theta)$ to be tangent to $\bigcup_i I_{\text{curr}}(t_i)$, where the union is taken over all subepochs prior to the current one. (During the first subepoch $L_{\text{hist}}(\theta)$ is chosen to pass through the origin vertex for all θ.)

Lemma 6. *For every active direction θ, $H_{\text{hist}}(\theta)$ contains all points of S_{hist}. If $v(\theta)$ is the current extreme vertex (sample point) in direction θ, then $\langle L_{\text{hist}}(\theta) - v(\theta), \theta\rangle$ is either negative or is $O(E(\theta)/r^2)$.*

By Theorem 2, $S_{\text{curr}} \subset I_{\text{curr}}$, and by Lemma 6, $S_{\text{hist}} \subset I_{\text{hist}}$; hence by definition $S \subset I$. Because $\langle L(\theta) - v(\theta), \theta\rangle = O(E(\theta)/r^2)$ for every active θ, we can use the fact that the active sample directions define a valid adaptive convex hull, plus an argument like that of Lemma 4, to show that if P is the current set of sample points, $err(P, I, \theta) = O(E(\theta)/r^2)$ for any θ. This in turn implies $err(P, S, \theta) = O(E(\theta)/r^2)$, since $S \subset I$. We have established our main theorem.

Theorem 3. *Let S be a set of points in the plane that arrives as a data stream, and let r be a given integer parameter. The adaptive convex hull algorithm described above constructs and maintains an extent coreset for S of size $O(r)$ in $O(\log r)$ amortized time per point. At all times and for any direction θ, the relative extent error of the coreset is $O(1/r^2)$.*

5 Conclusion

We have described a simple algorithm for maintaining an extent coreset of a data stream of two-dimensional points. The algorithm uses optimal space and amortized time. Although the proofs of correctness and error bounds are nontrivial, the data structure itself is very easy to define and implement. In particular, it consists of a single structure parameterized by the current bbox, rather than multiple structures dependent on a polylogarithmic number of prior bboxes, as is the case with previous coreset algorithms.

References

1. Agarwal, P., Har-Peled, S., Varadarajan, K.: Geometric approximations via coresets. Combinatorial and Computational Geometry – MSRI Pub. 52, 1–30 (2005)
2. Agarwal, P.K., Har-Peled, S., Varadarajan, K.R.: Approximating extent measures of points. J. ACM 51(4), 606–635 (2004)
3. Agarwal, P.K., Yu, H.: A space-optimal data-stream algorithm for coresets in the plane. In: Proc. 23rd Ann. Symp. Comput. Geometry, pp. 1–10. ACM, New York (2007)
4. de Berg, M., van Kreveld, M., Overmars, M., Schwarzkopf, O.: Computational Geometry: Algorithms and Applications, 2nd edn. Springer, Berlin (2000)
5. Chan, T.M.: Faster core-set constructions and data-stream algorithms in fixed dimensions. Comput. Geom. Theory Appl. 35(1), 20–35 (2006)
6. Cormode, G., Muthukrishnan, S.: An improved data stream summary: The count-min sketch and its applications. J. Algorithms 55(1), 58–75 (2005)
7. Flajolet, P., Martin, G.N.: Probabilistic counting algorithms for data base applications. Journal of Computer and System Sciences 31(2), 182–209 (1985)
8. Greenwald, M., Khanna, S.: Space-efficient online computation of quantile summaries. In: Proc. ACM SIGMOD Intl. Conf. Mgmt. of Data, pp. 58–66. ACM Press, New York (2001)
9. Hershberger, J., Suri, S.: Adaptive sampling for geometric problems over data streams. In: Proc. 23rd ACM Symp. Principles of Database Syst., pp. 252–262 (2004)
10. Muthukrishnan, S.: Data Streams: Algorithms and Applications. In: Foundations and Trends in Theoretical Computer Science, vol. 1(2), Now Publishers, Delft, Netherlands (2005)
11. Preparata, F.P., Shamos, M.I.: Computational Geometry. Springer, Heidelberg (1985)
12. Shrivastava, N., Buragohain, C., Agrawal, D., Suri, S.: Medians and beyond: New aggregation techniques for sensor networks. In: SenSys 2004, pp. 239–249. ACM, New York (2004)

Uniquely Represented Data Structures for Computational Geometry

Guy E. Blelloch[⋆], Daniel Golovin[⋆⋆], and Virginia Vassilevska[⋆⋆⋆]

Computer Science Department, Carnegie Mellon University, Pittsburgh, PA
{blelloch, dgolovin, virgi}@cs.cmu.edu

Abstract. We present new techniques for the construction of uniquely represented data structures in a RAM, and use them to construct efficient uniquely represented data structures for orthogonal range queries, line intersection tests, point location, and 2-D dynamic convex hull. Uniquely represented data structures represent each logical state with a unique machine state. Such data structures are *strongly history-independent*. This eliminates the possibility of privacy violations caused by the leakage of information about the historical use of the data structure. Uniquely represented data structures may also simplify the debugging of complex parallel computations, by ensuring that two runs of a program that reach the same logical state reach the same physical state, even if various parallel processes executed in different orders during the two runs.

1 Introduction

Most computer applications store a significant amount of information that is hidden from the application interface—sometimes intentionally but more often not. This information might consist of data left behind in memory or disk, but can also consist of much more subtle variations in the state of a structure due to previous actions or the ordering of the actions. For example a simple and standard memory allocation scheme that allocates blocks sequentially would reveal the order in which objects were allocated, or a gap in the sequence could reveal that something was deleted even if the actual data is cleared. Such location information could not only be derived by looking at the memory, but could even be inferred by timing the interface—memory blocks in the same cache line (or disk page) have very different performance characteristics from blocks in different lines (pages). Repeated queries could be used to gather information about relative positions even if the cache is cleared ahead of time. As an example of where this could be a serious issue consider the design of a voting machine. A careless design might reveal the order of the cast votes, giving away the voters' identities.

To address the concern of releasing historical and potentially private information various notions of *history independence* have been derived along with data structures that support these notions [14,18,13,7,1]. Roughly, a data structure is history independent if

⋆ Supported in part by NSF ITR grants CCR-0122581 (The Aladdin Center).

⋆⋆ Supported in part by NSF ITR grants CCR-0122581 (The Aladdin Center) and IIS-0121678.

⋆⋆⋆ Supported in part by NSF ITR grants CCR-0122581 (The Aladdin Center) and a Computer Science Department Ph.D. Scholarship.

J. Gudmundsson (Ed.): SWAT 2008, LNCS 5124, pp. 17–28, 2008.

someone with complete access to the memory layout of the data structure (henceforth called the "observer") can learn no more information than a legitimate user accessing the data structure via its standard interface (*e.g.*, what is visible on screen). The most stringent form of history independence, *strong history independence*, requires that the behavior of the data structure under its standard interface along with a collection of randomly generated bits, which are revealed to the observer, uniquely determine its memory representation. We say that such structures have a *unique representation.*

The idea of unique representations had also been studied earlier [24,25,2] largely as a theoretical question to understand whether redundancy is required to efficiently support updates in data structures. The results were mostly negative. Anderson and Ottmann [2] showed, for example, that ordered dictionaries require $\Theta(n^{1/3})$ time, thus separating unique representations from redundant representations (redundant representations support dictionaries in $\Theta(\log n)$ time, of course). This is the case even when the representation is unique only with respect to the pointer structure and not necessarily with respect to memory layout. The model considered, however, did not allow randomness or even the inspection of secondary labels assigned to the keys.

Recently Blelloch and Golovin [4] described a uniquely represented hash table that supports insertion, deletion and queries on a table with n items in $O(1)$ expected time per operation and using $O(n)$ space. The structure only requires $O(1)$-wise independence of the hash functions and can therefore be implemented using $O(\log n)$ random bits. The approach makes use of recent results on the independence required for linear probing [20] and is quite simple and likely practical. They also showed a perfect hashing scheme that allows for $O(1)$ worst-case queries, it although requires more random bits and is probably not practical. Using the hash tables they described efficient uniquely represented data structures for ordered dictionaries and the order maintenance problem [10]. This does not violate the Anderson and Ottmann bounds as it allows random bits to be part of the input.

In this paper we use these and other results to develop various uniquely represented structures in computational geometry. We show uniquely represented structures for the well studied dynamic versions of orthogonal range searching, horizontal point location, and orthogonal line intersection. All our bounds match the bounds achieved using fractional cascading [8], except that our bounds are in expectation instead of worst-case bounds. In particular for all problems the structures support updates in $O(\log n \log\log n)$ expected time and queries in $O(\log n \log\log n + k)$ expected time, where k is the size of the output. They use $O(n \log n)$ space and use $O(1)$-wise independent hash functions. Although better redundant data structures for these problems are known [15,17,3] (an $O(\log\log n)$-factor improvement), our data structures are the first to be uniquely represented. Furthermore they are quite simple, arguably simpler than previous redundant structures that match our bounds.

Instead of fractional cascading our results are based on a uniquely represented data structure for the ordered subsets problem (OSP). This problem is to maintain subsets of a totally ordered set under insertions and deletions to either the set or the subsets, as well as predecessor queries on each subset. Our data structure supports updates or comparisons on the totally ordered set in expected $O(1)$ time, and updates or queries to the subsets in expected $O(\log \log m)$ time, where m is the total number of element occurrences in subsets. This structure may be of independent interest.

We also describe a uniquely represented data structure for 2-D dynamic convex hull. For n points it supports point insertions and deletions in $O(\log^2 n)$ expected time, outputs the convex hull in time linear in the size of the hull, takes expected $O(n)$ space, and uses only $O(\log n)$ random bits. Although better results for planar convex hull are known ([6]) , we give the first uniquely represented data structure. Due to space considerations, the details of our results on horizontal point location and dynamic planar convex hull appear in the full version of the paper [5].

Our results are of interest for a variety of reasons. From a theoretical point of view they shed some light on whether redundancy is required to efficiently support dynamic structures in geometry. From the privacy viewpoint range searching is an important database operation for which there might be concern about revealing information about the data insertion order, or whether certain data was deleted. Unique representations also have potential applications to concurrent programming and digital signatures [4].

2 Preliminaries

Let $\mathbb{R}$ denote the real numbers, $\mathbb{Z}$ denote the integers, and $\mathbb{N}$ denote the naturals. Let $[n]$ for $n \in \mathbb{Z}$ denote $\{1, 2, \ldots, n\}$.

Unique Representation. Formally, an *abstract data type* (ADT) is a set V of logical states, a special starting state $v_0 \in V$, a set of allowable operations $\mathcal{O}$ and outputs $\mathcal{Y}$, a transition function $t : V \times \mathcal{O} \to V$, and an output function $y : V \times \mathcal{O} \to \mathcal{Y}$. The ADT is initialized to v_0, and if operation $O \in \mathcal{O}$ is applied when the ADT is in state v, the ADT outputs $y(v, O)$ and transitions to state $t(v, O)$. A *machine model* $\mathcal{M}$ is itself an ADT, typically at a relatively low level of abstraction, endowed with a programming language. Example machine models include the *random access machine* (RAM), the *Turing machine* and various *pointer machines*. An *implementation* of an ADT $\mathcal{A}$ on a machine model $\mathcal{M}$ is a mapping f from the operations of $\mathcal{A}$ to programs over the operations of $\mathcal{M}$. Given a machine model $\mathcal{M}$, an implementation f of some ADT (V, v_0, t, y) is said be *uniquely represented* (UR) if for each $v \in V$, there is a unique machine state $\sigma(v)$ of $\mathcal{M}$ that encodes it. Thus, if we run $f(O)$ on $\mathcal{M}$ exactly when we run O on (V, v_0, t, y), then the machine is in state $\sigma(v)$ iff the ADT is in logical state v.

Model of Computation & Memory allocation. Our model of computation is a unit cost RAM with word size at least $\log |U|$, where U is the universe of objects under consideration. As in [4], we endow our machine with an infinite string of random bits. Thus, the machine representation may depend on these random bits, but our strong history independence results hold no matter what string is used. In other words, a computationally unbounded observer with access to the machine state and the random bits it uses can learn no more than if told what the current logical state is. We use randomization solely to improve performance; in our performance guarantees we take probabilities and expectations over these random bits.

Our data structures are based on the solutions of several standard problems. For some of these problems UR data structures are already known. The most basic structure that is required throughout this paper is a hash table with insert, delete and search. The most common use of hashing in this paper is for memory allocation. Traditional memory allocation depends on the history since locations are allocated based on the ordering in which they are requested. We maintain data structures as a set of *blocks*. Each block has its own unique integer label which is used to hash the block into a unique *memory cell*. It is not too hard to construct such block labels if the data structures and the basic elements stored therein have them. For example, we can label points in $\mathbb{R}^d$ using their coordinates and if a point p appears in multiple structures, we can label each copy using a combination of p's label, and the label of the data structure containing that copy. Such a representation for memory contains no traditional "pointers" but instead uses labels as pointers. For example for a tree node with label l_p, and two children with labels l_1 and l_2, we store a cell containing (l_1, l_2) at label l_p. This also allows us to focus on the construction of data structures whose *pointer structure* is UR; such structures together with this memory allocation scheme yield UR data structures in a RAM. Note that all of the tree structures we use have pointer structures that are UR, and so the proofs that our structures are UR are quite straightforward. We omit the details due to lack of space.

Trees. Throughout this paper we make significant use of tree-based data structures. We note that none of the deterministic trees (e.g. red-black, AVL, splay-trees, weight-balanced trees) have unique representations, even not accounting for memory layout. We therefore use randomized treaps [22] throughout our presentation. We expect that one could also make use of skip lists [21] but we can leverage the elegant results on treaps with respect to limited randomness. For a tree T, let $|T|$ be the number of nodes in T, and for a node $v \in T$, let T_v denote the subtree rooted at v, and let $\mathrm{depth}(x)$ denote the length of the path from x to the root of T.

Definition 1 (k-Wise Independence). *Let $k \in \mathbb{Z}$ and $k \geq 2$. A set of random variables is k-*wise independent *if any k-subset of them is independent. A family $\mathcal{H}$ of hash functions from set A to set B is k-*wise independent *if the random variables in $\{h(x)\}_{x \in A}$ are k-wise independent and uniform on B when h is picked at random from $\mathcal{H}$.*

Unless otherwise stated, all treaps in this paper use 8-wise independent hash functions to generate priorities. We use the following properties of treaps.

Theorem 1 (Selected Treap Properties [22]). *Let T be a random treap on n nodes with priorities generated by an 8-wise independent hash function from nodes to $[p]$, where $p \geq n^3$. Then for any $x \in T$,*

(1) $\mathbf{E}[\mathrm{depth}(x)] \leq 2\ln(n) + 1$, *so access and update times are expected $O(\log n)$*
(2) $\mathbf{Pr}[|T_x| = k] = O(1/k^2)$ *for all $1 \leq k < n$*
(3) Given a predecessor handle, the expected insertion or deletion time is $O(1)$
(4) If the time to rotate a subtree of size k is $f(k)$ for some $f : \mathbb{N} \to \mathbb{R}_{\geq 1}$, the total time due to rotations to insert or delete an element is $O\left(\frac{f(n)}{n} + \sum_{0<k<n} \frac{f(k)}{k^2}\right)$ in expectation. Thus even if the cost to rotate a subtree is linear in its size (e.g., $f(k) = \Theta(k)$), updates take expected $O(\log n)$ time.

Dynamic Ordered Dictionaries. The dynamic ordered dictionary problem is to maintain a set $S \subset U$ for a totally ordered universe $(U, <)$. In this paper we consider supporting insertion, deletion, predecessor ($\text{Pred}(x, S) = \max\{e \in S | e < x\}$) and successor ($\text{Succ}(x, S) = \min\{e \in S | e > x\}$). Henceforth we will often skip successor since it is a simple modification to predecessor. If the keys come from the universe of integers $U = [m]$ a simple variant of the Van Emde Boas *et. al.* structure [26] is UR and supports all operations in $O(\log \log m)$ expected time [4] and $O(|S|)$ space. Under the comparison model we can use treaps to support all operations in $O(\log |S|)$ time and space. In both cases $O(1)$-wise independence of the hash functions is sufficient. We sometimes associate data with each element.

Order Maintenance. The *Order-Maintenance* problem [10] (OMP) is to maintain a total ordering L on n elements while supporting the following operations:

- Insert(x, y): insert new element y right after x in L.
- Delete(x): delete element x from L.
- Compare(x, y): determine if x precedes y in L.

In previous work [4] the first two authors described a randomized UR data structure for the problem that supports compare in $O(1)$ worst-case time and updates in $O(1)$ expected time. It is based on a three level structure. The top two levels use treaps and the bottom level uses state transitions. The bottom level contains only $O(\log \log n)$ elements per structure allowing an implementation based on table lookup. In this paper we use this order maintenance structure to support ordered subsets.

Ordered Subsets. The *Ordered-Subset* problem (OSP) is to maintain a total ordering L and a collection of subsets of L, denoted $\mathcal{S} = \{S_1, \ldots, S_q\}$ with $m = |L| + \sum_{i=1}^{q} |S_i|$ while supporting the OMP operations on L and the following ordered dictionary operations on each S_k:

- $\text{Insert}(x, S_k)$: insert $x \in L$ into set S_k.
- $\text{Delete}(x, S_k)$: delete x from S_k.
- $\text{Pred}(x, S_k)$: For $x \in L$, return $\max\{e \in S_k | e < x\}$.

Dietz [11] first describes this problem in the context of fully persistent arrays, and gives a solution yielding $O(\log \log m)$ expected amortized time operations. Mortensen [16] describes a solution that supports updates to the subsets in expected $O(\log \log m)$ time, and all other operations in $O(\log \log m)$ worst case time, where m is the total number of element occurrences in subsets. In section 3 we describe a UR version.

3 Uniquely Represented Ordered Subsets

Here we describe a UR data structure for the ordered-subsets problem. It supports the OMP operations on L in expected $O(1)$ time and the dynamic ordered dictionary problems on the subsets in expected $O(\log \log m)$ time, where $m = |L| + \sum_{i=1}^{q} |S_i|$. We use a somewhat different approach than Mortensen [16], which relied heavily on the solution of some other problems which we do not know how to make UR. Our solution is more self-contained and is therefore of independent interest beyond the fact that it is UR. Furthermore, our results improve on Mortensen's results by supporting insertion into and deletion from L in $O(1)$ instead of $O(\log \log m)$ time.

Theorem 2. *Let $m := |\{(x,k) : x \in S_k\}| + |L|$. There exists a UR data structure for the ordered subsets problem that uses $O(m)$ space, supports all OMP operations in expected $O(1)$ time, and all other operations in expected $O(\log\log m)$ time.*

We devote the rest of this section to proving Theorem 2. To construct the data structure, we start with a UR *order maintenance* data structure on L, which we will denote by D (see Section 2). Whenever we are to compare two elements, we simply use D.

We recall an approach used in constructing D [4], ***treap partitioning***: Given a treap T and an element $x \in T$, let its *weight* $w(x,T)$ be the number of descendants, including itself. For a parameter s, let $\mathcal{L}_s[T] = \{x \in T : w(x,T) \geq s\} \cup \{\mathrm{root}(T)\}$ be the *weight s partition leaders* of T[1]. For every $x \in T$ let $\ell(x,T)$ be the least (deepest) ancestor of x in T that is a partition leader. Here, each node is considered an ancestor of itself. The weight s partition leaders partition the treap into the sets $\{\{y \in T : \ell(y,T) = x\} : x \in \mathcal{L}_s[T]\}$, each of which is a contiguous block of keys from T.

In the construction of D [4] the elements of the order are treap partitioned twice, at weight $s := \Theta(\log|L|)$ and again at weight $\Theta(\log\log|L|)$. The partition sets at the finer level of granularity are then stored in UR hash tables. In the rest of the exposition we will refer to the treap on all of L as $T(D)$. The set of weight s partition leaders of $T(D)$ is denoted by $\mathcal{L}[T(D)]$, and the treap on these leaders by $T(\mathcal{L}[D])$.

The other main structure that we use is a treap $\mathcal{T}$ containing all elements from the set $\hat{L} = \{(x,k) : x \in S_k\} \cup \{(x,0) : x \in \mathcal{L}[T(D)]\}$. Treap $\mathcal{T}$ is partitioned by weight $\log m$ partition leaders. These leaders are labeled with the path from the root to their node (0 for left, 1 for right), so that label of each v is the binary representation of the root to v path. We keep a hash table H that maps labels to nodes, so that the subtreap of $\mathcal{T}$ on $\mathcal{L}[\mathcal{T}]$ forms a trie. It is important that only the leaders are labeled since otherwise insertions and deletions would require $O(\log m)$ time. We maintain a pointer from each node of $\mathcal{T}$ to its leader. In addition, we maintain pointers from each $x \in \mathcal{L}[T(D)]$ to $(x,0) \in \mathcal{T}$.

We store each subset S_k in its own treap T_k, also partitioned by weight $\log m$ leaders. When searching for the predecessor in S_k of some element x, we use $\mathcal{T}$ to find the leader ℓ in T_k of the predecessor of x in S_k. Once we have ℓ, the predecessor of x can easily be found by searching in the $O(\log m)$-size subtree of T_k rooted at ℓ. To guide the search for ℓ, we store at each node v of $\mathcal{T}$ the minimum and maximum T_k-leader labels in the subtree rooted at v, if any. Since we have multiple subsets we need to find predecessors in, we actually store at each v a *mapping* from each subset S_k to the minimum and maximum leader of S_k in the subtree rooted at v. For efficiency, for each leader $v \in \mathcal{T}$ we store a hash table H_v, mapping $k \in [q]$ to the tuple $(\min\{u : u \in \mathcal{L}[T_k]$ and $(u,k) \in \mathcal{T}_v\}, \max\{u : u \in \mathcal{L}[T_k]$ and $(u,k) \in \mathcal{T}_v\})$, if it exists. Recall $\mathcal{T}_v$ is the subtreap of $\mathcal{T}$ rooted at v. The high-level idea is to use the hash tables H_v to find the right "neighborhood" of $O(\log m)$ elements in T_k which we will have to update (in the event of an update to some S_k), or search (in the event of a predecessor or successor query). Since these neighborhoods are stored as treaps, updating and searching them takes expected $O(\log\log m)$ time. We summarize these definitions, along with some others, in Table 1.

We use the following Lemma to bound the number of changes on partition leaders.

[1] For technical reasons we include $\mathrm{root}(T)$ in $\mathcal{L}_s[T]$ ensuring that $\mathcal{L}_s[T]$ is nonempty.

Table 1. Some useful notation and definitions of various structures we maintain

H	hash table mapping label $i \in \{0,1\}^m$ to a pointer to the leader of $\mathcal{T}$ with label i
H_v	hash table mapping $k \in [q]$ to the tuple (if it exists) $(\min\{u : u \in \mathcal{L}[T_k] \wedge (u,k) \in \mathcal{T}_v\}, \max\{u : u \in \mathcal{L}[T_k] \wedge (u,k) \in \mathcal{T}_v\})$
$w(x,T)$	number of descendants of node x of treap T
$\mathcal{L}[T]$	weight $s = \Theta(\log m)$ partition leaders of treap T
$\ell(x,T)$	the partition leader of x in T
T_k	treap containing all elements of the ordered subset S_k, $k \in [q]$
$T(D)$	the treap on L
$T(\mathcal{L}[D])$	the subtreap of $T(D)$ on the weight $s = \Theta(\log m)$ leaders of $T(D)$
J_x	for $x \in \mathcal{L}[T(D)]$, a treap containing $\{u \in L : \ell(u, T(D)) = x \text{ and } \exists i : u \in S_i\}$
$\hat{L}$	the set $\{(x,k) : x \in S_k\} \cup \{(x,0) : x \in \mathcal{L}[T(D)]\}$
$\mathcal{T}$	a treap storing $\hat{L}$
I_x	for $x \in L$, a fast ordered dictionary [4] mapping each $k \in \{i : x \in S_i\}$ to (x,k) in $\mathcal{T}$

Lemma 1. *[4] Let $s \in \mathbb{Z}^+$ and let T be a treap of size at least s. Let T' be the treap induced on the weight s partition leaders in T. Then the probability that inserting a new element into T or deleting an element from T alters the structure of T' is c/s for some absolute constant c.*

Note that each partition set has size at most $O(\log m)$. The treaps T_k, J_x and $\mathcal{T}$, and the dictionaries I_x from Table 1 are stored explicitly. We also store the minimum and maximum element of each $\mathcal{L}[T_k]$ explicitly. We use a total ordering for $\hat{L}$ as follows: $(x,k) < (x',k')$ if $x < x'$ or $x = x'$ and $k < k'$.

OMP Insert & Delete Operations: These operations remain largely the same as in the order maintenance structure of [4]. We assume that when $x \in L$ is deleted it is not in any set S_k. The main difference is that if the set $\mathcal{L}[T(D)]$ changes we will need to update the treaps $\{J_v : v \in \mathcal{L}[T(D)]\}$, $\mathcal{T}$, and the tables $\{H_v : v \in \mathcal{L}[\mathcal{T}]\}$ appropriately.

Note that we can easily update H_v in time linear in $|\mathcal{T}_v|$ using in-order traversal of $\mathcal{T}_v$, assuming we can test if x is in $\mathcal{L}[T_k]$ in $O(1)$ time. To accomplish this, for each k we can store $\mathcal{L}[T_k]$ in a hash table. Thus using Theorem 1 we can see that all necessary updates to $\{H_v : v \in \mathcal{T}\}$ take expected $O(\log m)$ time. Clearly, updating $\mathcal{T}$ itself requires only expected $O(\log m)$ time. Finally, we bound the time to update the treaps J_v by the total cost to update $T(\mathcal{L}[D])$ if the rotation of subtrees of size k costs $k + \log m$, which is $O(\log m)$ by Theorem 1. This bound holds because $|J_v| = O(\log m)$ for any v, and any tree rotation on $T(D)$ causes at most $3s$ elements of $T(D)$ to change their weight s leader. Therefore only $O(\log m)$ elements need to be added or deleted from the treaps $\{J_v : v \in T(\mathcal{L}[D])\}$, and we can batch these updates in such a way that each takes expected amortized $O(1)$ time. However, we need only make these updates if $\mathcal{L}[T(D)]$ changes, which by Lemma 1 occurs with probability $O(1/\log m)$. Hence the expected overall cost is $O(1)$.

Predecessor & Successor: Suppose we wish to find the predecessor of x in S_k. (Finding the successor is analogous.) If $x \in S_k$ we can test this in expected $O(\log \log m)$ time using I_x. So suppose $x \notin S_k$. We will first find the predecessor w of (x,k) in $\mathcal{T}$ as follows. (We can handle the case that w does not exist by adding a special element to L that is smaller than all other elements and is considered to be part of $\mathcal{L}[T(D)]$). First search I_x for the predecessor k_2 of k in $\{i : x \in S_i\}$ in $O(\log\log m)$ time. If k_2 exists, then $w = (x, k_2)$. Otherwise, let y be the leader of x in $T(D)$, and let y' be the predecessor of y in $\mathcal{L}[T(D)]$. Then either $w \in \{(y', 0), (y, 0)\}$ or else $w = (z, k_3)$, where $z = \max\{u : u < x \text{ and } u \in J_y \cup J_{y'}\}$ and $k_3 = \max\{i : z \in S_i\}$. Thus we can find w in expected $O(\log\log m)$ time using fast finger search for y', treap search on the $O(\log m)$ sized treaps in $\{J_v : v \in \mathcal{L}[T(D)]\}$, and the fast dictionaries $\{I_x : x \in L\}$.

Once we have found the predecessor w of (x,k) in $\mathcal{T}$, we search for the predecessor w' of x in $\mathcal{L}[T_k]$. (If w' does not exist, we simply use $\min\{u \in \mathcal{L}[T_k]\}$). To find w', we first use w to search for a node u', defined as the leader (x,k) would have had in $\mathcal{T}$, had it been given a priority of $-\infty$. Note that with priority $-\infty$, (x,k) would be the leftmost leaf of the right subtree of w in $\mathcal{T}$. Hence its leader would either be the leader of w, or the deepest leader on the leftmost path starting from the right child of w. Hence u' can be found in expected $O(\log\log m)$ time, by binary searching on its label (i.e., if the label of w is α, then find the maximum k such that $\alpha \cdot 1 \cdot 0^k$ is an label in H).

Let P be the path from u' to the root of $\mathcal{T}$. We use the label of u' and H to binary search on P for the deepest node $v \in P$ for which $\min\{u : u \in \mathcal{L}[T_k] \text{ and } (u,k) \in \mathcal{T}_v\} < x$. This takes $O(\log|P|) = O(\log\log m)$ time in expectation. If $v \neq u'$, then u' is in the right subtree of v in $\mathcal{T}$, and (w', k) is in the left subtree of v. So let v_l be the left child of v and note that $w' = \max\{u : u \in \mathcal{L}[T_k] \text{ and } (u,k) \in \mathcal{T}_{v_l}\}$, which we can look up in $O(1)$ time after finding v by using H_v. Otherwise $v = u'$. In this case, lookup $a := \min\{u : u \in \mathcal{L}[T_k] \text{ and } (u,k) \in \mathcal{T}_v\}$ and $b := \max\{u : u \in \mathcal{L}[T_k] \text{ and } (u,k) \in \mathcal{T}_v\}$, find the least common ancestor c of $\{a,b\}$ in T_k, and starting from c search T_k for w'. Since a and b are both descendants of u', their distance (i.e., one plus the number of nodes between them in the order) in $\hat{L}$ is at most $s = \Theta(\log m)$, and thus their distance in T_k is at most $O(\log m)$. However, in random treaps the expected length of a path between nodes at distance d is $O(\log(d))$, even if priorities are generated using only 8-wise independent hash functions [22]. Thus we can find c in expected $O(\log\log m)$ time. Note c has at most $O(\log^2 m)$ descendants between a and b in T_k, since there are at most $O(\log m)$ partition leaders between a and b and each has at most $O(\log m)$ "followers" in its partition set, and we can find w' in expected $O(\log\log m)$ time starting from c. Once we have found w', the predecessor of x in $\mathcal{L}[T_k]$, we can simply find the successor of w' in $\mathcal{L}[T_k]$, say w'', via fast finger search, and then search the subtreaps rooted at w' and w'' for the actual predecessor of x in S_k in expected $O(\log\log m)$ time.

OSP-Insert and OSP-Delete: *OSP-Delete* is analogous to *OSP-Insert*, hence we focus on *OSP-Insert*. Suppose we wish to add x to S_k. First, if x is not currently in any sets $\{S_i : i \in [q]\}$, then find the leader of x in $T(D)$, say y, and insert x into J_y in expected $O(\log\log m)$ time. Next, insert x into T_k as follows. Find the predecessor w of x in S_k, then insert x into T_k in expected $O(1)$ time starting from w to speed up the insertion.

Find the predecessor w' of (x,k) in $\mathcal{T}$ as in the predecessor operation, and insert (x,k) into $\mathcal{T}$ using w' as a starting point. If neither $\mathcal{L}[T_k]$ nor $\mathcal{L}[\mathcal{T}]$ changes, then no modifications to $\{H_v : v \in \mathcal{L}[\mathcal{T}]\}$ need to be made. If $\mathcal{L}[T_k]$ does not change but $\mathcal{L}[\mathcal{T}]$ does, as happens with probability $O(1/\log m)$, we can update $\mathcal{T}$ and $\{H_v : v \in \mathcal{L}[\mathcal{T}]\}$ appropriately in expected $O(\log m)$ time. If $\mathcal{L}[T_k]$ changes, we must be careful when updating $\{H_v : v \in \mathcal{L}[\mathcal{T}]\}$. Let $\mathcal{L}[T_k]$ and $\mathcal{L}[T_k]'$ be the leaders of T_k immediately before and after the addition of x to S_k, and let $\Delta_k := (\mathcal{L}[T_k] - \mathcal{L}[T_k]') \cup (\mathcal{L}[T_k]' - \mathcal{L}[T_k])$. Then we must update $\{H_v : v \in \mathcal{L}[\mathcal{T}]\}$ appropriately for all nodes $v \in \mathcal{L}[\mathcal{T}]$ that are descendants of (x, k) as before, but must also update H_v for any node $v \in \mathcal{L}[\mathcal{T}]$ that is an ancestor of some node in $\{(u, k) : u \in \Delta_k\}$. It is not hard to see that these latter updates can be done in $O(|\Delta_k| \log m)$ time. Moreover, $\mathbf{E}\left[|\Delta_k| \mid x \in \mathcal{L}[T_k]'\right] = O(1)$, since $|\Delta_k|$ can be bounded by $2(R+1)$, where R is the number of rotations necessary to rotate x down to a leaf node in a treap on $\mathcal{L}[T_k]'$. Since it takes $\Theta(R)$ time to delete x given a handle to it, from Theorem 1 we easily infer $\mathbf{E}[R] = O(1)$. Since the randomness for T_k is independent of the randomness used for $\mathcal{T}$, these expectations multiply, for a total expected time of $O(\log m)$, conditioning on the fact that $\mathcal{L}[T_k]$ changes. Since $\mathcal{L}[T_k]$ only changes with probability $O(1/\log m)$, this part of the operation takes expected $O(1)$ time. Finally, insert k into I_x in expected $O(\log \log m)$ time, with a pointer to (x, k) in $\mathcal{T}$.

4 Uniquely Represented Range Trees

Let $P = \{p_1, p_2, \ldots, p_n\}$ be a set of points in $\mathbb{R}^d$. The well studied *orthogonal range reporting* problem is to maintain a data structure for P while supporting queries which given an axis aligned box B in $\mathbb{R}^d$ returns the points $P \cap B$. The dynamic version allows for the insertion and deletion of points. Chazelle and Guibas [8] showed how to solve the two dimensional dynamic problem in $O(\log n \log \log n)$ update time and $O(\log n \log \log n + k)$ query time, where k is the size of the output. Their approach used fractional cascading. More recently Mortensen [17] showed how to solve it in $O(\log n)$ update time and $O(\log n + k)$ query time using a sophisticated application of Fredman and Willard's q-heaps [12]. All of these techniques can be generalized to higher dimensions at the cost of replacing the first $\log n$ term with a $\log^{d-1} n$ term [9].

Here we present a uniquely represented solution to the problem. It matches the bounds of the Chazelle and Guibas version, except ours are in expectation instead of worst-case bounds. Our solution does not use fractional cascading and is instead based on ordered subsets. One could probably derive a UR version based on fractional cascading, but making dynamic fractional cascading UR would require significant work[2] and is unlikely to improve the bounds. Our solution is simple and avoids any explicit discussion of weight balanced trees (the required properties fall directly out of known properties of treaps).

Theorem 3. *Let P be a set of n points in $\mathbb{R}^d$. There exists a UR data structure for the orthogonal range query problem that uses $O(n \log^{d-1} n)$ space and $O(d \log n)$ random bits, supports point insertions or deletions in expected $O(\log^{d-1} n \cdot \log \log n)$ time, and queries in expected $O(\log^{d-1} n \cdot \log \log n + k)$ time, where k is the size of the output.*

If $d = 1$, simply use the dynamic ordered dictionaries solution [4] and have each element store a pointer to its successor for fast reporting. For simplicity we describe the

[2] We expect a variant of Sen's approach [23] could work.

two dimensional case. The remaining cases with $d \geq 3$ can be implemented using standard techniques [9] if treaps are used for the underlying hierarchical decomposition trees. The description will be deferred to the full paper. We will assume that the points have distinct coordinate values; thus, if $(x_1, x_2), (y_1, y_2) \in P$, then $x_i \neq y_i$ for all i. (There are various ways to remove this assumption, e.g., the composite-numbers scheme or symbolic perturbations [9].) We store P in a random treap T using the ordering on the first coordinate as our BST ordering. We additionally store P in a second random treap T' using the ordering on the second coordinate as our BST ordering, and also store P in an ordered subsets instance D using this same ordering. We cross link these and use T' to find the position of any point we are given in D. The subsets of D are $\{T_v : v \in T\}$, where T_v is the subtree of T rooted at v. We assign each T_v a unique integer label k using the coordinates of v, so that T_v is S_k in D. The structure is UR as long as all of its components (the treap and ordered subsets) are uniquely represented.

To insert a point p, we first insert it by the second coordinate in T' and using the predecessor of p in T' insert a new element into the ordered subsets instance D. This takes $O(\log n)$ expected time. We then insert p into T in the usual way using its x coordinate. That is, search for where p would be located in T were it a leaf, then rotate it up to its proper position given its priority. As we rotate it up, we can reconstruct the ordered subset for a node v from scratch in time $O(|T_v| \log \log n)$. Using Theorem 1, the overall time is $O(\log n \log \log n)$ in expectation. Finally, we must insert p into the subsets $\{T_v : v \in T \text{ and } v \text{ is an ancestor of } p\}$. This requires expected $O(\log \log n)$ time per ancestor, and there are only $O(\log n)$ of them in expectation. Since these expectations are computed over independent random bits, they multiply, for an overall time bound of $O(\log n \cdot \log \log n)$ in expectation. Deletion is similar.

To answer a query $(p, q) \in \mathbb{R}^2 \times \mathbb{R}^2$, where $p = (p_1, p_2)$ is the lower left and $q = (q_1, q_2)$ is the upper right corner of the box B in question, we first search for the predecessor p' of p and the successor q' of q in T (i.e., with respect to the first coordinate). We also find the predecessor p'' of p and successor q'' of q in T' (i.e., with respect to the second coordinate). Let w be the least common ancestor of p' and q' in T, and let $A_{p'}$ and $A_{q'}$ be the paths from p' and q' (inclusive) to w (exclusive), respectively. Let V be the union of right children of nodes in $A_{p'}$ and left children of nodes in $A_{q'}$, and let $\mathcal{S} = \{T_v : v \in V\}$. It is not hard to see that $|V| = O(\log n)$ in expectation, that the sets in $\mathcal{S}$ are disjoint, and that all points in B are either in $W := A_{p'} \cup \{w\} \cup A_{q'}$ or in $\cup_{S \in \mathcal{S}} S$. Compute W's contribution to the answer, $W \cap B$, in $O(|W|)$ time by testing each point in turn. Since $\mathbf{E}[|W|] = O(\log n)$, this requires $O(\log n)$ time in expectation. For each subset $S \in \mathcal{S}$, find $S \cap B$ by searching for the successor of p'' in S, and doing an in-order traversal of the treap in D storing S until reaching a point larger than q''. This takes $O(\log \log n + |S \cap B|)$ time in expectation for each $S \in \mathcal{S}$, for a total of $O(\log n \cdot \log \log n + k)$ expected time.

5 Horizontal Point Location and Orthogonal Segment Intersection

Let $S = \{(x_i, x'_i, y_i) : i \in [n]\}$ be a set of n horizontal line segments. In the *horizontal point location problem* we are given a point $(\hat{x}, \hat{y})$ and must find $(x, x', y) \in S$ maximizing y subject to the constraints $x \leq \hat{x} \leq x'$ and $y < \hat{y}$. In the related *orthogonal segment intersection problem* we are given a vertical line segment $s = (x, y, y')$, and must report all segments in S intersecting it, namely $\{(x_i, x'_i, y_i) : x_i \leq x \leq x'_i \text{ and } y \leq y_i \leq y'\}$. In the dynamic version we must additionally support updates to S. As with the orthogonal range reporting problem, both of these problems can be solved using fractional cascading and in the same time bounds [8] ($k = 1$ for point location and is the number of lines reported for segment intersection). Mortensen [15] improved orthogonal segment intersection to $O(\log n)$ updates and $O(\log n + k)$ queries.

We extend our ordered subsets approach to obtain the following results for horizontal point location and range reporting.

Theorem 4. *Let S be a set of n horizontal line segments in $\mathbb{R}^2$. There exists a uniquely represented data structure for the point location and orthogonal segment intersection problems that uses $O(n \log n)$ space, supports segment insertions and deletions in expected $O(\log n \cdot \log\log n)$ time, and supports queries in expected $O(\log n \cdot \log\log n + k)$ time, where k is the size of the output. The data structure uses $O(\log n)$ random bits.*

6 Uniquely Represented 2-D Dynamic Convex Hull

Using similar techniques we obtain a uniquely represented data structure for maintaining the convex hull of a dynamic set of points $S \subset \mathbb{R}^2$. Our approach builds upon the work of Overmars & Van Leeuwen [19]. Overmars & Van Leeuwen use a standard balanced BST T storing S to partition points along one axis, and likewise store the convex hull of T_v for each $v \in T$ in a balanced BST. In contrast, we use treaps in both cases, together with the hash table in [4] for memory allocation. Our main contribution is then to analyze the running times and space usage of this new uniquely represented version, and to show that even using only $O(\log n)$ random bits to hash and generate treap priorities, the expected time and space bounds match that of the original version up to constant factors. Specifically, we prove the following.

Theorem 5. *Let $n = |S|$. There exists a uniquely represented data structure for 2-D dynamic convex hull that supports point insertions and deletions in $O(\log^2 n)$ expected time, outputs the convex hull in $O(k)$ time, where k is the size of the convex hull, requires $O(n)$ space in expectation, and uses only $O(\log n)$ random bits.*

7 Conclusions

We have introduced uniquely represented data structures for a variety of problems in computational geometry. Such data structures represent every logical state by a unique machine state and reveal no history of previous operations, thus protecting the privacy of their users. For example, our uniquely represented range tree allows for efficient orthogonal range queries on a database containing sensitive information (e.g., viral load in the blood of hospital patients) without revealing any information about what order the current points were inserted into the database, whether points were previously deleted, or what queries were previously executed. Uniquely represented data structures have other benefits as well. They make equality testing particularly easy. They may also simplify the debugging of parallel processes by eliminating the conventional dependencies upon the specific sequence of operations that led to a particular logical state.

References

1. Acar, U.A., Blelloch, G.E., Harper, R., Vittes, J.L., Woo, S.L.M.: Dynamizing static algorithms, with applications to dynamic trees and history independence. In: Proc. SODA, pp. 531–540 (2004)
2. Andersson, A., Ottmann, T.: New tight bounds on uniquely represented dictionaries. SIAM Journal of Computing 24(5), 1091–1103 (1995)
3. Blelloch, G.E.: Space-efficient dynamic orthogonal point location, segment intersection, and range reporting. In: Proc. SODA (2008)
4. Blelloch, G.E., Golovin, D.: Strongly history-independent hashing with applications. In: Proc. FOCS, pp. 272–282 (2007)
5. Blelloch, G.E., Golovin, D., Vassilevska, V.: Uniquely represented data structures for computational geometry. Technical Report CMU-CS-08-115, Carnegie Mellon University (April 2008)
6. Brodal, G.S., Jacob, R.: Dynamic planar convex hull. In: Proc. FOCS, pp. 617–626 (2002)
7. Buchbinder, N., Petrank, E.: Lower and upper bounds on obtaining history independence. In: Boneh, D. (ed.) CRYPTO 2003. LNCS, vol. 2729, pp. 445–462. Springer, Heidelberg (2003)
8. Chazelle, B., Guibas, L.J.: Fractional cascading. Algorithmica 1, 133–196 (1986)
9. de Berg, M., van Kreveld, M., Overmars, M., Schwarzkopf, O.: Computational geometry: algorithms and applications. Springer, New York (1997)
10. Dietz, P., Sleator, D.: Two algorithms for maintaining order in a list. In: Proc. STOC, pp. 365–372 (1987)
11. Dietz, P.F.: Fully persistent arrays. In: Dehne, F., Santoro, N., Sack, J.-R. (eds.) WADS 1989. LNCS, vol. 382, pp. 67–74. Springer, Heidelberg (1989)
12. Fredman, M.L., Willard, D.E.: Trans-dichotomous algorithms for minimum spanning trees and shortest paths. Journal of Computer and System Sciences 48(3), 533–551 (1994)
13. Hartline, J.D., Hong, E.S., Mohr, A.E., Pentney, W.R., Rocke, E.: Characterizing history independent data structures. Algorithmica 42(1), 57–74 (2005)
14. Micciancio, D.: Oblivious data structures: applications to cryptography. In: Proc. STOC, pp. 456–464 (1997)
15. Mortensen, C.W.: Fully-dynamic two dimensional orthogonal range and line segment intersection reporting in logarithmic time. In: Proc. SODA, pp. 618–627 (2003)
16. Mortensen, C.W.: Fully-dynamic two dimensional orthogonal range and line segment intersection reporting in logarithmic time. In: Technical report TR-2003-22 IT University Technical Report Series (2003)
17. Mortensen, C.W.: Fully dynamic orthogonal range reporting on a RAM. SIAM J. Comput. 35(6), 1494–1525 (2006)
18. Naor, M., Teague, V.: Anti-presistence: history independent data structures. In: Proc. STOC, pp. 492–501 (2001)
19. Overmars, M.H., van Leeuwen, J.: Maintenance of configurations in the plane. J. Comput. Syst. Sci. 23(2), 166–204 (1981)
20. Pagh, A., Pagh, R., Ruzic, M.: Linear probing with constant independence. In: Proc. STOC, pp. 318–327 (2007)
21. Pugh, W.: Skip lists: A probabilistic alternative to balanced trees. In: Dehne, F., Santoro, N., Sack, J.-R. (eds.) WADS 1989. LNCS, vol. 382, pp. 437–449. Springer, Heidelberg (1989)
22. Seidel, R., Aragon, C.R.: Randomized search trees. Algorithmica 16(4/5), 464–497 (1996)
23. Sen, S.: Fractional cascading revisited. Journal of Algorithms 19(2), 161–172 (1995)
24. Snyder, L.: On uniquely representable data structures. In: Proc. FOCS, pp. 142–146 (1977)
25. Sundar, R., Tarjan, R.E.: Unique binary search tree representations and equality-testing of sets and sequences. In: Proc. STOC, pp. 18–25 (1990)
26. van Boas, P.E., Kaas, R., Zijlstra, E.: Design and implementation of an efficient priority queue. Mathematical Systems Theory 10, 99–127 (1977)

I/O Efficient Dynamic Data Structures for Longest Prefix Queries*

Moshe Hershcovitch[1] and Haim Kaplan[2]

[1] Faculty of Electrical Engineering
moshik1@gmail.com
[2] School of Computer Science
haimk@cs.tau.ac.il
Tel Aviv University, Tel Aviv 69978, Israel

Abstract. We present an efficient data structure for finding the longest prefix of a query string p in a dynamic database of strings. When the strings are IP-addresses then this is the IP-lookup problem. Our data structure is I/O efficient. It supports a query with a string p using $O(\log_B(n) + \frac{|p|}{B})$ I/O operations, where B is the size of a disk block. It also supports an insertion and a deletion of a string p with the same number of I/O's. The size of the data structure is linear in the size of the database and the running time of each operation is $O(\log(n) + |p|)$.

1 Introduction

We consider the *longest prefix problem* which is defined as follows. The input consists of a set of strings $S = \{p_1 \dots p_n\}$ which we shall refer to as *prefixes*. We want to preprocess S into a data structure such that given a query string q we can efficiently find the longest prefix in S which is a prefix of q or report that no prefix in S is a prefix of q. We focus on the dynamic version of the problem where we want to be able to insert and delete prefixes to and from S, respectively.

The main application of this problem is for packet forwarding in IP networks. In this application a router maintains a set of prefixes of IP addresses according to some routing protocol such as BGP (Border Gateway Protocol). When a packet arrives, the router finds the longest prefix of the destination address of the packet and sends the packet on the outgoing link associated with this longest prefix. A longest prefix match query in this settings is often called *IP-lookup*.

The rapid growth of the Internet has brought the need for routers to maintain large sets of prefixes, and to perform longest prefix match queries at high speeds [7]. A main issue in the design of routers is the size of the expensive high speed memory used by the router for packet forwarding. One can reduce the size of this expensive memory by using external memory components. Therefore the I/O efficiency of the algorithm we use is very important. In software implementations the entire data structure may not fit into the cache and we may flip parts of it

* This work is partially supported by United states - Israel Binational Science Foundation, project number 2006204.

J. Gudmundsson (Ed.): SWAT 2008, LNCS 5124, pp. 29–40, 2008.

back and forth from the main memory or the disk. In hardware implementations it may be too expensive to fit the entire data structure in the memory which is integrated in the chip that implements the forwarding algorithm itself. So parts of the data structure may reside in external memory devices (such as DRAM). In such designs the communication between the main chip and the external memory becomes the performance bottleneck of the system.

In addition to good I/O performance an efficient data structure for IP-lookup should be able to perform queries at the line rate (which is about 40 Gbps and more today). The data structure has to be scalable since the number of prefixes a router has to maintain is growing rapidly as well as the length of these prefixes. Finally, we need to support fast updates mainly due to instabilities in backbone routing protocols and security issues.

Our computational model. Our algorithm works in the classical *pointer machine model*[19] using only *comparisons* to manipulate prefixes. This is in contrast with many other algorithms for IP-lookup that use bit manipulations on IP addresses. See Section 1. This restriction which we obey does not come at the cost of a complicated data structure. On the contrary, our data structure is much simpler than previously known data structure with the same guarantees.

We develop our data structures in two steps. First we reduce the longest prefix problem to the problem of finding the shortest segment containing a query point. For this data structure we assume that each prefix fits into a constant number of computer words so that we can compare two endpoints of segments in $O(1)$ time. In the second step we relax this assumption and deal with variable length strings with no apriori upper bound on their length. We assume that the strings are over an ordered alphabet Σ and that we can compare two characters of Σ in $O(1)$ time. We do not require direct access to the characters of each prefix.

To analyze I/O performance we use the standard external memory model where memory is partitioned into blocks of size B, and we count the number of blocks that we have to transfer from slow to fast memory in order to perform the operation. This quantity is *the number of I/O operations* performed by the operation [20].

Overview of our results. We first consider the problem of maintaining a dynamic set of segments for point stabbing queries. Specifically, we consider a dynamic *nested family of segments* where each pair of segments are either disjoint or one contains the other. We develop a data structure that given a point p can efficiently find the shortest segment containing p.

Our data structure for this problem which is based on a segment tree (with large fan-out) is particularly simple. In a segment tree we map a segment s to every node v, such that s contains[1] v, and does not contain the parent of v. Typically one maintains at each node v all the segments that map to v in some secondary structure [18,13,12]. We make the crucial observation that for our point stabbing query it is sufficient to maintain for each node v only the shortest segment which maps to v.

[1] A segment contains a node if it contains all points in its subtree.

Our data structure performs a query, and an insertion or a deletion of a segment in $O(\log n)$ time and $O(\log_B(n))$ I/O operations. We manipulate the segments only via comparisons of their endpoints.

We use this result to solve the longest prefix problem as follows. We associate a segment $[pL, pR]$ with each string p, where L and R are two new characters, smaller and larger than all other characters, respectively. We then apply the previous data structure to this set of segments. This gives the data structure for the case where we assume that two prefixes can be compared in $O(1)$ time.

To handle strings with no fixed bound on their lengths we combine this idea with the powerful string B-tree of Ferragina and Grossi [9]. This data structure is a B-tree carefully designed for storing strings. For efficient searches and updates it uses a *Patricia trie* [14] in each node. We show how to maintain the information which we need for longest prefix queries using the string B-tree.

Since our data structure is based on a B-tree it is also I/O efficient. If we pick the size of a node so that it fits in a disk block of size B, we obtain that a query or update with a string q performs $O(\log_B(n)+|q|/B)$ I/O operations. The data structure requires $O(n/B)$ disk blocks in addition to the blocks required to store the strings themselves. The time for query or update with a string q is $O(\log(n) + |q|)$. (This is as efficient as with tries implemented carefully [16], but tries cannot be implemented I/O efficiently [6]).

Previous related results. There has been a lot of work mainly in the networking community on the IP-lookup problem. The different data structures can be classified into three families: trie based structures (See for example [7] and the references there), hash based structures (See for example [11] and the references there), and tree based structures. In the rest of this section we focus on dynamic tree based solutions with worst case guarantees that are related to our approach.

Sahni and Kim [15] describe a solution based on a collection of red-black trees that requires linear space and logarithmic time per operation. Feldmann and Muthukrishnan [8] proposed Fat Inverted Segment tree (FIS). This data structure supports queries in $O(\log\log n + \ell)$ time, where ℓ is the number of levels in the segment tree. The space requirement is $O(n^{1+1/\ell})$, and insert and delete take $O(n^{1/\ell} \log n)$ time, but there is an upper bound on the total number of insertions and deletions allowed. Suri et al. [18] proposed a data structure which is similar to ours in the sense that it is both a segment tree and a B-tree. But they store in each node all the segments which are mapped to it and therefore achieve logarithmic worst case time bound per operation and linear space only for IP-addresses. Lu and Sahni [13] suggested an improvement of the segment tree of Suri that stores each prefix only in one place. They maintain other bit vectors in internal nodes and their update operations are quite complicated. Our structure, which can be extended to general strings (and general segments) and uses only comparisons, is simpler than all the solutions mentioned above. In particular it is much cleaner than the latter two B-tree based implementations when applied to IP addresses.

Kaplan, Molad, and Tarjan [12] considered the problem of point stabbing a dynamic nested set of segments. In their setting, which is more general than ours, each segment has a priority associated with it and we want to find the segment of minimum priority containing a query point. They present a data structure performing query and update in $O(\log n)$ time that requires linear space. It uses both a balanced search tree and a dynamic tree [17] and thereby more complicated than ours (when applied to the special case where the priority of an interval is its length).

In recent years, external memory data structures have been developed for a wide range of applications [20]. A classical I/O efficient data structure is the B-tree [2]. This is a search tree in which we choose the degree of a node so that it occupies a single block. The *string B-tree* of Ferragina and Grossi [9] is a fundamental extension of the B-tree for storing unbounded strings. The main idea is to use a *Patricia trie* [14] in each node to direct the search. Unfortunately this data structure by itself does not solve the longest prefix problem.

Agarwal, Arge, and Yi [1] improved a more general data structure of Kaplan, Molad, and Tarjan [12] for stabbing-min queries against general segments (not necessarily nested). This data structure is based on a *B-tree* and can be implemented so that it is I/O efficient. Specifically, a data structure for n intervals uses $O(n/B)$ disk blocks and $O(\log_B(n))$ I/O operations for query and update. This data structure is quite complicated and assumes that endpoints of intervals can be compared in $O(1)$ time. Therefore it is not directly applicable for longest prefix queries in a collection of unbounded strings.

Brodal and Fagerberg [5] obtained a cache oblivious (see Section 4) data structure for manipulating strings. This data structure, which is essentially a trie can be used to obtain an I/O efficient (though complicated) solution for the static version of the longest prefix problem.

The outline of the rest of the paper is as follows. In section 2 we present our basic ideas using the assumption that strings are of constant size. In section 3 we combine our ideas with the string B-tree to obtain a general I/O efficient solution. In section 4 we suggest a future research.

2 B Tree for Longest Prefix Queries

Our input is a set of prefixes $S = \{p_1 \dots p_n\}$ which are strings over the alphabet Σ. We think of each prefix p as a segment $I(p) = [pL, pR]$, where L and R are two special characters not in Σ, L is smaller than all characters in Σ and R is larger than all characters in Σ. The longest prefix p of a query q corresponds to the the shortest segment $I(p)$ containing q. We describe a dynamic data structure to maintain a set of nested segments such that we can find the smallest segment containing a query point. Although we can apply our data structure to any set of nested segments we present our result using the string terminology and the set of segments $\{I(p) \mid p \in S\}$.

Let $P = \{p_iL,\ p_iR \mid p_i \in S\}$ be the set of endpoints of the prefixes in S. We store P ordered lexicographically at the leaves of a B^+ *tree* T. Each internal node x of T has $n(x)$ children, where $b \leq n(x) \leq 2b$. If x is a leaf then it stores $n(x)$ endpoints of P, where $b \leq n(x) \leq 2b$. (See Figure 1.)

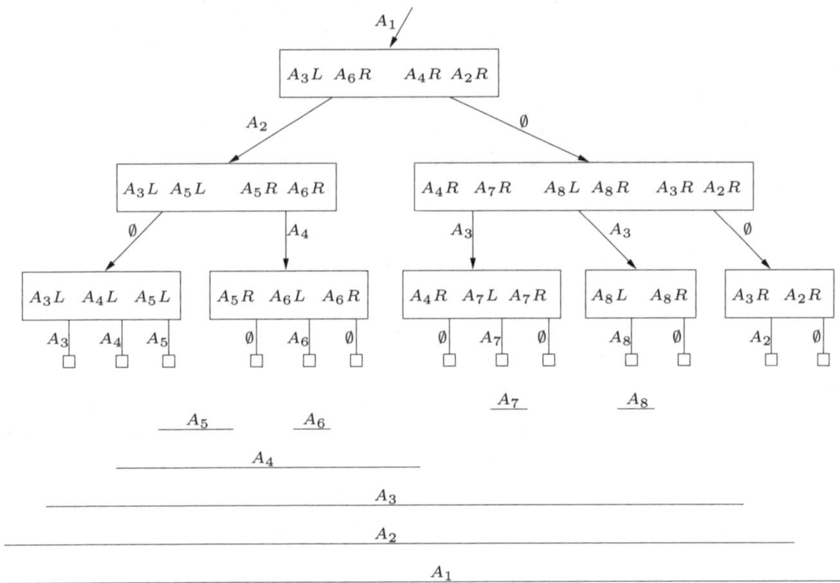

Fig. 1. A B^+ *tree* with $b = 2$ storing the prefixes $A_1, \ldots, A_8$. Rectangles correspond to internal nodes and squares correspond to dummy leaves. In each height-1 node we show the endpoints that it stores. In each internal node v of height > 1 we show the spans of its children which are also used as the keys which direct the search. (Note that when we use the spans as keys, a search can never reach the first dummy leaf in each height-1 node. Therefore we do not need to keep longest prefixes of these nodes and we do not show them in the figure.) The span of a child u of v is the closed interval from the point depicted to the left of the edge from v to u to the point depicted to the right of the edge from v to u. On each edge $(p(v), v)$ we show the longest prefix of v.

To simplify the presentation we assume that a leaf x with $n(x)$ endpoints has $n(x) + 1$ "dummy" children. From now on when we say a leaf of T, we refer to one of these dummy nodes, and we refer to x as a *height-1 node*. Each endpoint in a height-1 node plays the role of a key separating two consecutive dummy leaves. We associate each leaf with the open interval from the endpoint preceding the leaf to the endpoint following the leaf. We call this interval the *span* of the leaf. (Note that the last dummy leaf in a height-1 node and the first dummy leaf in the next height-1 node have the same span.) We define the *span* of an internal node v to be the smallest interval containing the endpoints which are descendents of v.[2] We denote the span of a node v by $span(v)$. We think of T as a segment tree and map each segment $[pL, pR]$ to every node v such that pL is to the left of $span(v)$ and pR is to the right of span(v), and either pL or pR are in $span(p(v))$. We define the longest prefix (or the shorter segment)of

[2] This is the interval that starts at the leftmost endpoint in the subtree of v, and ends at the rightmost endpoint in the subtree of v.

v and denote it $LP(v)$ to be the shortest segment mapped to v.[3] If there isn't any segment which is mapped to v we define $LP(v)$ to be empty. ($LP(v)$ is also defined if v is a dummy leaf.)

We store $span(v)$ and $LP(v)$ with the pointer to node v. Note that when we are at a node v we can use the span values of the children of v as the keys which direct the search. We denote by $B = O(b)$ the maximum size of a node. We pick b so that B is the size of a disk block.

2.1 Finding the Longest Prefix

Assume we want to find the longest prefix of a query string q. We search the B^+ *tree* with the string q[4] in a standard way and traverse a path A to a leaf of T. We return $LP(w)$, where w is the last node on A, such that $LP(w)$ is not empty.

The correctness of the query follows from the following observations. For each prefix p of q, $I(p)$ is mapped to some node u on A. Therefore p is $LP(u)$ unless some longer prefix is mapped to u. Furthermore, since for every v, $LP(p(v))$ is a prefix of $LP(v)$, it follows that the longest prefix of q must be $LP(w)$, where w is the last node on A for which $LP(v)$ is not empty.

2.2 Inserting a New Prefix

To insert a new prefix p we have to insert $I(p)$ into T. We insert pL and pR into the appropriate *height-1* nodes w and w', respectively, according to the lexicographic order of the endpoints. The endpoint pL is inserted into the span of a leaf y and the endpoint pR is inserted into the span of a leaf z. Assume first that $z \neq y$. The span of y is now split between two new leaves: y' that precedes pL and y'' that follows pL. We set the longest prefix of y' to be the longest prefix of y and the longest prefix of y'' to be p. The span of z is now split between two new leaves: z' that precedes pR and z'' that follows pR. We set the longest prefixes of z' to be p and the longest prefix of z'' to be the longest prefix of z. If $y = z$ then the span of y is split between three new leaves y^1, y^2, and y^3. We set the longest prefixes of y^1 and y^3 to be the longest prefix of y, and the longest prefix of y^2 to be p.

There may be nodes v in T, such that after adding p, we have to update $LP(v)$ to be p. Let y be the leaf preceding pL and let z be the leaf following pR. Let u be the lowest common ancestor of y and z. Let u' be the child of u which is an ancestor of y and let u'' be the child of u which is an ancestor of z. We may need to update $LP(v)$ if either:

Case 1: v is a child of a node w on the path from u' to y, which is right sibling of the child w' of w on this path.

Case 2: v is a child of a node w on the path from u'' to z, which is left sibling of the child w' of w on this path.

---

[3] Note that all segments mapped to v form a nested family of segments, so the shortest among them is unique.

[4] In fact we "pretend" to search with a string (not in the data structure) that immediately follows qL in the lexicographic order of the strings.

Case 3: v is a child of u which is a right sibling of u' and left sibling of u''.

For each such node v, we know that $span(v) \subset I(p)$ so we change $LP(v)$ to be p, if p is longer than the current $LP(v)$. Since the depth of T is $O(\log_B(n))$, we update $O(B \log_B(n))$ longest prefixes which are stored at $O(\log_B(n))$ nodes.

After inserting pL and pR, if the *height-1* node containing pL and the *height-1* node containing pR have no more than $2b$ children, we finish the insert. Otherwise we have to split at least one of these nodes. We split node v into two nodes v_1 and v_2. Node v_1 is the parent of the first b (or $b+1$) children of v and node v_2 is the parent of the last $b+1$ children of v. Both v_1 and v_2 replace v as consecutive children of $p(v)$. We compute $span(v_1)$ from the span of its first child and the span of its last child, and similarly for $span(v_2)$.

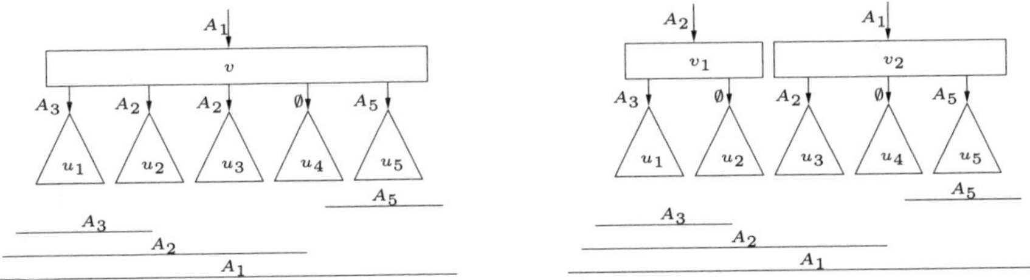

Fig. 2. A node v at the left which is split into nodes v_1 and v_2 to the right. Since $span(v_1) \subseteq I(A_2)$ the prefix A_2, which was the longest prefix of u_2 before the split, is the longest prefix of v_1 after the split. The longest prefix of u_2 after the split is empty.

Clearly we have to update $LP(v_1)$ and $LP(v_2)$. Furthermore, since a segment that was mapped to a child u of v may now be mapped to v_1 or v_2, we may also have to update $LP(u)$ for children u of v_1 and v_2. Other longest prefixes do not change. The following simple observations specify how to update the longest prefixes. In the following if u is a child of v prior the split, then $LP(u)$ refers to the longest prefix of u before the split. Note that since v exists only before the split then $LP(v)$ is the longest prefix of v before the split. Similarly, $LP(v_1)$ and $LP(v_2)$ are the longest prefix of v_1 and v_2, respectively, after the split.

Lemma 1. *Let u be a child of v_1 after the split. If $span(v_1) \subset I(LP(u))$ then after split $LP(u)$ should be empty.*

Proof. Let $p = LP(u)$ since $span(v_1) \subset I(LP(u))$ then the segment $I(p)$ is not mapped to u after the split. Since $I(p)$ was the shortest segment that was mapped to u no other segment is mapped to u after the split. □

Lemma 2. *Let u_1 and u_2 be children of v_1. If $span(v_1) \subset I(LP(u_1))$ and $span(v_1) \subset I(LP(u_2))$ then $LP(u_1) = LP(u_2)$.*

Proof. Since $span(v_1) \subset I(LP(u_1))$ then $span(u_2) \subset I(LP(u_1))$. So $LP(u_1)$ cannot be longer than $LP(u_2)$ since this would contradict the fact that $I(LP(u_2))$ is the shortest segment containing $span(u_2)$. Symmetrically, $LP(u_2)$ cannot be longer than $LP(u_1)$, so they must be equal. □

Lemma 3. *If there exist child u of v_1 such that $span(v_1) \subset I(LP(u))$ then $LP(v_1)$ is $LP(u)$.*

Proof. Obviously $LP(u)$ is mapped to v_1. Furthermore, $LP(u)$ is the longest prefix with this property, since if there is a longer prefix q with this property then q should have been $LP(u)$ before the split. □

Lemma 4. *If there isn't a child u of v_1 such that $span(v_1) \subset I(LP(u))$ then $LP(v_1)$ is equal $LP(v)$.*

Proof. We claim that there exists a child u of v_1 that $LP(u)$ is empty. From this claim the lemma follows since if there is a prefix q longer than $LP(v)$ such that $I(q)$ is mapped to v_1, then q is mapped to u before the split and $LP(u)$ couldn't have been empty.

We prove this claim as follows. Assume to the contrary that $LP(u)$ is not empty for every child u of v_1. Let w be a child of v_1 such that $I(LP(w))$ is not contained in $I(LP(w'))$ for any other child w' of v (w exists since segments do not overlap). From our assumption follows that $I(LP(w)) \subseteq span(v_1)$. Therefore at least one of the endpoints of $I(LP(w))$, say z is in the subtree of v_1. Let w'' be a child of v_1 whose subtree contains z. It is easy to see now that $I(LP(w''))$ and $I(LP(w))$ overlap which is a contradiction. □

A symmetric version of Lemmas 1, 2, 3, and 4 hold for v_2.

These observations imply the following straightforward algorithm to update longest prefixes when we perform a split. If there is a child u of v_1 such that $span(v_1) \subset I(LP(u))$ we set $LP(v_1)$ to be $LP(u)$, otherwise we set $LP(v_1)$ to be $LP(v)$. In addition we set $LP(u)$ to be empty for every child u of v_1 such that $span(v_1) \subset I(LP(u))$. We update the span of v_2 and its children analogously. See Figure 2.

After splitting v we recursively check if $p(v_1)$ or $p(v_2)$ has more than $2b$ children and if so we continue to split them until we reach a node that has no more than $2b$ children.

2.3 Deleting a Prefix

To delete a prefix p we need to delete $I(p)$ from T. We first find the longest prefix of p in S denoted by w.[5] Then we delete pL and pR from the *height-1* nodes containing them.

We have to change the longest prefix of every node v for which $LP(v) = p$ to w. Nodes v for which $LP(v)$ may be equal to p are of three kinds as specified in Cases (1), (2) and (3) of Section 2.2

As a result of deleting pL and pR from the *height-1* nodes containing them we may create nodes with less than b children. To fix such node v we either *borrow* a child from a sibling of v or *merge* v with one of its siblings. We omit

[5] We do that by a query with a string (not in the data structure) that immediately follows pR in the lexicographically order of strings.

the details of these rebalancing operations and their affect on longest prefixes from this abstract. The following theorem summarizes the results of this section.

Theorem 1. *Assuming each string occupies $O(1)$ words, the B-tree data structure which we described supports longest prefix queries, insertions, and deletions in $O(\log(n))$ time. Furthermore, it performs $O(\log_B(n))$ I/Os per operation, and requires linear space.*

3 String B-Tree for Longest Prefix Queries

In a *B-tree*, we assume that $\Theta(b)$ keys that reside at a single node fit into one disk block of size B. However if the keys are strings of variable sizes, which can be arbitrarily long, there may not be enough space to store $\Theta(b)$ strings in a single block. Instead, we can store $\Theta(b)$ pointers to strings in each node, but accessing these strings during the search requires more than a constant number of I/O operations per node. To reduce the number of I/Os, Ferragina and Grossi [9] developed an elegant generalization of a *B-tree* called the *string B-tree* or *SB-tree* for short.

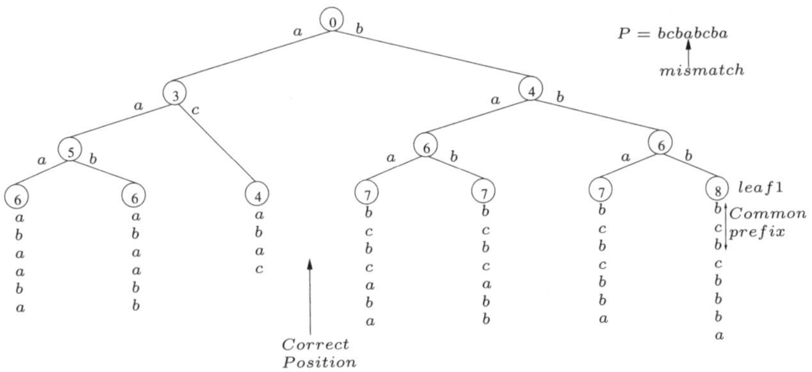

Fig. 3. A Patricia trie of a node in a string B-tree. The number in a node is its string depth. The character on an edge is the branching character of the edge.

An individual node v of an *SB-tree* is shown in Figure 3. Instead of storing the keys at a node v we store a *Patricia trie* [14] of the keys, denoted by $PT(v)$. Using this representation we can perform b-way branching using only $\Theta(b)$ characters that are stored in a constant number of disk blocks of size B. Each internal node ξ of the Patricia trie stores the length of the string corresponding the path from the root to ξ. We call this the *string depth* of ξ. We store with each edge e the first character of the string that corresponds to e. This character is called the *branching character* of e.

As an example Figure 3 shows a Patricia trie of a node in a string B-tree. The right child of the root has string depth 4 and it's outgoing edges have the branching characters "a" and "b", respectively. This means that the node's left subtrie consists of strings whose fifth character is "a" , and its right subtrie consists of strings whose fifth character is "*b*". The first four characters in all the strings in the right subtrie of the root are "*bcbc*". Let ξ be a node of the trie whose string depth is $d(\xi)$. To make a branching decision at ξ, we compare the $d(\xi)+1$ character of the string that we search, to the characters on the edges outgoing from ξ. For example, for the string "*bcbabcba*", the search in the trie in Figure 3 traverses the rightmost path of the Patricia trie, examining the characters 1, 5, and 7 of the string which we search.

Unfortunately, the leaf of the Patricia trie that we reach (in our example, the leaf at the far right, corresponding to "*bcbcbbba*") is not in general the correct branching point, from the node of the *SB-tree* represented by this trie, since we did not compare all characters of the string which we search. We fix this by sequentially comparing the string which we search with the key associated with the leaf of the trie which we reached. If they differ, we find the position in which they first differ. In the example the first character of the string "*bcbabcba*" that is not equal to the corresponding character of the key "*bcbcbbba*", is the fourth character. Since the fourth character of "*bcbabcba*" is smaller we know that the string which we search is lexicographically smaller than all keys in the right subtree of the root. It thus fits in between the leaves "*abac*" and "*bcbcaba*". For more details see [9].

Searching each Patricia trie requires constant number of I/O to load it into memory, plus additional I/Os to do the sequential scan of the key associated with the leaf we reached. Therefore our structure as defined so far does not guarantee that the total number of I/Os is $O(\log_B n + \ell/B)$, where ℓ is the length of the string that we search.

To further reduce the number of I/Os Ferragina and Grossi [9] used the leftmost and the rightmost strings in the subtree of a node v as keys at $p(v)$. Recall that we in fact did the same in our B-tree when we use the spans of the children of v as the keys at v. Having the keys defined this way, we can use information from the search in the trie $PT(v)$ of a node v to reduce the number of I/Os in the followings search of the trie $PT(u)$ of a child u of v. Specifically, let s be the string which we search, and let ℓ be the length of the longest common prefix of s and the key at the leaf of $PT(v)$, where the search ended. Then it is guaranteed that the length of the longest common prefix of s and the key at the leaf of $PT(u)$, where the search of s ends is at least ℓ. Thus, we can avoid the first ℓ comparisons and the I/Os associated with them. Ferragina and Grossi [9] also showed how to insert and delete a string in $O(\log_B n + \ell/B)$ time in the worst case.

We now describe how to combine the SB-tree with our algorithm for longest prefix queries so that our input prefixes $S = \{p_1 \dots p_n\}$ can be arbitrarily long. As Ferragina and Grossi [9], we use the endpoints of *span*(v) as keys at $p(v)$, and represent the keys of each node v in a Patricia trie $PT(v)$. Each leaf of the Patricia trie stores a pointer to the first block containing the key that it corresponds to. We use the same definition of the longest prefix of a node v, denoted by $LP(v)$, as in Section 2. Recall that from these definitions follow that if $LP(v)$ is not empty then $span(v) \subset I(LP(v))$ and therefore $LP(v)$ is a prefix of every key in the subtree of v. Let $span(v) = [KL(v), KR(v)]$. That is $KL(v)$ be the leftmost string in the subtree of v and $KR(v)$ is the rightmost string in the subtree of v. Clearly $LP(v)$ is a prefix of $KL(v)$ and $KR(v)$. The string $KL(v)$ is a key separating v from its sibling in $p(v)$ and therefore corresponds to a leaf in $PT(v)$. So we represent $LP(v)$ by storing its length, and pointer to it, in the leaf of $PT(v)$, that corresponds to $KL(v)$. If $LP(v)$ is empty we encode this by storing zero at the associated leaf.

Finding the longest prefix. We search the *SB-tree* and traverse a path A to a leaf of T. Let w be the last node on A for which $LP(w)$ is not empty. Together with the pointer to w in $p(w)$, we find $|LP(w)|$ and a pointer to $LP(w)$.

Inserting a new prefix. Assume we want to insert a new prefix $p \in S$ to the data structure. We insert pL and pR into the *SB-tree* using the insertion algorithm of the *SB-tree*. As in Section 2.2 there may be nodes v in T, such that after adding p, we need to update $LP(v)$ to be p. Nodes v for which $LP(v)$ may be equal to p are of three kinds as specified in Cases (1), (2) and (3) of Section 2.2. For each such node v we change $|LP(v)|$ to be $|p|$, if $|p|$ is longer than the current value $|LP(v)|$. This is correct since for each of these nodes v, we know that $span(v) \subset I(p)$. Note that all these changes are located at $O(\log_B(n))$ nodes of the *SB-tree*, and therefore we can perform them while doing $O(\log_B(n))$ I/O operations.

After inserting a prefix p we may split node v into two nodes v_1 and v_2. We split a node in the *SB-tree* using the algorithm of Ferragina and Grossi [9]. Splitting may change the longest prefixes. To perform these changes we use the same algorithm as in Section 2.2. To implement this algorithm we need to determine if there is a child u of v_1 such that $span(v_1) \subset I(LP(u))$.

Let u be a child of v_1. We decide if $span(v_1) \subset I(LP(u))$ as follows. Since $LP(u)$ is a prefix of $KL(u)$ and $KR(u)$, and $KL(u)$ and $KR(u)$ are keys in $PT(v_1)$ then there is a path in $PT(v_1)$ that corresponds to the string $LP(u)$. It follows that $LP(u)$ is a prefix of all the keys in $PT(v_1)$, and in particular of $KL(v_1)$ and $KR(v_2)$, if $|LP(u)|$ is not larger than the string depth of the root of $PT(v_1)$. We check if there is a child u of v_2 that $span(v_2) \subset I(LP(u))$ analogously.

Deletion of a prefix is similar, we omit the details from this abstract. The following theorem summarizes the results of this section.

Theorem 2. *The data structure which we described in this section supports longest prefix queries, insertions, and deletions in $O(\log(n) + |q|)$ time where q is the string which we perform the operation with. Furthermore, it performs $O(\log_B(n) + |q|/B)$ I/Os per operation, and requires linear space.*

4 Future Research

The cache oblivious model [10] is a generalization of the I/O model. In this model we seek I/O efficient algorithms which do not depend on the block size.

Among the state of the art in this model is a cache-oblivious B-tree [3], and an almost efficient cache-oblivious string B-tree [4] whose query time is optimal but updates are not. An obvious open question is to find a cache oblivious data structure for longest prefix queries.

References

1. Agarwal, P.K., Arge, L., Yi, K.: An Optimal Dynamic Interval Stabbing-Max Data Structure? In: Proceedings of SODA, pp. 803–812 (2005)
2. Bayer, R., McCreight, E.M.: Organization and Maintenance of Large Ordered Indexes. Acta Informtica 1(3), 173–189 (1972)
3. Bender, M.A., Demaine, E., Farach-Colton, M.: Cache-Oblivious B-Trees. SIAM Journal on Computing 35(2), 341–358 (2005)
4. Bender, M.A., Farach-Colton, M., Kusznaul, B.C.: Cache-Oblivious String B-Trees. In: Proceedings of PODS, pp. 233–242 (2006)
5. Brodal, G.S., Fagerberg, R.: Cache-oblivious string dictionaries. In: Proc. 17th Annual ACM-SIAM Symposium on Discrete Algorithms, pp. 581–590 (2006)
6. Demaine, E.D., Iacono, J., Langerman, S.: Worst-case optimal tree layout in a memory hierarchy (2004)
7. Eatherton, W., Dittia, Z., Varghese, G.: Tree Bitmap: Hardware/Software IP Lookups with Incremental Updates. ACM SIGCOMM Computer Communications Review 34(2), 97–122 (2004)
8. Feldmann, A., Muthukrishnan, S.: Tradeoffs for Packet Classification. In: Proceedings of INFOCOM, pp. 1193–1202 (2000)
9. Ferragina, P., Grossi, R.: The String B-Tree: A New Data Structure for String Search in External Memory and Its Applications. Journal of the ACM 46(2), 236–280 (1999)
10. Frigo, M., Leiserson, C.E., Prokop, H., Ramachandran, S.: Cache-Oblivious Algorithms. In: Proceedings of FOCS, pp. 285–297 (1999)
11. Hasan, J., Cadambi, S., Jakkula, V., Chakradhar, S.: Chisel: A Storage-Efficient, Collision-Free Hash- Based Network Processing Architecture. In: Proceedings of ISCA, May 2006, pp. 203–215 (2006)
12. Kaplan, H., Molad, E., Tarjan, R.E.: Dynamic Rectangular Intersection with Priorities. In: Proceedings of STOC, pp. 639–648 (2003)
13. Lu, H., Sahni, S.: A B-Tree Dynamic Router-Table Design. IEEE Transactions on Computers 54(7), 813–824 (2005)
14. Morrison, D.R.: Patricia: Practical Algorithm to Retrieve Information Coded in Alphanumeric. Journal of the ACM 15(4), 514–534 (1968)
15. Sahni, S., Kim, K.: O(log n) Dynamic Packet Routing. In: Proceedings of ISCC, pp. 443–448 (2002)
16. Sleator, D., Tarjan, R.E.: Self-adjusting binary search trees. Journal of the ACM 32, 652–686 (1985)
17. Sleator, D.D., Tarjan, R.E.: A Data Structure for Dynamic Trees. JCSS 26(3), 362–391 (1983)
18. Suri, S., Varghese, G., Warkhede, P.: Multiway Range Trees: Scalable IP Lookup with Fast Updates. In: Proceedings of GLOBECOM, pp. 1610–1614 (2001)
19. Tarjan, R.E.: A Class of Algorithms which Require Nonlinear Time to Maintain Disjoint Sets. Journal of Computing System Science 18, 110–127 (1979)
20. Vitter, J.S.: External Memory Algorithms and Data Structures: Dealing with Massive Data. ACM Computing Surveys 33(2), 209–271 (2001)

Guarding Art Galleries: The Extra Cost for Sculptures Is Linear*

Louigi Addario-Berry[1], Omid Amini[2], Jean-Sébastien Sereni[3], and Stéphan Thomassé[4]

[1] Department of Statistics, Oxford University, Oxford, UK
louigi@gmail.com
[2] Max-Planck-Institut für Informatik, Saarbrücken, Germany
amini@mpi-inf.mpg.de
[3] Institute for Theoretical Computer Science and Department of Applied Mathematics, Charles University, Prague, Czech Republic
sereni@kam.mff.cuni.cz
[4] LIRMM-Université Montpellier II, Montpellier, France
thomasse@lirmm.fr

Abstract. Art gallery problems have been extensively studied over the last decade and have found different type of applications. Normally the number of sides of a polygon or the general shape of the polygon is used as a measure of the complexity of the problem. In this paper we explore another measure of complexity, namely, the number of guards required to guard the boundary, or the walls, of the gallery. We prove that if n guards are necessary to guard the walls of an art gallery, then an additional team of at most $4n-6$ will guard the whole gallery. This result improves a previously known quadratic bound, and is a step towards a possibly optimal value of $n-2$ additional guards. The proof is algorithmic, uses ideas from graph theory, and is mainly based on the definition of a new reduction operator which recursively eliminates the simple parts of the polygon. We also prove that every gallery with c convex vertices can be guarded by at most $2c-4$ guards, which is optimal.

Keywords: Art Gallery, Pseudo-triangulation.

1 Introduction

Art gallery problems are, broadly speaking, the study of the relation between the shapes of regions in the plane and the number of points needed to *guard* them. The problem of determining how many guards are sufficient to see every point in the interior of an n-wall art gallery room was first posed by Klee [11]. Conceptually, the room is a simple polygon P with n vertices, and the guards are stationary points in P that can see any point of P connected to them by a straight line segment lying entirely within P. The first "art gallery theorem"

* This work was supported by the European project IST FET AEOLUS.

J. Gudmundsson (Ed.): SWAT 2008, LNCS 5124, pp. 41–52, 2008.

was obtained by Chvátal [3], who demonstrated that given any simple polygon with n sides, the interior of the polygon can be guarded with at most $\lfloor \frac{n}{3} \rfloor$ guards and that this number of guards is sometimes necessary. Fisk [8] later found a simpler proof which lends itself to an $O(n \log n)$-time algorithm developed by Avis and Toussaint [2] for locating these $\lfloor \frac{n}{3} \rfloor$ stationary guards. With some restriction on the shape of the polygon, for example if the polygon is rectilinear, that is, the edges of the polygon are either horizontal or vertical, Kahn *et al.* [12] have shown that $\lceil \frac{n}{4} \rceil$ guards are sufficient and sometimes necessary. Sack [20] and Edelsbrunner *et al.* [6] have devised an $O(n \log n)$-time algorithm to locate these $\lfloor \frac{n}{4} \rfloor$ guards. These classical results in the theory of art galleries have spawned a plethora of research (see the monograph by O'Rourke [18], and the surveys [21,23,25] for overviews of previous work). In particular, since then the art gallery problems have emerged as a research area that stresses complexity and algorithmic aspects of visibility and illumination in configurations comprising obstacles and guards.

In most of the reseach papers in the field, the number of sides of a polygon or restriction on the shape of the polygon is used as a very natural measure of the "complexity" of the polygon. The aim of this paper is to explore another measure of complexity, namely the number of guards required to guard the boundary, or the walls, of the gallery. As we will see in the next sections, this new complexity measure can be regarded as a mixture of the two named ones: the shape and the number of sides, but remains different and has its own characteristics. As shown in Figure 1, a team of guards inside a gallery can see the walls (where paintings are hung), without necessarily guarding the whole gallery (where sculptures are displayed), showing that these two notions of complexity are in general different. More precisely, the question we investigate in this paper is the following: given that the interior walls of a polygon can be guarded with at most n guards, how many *additional* guards may be needed to guard the whole interior? This question has been first explored by Aloupis *et all.* in [1] in their study of fat polygons. They proved that an additional number of at most $3n^2/2$ guards can guard the whole gallery.

Main Results. We prove the following linear bound.

Theorem 1. *Let M be a polygonal gallery. If the walls of M can be guarded by at most $n > 1$ guards, an additional set of $4n - 6$ guards is sufficient to guard the interior of M.*

Observe that when $n = 1$, the unique guard sees all the walls, hence sees the whole gallery. Most likely, the previous bound is not sharp. We offer the following conjecture.

Conjecture 1. *If the walls of a gallery can be guarded by $n > 1$ guards, then $n - 2$ additional guards are sufficient to guard the whole gallery.*

If Conjecture 1 is true then the given value would be optimal, as is shown by the example in Figure 2. In this example, there are $n-2$ "small rooms" attached by narrow entrances to a main room. Guarding the walls requires at most n guards

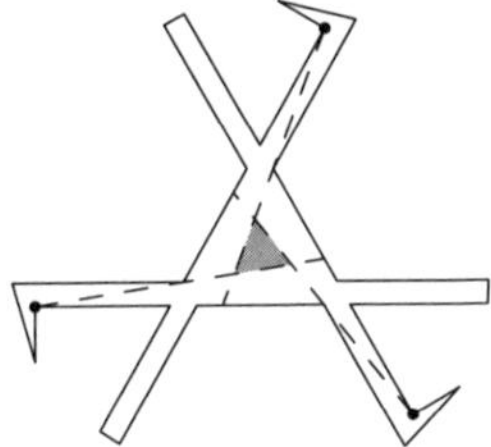

Fig. 1. Three guards are enough the guard the paintings (on the walls), but not the sculpture in the shaded area. Dashed lines are lines of sight of the guards.

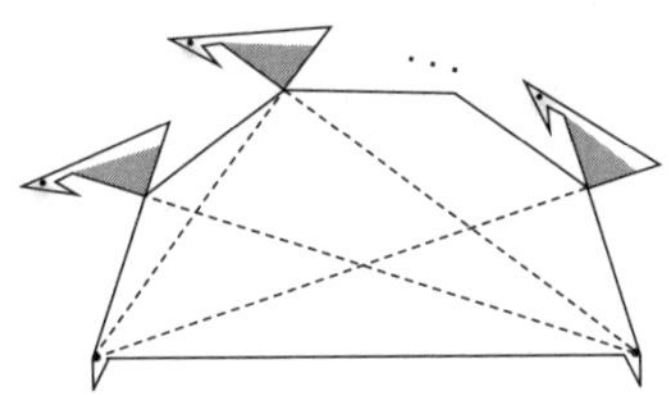

Fig. 2. Black dots indicate guards. The shaded areas indicate parts not seen by any of the current n guards, and dashed lines are lines of sight of the guards.

(as shown): one guard in each of the "small rooms" off the main room, and one guard each of the two far corners of the main room. These latter two guards each have a line of sight along one wall of each small room. However, with such a set of guards the parts of the gallery's interior shaded in dark grey are left unguarded. To guard the whole gallery requires *two* guards in each of the "small rooms", and an additional two guards in the main room.

The proof of Theorem 1 uses the fact that every gallery with c convex vertices can be guarded by at most $2c - 4$ guards. This latter result is optimal. In order to apply induction to bound the number of additional guards required to guard M, we first *reduce* our gallery to another gallery with certain guaranteed structural properties that make it easier to analyse. We do so by means of a new *transformation operator* $T(\cdot, \cdot)$, which takes as an argument a gallery N and a set of guards G that guards the walls of N, and returns another gallery N'. The operator T captures the complexity of the polygon by successfully deleting the parts which do not contribute to the main complexity. It has a nice definition and the general idea behind it may hopefully be applied to other contexts. On the complexity side, using the reduction operator T and earlier results [16,7], one can infer that the general problem of calculating the number of extra guards needed is NP-complete and does not admit a PTAS, i.e. is APX-hard.

Related Work. As we mentioned before, the literature on the art gallery problems is huge and different type of strategies and situations have been considered. Let us briefly review some of the works related to this paper. Laurentini [14] investigated the problem of covering the sides of the polygon and not necessarily the interior—related complexity questions being studied in another paper [9]. Efrat *et al.* [5] introduced the *link diagram* of a polygon. As we will see later, the last step in the proof of Theorem 1 is based on a certain kind of link diagram between the guards. It is interesting to explore the relations between the two notions. As the graph we use is based on the connectivity between guards, another related subject is that of the *guarded guard art gallery* problem [15,17]. In particular, one could investigate the guarded guard version of our problem.

Notations and Basic Definitions. Let us give some formal definitions. We let $\overline{S}$ be the closure of the set $S \subset \mathbf{R}^2$. A simply connected, compact set $M \subset \mathbf{R}^2$ is *polygonal* if its boundary ∂M is a simple closed polygon with finitely many vertices. The set M is *nearly* polygonal if M can be written as the union of polygonal galleries $M_1, \ldots, M_k$ such that

(i) for distinct $i, j \in \{1, \ldots, k\}$, letting $E_{i,j} = M_i \cap M_j$, either $E_{i,j} = \emptyset$ or $E_{i,j}$ contains a single point $e_{i,j}$; and
(ii) the connectivity graph with node set $\{v_1, \ldots, v_k\}$, where v_i represents the polygonal gallery M_i and nodes v_i and v_j are adjacent whenever $M_i \cap M_j \neq \emptyset$, is a tree.

We sometimes refer to the set ∂M as *the walls of* M, and to $M_1, \ldots, M_k$ as the *rooms* of M. A point $p \in M$ is a *cut-vertex* of M if $M \setminus \{p\}$ is not connected—so the cut-vertices of M are precisely the points $e_{i,j}$ defined above.

If M is a polygonal gallery, then we may describe M by simply listing the vertices of the polygon ∂M in their cyclic order, which we always assume is given in the "clockwise direction". Similarly, we may describe a nearly polygonal gallery M by listing the vertices of ∂M in cyclic order (again, in this paper always clockwise). If M is the nearly polygonal gallery described by $P = (p_1, \ldots, p_k, p_{k+1} = p_1)$, then M is polygonal precisely if P has no repeated points. Given points x and y of ∂M, by $\partial M[x, y]$ we mean the subset of ∂M starting at x and ending at y and following the cyclic order. These straightforward definitions and facts are depicted in Figures 3 and 4. We will also often abuse notation and write P or $P[x, y]$ in place of ∂M or $\partial M[x, y]$, respectively.

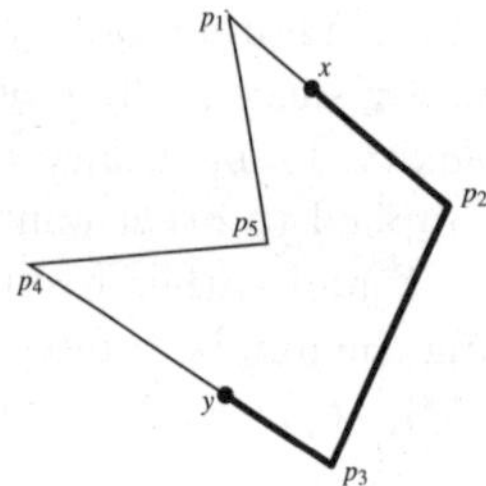

Fig. 3. (a) A polygonal gallery defined by the sequence $(p_1, p_2, p_3, p_4, p_5, p_1)$. The set $\partial M[x, y]$ is shown in bold.

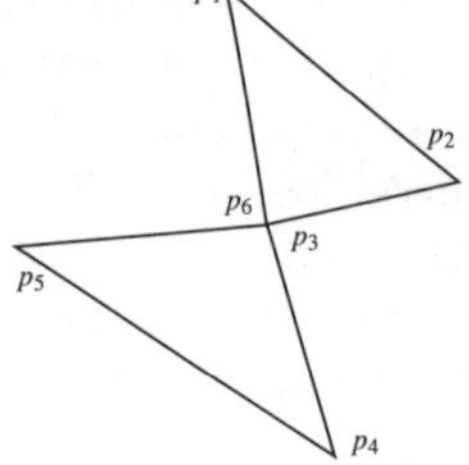

Fig. 4. (b) A nearly polygonal gallery defined by the sequence $(p_1, p_2, p_3, p_4, p_5, p_6, p_1)$, with $p_6 = p_3$

A *guard* is a point of M. A guard g *sees* a point p of M if the line segment $[g, p]$ is included in M. Unless otherwise stated, G is always a set of guards in M. We say that G *guards* M if every point of M is seen by a guard of G. Similarly, G *guards* ∂M (or G *guards the walls* of M) if every point of ∂M is seen by a guard of G. The *guarding number* of M is the minimum number of guards necessary to guard M.

2 Guards Versus Convex Vertices

Let $P = (p_1, \ldots, p_k, p_{k+1} = p_1)$ describe a nearly polygonal gallery M. The goal of this section is to prove that the guarding number of M is at most $2c - 4$, where c is the number of convex vertices of M. This bound is sharp: an example is given in Figure 5. The polygon shown in Figure 5 contains five convex vertices. To bound the number of guards required, consider the grey shaded region of the polygon. Regardless of how guards are placed outside the shaded region, the dark grey area remains unguarded, and no single guard can see all the dark grey area. Thus, the grey shaded region of the polygon must contain two guards. Similarly, the two other "concave triangular" areas must each contain two guards, for a total of six guards. This example can easily be generalised to show that for every $c \geq 5$, there is a polygon with c convex vertices requiring $2c - 4$ guards.

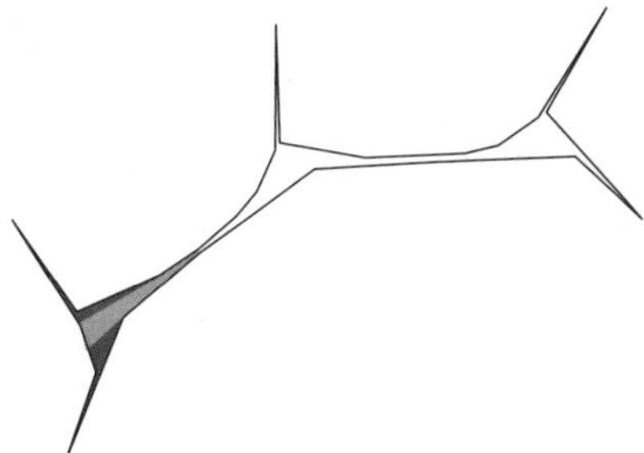

Fig. 5. A gallery with five convex vertices and guarding number six

We will use *pseudo-triangulations* of polygons to obtain our bound. A *pseudo-triangle* is a simple polygon with exactly three convex vertices. Given a simple polygon P, a *pseudo-triangulation* of P is a partition of the interior of P into non-overlapping pseudo-triangles whose vertices are all among vertices of P. We refer to the survey of Rote *et al.* [19] for further exposition about pseudo-triangulations. In our considerations, we need the following result.

Theorem 2. *Every simple polygon with k convex vertices admits a pseudo-triangulation consisting of $k - 2$ pseudo-triangles.*

It is easy to see that

Lemma 1. *The guarding number of a pseudo-triangle is at most 2. Moreover, it is one if the pseudo-triangle contains two consecutive convex vertices.*

The next theorem is a direct consequence of Theorem 2 and Lemma 1.

Theorem 3. *Let M be a polygonal gallery with c convex vertices for some integer $c \geq 3$. Then the guarding number of M is at most $2c - 4 - s$, where $s = 1$ if M contains two consecutive convex vertices, and $s = 0$ otherwise.*

More generally, using pseudo-triangulations, one can show [1]

Theorem 4. *Let M be a polygonal gallery with c convex vertices for some integer $c \geq 3$ such that these vertices appear in t chains of consecutive convex vertices. Then the guarding number of M is at most $c + t - 4$.*

As we don't use Theorem 4 in this full generality, we leave its proof to the full version of this paper.

3 Sculpture Galleries

We now turn our attention to the proof of Theorem 1, which we will prove inductively. For the purposes of our induction, we will in fact prove the following, stronger result.

Theorem 5. *Let M be a nearly polygonal gallery. If ∂M can be guarded with at most n guards, an additional set of $4n - 6$ guards is sufficient to guard M.*

In order to apply induction to bound the number of additional guards required to guard M, we first "reduce" M to another gallery M' with certain guaranteed structural properties that make it easier to analyse. We do so by means of a transformation operator $T(\cdot, \cdot)$, which takes as an argument a nearly polygonal gallery N and a set of guards G that guards the walls of N, and returns another nearly polygonal gallery N'.

Roughly speaking, the effect of T is to "trim off" a section of the polygon N that is unimportant to any of the lines of sight of the guards. Before defining T, then, we first formalise this notion of "importance". Let $U = U(N, G)$ be the set of points of N not seen by any guard $g \in G$. We say that a point p of N is *important* (with respect to N and G) if $p \in G$ or if $p \in \overline{U}$ or if p is a cut-vertex of N.

When there is no risk of confusion, we will write $T(N)$ instead of $T(N, G)$. We also remark that the operator T will be such that $T(N, G)$ is nearly polygonal, $U \subset T(N, G) \subset N$, $G \subset T(N, G)$, and G guards the walls of $T(N, G)$. We ask the reader to keep these properties in mind while reading the definition of $T(N, G)$, to which we now proceed.

3.1 The Definition of the Operator $T(N, G)$

Let N have rooms $N_1, \ldots, N_k$, and suppose that N is described by $P = (p_1, \ldots, p_m, p_1)$. We say N_i is a *leaf* if there is at most one $j \neq i$ such that $N_j \cap N_i \neq \emptyset$. By N_i^- we mean the set $N_i \setminus \cup_{j \neq i} N_j$. N_i is *empty* if $N_i^- \cap G = \emptyset$.

(A) If there is $1 \leq i \leq k$ such that N_i is an empty leaf then set $T(N) = N \setminus N_i^- = \overline{N \setminus N_i}$.

[1] We are very gratefull to the first referee for pointing out that the proof of Theorem 3 we presented in the first version of this paper, which didn't use the pseudo-triangulations, could be simplified and generalised to derive theorem 4 without using the pseudo-triangulations.

(B) Otherwise, if N contains a cut-vertex p such that $p \notin G$ and such that for each simply connected component N^* of $N \setminus \{p\}$, $|G \cap \overline{N^*}| < |G|$ and $G \cap \overline{N^*}$ guards $\overline{N^*}$, then set $T(N) = N$.

(C) Otherwise, if every convex vertex of ∂N is important, set $T(N) = N$.

If none of (A),(B), or (C) occur, then ∂N contains a convex vertex p_i that is not important (and in particular is not a cut-vertex). Choose $x \in [p_{i-1}, p_i[$ and $y \in]p_i, p_{i+1}]$ such that $\triangle_{xp_iy}$ contains no important points except perhaps p_{i-1} and or p_{i+1}, with x as close to p_{i-1} as possible subject to this, and with y as close to p_{i+1} as possible subject to the previous constraints. We note that if $[x, y] \cap \partial N$ contains an interval of positive length, then $[x, y]$ must contain at least one guard; for, letting $[a, b]$ be some interval in $[x, y] \cap \partial N$, no finite set of guards lying outside $\triangle_{xp_iy}$ can see all of $\triangle_{ap_ib}$. (This situation is depicted in Figure 6.)

(D) If $G \cap]x, y[$ is non-empty choose points $x' \in]x, p_i[$ and $y' \in]p_i, y[$ arbitrarily. Let g be some guard of $G \cap]x, y[$ and let $T(N) = \overline{(N \setminus \triangle_{xp_iy}) \cup \triangle_{xx'g} \cup \triangle_{y'yg}}$. (This case is shown in Figure 7)

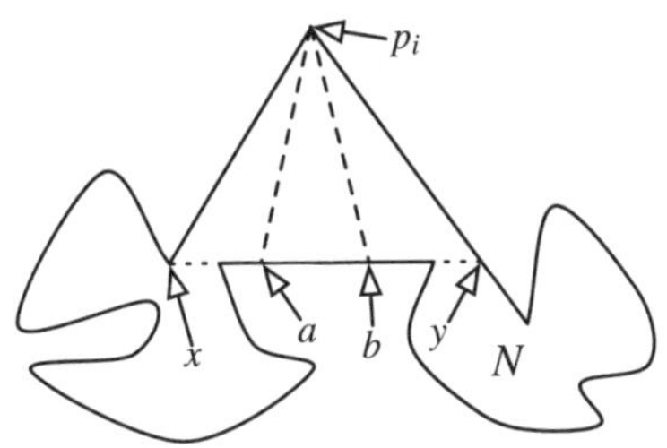

Fig. 6. No matter how guards are placed outside of $\triangle_{xp_iy}$, some part of $\triangle_{ap_ib}$ close to $[a, b]$ will not be seen by any guard

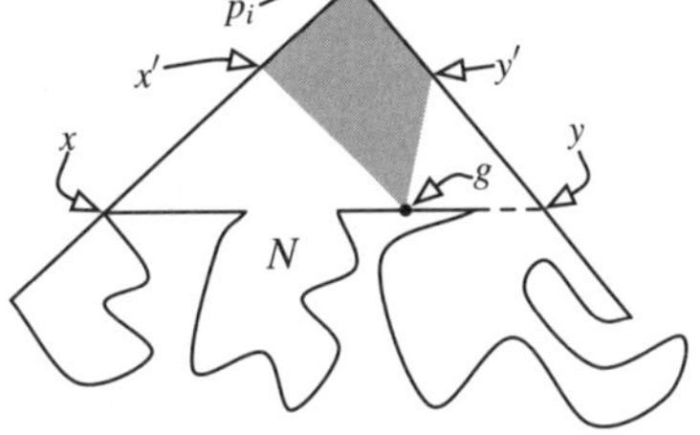

Fig. 7. The situation in case (D). The dark shaded region belongs to N but not to $T(N)$.

(E) Otherwise, suppose $x \in G$ or $y \in G$ – without loss of generality, we presume $x \in G$. Choose $z \in [x, y] \cap \partial N$ such that $[x, z]$ is not contained in ∂N, and as close to x as possible subject to this. Choose $z' \in \partial N$ very close to z and after z in the cyclic order. Finally, choose x' in $[x, y] \cap \partial N$ as far from x as possible such that $[x, x'] \subset \partial N$ (possibly $x' = x$), and set $T(N) = \overline{N \setminus \triangle_{x'zz'}}$. (This case is shown in Figure 8.)

If none of (A)-(E) occur then $[x, y] \cap G = \emptyset$, so $[x, y] \cap \partial N$ contains no interval of positive length.

(F) Otherwise, if $x = p_{i-1}$, $y = p_{i+1}$, or if $]x, y[\cap \partial N \neq \emptyset$, then set $T(N) = \overline{(N \setminus \triangle_{xp_iy})}$.

If (F) does not occur then we may assume without loss of generality that $x \neq p_{i-1}$. Since $[x,y] \cap G = \emptyset$ and $]x,y[\cap \partial M = \emptyset$, by our choice of x and y there must be $z \in [x,y[\cap \overline{U}$.

(G) Otherwise, if $x \in \overline{U}$, then set $T(N) = \overline{(N \setminus \triangle_{xp_iy})}$.
(H) Otherwise, let z be the point of $]x,y[$ that is closest to x such that $z \in \overline{U}$. Pick a point $z' \notin \triangle xp_iy$ chosen close to z in order to guarantee that $\triangle_{xzz'}$ does not contain a guard and is disjoint from U and from ∂N. Finally, set $T(N) = \overline{N \setminus (\triangle_{xp_iy} \cup \triangle_{x'zz'})}$. (This case is shown in Figure 9.)

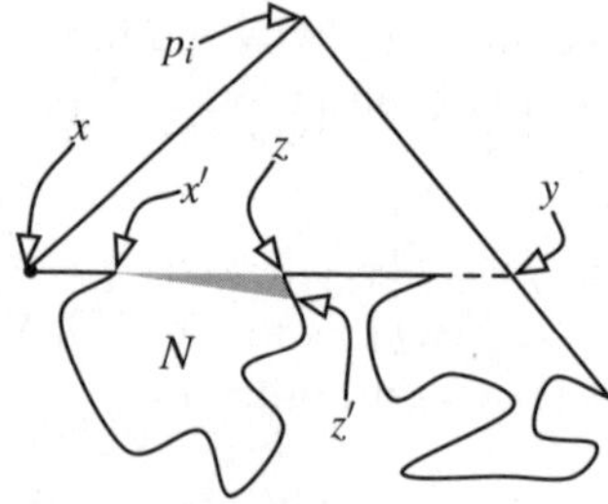

Fig. 8. The situation in case (E). The grey shaded region belongs to N but not to $T(N)$.

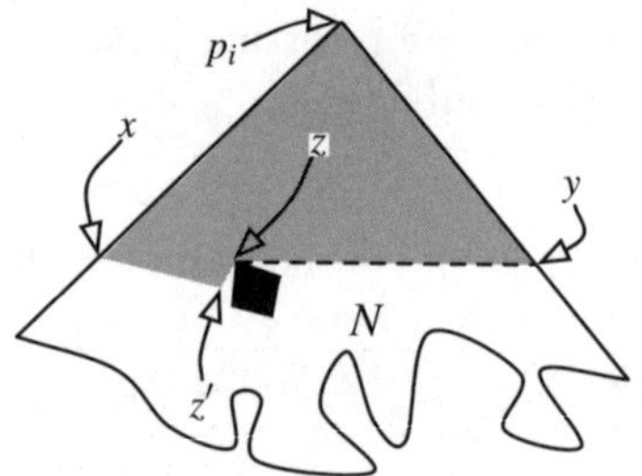

Fig. 9. The situation in case (H). The grey shaded region belongs to N but not to $T(N)$, and the black shaded region belongs to U.

When applying T repeatedly, we will usually write $T^2(N)$ in place of $T(T(N))$. As mentioned at the start of Section 3.1, a key property of this (rather cumbersome) transformation is that if N and G satisfy the hypotheses of Theorem 5 then $T(N)$ and G also satisfy the hypotheses of Theorem 5. Another important property of T is that its repeated application is guaranteed to increase the value of a certain bounded invariant that can be associated to gallery-guard set pairs N, G, and so by applying T to any such pair N, G enough times, we are guaranteed to reach a fixed point of the transformation T.

To define this invariant, we first introduce one additional piece of notation. Let C denote the set of cut-vertices of N, and, for $c \in C$, let $\kappa(c)$ be the number of simply connected components of $N \setminus \{c\}$. The invariant, which we denote $\Phi(N,G)$ (or $\Phi(N)$ when there is no risk of confusion), is equal to the number of convex vertices which are guards, plus the number of convex vertices in $\overline{U}$, plus $\sum_{c \in C \cap G}(\kappa(c) - 1)$.

We hereafter refer to vertices that are also guards as *occupied*, and vertices that are in $\overline{U}$ as *critical*. We observe that every occupied (resp. critical) convex vertex of N is an occupied (resp. critical) convex vertex of $T(N)$. It thus follows from the definition of Φ that $\Phi(T(N),G) \geq \Phi(N,G)$. The main property that makes this invariant useful is captured by the following lemma.

Lemma 2. *For any nearly polygonal gallery M with no set of empty leaves, and any finite set of guards G that see all of ∂M, if $T^2(M) \neq T(M) \neq M$ then either $\Phi(T(M)) > \Phi(M)$ or $T(M)$ has strictly fewer vertices than M.*

Let us postpone the proof of Lemma 2 to the end of the paper.

3.2 The Proof of Theorem 5

Let M and G be as in the statement of Theorem 5 and let $g = |G|$. If $g = 1$, the unique guard sees the whole gallery M. Next suppose that $g > 1$ and that the statement of Theorem 5 holds for all values $n < g$. As previously, we let $U = U(M, G)$ be the (open) set of points of the gallery M that are not seen by any guard of G.

Let $M_0 = M$; for $i \geq 1$ set $M_i = T(M_{i-1}, G)$, It turns out that the number of critical convex vertices in *all* of the galleries M_i can be bounded uniformly in terms of g; this is the substance of the following lemma.

Lemma 3. *For all $i \geq 0$, there are at most $g - 1$ critical convex vertices in M_i.*

We will prove this lemma along with Lemma 2, at the end of the paper.

We observe that M contains precisely $1 + \sum_{\{c \in C : \kappa(c) > 2\}} (\kappa(c) - 1)$ leaves (this can be seen by a straightforward induction). Since M contains no empty leaves, it follows that

$$g - 1 \geq \sum_{\{c \in C : \kappa(c) > 2\}} (\kappa(c) - 1) \geq \sum_{\{c \in C \cap G : \kappa(c) > 2\}} (\kappa(C) - 1),$$

so

$$\sum_{c \in C \cap G} (\kappa(C) - 1) \leq g + \sum_{\{c \in C \cap G : \kappa(c) > 2\}} (\kappa(C) - 1) \leq 2g - 1 .$$

Furthermore, there are at most g occupied convex vertices. It follows by Lemma 3, the above inequalities and the definition of Φ that $\Phi(M_i) \leq 4g - 1$ for all i. By the observation immediately preceding Lemma 2, $\Phi(M_{i+1}) \geq \Phi(M_i)$ for all i, and Lemma 2 then implies that there exists an integer j such that $T(M_j) = M_j$. In this case, by the definition of the operator $T(\cdot)$, one of (B) or (C) occurs for M_j.

We now show that an additional set of at most $4g-6$ guards suffices to guard M_j. Since $U \subset M_j$, these guards also guard U in M_j; since $M_j \subseteq M$, these guards also guard U in M; so together with G, they guard all of M, as claimed. We now assume, purely for the ease of exposition, that $j = 0$, i.e., that $P_j = P$ and $M_j = M$; this eases the notational burden without otherwise changing the proof.

If (B) occurs then we let p be a cut-vertex as described in (B). Let $N_1^-, \ldots, N_r^-$ be the simply connected components of $M - \{p\}$, and for $i \in \{1, \ldots, r\}$ let $N_i^+ = \overline{N_i^-}$, let $G_i = N_i^+ \cap G$ and let $g_i = |G_i|$. Since $p \notin G$, we have $\sum_{i=1}^r g_i = g$. Furthermore, since G_i guards N_i^+ and $g_i < g$ for all $i \in \{1, \ldots, r\}$, the induction hypothesis implies the existence of $H_i \subset N_i^+$ such that $|H_i| \leq 4g_i - 6$ and $G_i \cup H_i$ guards N_i^+. In this case $\bigcup_{i=1}^r (G_i \cup H_i)$ guards M, and

$$\left| \bigcup_{i=1}^r H_i \right| \leq \sum_{i=1}^r (4g_i - 6) \leq 4 \sum_{i=1}^r g_i - 6r \leq 4r - 8 < 4r - 4 .$$

If (C) occurs then since the number of critical convex vertices is at most $g-1$ by Lemma 3, and the number of occupied vertices is at most g, the total number of convex vertices of our gallery M is at most $2g-1$. It follows from Theorem 3 that there is a set H of at most $2(2g-1)-4 = 4g-6$ additional guards that guard M, and hence guard U. □

3.3 Proofs of Lemmas 2 and 3

Having established Theorem 1 assuming that Lemmas 2 and 3 hold, we now turn to the proofs of these lemmas.

Proof (of Lemma 2). Let M and G be as in the statement of the lemma, and suppose that $T(M) \neq M$. Let $M_1 = T(M)$ and let $M_2 = T^2(M)$. For any of the conditions (A)-(H) in the description of T – say (E), for example – we will use the shorthand "(E) holds for M (resp. M_1, M_2)" if, letting $N = M$ (resp. M_1, M_2) in the definition of $T(\cdot)$, the condition described in (E) holds and none of the earlier conditions hold.

By the definition of $T(\cdot)$, if $M_1 \neq M$, then one of (A) or (D)-(H) must hold for M. We now show that in each case, either M_1 has strictly fewer vertices than P, $\Phi(M_1) > \Phi(M)$, or $M_2 = M_1$.

- If (A) holds for M then $T(N)$ has strictly fewer vertices than N.
- If (D) holds for M then g is a cut-vertex in M_1 and, more strongly, $M_1 \setminus \{g\}$ has strictly more connected components than $M \setminus \{g\}$. Thus $\Phi(M_1) > \Phi(M)$.
- If (E) holds for M then since $\mathrm{int}(\triangle_{xp_iy}) \cap G = \emptyset$, x' is a cutvertex in M_1 and, more strongly, (B) holds for M_1 (with $p = x'$). Thus $M_2 = M_1$.
- If (F) holds for M and $x = p_{i-1}$, $y = p_{i+1}$, then M_1 contains strictly fewer vertices than M. If (F) holds for M and $]x, y[\cap \partial M \neq \emptyset$, we first observe that since $[x, y] \cap G = \emptyset$, no guard on the line through x and y can see $\mathrm{int}(\triangle_{xp_iy})$. As every point in $\mathrm{int}(\triangle_{xp_iy})$ is guarded, it follows that every point in $[x, y]$ is seen by some guard g *not* on the line through $[x, y]$. Now let z be a point in $]x, y[\cap \partial M$; then z is a cut-vertex in M_1. Furthermore, by the above comments it must be the case that (B) holds for M_1 (with $p = z$). Thus $M_2 = M_1$.
- If (G) holds for M then x is a critical convex vertex in M_1 but not in M, so $\Phi(M_1) > \Phi(M)$.
- Finally, if (H) holds for M then z is a critical convex vertex in M_1 but not in M, so $\Phi(M_1) > \Phi(M)$.

This completes the proof of Lemma 2.

Proof (of Lemma 3). Let $P = (p_1, \ldots, p_k, p_1)$ describe M. We form a graph $\mathcal{G}$ whose vertex set is the set G of guards of M. For every critical convex vertex p_j, we choose one guard g_i that sees some non-empty interval $]x_j, p_j[$ of $[p_{j-1}, p_j]$, and one guard g'_j that sees some non-empty interval $]p_j, y_j[$ of $[p_j, p_{j+1}]$, and add the edge $g_j g'_j$ to $\mathcal{G}$. By construction, the number of critical convex vertices of P

is at most the number of edges of $\mathcal{G}$. We shall show that $\mathcal{G}$ contains no cycles, from which the conclusion immediately follows.

We first observe that if p_j is a critical convex vertex of P, then the angle $p_{j-1}p_jp_{j+1}$ is strictly positive. Therefore, g_j is different from g_{j+1}, or else the quadrilateral $g_jx_jp_jy_j$ would be entirely seen by g_j, contradicting the fact that p_j is in the closure of U. It follows that $\mathcal{G}$ contains no loops (cycles of length 1).

Next, suppose that $\mathcal{G}$ contains a cycle $g_1, g_2, \ldots, g_k, g_{k+1} = g_1$ with $k \geq 2$ (in which case $g_{i+1} = g_i'$ and g_ig_{i+1} is the edge corresponding to some critical convex vertex p_i, for $i \in \{1, \ldots, k\}$). In this case, the polygonal line

$$\mathrm{PL} = (g_1, p_1, g_2, p_2, \ldots, g_k, p_k, g_1)\,,$$

which is not necessarily simple or even uncrossing, contains some simple, closed polygonal line $\mathrm{PL}_1 = (x, g_i, p_i, \ldots, x)$ or $\mathrm{PL}_2 = (x, p_i, g_{i+1}, \ldots, x)$. We emphasise that though a line segment $[g_i, g_{i+1}]$ may not be contained within M, PL is fully contained within M.

Given a critical convex vertex p_j, as the angle at p_j is convex, p_j can only appear in PL as the endpoint of a line segment. Furthermore, by definition there is no guard, so no vertex of $\mathcal{G}$, at position p_j. It follows that PL contains each of $p_1, \ldots, p_k$ exactly once, so the point x is not the point p_i of PL_1 or PL_2. Suppose first that PL contains a closed circuit such as PL_1. Since x is not p_i, p_i is proceeded by $g_{i+1} = g_i'$ in PL_1. Since PL_1 is simple, its interior (the bounded component of $\mathbf{R}^2 \setminus \mathrm{PL}_1$) lies entirely within M. Since p_i is convex, the guard g_i sees a non-empty interval $]p_i, y[$ for some $y \in]p_i, g_i'[$. This means that the triangle g_ip_iy is entirely seen by g_i, which contradicts the fact that p_i is in the closure of U. A similar contradiction occurs when considering PL_2 instead of PL_1. Therefore, $\mathcal{G}$ contains no cycles of length at least 2, so no cycles at all, and hence has at most $g - 1$ edges.

4 Conclusion

It would be interesting to consider the approximability of the problem. In particular, we do not know if the problem admits a constant factor approximation (the best approximation algorithm [10,4] for the general art gallery problems has ratio log(OPT)). The generalisation of the problem to three dimensions is also another natural question to investigate.

References

1. Aloupis, G., Bose, P., Dujmović, V., Gray, C., Langerman, S., Speckmann, B.: Guarding Fat Polygons and Triangulating Guarded Polygons, September (preprint, 2007)
2. Avis, D., Toussaint, G.T.: An efficient algorithm for decomposing a polygon into star-shaped polygons. Pattern Recognition 13(6), 395–398 (1981)
3. Chvatal, V.: A combinatorial theorem in plane geometry. J. Combin. Theory Ser. B 18, 39–41 (1975)

4. Deshpande, A., Taejung, K., Demaine, E.D., Sarma, S.E.: A Pseudopolynomial Time O(log $c_{opt)}$-Approximation Algorithm for Art Gallery Problems. In: Dehne, F., Sack, J.-R., Zeh, N. (eds.) WADS 2007. LNCS, vol. 4619, pp. 163–174. Springer, Heidelberg (2007)
5. Efrat, A., Guibas, L.J., Har-Peled, S., Lin, D.C., Mitchell, J.S.B., Murali, T.M.: Sweeping simple polygons with a chain of guards. In: Proc. 11th annual ACM-SIAM symposium on Discrete algorithms, pp. 927–936 (2000)
6. Edelsbrunner, H., O'Rourke, J., Welzl, E.: Stationing guards in rectilinear art galleries. Computer vision, graphics, and image processing 27(2), 167–176 (1984)
7. Eidenbenz, S., Stamm, C., Widmayer, P.: Inapproximability Results for Guarding Polygons and Terrains. Algorithmica 31(1), 79–113 (2001)
8. Fisk, S.: A short proof of Chvatals watchman theorem. J. Combin. Theory Ser. B 24(3), 374 (1978)
9. Fragoudakis, C., Markou, E., Zachos, S.: Maximizing the guarded boundary of an Art Gallery is APX-complete. Comput. Geom. 38(3), 170–180 (2007)
10. Ghosh, S.: Approximation algorithms for art gallery problems. In: Proc. Canadian Inform. Process. Soc. Congress, pp. 429–434 (1987)
11. Honsberger, R.: Mathematical Gems II. Dolciani Mathematical Expositions No. 2. Mathematical Association of America, Washington, pp. 104–110 (1976)
12. Kahn, J., Klawe, M., Kleitman, D.: Traditional Galleries Require Fewer Watchmen. SIAM J. Algebraic Discrete Methods 4, 194–206 (1983)
13. Kranakis, E., Pocchiola, M.: Camera placement in integer lattices. Discrete and Comput. Geom. 12(1), 91–104 (1994)
14. Laurentini, A.: Guarding the walls of an art gallery. The Visual Computer 15(6), 265–278 (1999)
15. Liaw, B.C., Huang, N.F., Lee, R.C.T.: The minimum cooperative guards problem on k-spiral polygons. In: Proc. of 5th Canadian Conference on Computational Geometry, pp. 97–101 (1993)
16. Lee, D., Lin, A.: Computational complexity of art gallery problems. IEEE Trans. Inform. Theory 32(2), 276–282 (1986)
17. Michael, T., Pinciu, V.: Art gallery theorems for guarded guards. Comput. Geom. 26(3), 247–258 (2003)
18. O'Rourke, J.: Art gallery theorems and algorithms. Oxford University Press, New York (1987)
19. Rote, G., Santos, F., Streinu, I.: Pseudo-triangulations: A survey, Arxiv preprint math/0612672v1 (December 2006)
20. Sack, J.R.: An $O(n \log n)$ algorithm for decomposing simple rectilinear polygons into convex quadrilaterals. In: Proc. 20th Conference on Communications, Control, and Computing, pp. 64–74 (1982)
21. Shermer, T.C.: Recent results in art galleries [geometry]. Proc. IEEE 80(9), 1384–1399 (1992)
22. Speckmann, B., Tóth, C.D.: Allocating Vertex p-Guards in Simple Polygons via Pseudo-Triangulations. Discrete Comput. Geom. 33(2), 345–364 (2005)
23. Szabo, L.: Recent results on illumination problems. Bolyai Soc. Math. Stud. 6, 207–221 (1997)
24. Tóth, L.F.: Illumination of convex discs. Acta Math. Hungar. 29(3-4), 355–360 (1977)
25. Urrutia, J.: Art gallery and illumination problems. Handbook of Computational Geometry, 973–1027 (2000)

Vision-Based Pursuit-Evasion in a Grid

Adrian Dumitrescu[1,⋆], Howi Kok[1], Ichiro Suzuki[1], and Paweł Żyliński[3,⋆⋆]

[1] Department of Computer Science
University of Wisconsin-Milwaukee
WI 53201-0784, USA
{ad,suzuki}@cs.uwm.edu

[2] Institute of Computer Science
University of Gdańsk
80-952 Gdańsk, Poland
pz@inf.univ.gda.pl

Abstract. We revisit the problem of pursuit-evasion in the grid introduced by Sugihara and Suzuki in the line-of-sight vision model. Consider an arbitrary evader Z with the maximum speed of 1 who moves (in a continuous way) on the streets and avenues of an $n \times n$ grid G_n. The cunning evader is to be captured by a group of pursuers, possibly only one. The maximum speed of the pursuers is $s \geq 1$ (s is a constant for each pursuit-evasion problem considered, but several values for s are studied). We prove several new results; no such algorithms were available for capture using one, two or three pursuers having a constant maximum speed limit:

(i) A randomized algorithm through which one pursuer A with a maximum speed of $s \geq 3$ can capture an arbitrary evader Z in G_n in expected polynomial time. For instance, the expected capture time is $O(n^{1+\log_{6/5} 16}) = O(n^{16.21})$ for $s = 3$, $O(n^{1+\log 12}) = O(n^{4.59})$ for $s = 4$, $O(n^{1+\log 60/13}) = O(n^{3.21})$ for $s = 6$, and it approaches $O(n^3)$ with the further increase of s.

(ii) A randomized algorithm for capturing an arbitrary evader in $O(n^3)$ expected time using two pursuers who can move slightly faster than the evader ($s = 1 + \varepsilon$, for any $\varepsilon > 0$).

(iii) Randomized algorithms for capturing a certain "passive" evader using either a single pursuer who can move slightly faster than the evader ($s = 1 + \varepsilon$, for any $\varepsilon > 0$), or two pursuers having the same maximum speed as the evader ($s = 1$).

(iv) A deterministic algorithm for capturing an arbitrary evader in $O(n^2)$ time, using three pursuers having the same maximum speed as the evader ($s = 1$).

1 Introduction

An $n \times n$ grid G_n, $n \geq 2$, is the set of points with integer coordinates in $[0, n-1] \times [0, n-1]$ together with their connecting edges viewed as a connected planar set.

⋆ Supported in part by NSF CAREER grant CCF-0444188.

⋆⋆ Work by Paweł Żyliński was done while he was a visitor at UWM, and with support from NSF grant CCF-0444188.

J. Gudmundsson (Ed.): SWAT 2008, LNCS 5124, pp. 53–64, 2008.

Alternatively, G_n can be viewed as the union of the following $2n$ line segments: (a) the line segment between $(i, 0)$ and $(i, n-1)$, called *column* i, $0 \leq i \leq n-1$, and (b) the line segment between $(0, j)$ and $(n-1, j)$, called *row* j, $0 \leq j \leq n-1$. A point (x, y) in G_n is called a *vertex* if both x and y are integers.

We consider a vision-based pursuit-evasion problem in G_n in which a group of *pursuers* (*searchers*) are required to search for and capture an *evader* (*fugitive*). Both the pursuers and evader are represented by a (moving) point in G_n (two players can be at the same point at one time). The vision of the players is limited to a straight line of sight (i.e. a row or column): a player at a vertex can see both the corresponding row and column, while one located in the interior of an edge can see only the row or column containing that edge.[1] A player is said to have a *direction detection capability* if he can see in which direction an opponent moves (left or right) when disappearing from the line of sight. A *distance detection capability* is one that allows a player to know either the exact or an approximate distance between his current location and that of an opponent in sight. Generally, we assume that the pursuers have no direction detection capability, and their distance detection capability is limited—a pursuer can tell only whether or not an evader in sight is within distance 1 of his current position, or they have an approximate distance detection capability with a constant relative error.[2] In contrast, the evader may have both direction detection and exact distance detection capabilities. The pursuers can communicate with one another in real time (without delay). The players can also initiate and start executing any movement without delay. The evader may know the algorithm of the pursuers, but he does not get to know the outcomes of their random choices, in case they use a randomized algorithm. The evader is considered captured if there exists a time during the pursuit when his position coincides with the position of one of the pursuers.

We discuss this problem in the *continuous model*: any move in G_n is allowed within the speed limit constraint, which is 1 for the evader w.l.o.g., and some constant s for the pursuers. In this paper we consider the case $s \geq 1$, and present algorithms for capturing the evader using a constant number of pursuers having either $s = 1$ (the pursuers have the same maximum speed as the evader), $s = 1+\varepsilon$ for any small $\varepsilon > 0$ (the pursuers are slightly faster than the evader), $s = 2 + \varepsilon$ for any small $\varepsilon > 0$, or $s \geq 3$.

The vision-based continuous pursuit-evasion problem in a grid described above was first introduced by Sugihara and Suzuki [10] as a variant of the well-known *graph search problem* [5,6,8], which is essentially the same problem except it is played in an arbitrary connected graph by "blind" pursuers and an evader having an unbounded speed. In [10] it is shown that it is possible to capture an arbitrary evader using four pursuers having a maximum speed of $s = 1$. Subsequently, Dawes [1] showed that a single pursuer having a speed of n can *find* (i.e., see) in G_n an arbitrary evader having full knowledge about the pursuer's move, and later Neufeld [7] improved the speed bound to $\lceil 2(n-1)/3 \rceil + 2$; no direction and

[1] A player does not block the view of another.

[2] Sections 3 and 4 present some results on pursuers capable of exact distance detection.

distance detection capabilities are needed for the pursuer, since the game ends as soon as he finds the evader. A variant of the problem in which all players have "full vision" and thus know the positions of the others at all times has been considered in [9]. See [3] for a survey of other known results on the relation between the pursuers' maximum speed and the possibility of capturing an evader in various graphs.

Table 1. Summary of the main results. EDD denotes the exact distance detection capability, and s denotes the maximum speed of the pursuer(s). For 2nd row in the table: $p(s) = \frac{s^2-4s+1}{4(s-1)(s-3)}$.

Number of pursuers	s	Evader	Duration of iterative step	Prob. of capture	Expected time to capture
1 (EDD)	4	arbitrary	$O(n)$	$\frac{1}{n^{\log 12}}$	$O(n^{1+\log 12})$
1 (EDD)	≥ 4	arbitrary	$O(n)$	$\frac{1}{n^{\log 1/p(s)}}$	$O(n^{1+\log 1/p(s)})$
1	$1+\varepsilon$	K-passive	$O(n+K)$	$\frac{2}{5n-4}$	$O(n^2+nK+\frac{1}{\varepsilon})$
1 (EDD)	$2+\varepsilon$	K-passive, $K \leq \frac{n}{2}$	$O(n)$	$\frac{1}{5K}$	$O(nK+\frac{1}{\varepsilon})$
2	$1+\varepsilon$	arbitrary	$O(n^2)$	$\frac{1}{n-1}$	$O(n^3)$
2	1	K-passive	$O(n+K)$	$\frac{4}{9n-6}$	$O(n^2+nK)$
1 (EDD)	1	K-passive, $K \leq \frac{n}{2}$	$O(n)$	$\frac{2}{9K+1}$	$O(n^2)$
3	1	arbitrary	$O(n^2)$	1	$O(n^2)$

Our results. We first present a randomized algorithm through which one pursuer A with a maximum speed of $s \geq 3$ can capture an arbitrary evader Z in G_n in expected polynomial time (Section 3). The expected capture time is $O(n^{1+\log_{6/5} 16}) = O(n^{16.21})$ for $s = 3$, $O(n^{1+\log 12}) = O(n^{4.59})$ for $s = 4$, $O(n^{1+\log 60/13}) = O(n^{3.21})$ for $s = 6$, and it approaches $O(n^3)$ with the further increase of s. Next, we present a randomized algorithm through which two pursuers having a speed of $s = 1+\varepsilon$ can capture an arbitrary evader in $O(n^3)$ expected time (i.e., we only require that the pursuers can move slightly faster than the evader). We also present a three-pursuer deterministic algorithm for capturing an arbitrary evader in G_n in $O(n^2)$ time, using three pursuers with a maximum speed of $s = 1$. No such algorithms for capture were known using one, two or three pursuers having a constant maximum speed limit. In particular, the latter result improves upon the four-searcher algorithm of [10], by using only three pursuers. It is worth noting that with arbitrary grids (i.e., connected subgraphs of G_n), it is easy to construct examples which require arbitrarily many searchers for capturing a fugitive.

Furthermore, under the additional assumption that the evader is K*-passive* for some known K, that is, he will stop moving after not seeing any pursuer for K time units, the expected capture time can be reduced, with even a smaller maximum speed requirement for the pursuers. Namely, we show that it is possible to capture a K-passive evader using a single pursuer having a maximum speed of $s = 1 + \varepsilon$, in expected $O(n^2 + nK + \frac{1}{\varepsilon})$ time. With two pursuers, we only need $s = 1$ for both, and the expected capture time becomes $O(n^2 + nK)$. If $K \leq \frac{n}{2}$, further improvements are possible, provided that the pursuers have the exact distance detection capability. See Table 1 for a summary of these results. Due to space limits, most proofs are omitted. The missing proofs can be found in the full version [2].

2 Preliminaries

In this section, we present useful schemes for pursuers to find and possibly capture the evader. We also introduce the concept of K-passiveness that determines the length of period in which an evader can remain active after seeing a pursuer. In the rest of the paper, we denote the pursuers by A, B, ... and the evader by Z. We use shortest path L_1-distances in G_n in measuring distances between the pursuer(s) and evader; e.g., the distance between $(0, 2/3)$ and $(1, 2/3)$ is $5/3$. Let $B(p, r)$ be the ball of radius r centered at a point p in the L_1 norm. We denote by $|I|$ the length of an interval I on the line.

2.1 Searching for Z

The first goal in the process of capturing Z is seeing Z. The algorithm used by A is essentially a repeated walk in a random direction on the boundary of G_n, and with random delays.

Lemma 1. *Using a randomized algorithm, one pursuer A with a maximum speed of $s \geq 1$ can find (see) Z in $O(n)$ expected time.*

Proof (Sketch). A executes the following algorithm to search for Z:

Search algorithm. Parameter: s.

0. Set $W := 2n$.
1. A goes to $(0, 0)$ from its current location at maximum speed s.
2. Time is reset to 0. A selects a waiting time $w \in [0, W]$ uniformly at random, and waits for time w at $(0, 0)$.
3. Uniformly at random, A selects one of the two axis directions $x+$ or $y+$, and starts moving from $(0, 0)$ in that direction at maximum speed s (towards $(n-1, 0)$ or resp. $(0, n-1)$). If A has not seen Z, repeat the algorithm from step 2 (by symmetry, the step can be iterated by starting at any of the four corners of G_n instead of $(0, 0)$). □

2.2 Chasing

We say that pursuer A with the maximum speed $s \geq 1$ *chases* Z if he continuously moves towards Z at a speed of s, after seeing Z within distance at most s. Notice that chasing by a pursuer forces Z to continue to move forward to avoid an immediate capture, and—because of the assumption of no direction-detection—A may not know temporarily where Z is when Z turns left or right at a vertex v during a chase. However, due to the initial distance of s assumption, we ensure that A will see Z again within distance 1 when he reaches v, thus A will always be able to continue to chase. If $s > 1$, or if $s = 1$ but there are at least two pursuers, the chase always terminates with a capture.

Lemma 2

(i) *If $s = 1+\varepsilon$, then a single pursuer can capture Z within $\frac{s}{\varepsilon} = O(\frac{1}{\varepsilon})$ time after he starts chasing Z.*
(ii) *If $s = 1$, then two pursuers can capture Z within $O(n^2)$ time after one of them starts chasing Z.*

2.3 K-Passiveness and Guessing

For any integer $K \geq 1$, evader Z is said to be *K-passive* if he can move only for K time units after seeing a pursuer; thereafter such Z becomes stationary, until he sees a pursuer again. A 1-passive evader is analogous to a "reactive rabbit" considered in [4] that can move (in a discrete model)[3] only when a hunter is in sight, that is, a hunter is adjacent to a rabbit (one-edge visibility). An arbitrary evader may be thus considered as ∞-passive.

It sometimes happens that the pursuers have not seen a K-passive Z for K time units, and hence they know that he is stationary. (Actually, we often let the pursuers attempt to "hide" from a K-passive Z till he becomes stationary.) Once this happens, the pursuers can guess the location of Z, approach and start chasing him[4] with a probability of success of $\Omega(\frac{1}{n})$. Specifically, we have the following lemma:

Lemma 3. *Assume that the pursuer(s) (one, resp. two) having maximum speed $s \geq 1$, currently located in column 0, know(s) that a K-passive Z is stationary (somewhere out of their sight in G_n). Then:*

(i) *With probability at least $\frac{2}{5n-4}$, a single pursuer can start chasing Z within $O(n)$ time.*
(ii) *With probability at least $\frac{4}{9n-6}$, two pursuers can start chasing Z within $O(n)$ time.*

---

[3] In the discrete model, the moves are restricted to the vertices of G_n and executed at discrete time steps $t = 0, 1, \ldots$, simultaneously, by each player. A move consists of either moving to an adjacent vertex, or staying at the current vertex.

[4] For convenience, we say "pursuers start chasing Z" to mean "one of the pursuers starts chasing Z".

2.4 Hiding

In most of our randomized algorithms, pursuers use variations of the following general scheme to "hide" in column i, so that they can start chasing Z with probability $\Omega(\frac{1}{n})$ if Z appears in column i later. The general idea is for a pursuer, say A, to choose one of the $n-1$ edges in column i uniformly at random, say edge e, and hide in the interior of e without letting Z know which edge it is. If subsequently Z reaches column i at some vertex v, then with a probability at least $\frac{1}{n-1}$, v will be an endpoint of e, and thus A will be within distance 1 of Z, which will allow A to start chasing Z at that moment.

Specifically, assume that a single pursuer A, with maximum speed $s > 1$, is at vertex $(0, 0)$ and sees Z in row 0 at time t at distance > 1. A uses the following procedure $\mathcal{H}$ to hide in column 0. When it is finished, we say that column 0 is *guarded* by A. An illustration is in Fig. 1.

Procedure $\mathcal{H}$: A chooses one of the $n-1$ edges in column 0 uniformly at random, and immediately starts moving from vertex $(0, 0)$ straight to the midpoint of that edge at speed s.

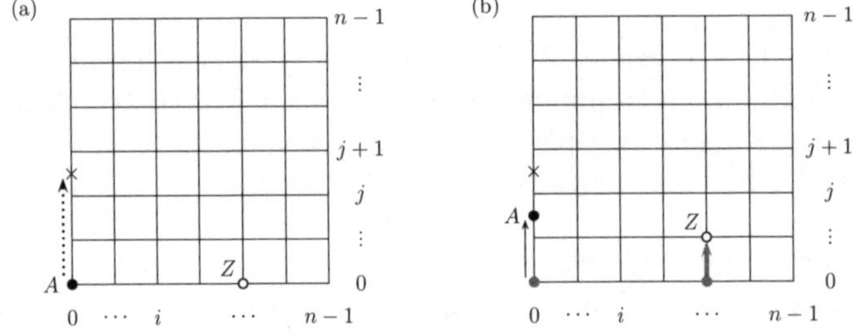

Fig. 1. Procedure $\mathcal{H}$: (a) A chooses a target position and (b) moves towards it at speed $s > 1$

Lemma 4. *Suppose pursuer A, whose maximum speed is $s > 1$, starts procedure $\mathcal{H}$ at time t at vertex $(0, 0)$ and hides in column 0. If Z (initially located in row 0) reaches column 0 for the first time after t (if any), A is within distance one of Z with probability of at least $\frac{1}{n-1}$.*

Note that the direction detection capability of Z does not help him to gain any knowledge about the edge in which A hides in the above scenario, since A always moves north from $(0, 0)$. Clearly, by interchanging the rows and columns, we can use procedure $\mathcal{H}$ to guard a column by hiding a pursuer in it.

3 One-Pursuer Randomized Algorithms

In the first part of this section, we show how to capture an arbitrary evader with one pursuer A in expected polynomial time, provided the maximum speed s of

A is about three times the maximum speed of Z (which is 1). We present specific results for $s \geq 3$. The method essentially breaks down for s approaching $s_0 = \frac{3+\sqrt{5}}{2} \approx 2.62$, with the increase of the expected capture time, as this becomes unbounded as a polynomial in n. In this section we assume the players (in particular, A) have an exact distance detection capability. However, in the end, we point out that similar results can be obtained under a weaker approximate distance detection capability for the players. In fact, such results hold even if Z has the exact distance detection capability, while A has only an approximate one.

Theorem 1. *Using a randomized algorithm, one pursuer A with a maximum speed of $s \geq 3$ and the exact distance detection capability can capture an arbitrary Z in G_n in expected polynomial time. More precisely:*

(i) *Let $p(s) = \frac{s^2-4s+1}{4(s-1)(s-3)}$. For $s \geq 4$, the expected capture time is $O(n^{1+\log 1/p(s)})$. Specifically, the expected capture time is $O(n^{1+\log 12}) = O(n^{4.59})$ for $s = 4$, $O(n^{1+\log 60/13}) = O(n^{3.21})$ for $s = 6$, and it approaches $O(n^3)$ with the further increase of s ($s \geq 6$).*
(ii) *For $s \in [3.2, 4)$, the expected capture time is $O(n^{1+\log_{s-2} 4(s-1)^2})$.*
(iii) *For $s \in [3, 3.2)$, the expected capture time is $O(n^{1+\log_{6/5} 16}) = O(n^{16.21})$.*

Proof. We only discuss the case $s = 4$. The algorithm for A is composed of *phases*. A succeeds in capturing as soon as one phase is successful. Each phase is composed of several (a logarithmic number) of *rounds*. If each round in the current phase is successful, the current phase is declared successful. After each successful round the distance between A and Z is reduced by a constant factor (this distance is measured on the line, since after each successful round, A succeeds in seeing Z). We will impose the condition that the distance reduction is bounded from below by a constant fraction larger than 1 (say, 2) after each successful round. We will show that the probability that each round is successful is bounded from below by another constant (say, $1/12$). Putting these together will ensure that A can capture Z in expected polynomial time.

One phase. Parameter: $s = 4$.

1. A executes the search algorithm (Lemma 1) until he sees Z; this takes $O(n)$ expected time.
2. A repeatedly executes one round (details below) until the current round terminates in failure or the distance between A and Z is less than $s = 4$ (this latter case happens after at most $O(\log n)$ successful rounds, and this makes current phase successful). If the current round terminates in failure, A starts a new phase. If the current phase is successful, A captures Z.

One round. Parameters: $s = 4$, $x = 2k$.

0. Let $d = x = 2k$ be the distance between A and Z at the start of the current round ($t = 0$). W.l.o.g., assume that A and Z are on the same row. Let $W := k/4$. Let $o = (x_0, y_0)$ be the initial position of Z at $t = 0$. (Refer to Fig 2.)

1. A moves at maximum speed s towards o. As explained below, we can assume that A does not see Z during this time interval, namely $[0, k/2]$. Note that when A reaches o at $t = k/2$, Z is somewhere in $B(o, k/2)$.
2. A selects a waiting time $w \in [0, W]$ uniformly at random, and waits for time w at o.
3. A selects one of the four axis directions $x+$, $x-$, $y+$, $y-$, and starts moving from o in that direction at maximum speed s. If A hits the boundary of G_n (when $B(o, k)$ is not entirely contained in G_n), it stays there until time k. Note that A reaches the boundary of the ball $B(o, k)$ or that of G_n latest at time $t = k$ (by our choice of parameters); while Z is also confined to the same ball $B(o, k)$ on the time interval $[0, k]$. If A sees Z during this last segment of his move, the current round is successful, and otherwise it is not.

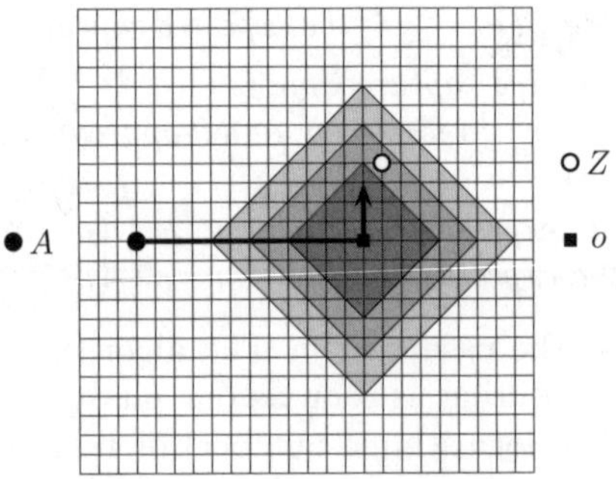

Fig. 2. One round of A's algorithm for $s = 4$; here $k = 8$, $x = 2k = 16$, $W = k/4 = 2$. The three concentric balls around o, $B(o, k/2)$, $B(o, 3k/4)$ and $B(o, k)$, are shown.

We first observe that in a successful round, the distance reduction is at least 2: $\frac{d'}{d} \le \frac{k}{d} = \frac{k}{2k} = \frac{1}{2}$, as A and Z are within distance k when A sees Z in $B(o, k)$. For one round: let o be the *last* position of Z when A sees Z; w.l.o.g., o is a grid point of G_n, because once A sees Z (say, along a row), A will force Z out of this row, and this happens at an integer grid point of G_n. The time is reset to 0 when o is chosen. Consider the ball $B(o, k)$ of radius k centered at o (the position of Z at time $t = 0$ when A sees Z). If it happens that A sees Z before reaching o, the time is reset, a new origin is chosen, and the current round (not a new one) is restarted from the smaller current distance between A and Z.

Consider the time interval $I = [2k/3, 3k/4]$. Note that $I \subset [k/2, k/2 + W]$. Let h^+ (resp. h^-) be the total time in I during which Z is visible along a horizontal line (row) with y-coordinate $\ge y_0$ (resp. $< y_0$). Similarly, let v^+ (resp. v^-) be the total time in I during which Z is visible along a vertical line (column) with x-coordinate $\ge x_0$ (resp. $< x_0$). Obviously $h^+ + h^- + v^+ + v^- \ge |I|$ (when Z is at integer grid points, he is visible along both a horizontal and a vertical line). Recall that the current round is successful if A sees Z in the time interval $[k/2, k]$.

Assume that Z is visible along a vertical (resp. horizontal) line at coordinate $x = x_0 + z$ (resp. $y = y_0 + z$) at time $t \in I$. W. l. o. g., we can assume that

$z \geq 0$. Note that if A starts its last segment (step 3 of the round) from o at $t_0 = t - z/4$ in the direction $x+$ (resp. $y+$), then A would see Z at t as indicated above. We also need to argue that A can see Z at most one time during his last segment of move; to put it differently, if Z is visible by A along the same direction (vertical or horizontal) at two different moments $t_1 < t_2$, $t_1, t_2 \in I$, the corresponding As must have different starting times. This translates in the fact that disjoint intervals in which Z is seen during the time interval I generate disjoint time intervals in $[k/2, k/2 + W]$ in which corresponding As can start. Specifically we prove the following two claims regarding the interval I we have chosen. In general, we say that $I \subset [k/2, k/2 + W]$ is a *good interval* if it satisfies these two claims.

Claim. If $t \in I = [2k/3, 3k/4]$, then $t_0 := t - z/4 \in [k/2, 3k/4] = [k/2, k/2 + W]$.

Proof. Upper bound: Since $t \leq 3k/4$, we have $t_0 = t - z/4 \leq t \leq 3k/4$. For the lower bound:

$$t_0 = t - z/4 = \frac{t - z}{4} + \frac{3t}{4} \geq \frac{3}{4} \cdot \frac{2k}{3} = \frac{k}{2},$$

where $z \leq t$ follows from the maximum unit speed assumption for Z. □

Claim. Let Z be visible along a vertical (resp. horizontal) line at moments $t_1 \leq t_2$, $t_1, t_2 \in I$, so that Z is at $x = x_0 + z_1$ (resp. $y = y_0 + z_1$) at t_1, and Z is at $x = x_0 + z_2$ (resp. $y = y_0 + z_2$) at t_2. Then $t_1 = t_2$.

Proof. The corresponding starting time for A in his last segment of move would be $t_1 - z_1/4 = t_2 - z_2/4$. This readily implies $z_2 - z_1 = 4(t_2 - t_1)$, which leads to a contradiction since $|z_2 - z_1| \leq t_2 - t_1$, by the maximum unit speed assumption for Z, unless $t_1 = t_2$ and $z_1 = z_2$. □

We now show that the probability that any given round is successful is at least 1/12. By conditioning on the four possible choices (axis directions) followed by A in his last segment, and by implicitly using the above claims, we get:

$$\text{Prob(success in one round)} \geq \frac{\frac{1}{4}h^+ + \frac{1}{4}h^- + \frac{1}{4}v^+ + \frac{1}{4}v^-}{W} \geq \frac{|I|/4}{W} = \frac{|I|/4}{k/4} = \frac{|I|}{k} = \frac{1}{12}.$$

Observe that the bound remains valid even if A hits the boundary of G_n during the last segment of his move (step 3), since we can imagine that A continues his move beyond this boundary. Since the initial distance between A and Z (after step 1 of a phase) is $x \leq n$, after at most $\log n$ successful rounds, the distance between A and Z becomes less than 4 (recall, the distance reduction after successful round is at least 2 : $(d'/d \leq 1/2)$. The current phase is then successful since A can chase and then capture Z within another 4/3 time by Lemma 2.

$$\text{Prob(success in one phase)} \geq \left(\frac{1}{12}\right)^{\log n} = \frac{1}{n^{\log 12}}.$$

The execution time for one phase is bounded by $O(n) + \frac{1}{4}\left(n + \frac{n}{2} + \frac{n}{4} + \ldots\right) = O(n)$. It follows that the expected number of phases until a successful one occurs is $O(n^{\log 12})$, and the expected capture time is consequently $O(n^{1+\log 12}) = O(n^{4.59})$, as claimed. This concludes the analysis for the case $s = 4$.

The extension of our result for both (constant) $s \geq 4$ and the interval $s \in [3, 4)$ appears in the full version [2]. Our method can be further pushed for values $s < 3$ (provided $s > \frac{3+\sqrt{5}}{2} \approx 2.62$), however, the bound on the expected capture time becomes prohibitive already for $s = 3$. □

Remark. The following assumption of approximate distance detection capability for the players is a natural one. If the players (A and Z) are visible to each other at some distance d, the distance $\widetilde{d}$ observed by some player satisfies $1 - \rho \leq \widetilde{d}/d \leq 1 + \rho$, for a small constant ρ (e.g., $\rho = 1/10$ or $\rho = 1/100$). (Of course, the distances observed by A and Z may differ). The same algorithm of A for capturing Z can be used, so that similar results hold under this assumption of approximate distance detection capability as well. Essentially, A proceeds according to the estimated distance perceived, with the effect that the probability of success per round is slightly reduced. We omit the calculations.

3.1 One-Pursuer Randomized Algorithm for a K-Passive Z

Here we show that for any $K \geq 1$, a single pursuer A can capture a K-passive Z in G_n in expected time $O(n^2 + nK + \frac{1}{\varepsilon})$, provided he can move slightly faster than Z, i.e., $s = 1 + \varepsilon$ for an arbitrary small $\varepsilon > 0$. We assume that A knows the value of K.

The algorithm works as follows. A goes to $(0, 0)$ and waits for Z to appear in row 0 or column 0 for up to K time units. If Z appears in row 0 within K time (w.l.o.g. at distance greater than 1; otherwise A immediately starts chasing Z), then A immediately hides in column 0 using procedure $\mathcal{H}$. Once this is done, if Z appears in column 0 within the next K time, then A will be within distance 1 of Z with probability of at least $\frac{1}{n-1}$ (by Lemma 4), and then can start chasing Z if this happens. On the other hand, if Z does not appear in column 0 for K time, either while A waits at $(0, 0)$ in the beginning or after A hides in column 0, then A knows K is stationary. Then A guesses Z's location, approaches him, and starts chasing him if he is indeed there. The probability of success is at least $\frac{2}{5n-4}$ by Lemma 3. The total time so far is $O(n+K)$, and if A successfully starts chasing Z, then a capture occurs in additional $\frac{1}{\varepsilon}$ time by Lemma 2. The case in which Z appears in row 0 is handled in a similar manner. To summarize:

Theorem 2. *Using the above approach, with probability at least $\frac{2}{5n-4}$, a single pursuer with maximum speed of $s = 1 + \varepsilon$ can start chasing a K-passive Z within $O(n + K)$ time. The expected time to capture Z by repeating this process is $O(n^2 + nK + \frac{1}{\varepsilon})$.*

Using a similar approach, for small values of K, namely, $K \leq \frac{n}{2}$, and with the exact distance detection capability, we obtain the following theorem.

Theorem 3. *With probability at least $\frac{1}{5K}$, a single pursuer with maximum speed of $s = 2 + \varepsilon$, $\varepsilon > 0$, and the exact distance detection capability can start chasing a K-passive Z, $K \leq \frac{n}{2}$, within $O(n)$ time. The expected time to capture Z is $O(nK + \frac{1}{\varepsilon})$.*

4 Other Results and Concluding Remarks

Due to space limits, we mention here only briefly further results (see [2]).

Theorem 4. *With probability at least $\frac{1}{n-1}$, two pursuers with a maximum speed of $s = 1 + \varepsilon$ can start chasing an arbitrary evader Z within $O(n^2)$ time. The expected time to capture Z by repeating this process is $O(n^3)$.*

Theorem 5. *With probability at least $\frac{4}{9n-6}$, two pursuers with a maximum speed of $s = 1$ can start chasing a K-passive Z within $O(n + K)$ time. The expected time to capture Z by repeating the process is $O(n^2 + nK)$.*

Theorem 6. *Using a deterministic algorithm, three pursuers with a maximum speed of $s = 1$ can capture an arbitrary evader Z within $O(n^2)$ time.*

In most of our results we have only assumed a limited distance detection capability for the pursuers—they can tell only whether or not Z in sight is within distance 1 of their locations, or have an approximate distance detection capability with constant ratio. This feature is desirable in practical applications. In addition, for some results requiring exact distance detection capability (the one-pursuer randomized algorithm for capturing an arbitrary Z), we pointed out that similar results hold under the assumption of approximate distance detection capability. Our algorithms are applicable in many scenarios with autonomous robots in locating and capturing a hostile or uncontrollable robot moving on the ground in a grid-like city environment, or in a contaminated environment not suitable for human intervention. A similar pursuit-evasion problem has recently been considered in a 3D grid [11].

Some questions remain for future study. Most likely, the maximum speed requirements and the capture times in our algorithms can be further reduced. Particularly interesting are: Can a single pursuer with a maximum speed of $s = 1 + \varepsilon$ capture an arbitrary evader in G_n in a polynomial (in n) number of steps? Is it possible to capture an arbitrary evader in polynomial time using two pursuers with a maximum speed of $s = 1$? Under what conditions can a group of searchers with a maximum speed $s < 1$ make a capture?

As a final remark, we show that two deterministic pursuers are indeed sufficient in the discrete model, despite the fact that all players have the same "speed" of one edge per step.[5] It is obvious that one pursuer is not enough even in the discrete model.

[5] In the discrete model, the evader is considered captured if he and a pursuer are either at the same grid point at a discrete time step, or if they traverse the same edge from opposite directions in the interval between two consecutive time steps.

Theorem 7. *In the discrete model, with no direction and exact distance detection capabilities for pursuers, two pursuers can capture Z in $O(n^2)$ time steps using a deterministic algorithm.*

Proof. A and B always maintain a tandem formation in which A is exactly at the adjacent vertex west of B. A and B start by moving in tandem along the south boundary of the grid (row 0) eastward from the west end. When one of them sees Z, they move north in tandem, and continue to do so until one of them starts chasing Z or they lose sight of Z. In the latter case, Z must be in a row to the north of A and B, and depending on whether Z was visible to A or B in the previous step, he is either on the column immediately to the west A or immediately to the east of B. Then A and B repeatedly move in the direction of Z (either west or east) within their current row, until one of them sees Z again. They then repeat the same procedure, by first moving north towards Z. As a result, Z is forced to remain to the north of A and B, and chasing starts in $O(n^2)$ time steps. The pursuers in the tandem formation can then capture Z within $O(n^2)$ steps after chasing starts. □

Acknowledgment. We wish to thank Amol Mali and Kazuo Sugihara for interesting discussions on these topics.

References

1. Dawes, R.W.: Some pursuit-evasion problems on grids. Information Processing Letters 43, 241–247 (1992)
2. Dumitrescu, A., Kok, H., Suzuki, I., Żyliński, P.: Vision-based pursuit-evasion in a grid. Manuscript submitted for publication (2007)
3. Fomin, F.V., Thilikos, D.: An annotated bibliography on guaranteed graph searching. Theoretical Computer Science (to appear, 2008)
4. Isler, V., Kannan, S., Khanna, S.: Randomized pursuit-evasion with local visibility. SIAM Journal of Discrete Mathematics 20(1), 26–41 (2006)
5. Kirousis, L.M., Papadimitriou, C.H.: Searching and pebbling. Theoretical Computer Science 47, 205–218 (1986)
6. Megiddo, N., Hakimi, S.L., Garey, M.R., Johnson, D.S., Papadimitriou, C.H.: The complexity of searching a graph. Journal of the ACM 35(1), 18–44 (1986)
7. Neufeld, S.W.: A pursuit-evasion problem on a grid. Information Processing Letters 58, 5–9 (1996)
8. Parsons, T.D.: Pursuit-evasion in a graph. Lecture Notes in Mathematics, vol. 642, pp. 426–441. Springer, Heidelberg (1992)
9. Sugihara, K., Suzuki, I.: On a pursuit-evasion problem related to motion coordination of mobile robots. In: Proc. the 21st Hawaii Int. Conf. on System Sciences, Kailua-Kona, Hawaii, pp. 218–226 (1988)
10. Sugihara, K., Suzuki, I.: Optimal algorithms for a pursuit-evasion problem in grids. SIAM Jorunal of Discrete Mathematics 2(1), 126–143 (1989)
11. Suzuki, I., Żyliński, P.: Strategies for capturing an evader in a building by mobile robots. IEEE Robotics and Automation Magazine (to appear, 2008)

Angle Optimization in Target Tracking

Beat Gfeller[1,**], Matúš Mihalák[1], Subhash Suri[2,*], Elias Vicari[1,**], and Peter Widmayer[1,**]

[1] Department of Computer Science, ETH Zurich, Zurich, Switzerland
{gfeller,mmihalak,vicariel,widmayer}@inf.ethz.ch
[2] Department of Computer Science, University of California, Santa Barbara, USA
suri@cs.ucsb.edu

Abstract. We consider the problem of tracking n targets in the plane using $2n$ cameras, where tracking each target requires two distinct cameras. A single camera (modeled as a point) sees a target point in a certain direction, ideally with unlimited precision, and thus two cameras (not collinear with the target) unambiguously determine the position of the target. In reality, due to the imprecision of the cameras, instead of a single viewing direction a target defines only a viewing cone, and so two cameras localize a target only within the intersection of two such cones. In general, the true localization error is a complicated function of the angle subtended by the two cameras at the target (the tracking angle), but a commonly accepted tenet is that an angle of 90° is close to the ideal. In this paper, we consider several algorithmic problems related to this so-called "focus of attention" problem. In particular, we show that the problem of deciding whether each of n given targets can be tracked with 90° is NP-complete. For the special case where the cameras are placed along a single line while the targets are located anywhere in the plane, we show a 2-approximation both for the sum of tracking angles and the bottleneck tracking angle (i.e., the smallest tracking angle) maximization problems (which is a natural goal whenever targets and cameras are far from each other). Lastly, for the uniform placement of cameras along the line, we further improve the result to a PTAS.

1 Introduction

We study the problem of tracking targets by a set of cameras in the plane. The position of a target can be estimated if two distinct cameras are dedicated to tracking the target. We consider simple low-resolution cameras with very limited

* Work done while the author was a visiting professor at the Institute of Theoretical Computer Science, ETH, Zurich. The author wishes to acknowledge the support provided by the National Science Foundation under grants CNS-0626954 and CCF-0514738.

** Work partially supported by the National Competence Center in Research on Mobile Information and Communication Systems NCCR-MICS, a center supported by the Swiss NSF under grant number 5005 – 67322, and by the Swiss SBF under contract no. C05.0047 within COST-295 (DYNAMO) of the European Union.

J. Gudmundsson (Ed.): SWAT 2008, LNCS 5124, pp. 65–76, 2008.

image processing capabilities. The quality of the estimated position of a target depends mainly on the relative position of the target and the two cameras [1]. Fig. 1 depicts two different tracking situations of target t with two cameras to illustrate this phenomenon. The field of view of a camera is a cone, a target can be tracked by the camera if it lies in that cone, and therefore any target to be tracked by two cameras needs to lie in the intersection of the two respective cones. The geometry of the situation indicates that tracking accuracy is best if the angle at the target is closest to 90°.

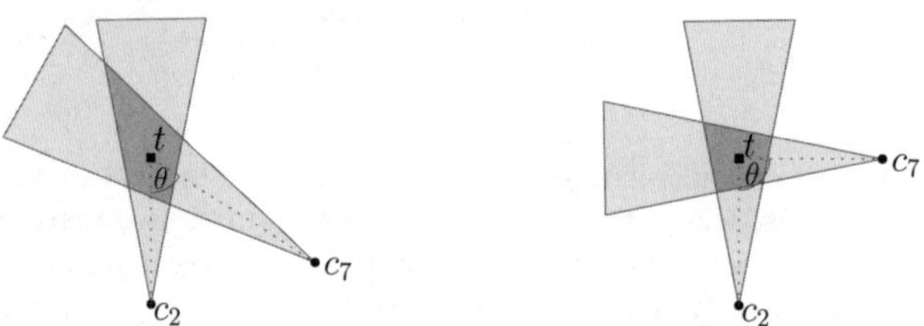

Fig. 1. Two scenarios of tracking target t with tracking angle θ: less than 90° (left), and exactly 90° (right)

The cameras in our setting cannot move, but can freely choose their viewing direction. A pair of cameras can be dedicated to track one target. Thus, tracking n targets requires $2n$ cameras. We assume for the moment that not only the cameras, but also the targets are points in the plane in fixed positions. The *Focus of Attention* problem (FoA) for n targets and $2n$ cameras is to find a pairing of cameras and an assignment of camera pairs to targets that is optimum for some measure of tracking quality. In such an assignment, where each camera is assigned to exactly one target, and each target is assigned to two cameras, each target forms a triangle with its two assigned cameras. We evaluate the quality of the assignment as a function of the *tracking angles*, i.e., the angles that the triangles form at the targets. We consider three specific problems which belong to the following general family of combinatorial optimization problems:

*Problem Family: Focus of Attention (*FoA*).*

Input: A set T of n targets and a set C of $2n$ cameras, given as points in the plane.

Feasible solution: A camera assignment where each target is assigned two cameras, such that each camera is assigned to exactly one target.

Measure: A tracking angle for every triple consisting of a target and two assigned cameras.

Goal: Find a feasible solution which is optimal for one from a collection of objective functions defined on the set of tracking angles.

In this paper we consider the following objectives for the Focus of Attention problem. In the problem SumOfAngleDeviations the objective is to minimize the sum of the deviations of tracking angles from ninety degrees (90°);

for SumOfAngles it is to maximize the sum of tracking angles, and for BottleneckAngle it is to maximize the minimum tracking angle. The latter two problems are interesting whenever targets and cameras tend to form small tracking angles, e.g. because the targets are far from cameras.

Notice that we assume the algorithms to know the exact position of the targets. In reality this is not possible for all scenarios. Nonetheless, the assumption is not a severe modeling simplification, if we assume that the targets are labeled and every camera can recognize the label of the tracked target. In such a situation every camera can look around (rotate the viewing focus) and for each target keep track of the angle under which the target is tracked. This information, together with the known positions of the cameras, allows to compute the tracking angle of any target and two cameras.

Related Work. Object (or target) tracking is an important task for environment surveillance and monitoring applications. It is a well established research subject [2] in the field of computer vision and image processing. Currently, multi-camera systems are being developed, where a certain depth information of objects needs to be computed for a given scene, ideally at low cost.

Isler et al. [3] were the first to consider this task as a combinatorial optimization problem. They defined the *Focus of Attention* problem as a theoretical abstraction of the problem of lowering the computational costs of the depth estimation by assigning cameras to targets in an "optimal" way, and pointed out that for a very general problem setting (not in the plane) this comprises the classical NP-hard *3-Dimensional Matching* (3DM) problem as a special case. Therefore, the focus in [3] is on the problem version in which all cameras are restricted to lie on a single line ℓ. The objective is the aspect ratio $z_t/d(c_i, c_j)$, where z_t is the distance of target t from ℓ, and $d(c_i, c_j)$ is the distance between the two cameras c_i and c_j that are assigned to t. They give a 2-approximation for the problem of minimizing the sum of aspect ratios and for the problem of minimizing the maximum aspect ratio. Also, if the cameras are placed equidistantly on the line, they present a PTAS for the problem of maximizing the sum of aspect ratios. They also consider cameras on a circle and targets inside the circle with tracking cost being $1/\sin\theta$, where θ is the tracking angle, and deliver a 1.42-approximation for the problem of minimizing the sum of tracking costs, and for minimizing the maximum tracking cost.

Naturally, it is the powerful geometric structure that sets the Focus of Attention problems apart from more general assignment problems and makes it particularly interesting. If we would abandon the constraints that geometry imposes, Focus of Attention would belong to the class of *Multi-Index Assignment Problems* [4], where the well known NP-hard *3-Dimensional Matching* problem is a special problem in this class. Other NP-hard versions of multi-index assignment problems also focus on geometry, such as those aiming at the circumference or the area of a triangle formed by three assigned points in the plane [5]. An easy modification of the NP-hardness proof of the latter problem implies NP-hardness of SumOfAngles and BottleneckAngle [6].

Our Contribution. While aspect ratios of rectangles as in [3] might capture tracking quality very well in some cases, we believe that it will in general be more useful to optimize the most influential component of tracking quality directly, namely the tracking angles at the targets, even though this turns out to be surprisingly complicated.

We first show that the problem of minimizing the sum of the deviations of tracking angles from 90° (SUMOFANGLEDEVIATIONS) is NP-hard, and that it admits no (multiplicative) approximation. We then consider FOA that asks for a camera assignment with maximum sum of tracking angles (SUMOFANGLES), and FOA that asks for a camera assignment where the minimum tracking angle is maximized (BOTTLENECKANGLE). For cameras on a line, we present an algorithm that is a 2-approximation for both SUMOFANGLES and BOTTLENECKANGLE at the same time. This is the first constant approximation for the BOTTLENECKANGLE on a line, and for the case on a line, it also improves upon the previous $2+1/t$ approximation [7] (t is the size of the local neighborhood in the local-search algorithm) for SUMOFANGLES. For the special case where the spacing of the cameras on the line is totally regular, we present a PTAS for SUMOFANGLES.

2 NP-Hardness of SumOfAngleDeviations

In this section we consider the minimization FOA problem where the objective is the deviation of the tracking angle from 90°, and state that the problem is NP-hard, by showing that the corresponding decision problem ORTHOGONALASSIGNMENT is NP-complete.

ORTHOGONALASSIGNMENT: For a FOA problem, where every point has integer coordinates, decide whether there exists a camera assignment where every tracking angle is exactly 90°.

We reduce the following NP-complete RESTRICTEDTHREEDM problem to our ORTHOGONALASSIGNMENT problem.

RESTRICTEDTHREEDM: Given three disjoint sets X, Y, and Z, each with q elements, and a set $S \subseteq X \times Y \times Z$ such that every element of $X \cup Y \cup Z$ appears in at most three triples from S, decide whether there exists a subset $S' \subseteq S$ of size q such that each element of $X \cup Y \cup Z$ occurs in precisely one triple from S'.

Our proof is in the tradition of an NP-hardness proof of Spieksma and Woeginger [5] who showed that the following problem related to ORTHOGONALASSIGNMENT is NP-complete: Given sets of planar points A_1, A_2 and A_3, does there exist an assignment $X \subset A_1 \times A_2 \times A_3$ such that for every $x \in X$, the three points of x lie on a line? One can easily see that this result immediately implies that a "degenerate tracking of targets", where each target is collinear with the two cameras that track it (creating an angle of 0° or 180°) is NP-complete. However, the problem of tracking with 90° angles that we consider does not follow easily, but requires several new gadgets, constructions, and proof ideas. The complete proof of the following statement can be found in [8].

Theorem 1. ORTHOGONALASSIGNMENT *is* NP-*complete.*

Theorem 1 implies NP-hardness of all those FoA problems for which the objective function is optimum when the tracking angles equal 90°. These include the objectives of minimizing the sum/maximum of deviations of the tracking angles from 90°, or the goal of maximizing the sum/minimum of $\sin \theta_t$ over all tracking angles θ_t. Furthermore, the maximization FoA problem with the deviation of the tracking angle from 90° as the objective cannot be approximated, unless P = NP. We summarize this discussion:

Corollary 1. *Every problem from the family* FoA *for which the only optimum solution is a camera assignment with all tracking angles equal* 90° *is* NP*-hard.*

3 Maximizing the Sum/Min of Tracking Angles

In this section we consider the maximization FoA problems, where the objective is to obtain large tracking angles. In SumOfAngles we ask for a camera assignment such that the sum of the tracking angles is maximized. We also consider a bottleneck variant of the problem, BottleneckAngle, which asks for a camera assignment where the minimum tracking angle is maximized.

The approach to maximize tracking angles appears unreasonable whenever these angles get close to 180 degrees. However, it makes sense whenever targets are fairly far from cameras, i.e., for any assignment the tracking angle is always at most 90° (in other words, each target lies outside the Thales circle formed by any two cameras).

3.1 Cameras on a Line

We consider the scenario where the cameras are positioned on a horizontal line, and the targets are placed freely in the plane. We may assume, without loss of generality, that all targets lie above the line with cameras (otherwise we mirror the targets from below to the part above the line, with no change in the resulting assignment). An example of such a scenario is frontier monitoring, where the shape of the border can be approximated by a line. We present a 2-approximation algorithm for both SumOfAngles and BottleneckAngle. Note that the previous best approximation ratio for SumOfAngles was $(2+\epsilon)$ which was implied by the result of Arkin and Hassin [7], and nothing was known about the bottleneck version.

In the following, we denote by c_i both the i-th camera on the line, and its position, and assume that $c_1 < c_2 < \ldots < c_{2n}$. Note that we assume that no two cameras have the same position. This avoids complicated special cases, but our results still hold without this assumption.

We call the interval between two paired cameras the *baseline* of these cameras. Furthermore, we call a pairing of cameras *all-overlapping* if the baselines of any two camera pairs (of the chosen pairing) intersect. Observe that there always exists an optimum solution which uses an all-overlapping pairing – any two non-intersecting pairs can exchange their closest endpoints to create two intersecting pairs with larger baselines and thus larger tracking angles.

Lemma 1. *For both* SumOfAngles *and* BottleneckAngle *with cameras on a line, there exists an optimal solution with an all-overlapping pairing.*

Note that for the objective of maximizing the sum of tracking angles, *every* optimal solution must have this property. We will make heavy use of the following consequence:

Corollary 2. *For both* SumOfAngles *and* BottleneckAngle *on a line, there exists an optimal solution where every left camera of a camera pair is among the leftmost n cameras, and every right camera of a camera pair is among the rightmost n cameras. These two groups of cameras can be separated by some point M on the line, such that $c_1 < c_2 < \ldots < c_n < M < c_{1+n} < c_{2+n} < \ldots < c_{2n}$.*

Approximation Algorithm. Our approximation algorithm uses the structural properties of Lemma 1. We first create a simple *interleaved camera pairing* that proved its value in earlier work [3]: For $i = 1, \ldots, n$, pair camera c_i with camera c_{i+n}. Then we assign the interleaved camera pairs to the targets in an optimum way.

For SumOfAngles an optimum assignment of the camera pairs to the targets can be found by computing a maximum-weight perfect matching in a weighted complete bipartite graph where the camera pairs and the targets are the two vertex sets, and the weight of an edge between a pair and a target is the tracking angle of the triangle formed by the target and the pair of cameras. This matching can be computed in $O(|V|(|E|+|V|\log|V|))$ time [9], which is $O(n^3)$ in our case, as we have a complete bipartite graph.

For BottleneckAngle, we can find an optimum assignment by a binary search for the maximum tracking angle (in the set of at most n^2 different tracking angles) for which a perfect matching exists. This means that for a tracking angle θ considered by binary search, all edges with value less than θ are discarded from the complete bipartite graph, and a maximum cardinality matching is computed. If the computed cardinality is less than n, we know that there is no camera assignment with interleaved cameras where the minimum tracking angle is at least θ, and the binary search proceeds with an angle smaller than θ, otherwise it proceeds with an angle larger than θ. The exact procedure is described in [8].

The aforementioned discussion shows that for any given camera pairing, we are able to efficiently find the best assignment of camera pairs to targets. Our algorithm, which we call Interleave, uses the interleaved pairing. In the following we analyze its approximation ratio.

Theorem 2. *For any solution O using an all-overlapping pairing, there exists a camera assignment with the interleaved pairing where each tracking angle is at least half of the corresponding tracking angle of the solution O.*

Proof. Let H denote the interleaved pairing, and let O be a target assignment using an all-overlapping pairing. We will show that any solution which uses an all-overlapping pairing can be transformed into a solution which uses the interleaved pairing H by a sequence of at most n steps of a particular nature:

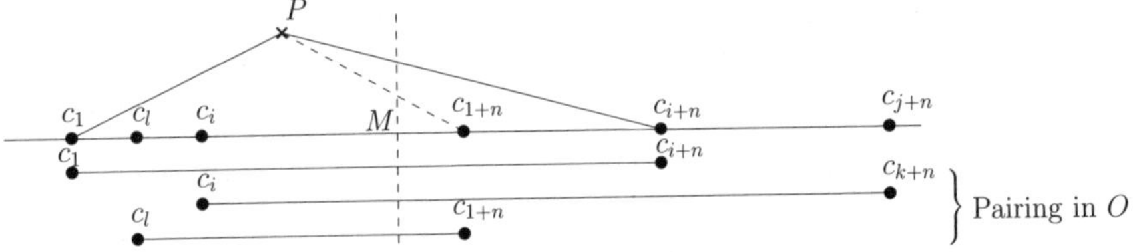

Fig. 2. Comparing the camera pairings in O with those in H

In each step, two pairs of cameras are chosen to create two new pairs under the following constraints: (1) One of the tracking angles may decrease by a factor of at most two, but from then on it stays the same throughout the rest of the transformation. (2) All other angles either stay the same or increase. The claim then follows from the constraints of the transformation. We will now show that for any solution O using an all-overlapping pairing, there exists a solution T using the interleaved pairing H, such that each tracking angle in T at any target is at least half of the corresponding tracking angle in O.

As O contains an all-overlapping pairing, there exists a point M on the line which separates the left ends of all baselines from the right ends of all baselines in O (see Fig. 2). We prove the existence of the transformation by induction on the number k of targets in the instance. If $k = 1$, the claim is trivially true because O's pairing is equal to H (and thus no transformation is needed). For $k \geq 2$, let (c_1, c_{i+n}) be the baseline with leftmost starting point in O. If $i = 1$, then this camera pair is present in both O's pairing and in H, and by the induction hypothesis the desired transformation exists for the $k - 1$ other targets and camera pairs. In the following, assume that $i \neq 1$. In H, c_1 and c_i will be paired with c_{1+n} and c_{i+n}, respectively. Considering O, let c_l be the camera paired with c_{1+n}, and let c_{j+n} be the camera paired with c_i.

Let P be the target which O assigns to the pair (c_1, c_{i+n}). We now distinguish two cases:

(A) The angle c_1, P, c_{1+n} is at least half of the angle c_1, P, c_{i+n}. Let Q be the target which O assigns to the pair (c_l, c_{1+n}). We create from O a new camera assignment O' by transforming the pairs (c_1, c_{i+n}), (c_l, c_{1+n}) into pairs (c_1, c_{1+n}), (c_l, c_{i+n}) and assigning these pairs to targets P and Q, respectively. Thus, the tracking angle at P in O' is at most cut in half, and the tracking angle at Q in O' increases, as the baseline (c_l, c_{i+n}) extends to the right (compared to (c_l, c_{1+n})).

(B) The angle c_i, P, c_{i+n} is at least half of the angle c_1, P, c_{i+n}. Let Q be the target which O assigns to the pair (c_i, c_{j+n}). We create from O a new camera assignment O' by transforming the pairs (c_1, c_{i+n}), (c_i, c_{j+n}) into pairs (c_1, c_{j+n}), (c_i, c_{i+n}), and assigning them to targets Q and P, respectively. The tracking angle at Q increases (as the baseline lengthens in the new assignment), and the tracking angle at P is at most cut in half.

As c_i and c_{1+n} both lie between c_1 and c_{i+n}, and c_i is to the left of c_{1+n}, at least one of these cases applies. In both cases, one angle increased, and one angle decreased by a factor of at most 2, but this angle uses a pair from the interleaved pairing. Thus, there remain $k-1$ camera pairs which are potentially not assigned to a camera pair from the interleaved pairing. By the induction hypothesis (applied on the $k-1$ targets and the cameras assigned to them) these targets can be assigned to camera pairs from the interleaved pairing such that each of these $k-1$ other angles are at most cut in half during the transformation. □

This theorem directly proves that our algorithm computes a 2-approximation for both considered objectives.

Corollary 3. *The algorithm* Interleave *computes a 2-approximation for both* SumOfAngles *and* BottleneckAngle.

It can be showed that the analysis of the approximation ratio of the algorithm is tight. We refer to [8] for the missing details.

3.2 Equidistant Cameras on a Line

We consider the special setting where the cameras lie on a (horizontal) line ℓ and the distance between any two neighboring cameras on the line is the same. Without loss of generality we assume unit distance. We consider the problem of maximizing the sum of tracking angles and present a PTAS for this problem.

We consider the cameras in the order as they appear on the line ℓ (from left to right). According to Corollary 2 we know that in every optimum camera assignment the first n cameras are paired with the last n cameras. Let L denote the first n cameras and R the last n cameras. We denote the cameras as they appear in the order on ℓ as $l_1, l_2, \ldots, l_n$ for cameras in L, and $r_1, r_2, \ldots, r_n$ for cameras in R. Hence, the distance between l_1 and r_n is $2n-1$.

The main idea of the algorithm is to partition L and R into k equally-sized sets $L_1, L_2, \ldots, L_k$ and $R_1, R_2, \ldots, R_k$, and to correctly guess what types of paired cameras an optimum solution OPT contains with respect to the partition, i.e., we want to know for every s and t how many pairs of OPT have a camera from L_s and a camera from R_t. Camera pair $\{l_i, r_j\}$ is called a *pair of type* (s,t) (with respect to the partition), if $l_i \in L_s$ and $r_j \in R_t$. There are k^2 different types of pairs. We can characterize every camera assignment by its types of the camera pairs – for each type (s,t) we know how many pairs are of this type. Let $m_{s,t}$ denote this number. For these k^2 numbers $m_{s,t}$, each $m_{s,t}$ is in the range $\{0, \ldots, \frac{n}{k}\}$. According to this classification, there are at most $(\frac{n}{k})^{k^2}$ different classes of camera pairings, which we call *camera-pairing types*. Thus, if k is a constant, the algorithm can enumerate all camera-pairing types in polynomial time. For each enumerated type of camera pairing the algorithm constructs some camera pairing of that type (if that is possible, otherwise it reports that no camera assignment using such a camera pairing type exists) and optimally assigns the targets to the camera pairs. At the end the algorithm outputs the best solution among all constructed camera assignments. The algorithm can create a camera pairing of a type specified by values $m_{s,t}$ in the following way: for every camera-pair type (s,t) it creates $m_{s,t}$ pairs of type (s,t) by pairing $m_{s,t}$ cameras from L_s with $m_{s,t}$ cameras from R_s. Clearly, if a camera pairing of the considered camera-pairing type exists, the algorithm finds one, otherwise the algorithm fails to create one, in which case it continues with the next enumerated camera-pairing.

In the following, we concentrate on the situation where the algorithm considers the same camera-pairing type as the OPT solution has. The algorithm creates some camera pairing of that type, and assigns the camera pairs to the targets in an optimum way by computing a maximum weight matching between the pairs and the targets. Let A denote the resulting camera assignment. We show that the sum of the tracking angles of A is a good approximation of the sum of tracking angles of OPT. We say that a camera pair $\{l_i, r_j\}$ is a *short pair*, if the distance between l_i and r_j is at most $\frac{n}{\sqrt{k}}$, otherwise we say it is a *long pair*. Observe that in any camera assignment there are a lot more long pairs than short ones, as there are at most $n/\sqrt{k}$ short pairs (every short pair has to have its left camera among the last $n\sqrt{k}$ cameras in L), and thus at least $n-n/\sqrt{k}$ long pairs. We show that the algorithm guarantees good tracking angles at long pairs. We further show that there exists a solution Q_{OPT} (quasi-optimum) which incurs most of the tracking profit at the long pairs, and which is not much worse than the optimum solution OPT. This then implies that the solution A computed by the algorithm is a good approximation of Q_{OPT}, and therefore it is a good approximation of OPT, too. In the following, if X is a camera assignment, we use $X|\mathrm{LONG}$ to denote the subset of X which consists of long pairs only. By $w(X)$ we denote the weight of X, i.e., the sum of tracking angles arising in X.

Let OPT$'$ denote an optimum solution for the problem of assigning all pairs of OPT to targets, and maximizing the sum of tracking angles at long pairs. Thus, OPT$'$, OPT, and A use the same type of camera-pairing. Consider a long pair $\{l_{\mathrm{O}}, r_{\mathrm{O}}\}$ of type (i,j) from OPT$'$. Let t be the target to which that pair is assigned, and let $\theta_{\mathrm{OPT}'}$ be the tracking angle of t in OPT$'$. See Fig. 3 for illustration. Let $\{l_A, r_A\}$ be a (long) pair of type (i,j) that is created by the algorithm. In the camera assignment of A, $\{l_A, r_A\}$ is assigned to some target t'. We create a new camera assignment A' that uses the pairing of A, and the targets of long pairs of OPT$'$ in the following way: we match every long pair $\{l_A, r_A\}$ from A of type (i,j) with a long pair $\{l_{\mathrm{O}}, r_{\mathrm{O}}\}$ from OPT$'$ of the same type. Then, let $\{l_{\mathrm{O}}, r_{\mathrm{O}}\}$ be assigned to target t. We create A' by assigning the matched pair $\{l_A, r_A\}$ to t. Clearly, A' is a camera assignment of the same type as OPT' and A. Clearly, as A and A' are using the same pairs, the solution of A is at least as good as A', i.e., $w(A) \geq w(A')$. We now show that A' is a good approximation for OPT' on long pairs.

Let $\theta_{A'}$ denote the tracking angle of t in A'. We show that $\theta_{A'}$ is not much smaller (if at all) than $\theta_{\mathrm{OPT}'}$. Clearly, the worst case for the difference between the two angles is when l_{O} is the leftmost vertex in L_i, r_{O} is the rightmost vertex in R_j, and l_A is the rightmost vertex in L_i and r_A is the leftmost vertex in R_j. The distance between l_{O} and l_A is at most n/k (the size of L_i). Similarly for r_{O}

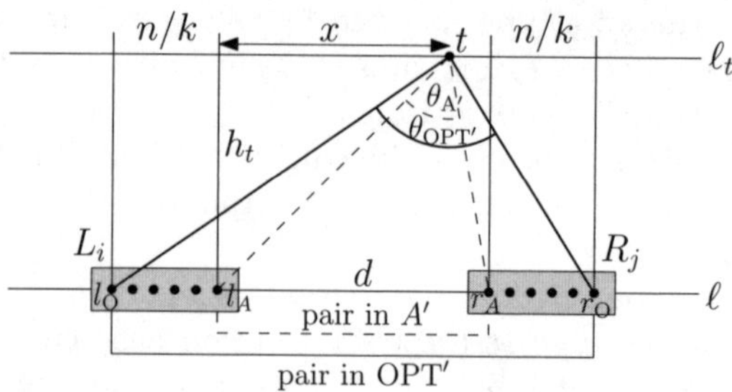

Fig. 3. A long pair of type (i, j) in OPT', and a long pair of type (i, j) in solution A'

and r_A, the distance between these two cameras is at most n/k. On the other hand, the distance between l_A and r_A is at least $n/\sqrt{k}$, because $\{l_A, r_A\}$ is a long pair. We can express the ratio $\theta_{A'}/\theta_{\mathrm{OPT}'}$ in the following way. Let x express the position of t on the line ℓ_t parallel to ℓ through t. We assume that $x = 0$ when t is exactly above l_A. Further, we denote by h_t the distance between ℓ and ℓ_t and by d the distance between l_A and r_A (cf. Fig. 3). We can then express the ratio $\theta_{A'}/\theta_{\mathrm{OPT}}$ as follows:

$$\frac{\theta_{A'}}{\theta_{\mathrm{OPT}'}} = \frac{\arctan\left(\frac{x}{h_t}\right) + \arctan\left(\frac{d-x}{h_t}\right)}{\arctan\left(\frac{n/k+x}{h_t}\right) + \arctan\left(\frac{d+n/k-x}{h_t}\right)} \tag{1}$$

The analysis of the first and second derivative with respect to x of the previous function shows that $\theta_{A'}/\theta_{\mathrm{OPT}'}$ is minimal for $x = d/2$, i.e., when the target lies in the middle point of the segment (l_A, r_A). Hence, by setting $x = d/2$ in (1) we get the following lower bound: $\frac{\theta_{A'}}{\theta_{\mathrm{OPT}'}} \geq \frac{\arctan\left(\frac{d}{2h_t}\right)}{\arctan\left(\frac{n}{kh_t} + \frac{d}{2h_t}\right)}$ We now distinguish two cases. First, if $h_t = o(n)$, then the two arguments inside the arctan functions of the previous term are unbounded as n grows (remember that $d \geq n/\sqrt{k}$). As arctan is a bounded function, $\theta_{A'}/\theta_{\mathrm{OPT}'}$ approaches 1 as n gets large. Therefore, given any ϵ and k, we can find n large enough, such that $\theta_{A'}/\theta_{\mathrm{OPT}'} \geq 1 - \epsilon$, as desired. Second, if $h_t = \Omega(n)$, the arguments inside the arctan functions of the previous term are bounded from above and may even approach zero as n goes to infinity (remember that $d \leq 2n - 1$), so we have to examine the behavior of the term in this case, and, as we will see, we also have to involve k. We denote $\alpha := \frac{d}{2h_t}$, and thus obtain $\theta_{A'}/\theta_{\mathrm{OPT}'} \geq \frac{\arctan(\alpha)}{\arctan\left(\frac{n}{kh_t} + \alpha\right)}$. We express the term $\frac{n}{kh_t}$ in terms of α (remember that $d \geq n/\sqrt{k}$): $\frac{n}{kh_t} = \frac{2dn}{2dkh_t} = \frac{2n}{dk}\alpha \leq \frac{2}{\sqrt{k}}\alpha$. Thus we have $\theta_{A'}/\theta_{\mathrm{OPT}'} \geq \frac{\arctan(\alpha)}{\arctan\left((1+\frac{2}{\sqrt{k}})\alpha\right)}$. The derivative of the last fraction with respect to α is positive for any $\alpha \geq 0$, and thus the last fraction is minimized when α approaches zero. The limit of that fraction, when α approaches zero, is $1/(1 + \frac{2}{\sqrt{k}}) = \frac{\sqrt{k}}{\sqrt{k}+2}$, and thus, for any fixed ϵ, setting k appropriately, and for n large enough, we get that $\theta_{A'}/\theta_{\mathrm{OPT}'} \geq 1 - \epsilon$.

In the remaining we show that there exists a camera assignment Q_{OPT} with the same camera-pairing type as OPT, which gains most of its profit (i.e., a $(1-\delta)$ fraction of the total profit, $0 < \delta < 1$) on long pairs, and which is a good approximation to OPT. The optimum solution OPT has at most $n/\sqrt{k}$ short pairs. Let S denote the subset of OPT that contains the short pairs only. Let us consider those long pairs of OPT, where every long pair, considered as an interval on ℓ, fully contains every short pair. Let G denote the subset of OPT on such pairs. Observe that there are at least $n - 2n/\sqrt{k}$ pairs in G (there are at least $n - n/\sqrt{k}$ long pairs that contain the left camera of any short pair – the pairs formed by cameras $l_1, l_2, \ldots, n - n/\sqrt{k}$; among those pairs, at most $n/\sqrt{k}$ can be formed by cameras $r_1, r_2, \ldots, r_{n/\sqrt{k}}$). We split G into subsets of size $|S|$. Let $G_1, G_2, \ldots, G_z$ denote these sets, where $z = |G|/|S| \geq \sqrt{k} - 2$. For simplicity we assume that $|G|$ is divisible by $|S|$. As one can check, this is not a crucial assumption in our analysis. Let us order the sets G_i such that $w(G_1) \geq w(G_2) \geq \ldots w(G_z)$.

Observe first that if $w(G_z) \geq w(S)$, we get $w(\text{OPT}) \geq \sum_i w(G_i) \geq z \cdot w(S)$, and thus $\frac{w(\text{OPT})}{z} \geq w(S)$. As $w(\text{OPT}|\text{LONG})$ denotes the contribution of all long pairs to the total weight $w(\text{OPT})$, we obtain $w(\text{OPT}|\text{LONG}) = w(\text{OPT}) - w(S) \geq w(\text{OPT}) - \frac{w(\text{OPT})}{z} \geq \frac{z-1}{z} w(\text{OPT}) \geq \frac{\sqrt{k}-3}{\sqrt{k}-2} w(\text{OPT})$. Hence, setting k appropriately, OPT gains a $(1-\delta)$ fraction of its profit on long pairs, and thus we can set $Q_{\text{OPT}} := \text{OPT}$.

Assume now that $w(S) > w(G_z)$. We create a new solution Q_{OPT}: we (arbitrarily) assign the pairs of G_z to targets of S and the pairs of S to targets of G_z. Let S' and G'_z denote the new assignments. Clearly, as $w(Q_{\text{OPT}}) \leq w(\text{OPT})$, we have $w(G'_z) + w(S') \leq w(G_z) + w(S)$, as only pairs in G_z and S have possibly been assigned to different targets. Observe now that $w(G'_z) > w(S)$, because in Q_{OPT} the pairs of G_z are assigned to the same targets as the pairs of S in OPT, and every pair from G_z fully contains every pair from S (if imagined as an interval), which makes every tracking angle of the respective target bigger in Q_{OPT}. Thus, using the last two inequalities, $w(S') < w(G_z)$. Hence, we also have that $w(S') < w(G_i)$, $i = 1, \ldots, z-1$, and thus $w(S') \leq \frac{w(\text{OPT})}{z}$. Applying the argumentation from above we get $w(Q_{\text{OPT}}|\text{LONG}) = w(Q_{\text{OPT}}) - w(S') \geq \frac{z-1}{z} w(Q_{\text{OPT}})$. Observe also that $w(Q_{\text{OPT}}) \geq w(\text{OPT}) - w(G_z)$. Thus, since $w(\text{OPT}) - w(G_z) \geq \frac{z-2}{z-1} w(\text{OPT})$, we obtain $w(Q_{\text{OPT}}) \geq \frac{z-2}{z-1} w(\text{OPT})$, and hence $w(Q_{\text{OPT}}|\text{LONG}) \geq \frac{z-2}{z} w(\text{OPT})$.

Putting the previously derived inequalities together, we obtain $w(A) \geq w(A') \geq (1-\epsilon) w(\text{OPT}'|\text{LONG}) \geq (1-\epsilon) w(Q_{\text{OPT}}|\text{LONG}) \geq (1-\epsilon)\frac{z-2}{z} w(\text{OPT}) \geq (1-\epsilon)\frac{\sqrt{k}-4}{\sqrt{k}-2} w(\text{OPT})$. Hence, for any given ϵ^*, we can find ϵ and k such that $(1-\epsilon)\frac{\sqrt{k}-4}{\sqrt{k}-2} \geq 1 - \epsilon^*$, and thus $w(A) \geq (1-\epsilon^*)\text{OPT}$. This yields:

Theorem 3. *There is a PTAS for* SumOfAngles *with equidistant cameras on a line.*

4 Conclusions

We have considered different variants of the "focus of attention" problem, where the objective function depends on the tracking angles. We have shown that the natural goal of assigning targets under 90° is (in general) an NP-complete problem. It remains an open problem whether the more restricted instances where the cameras are placed on a line can be solved in polynomial time. The hardness result shows that there is no approximation algorithm (unless P = NP) for the problem of minimizing the sum of deviations of tracking angles from 90°. In this context it would be interesting to consider different optimization goals which capture the optimality of tracking angles θ_i being 90° and which would allow a good approximation. The first candidate for such an objective could be $\sin \theta_i$. Any results for this or similar objective function would be interesting. Also for the objective functions considered in this paper there are unresolved questions. For example, we have only considered SumOfAngles on a line. For the general case, a simple greedy algorithm achieves a 3-approximation [4], and it remains open whether one could do better.

References

1. Hartley, R.I., Zisserman, A.: Multiple View Geometry in Computer Vision, 2nd edn. Cambridge University Press, Cambridge (2004)
2. Yilmaz, A., Javed, O., Shah, M.: Object tracking: A survey. ACM Computing Surveys 38(4) (December 2006)
3. Isler, V., Khanna, S., Spletzer, J., Taylor, C.J.: Target tracking with distributed sensors: the focus of attention problem. Computer Vision and Image Understanding 100(1-2), 225–247 (2005)
4. Spieksma, F.: Multi index assignment problems: complexity, approximation, applications. In: Nonlinear Assignment Problems, pp. 1–12. Kluwer, Dordrecht (2000)
5. Spieksma, F., Woeginger, G.J.: Geometric three-dimensional assignment problems. European Journal of Operational Research 91, 611–618 (1996)
6. Goossens, D., Spieksma, F.: On the FOA-problem. Manuscript (January 2004)
7. Arkin, E.M., Hassin, R.: On local search for weighted k-set packing. Mathematics of Operations Research 23(3), 640–648 (1998)
8. Gfeller, B., Mihalák, M., Suri, S., Vicari, E., Widmayer, P.: Angle optimization in target tracking. Technical Report 592, Department of Computer Science, ETH Zurich (2008), `http://www.inf.ethz.ch/research/disstechreps/techreports`
9. Galil, Z.: Efficient algorithms for finding maximum matching in graphs. ACM Comput. Surv. 18(1), 23–38 (1986)

Improved Bounds for Wireless Localization

Tobias Christ[1], Michael Hoffmann[1], Yoshio Okamoto[2], and Takeaki Uno[3]

[1] Institute for Theoretical Computer Science, ETH Zürich, Switzerland
[2] Tokyo Institute of Technology, Japan
[3] National Institute of Informatics, Tokyo, Japan

Abstract. We consider a novel class of art gallery problems inspired by wireless localization. Given a simple polygon P, place and orient guards each of which broadcasts a unique key within a fixed angular range. Broadcasts are not blocked by the edges of P. The interior of the polygon must be described by a monotone Boolean formula composed from the keys. We improve both upper and lower bounds for the general setting by showing that the maximum number of guards to describe any simple polygon on n vertices is between roughly $\frac{3}{5}n$ and $\frac{4}{5}n$. For the natural setting where guards may be placed aligned to one edge or two consecutive edges of P only, we prove that $n-2$ guards are always sufficient and sometimes necessary.

1 Introduction

Art gallery problems are a classic topic in discrete and computational geometry, dating back to the question posed by Victor Klee in 1973: "How many guards are necessary, and how many are sufficient to patrol the paintings and works of art in an art gallery with n walls?" Chvátal [2] was the first to show that $\lfloor n/3 \rfloor$ guards are always sufficient and sometimes necessary, while the beautiful proof of Fisk [6] made it into "the book" [1]. Nowadays there is a vast literature [12,14,16] about variations of this problem, ranging from optimization questions (minimizing the number of guards [10] or maximizing the guarded boundary [7]) over special types of guards (mobile guards [11] or vertex pi-guards [15]) to special types of galleries (orthogonal polygons [8] or curvilinear polygons [9]).

A completely different direction has recently been introduced by Eppstein, Goodrich, and Sitchinava [5]. They propose to modify the concept of visibility by not considering the edges of the polygon/gallery as blocking. The motivation for this model stems from communication in wireless networks where the signals are not blocked by walls, either. For illustration, suppose you run a café (modeled, say, as a simple polygon P) and you want to provide wireless Internet access to your customers. But you do not want the whole neighborhood to use your infrastructure. Instead, Internet access should be limited to those people who are located within the café. To achieve this, you can install a certain number of devices, let us call them guards, each of which broadcasts a unique (secret) key in an arbitrary but fixed angular range. The goal is to place guards and adjust their angles in such a way that everybody who is inside the café can

J. Gudmundsson (Ed.): SWAT 2008, LNCS 5124, pp. 77–89, 2008.

prove this fact just by naming the keys received and nobody who is outside the café can provide such a proof. Formally this means that P can be described by a monotone Boolean formula over the keys, that is, a formula using the operators AND and OR only, negation is not allowed. It is convenient to model a guard as a subset of the plane, namely the area where the broadcast from this guard can be received. This area can be described as an intersection or union of at most two halfplanes. Using this notation, the polygon P is to be described by a combination of the operations union and intersection over the guards. For example, the first polygon to the right can be described by $(a \cup b) \cap c \cap d$.

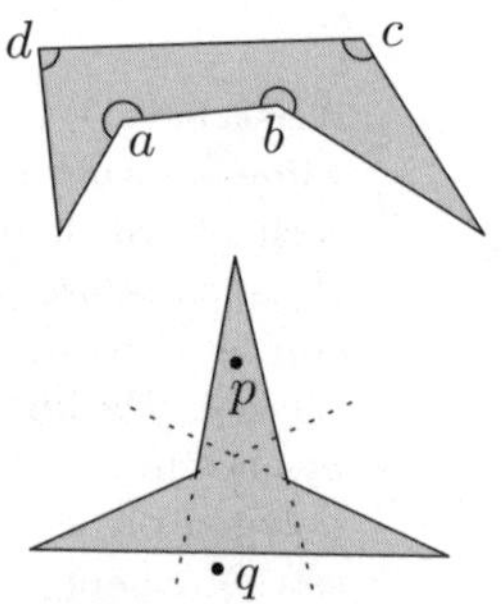

Natural guards. Natural locations for guards are the vertices and edges of the polygon. A guard which is placed at a vertex of P is called a *vertex guard.* A vertex guard is *natural* if it covers exactly the interior angle of its vertex. But natural vertex guards alone do not always suffice [5], as the second polygon P shown to the right illustrates: No natural vertex guard can distinguish the point p inside P from the point q outside of P. A guard placed anywhere on the line given by an edge of P and broadcasting within an angle of π to the inner side of the edge is called a *natural edge guard.* Dobkin, Guibas, Hershberger, and Snoeyink [4] showed that n natural edge guards are sufficient for any simple polygon with n edges.

Vertex guards. Eppstein et al. [5] proved that any simple polygon with n edges can be guarded using at most $n - 2$ (general, that is, not necessarily natural) vertex guards. More generally, they show that $n + 2(h - 1)$ vertex guards are sufficient for any simple polygon with n edges and h holes. This bound is not known to be tight. Damian, Flatland, O'Rourke, and Ramaswami [3] describe simple polygons with n edges which require at least $\lfloor 2n/3 \rfloor - 1$ vertex guards.

General guards. In the most general setting, we do not have any restriction on the placement and the angles of guards. So far the best upper bound known has been the same as for vertex guards, that is, $n - 2$. On the other hand, if the polygon does not have collinear edges then at least $\lceil n/2 \rceil$ guards are always necessary [5]. The lower bound construction of Damian et al. [3] for vertex guards does not provide an improvement in the general case, where these polygons can be guarded using at most $\lceil n/2 \rceil + 1$ guards. As O'Rourke wrote [13]: "The considerable gap between the $\lceil n/2 \rceil$ and $n - 2$ bounds remains to be closed."

Results. We provide a significant step in bringing the two bounds for general guards closer together by improving both on the upper and on the lower side. On one hand we show that for any simple polygon with n edges $\lfloor (4n-2)/5 \rfloor$ guards are sufficient. The result generalizes to polygons combined in some way by the operations intersection and/or union. Any simple polygon with h holes can be guarded using at most $\lfloor (4n - 2h - 2)/5 \rfloor$ guards. On the other hand we describe a family of polygons which require at least $\lceil (3n - 4)/5 \rceil$ guards. Furthermore,

Table 1. Number of guards needed for a simple polygon on n vertices. The mark $*$ indicates the results of this paper.

	natural		general	
	vertex guards	guards	vertex guards	guards
upper bound	does not exist [5]	$n-2$ [*]	$n-2$ [5]	$\lfloor (4n-2)/5 \rfloor$ [*]
lower bound	does not exist [5]	$n-2$ [*]	$\lfloor 2n/3 \rfloor - 1$ [3]	$\lceil (3n-4)/5 \rceil$ [*]

we extend the result of Dobkin et al. [4] to show that $n-2$ natural (vertex or edge) guards are always sufficient. It turns out that this bound is tight.

2 Notation and Basic Properties

We are given a simple polygon $P \subset \mathbb{R}^2$. A *guard* g is a closed subset of the plane, whose boundary ∂g is described by a vertex v and two rays emanating from v. The ray that has the interior of the guard to its right is called the *left ray*, the other one is called the *right ray*. The *angle* of a guard is the interior angle formed by its rays. For a guard with angle π, the vertex is not unique.

A guard g *covers* an edge e of P *completely* if $e \subseteq \partial g$ and their orientations match, that is, the inner side of e is on the inner side of g. We say e is covered *partly* by g if their orientations match and $e \cap \partial g$ is a proper sub-segment of e that is not just a single point. We call a guard a k-guard if it covers exactly k edges completely. As P is simple, a guard can cover at most one edge partly. If a guard covers an edge partly and k edges completely, we call it a k'-guard. Assuming there are no collinear edges, a guard can cover at most two edges; then a natural vertex guard is a 2-guard and a natural edge guard is a 1-guard. A *guarding* $\mathcal{G}(P)$ for P is a formula composed of a set of guards and the operators union and intersection that defines P. The *wireless localization problem* is to find a guarding with as few guards as possible. The same problem is sometimes referred to as *guard placement for point-in-polygon proofs* or the *sculpture garden problem* [5]. The following basic properties are restated without proof.

Observation 1. *For any guarding $\mathcal{G}(P)$ and for any two points $p \in P$ and $q \notin P$ there is a $g \in \mathcal{G}(P)$ which* distinguishes *p and q, that is, $p \in g$ and $q \notin g$.*

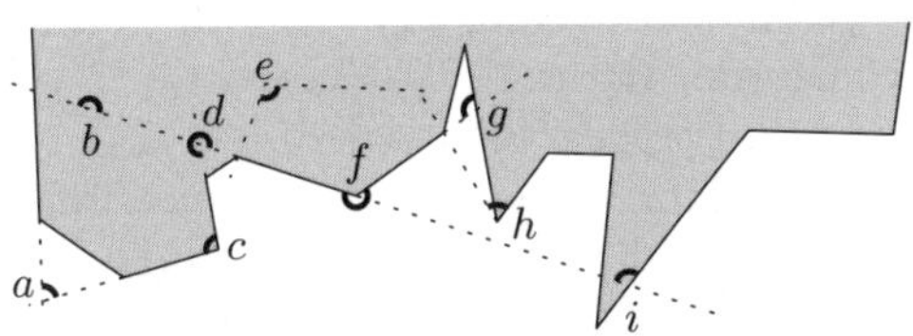

Fig. 1. (a) a 2-guard, (b) a 1-guard (and a natural edge guard), (c) a 2-guard (and a natural vertex guard), (d) a 2-guard, (e) a 0-guard, (f) a 0-guard, (g) a 1-guard (not a $1'$-guard), (h) a 1-guard (a non-natural vertex guard), (i) a $1'$-guard.

Lemma 1. [4] *Every edge of P must be covered by at least one guard or it must be covered partly by at least two guards*

3 Upper Bounds

Following Dobkin et al. [4] we use the notion of a *polygonal halfplane* which is a topological halfplane bounded by a *simple bi-infinite polygonal chain* with edges $(e_1, \ldots, e_n)$, for $n \in \mathbb{N}$. For $n = 1$, the only edge e_1 is a line and the polygonal halfplane is a halfplane. For $n = 2$, e_1 and e_2 are rays which share a common source but are not collinear. For $n \geq 3$, e_1 and e_n are rays, e_i is a line segment, for $1 < i < n$, and e_i and e_j, for $1 \leq i < j \leq n$, do not intersect unless $j = i + 1$ in which case they share an endpoint. For brevity we use the term *chain* in place of simple bi-infinite polygonal chain in the following. For a polygonal halfplane H define $\gamma(H)$ to be the minimum integer k such that there exists a guarding $\mathcal{G}(H)$ for H using k guards. Similarly, for a natural number n, denote by $\gamma(n)$ the maximum number $\gamma(H)$ over all polygonal halfplanes H that are bounded by a chain with n edges. Obviously $\gamma(1) = \gamma(2) = 1$. Dobkin et al. [4] show that $\gamma(n) \leq n$.

Lemma 2. *Any simple polygon P on $n \geq 4$ vertices is an intersection of two polygonal halfplanes each of which consists of at least two edges.*

Proof. Let p_- and p_+ be the vertices of P with minimal and maximal x-coordinate, respectively. If they are not adjacent along P, split the circular sequence of edges of P at both p_- and p_+ to obtain two sequences of at least two segments each. Transform each sequence into a chain by linearly extending the first and the last segment beyond p_- or p_+ to obtain a ray. As p_- and p_+ are opposite extremal vertices of P, the two chains intersect exactly at these two points. Thus, the polygon P can be expressed as an intersection of two polygonal halfplanes bounded by these chains. Now consider the case that p_- and p_+ are adjacent along P. Without loss of generality assume that P lies above the edge from p_- to p_+. Rotate clockwise until another point q has x-coordinate larger than p_+. If q and p_- are not adjacent along P, then split P at these points. Otherwise the convex hull of P is the triangle qp_-p_+. In particular, q and p_+ are opposite non-adjacent extremal vertices and we can split as described above. □

Theorem 3. *Any simple polygon P with $n \geq 4$ edges can be guarded using at most $n - 2$ natural (vertex or edge) guards.*

Proof. Dobkin et al. [4] showed that for any chain there is a Peterson-style formula, that is, a guarding using natural edge guards only in which each guard appears exactly once and guards appear in the same order as the corresponding edges appear along the chain. Looking at the expression tree of this formula there is at least one vertex both of whose children are leaves. In other words, there is an operation (either union or intersection) that involves only two guards. As these two guards belong to two consecutive edges of P, we can replace this operation in the formula by the natural vertex guard of the common vertex, thereby saving one guard. Doing this for both chains as provided by Lemma 2 yields a guarding for P using $n - 4$ natural edge guards and two natural vertex guards. □

The (closure of) the complement of a polygonal halfplane H, call it $\overline{H}$, is a polygonal halfplane as well.

Observation 2. *Any guarding for H can be transformed into a guarding for $\overline{H}$ using the same number of guards.*

Proof. Use de Morgan's rules and invert all guards (keep their location but flip the angle to the complement with respect to 2π). Note that the resulting formula is monotone. Only guards complementary to the original ones appear (in SAT terminology: only negated literals); a formula is not monotone only if both a guard and its complementary guard appear in it.

Corollary 4. *Let $P_1, \ldots, P_m$ be a collection of $m \geq 1$ simple polygons t of which are triangles, for $0 \leq t \leq m$. Let R be a region that can be described as a formula composed of the operations intersection, union, and complement over the variables $\{P_1, \ldots, P_m\}$ in which each P_i appears exactly once. Then R can be guarded using at most $n - 2m + t$ natural (vertex or edge) guards, where n is the total number of edges of the polygons P_i, for $1 \leq i \leq m$.* □

Corollary 5. *Any simple polygon with $n \geq 4$ edges and h non-triangular holes can be guarded using at most $n - 2(h+1)$ natural (vertex or edge) guards.* □

Our guarding scheme for chains is based on a recursive decomposition in which at each step the current chain is split into two or more subchains. At each split some segments are extended to rays and we have to carefully control the way these rays interact with the remaining chain(s). This is particularly easy if the split vertex lies on the convex hull because then the ray resulting from the segment extension cannot intersect the remainder of the chain at all. However, we have to be careful what we mean by convex hull. Instead of looking at the convex hull of a polygonal halfplane H we work with the convex hull of its bounding chain C. The convex hull $\mathrm{h}(C)$ of a chain $C = (e_1, \ldots, e_n)$, for $n \geq 2$, is either the convex hull of H or the convex hull of $\overline{H}$, whichever of these two is not the whole plane which solely depends on the direction of the two rays of C. The boundary of $\mathrm{h}(C)$ is denoted by $\partial\mathrm{h}(C)$. There is one degenerate case, when the two rays defining C are parallel and all vertices are contained in the strip between them; in this case, $\mathrm{h}(C)$ is a strip bounded by the two parallel lines through the rays and thus $\partial\mathrm{h}(C)$ is disconnected.

Theorem 6. *Any polygonal halfplane bounded by a simple bi-infinite polygonal chain with $n \geq 2$ edges can be guarded using at most $\lfloor (4n-1)/5 \rfloor$ guards.*

Proof. The statement is easily checked for $2 \leq n \leq 3$. We proceed by induction on n. Let C be any chain with $n \geq 4$ edges. Denote the sequence of edges along C by $(e_1, \ldots, e_n)$ and let v_i, for $1 \leq i < n$, denote the vertex of C incident to e_i and e_{i+1}. The underlying (oriented) line of e_i, for $1 \leq i \leq n$, is denoted by ℓ_i. For $2 \leq i \leq n-1$, let e_i^+ be the ray obtained from e_i by extending the segment linearly beyond v_i. Similarly e_i^- refers to the ray obtained from e_i by extending the segment linearly beyond v_{i-1}. For convenience, let $e_1^+ = \ell_1$ and $e_n^- = \ell_n$.

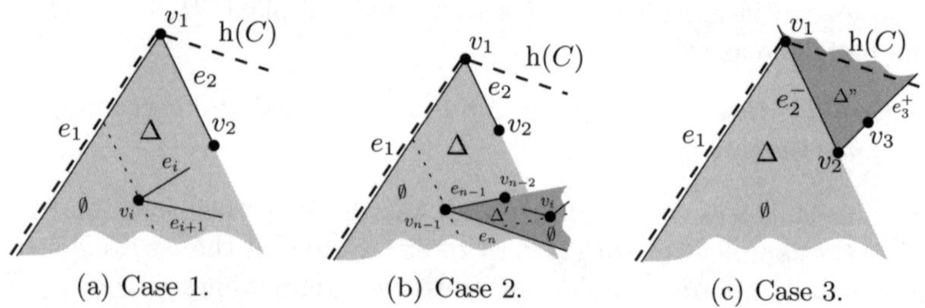

(a) Case 1. (b) Case 2. (c) Case 3.

Fig. 2. The chain C can interact with the shaded region Δ in three possible ways. The label $\emptyset$ marks an area which does not contain any vertex from C.

Without loss of generality (cf. Observation 2) suppose that v_1 is reflex, that is, the interior of the region bounded by C lies in the angle of C incident to v_1 which is larger than π. If there is any vertex v_i on $\partial\text{h}(C)$, for some $1 < i < n-1$, then split C into two chains $C_1 = (e_1, \ldots, e_i^+)$ and $C_2 = (e_{i+1}^-, \ldots, e_n)$. We obtain a guarding for C as $\mathcal{G}(C_1) \cup \mathcal{G}(C_2)$ and thus $\gamma(C) \leq \gamma(i) + \gamma(n-i)$, for some $2 \leq i \leq n-2$. As both $i \geq 2$ and $n-i \geq 2$, we can bound by the inductive hypothesis $\gamma(C) \leq \lfloor (4i-1)/5 \rfloor + \lfloor (4n-4i-1)/5 \rfloor \leq \lfloor (4i-1)/5 + (4n-4i-1)/5 \rfloor \leq \lfloor (4n-1)/5 \rfloor$. Else, if both e_1 and e_n are part of $\partial\text{h}(C)$ and ℓ_1 intersects ℓ_n then we place a guard g that covers both rays at the intersection of ℓ_1 and ℓ_n to obtain a guarding $g \cup \mathcal{G}(e_2^-, \ldots, e_{n-1}^+)$ for C. Therefore, in this case $\gamma(C) \leq 1 + \gamma(n-2)$. Observe that this is subsumed by the inequality from the first case with $i = 2$. Otherwise, either ℓ_1 does not intersect ℓ_n and thus v_1 and v_{n-1} are the only vertices of $\partial\text{h}(C)$ (the degenerate case where $\partial\text{h}(C)$ is disconnected) or without loss of generality (reflect C if necessary) v_1 is the only vertex of $\partial\text{h}(C)$. Let Δ denote the open wedge bounded by e_1 and e_2^+. We distinguish three cases.

Case 1. There is a vertex of C in Δ and among these, a vertex furthest from ℓ_2 is v_i, for some $3 \leq i \leq n-2$ (Fig. 2(a)). Split C into three chains, $C_1 = (\ell_1)$, $C_2 = (e_2^-, \ldots, e_i^+)$, and $C_3 = (e_{i+1}^-, \ldots, e_n)$. By the choice of v_i there is no intersection between C_2 and C_3 other than at v_i. A guarding for C can be obtained as $\mathcal{G}(C_1) \cup (\mathcal{G}(C_2) \cap \mathcal{G}(C_3))$. In this case $\gamma(C) \leq 1 + \gamma(j) + \gamma(n-j-1)$, for some $2 \leq j \leq n-3$. Since $j \geq 2$ and $n-j-1 \geq n-(n-3)-1 = 2$, we can apply the inductive hypothesis to bound $\gamma(C) \leq 1 + \lfloor (4j-1)/5 \rfloor + \lfloor (4n-4j-5)/5 \rfloor \leq \lfloor (4n-1)/5 \rfloor$.

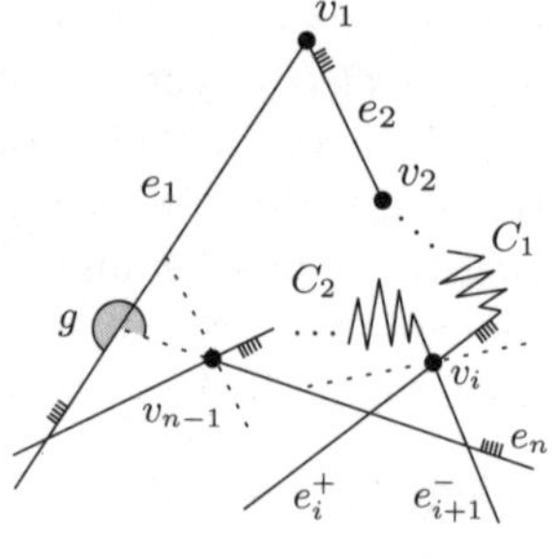

Case 2. There is a vertex of C in Δ and among these, the unique one furthest from ℓ_2 is v_{n-1} (Fig. 2(b)). We may suppose that ℓ_1 intersects ℓ_n; otherwise (in the degenerate case where $\partial\text{h}(C)$ is disconnected), exchange the roles of v_1 and v_{n-1}. We cannot end up in Case 2 both ways. Let Δ' denote the open (convex) wedge bounded by e_n and e_{n-1}^-. If there is any vertex of C in Δ',

let v_i be such a vertex which is furthest from ℓ_{n-1}. Let $C_1 = (e_1, \ldots, e_i^+)$ and $C_2 = (e_{i+1}^-, \ldots, e_{n-1}^+)$. Both C_1 and C_2 are simple, except that their first and their last ray may intersect (in that case split the resulting polygon into two chains). Put a guard g at the intersection of ℓ_n with e_1 such that g covers e_n completely and e_1 partially (see figure, the small stripes indicate the side to be guarded). A guarding for C can be obtained as $g \cap (\mathcal{G}(C_1) \cup \mathcal{G}(C_2))$. Again this yields $\gamma(C) \leq 1 + \gamma(i) + \gamma(n-i-1)$, for some $2 \leq i \leq n-3$, and thus $\gamma(C) \leq \lfloor (4n-1)/5 \rfloor$ as above in Case 1.

Otherwise there is no vertex of C in Δ'. We distinguish two sub-cases. If e_{n-1}^+ intersects e_1 then put two guards (see figure): a first guard g_1 at the intersection of ℓ_n with e_1 such that g_1 covers e_n completely and e_1 partially, and a second guard g_2 at the intersection of ℓ_{n-1} with e_1 such that g_2 covers e_{n-1} completely and e_1 partially. Together g_1 and g_2 cover e_1 and $g_1 \cap (g_2 \cup \mathcal{G}(C'))$ provides a guarding for C, with $C' = (e_2^-, \ldots, e_{n-2}^+)$. In this case we obtain $\gamma(C) \leq 2 + \gamma(n-3)$ and thus by the inductive hypothesis $\gamma(C) \leq 2 + \lfloor (4n-13)/5 \rfloor \leq \lfloor (4n-1)/5 \rfloor$.

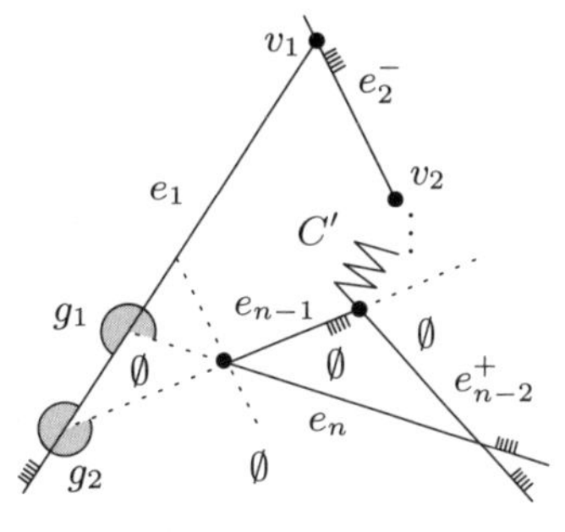

Finally, suppose that e_{n-1}^+ does not intersect e_1. Then for the chain $C' = (e_1, \ldots, e_{n-1}^+)$ there is some vertex other than v_1 on the convex hull boundary $\mathrm{h}(C')$. Thus we can obtain a guarding for C' as described above for the case that there is more than one vertex on the convex hull. Put a guard g at the intersection of ℓ_n with e_1 such that g covers e_n completely and e_1 partially (see figure). This yields a guarding $g \cap \mathcal{G}(C')$ for C with $\gamma(C) \leq 1 + \gamma(C') \leq 1 + \gamma(i) + \gamma(n-i-1)$, for some $2 \leq i \leq n-3$. As in Case 1 we conclude that $\gamma(C) \leq \lfloor (4n-1)/5 \rfloor$.

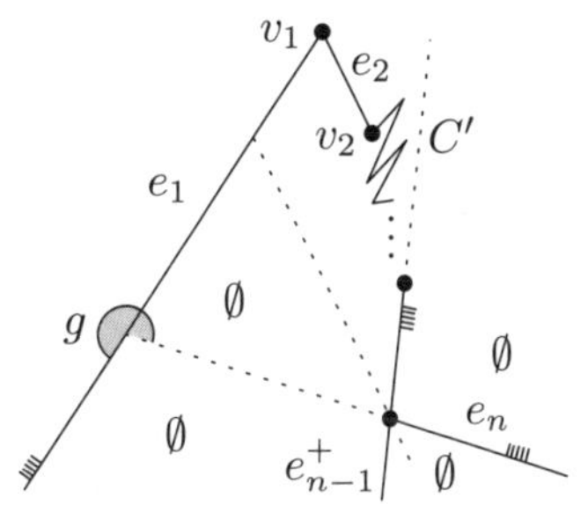

Case 3. There is no vertex of C in Δ (Fig. 2(c)). Let Δ'' denote the open (convex) wedge bounded by e_2^- and e_3^+. If e_3^- does not intersect e_1 then put a natural vertex guard g at v_1 to obtain a guarding $g \cap \mathcal{G}(C')$ for C, where $C' = (e_3^-, \ldots, e_n)$. This yields $\gamma(C) \leq 1 + \gamma(n-2)$ and thus by the inductive hypothesis $\gamma(C) \leq 1 + \lfloor (4n-9)/5 \rfloor \leq \lfloor (4n-1)/5 \rfloor$.

Now suppose that e_3^- intersects e_1. We distinguish two sub-cases. If there is no vertex of C in Δ'', then place two guards: a natural vertex guard g_1 at v_1 and a guard g_2 at the intersection of e_3^- with e_1 such that g_1 covers e_3 completely and e_1 partially. A guarding for C is provided by $g_1 \cap (g_2 \cup \mathcal{G}(C'))$, with $C' = (e_4^-, \ldots, e_n)$. In this case we obtain $\gamma(C) \leq 2 + \gamma(n-3)$ and thus in the same way as shown above $\gamma(C) \leq \lfloor (4n-1)/5 \rfloor$.

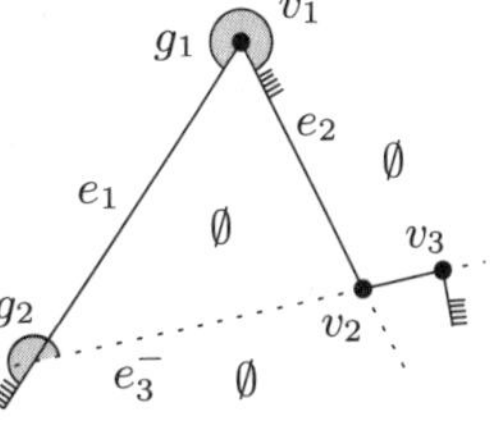

Otherwise there is a vertex of C in Δ''. Let v_i, for some $4 \le i \le n-1$, be a vertex of C in Δ'' which is furthest from ℓ_3. First suppose e_{i+1}^- does not intersect e_2. Then neither does e_i^+ and hence we can split at v_i in the same way as if v_i would be on $\partial \mathrm{h}(C)$. If $i = n-1$, e_n^- must intersect e_2 (otherwise, e_n would be on $\partial \mathrm{h}(C)$). Thus we have $i < n-1$ and both chains consist of at least two segments/rays.

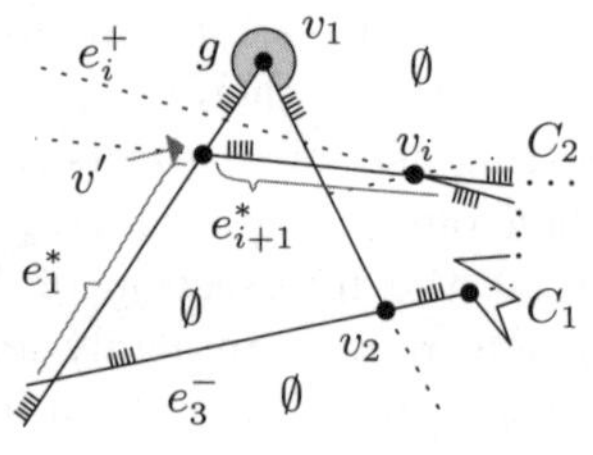

Now suppose that e_{i+1}^- intersects e_2 and thus e_1, and denote the point of intersection between e_{i+1}^- and e_1 by v'. Let e_1^* be the ray originating from v' in direction e_1, and let e_{i+1}^* denote the segment or ray (for $i = n-1$) originating from v' in direction e_{i+1}^-. Place a natural vertex guard g at v_1. Regardless of whether or not e_i^+ intersects e_2 and e_1, a guarding for C is provided by $g \cap (\mathcal{G}(C_1) \cup \mathcal{G}(C_2))$, with $C_1 = (e_3^-, \ldots, e_i^+)$ and $C_2 = (e_1^*, e_{i+1}^* \ldots, e_n)$ (if $i = n-1$ then $C_2 = (e_1^*, e_n^*)$). Observe that by the choice of v_i both C_1 and C_2 are simple and $\gamma(C) \le 1 + \gamma(j) + \gamma(n-j-1)$, for some $2 \le j \le n-3$. As above, this yields $\gamma(C) \le \lfloor (4n-1)/5 \rfloor$.

We have shown that in every case $\gamma(C) \le \lfloor (4n-1)/5 \rfloor$ and as C was arbitrary it follows that $\gamma(n) \le \lfloor (4n-1)/5 \rfloor$. □

Corollary 7. *Any simple polygon P with n edges can be guarded using at most $\lfloor (4n-2)/5 \rfloor$ guards.*

Corollary 8. *Let $P_1, \ldots, P_m$ be a collection of $m \ge 1$ simple polygons with n edges in total, and let R be a region that can be described as a formula composed of the operations intersection, union, and complement over the variables $\{P_1, \ldots, P_m\}$ in which each P_i appears exactly once. Then R can be guarded using at most $\lfloor (4n-2m)/5 \rfloor$ guards.*

Corollary 9. *Let P be any simple polygon P with h holes such that P is bounded by n edges in total. Then P can be guarded using at most $\lfloor (4n-2h-2)/5 \rfloor$ guards.*

4 Lower Bounds

For any natural number m we construct a polygon P_m with $2m$ edges which requires "many" guards. The polygon consists of spikes $S_1, S_2, ..., S_m$ arranged in such a way that the lines through both edges of a spike cut into every spike to the left (see Fig. 3). Denote the apex of S_i by w_i and its left vertex by v_i. The edge from v_i to w_i is denoted by e_i, the edge from w_i to v_{i+1} by f_i. We can construct P_m as follows: Consider the two hyperbolas $\{(x,y) \in \mathbb{R}^2 \mid x \ge 1, y = \frac{1}{x}\}$ and $\{(x,y) \in \mathbb{R}^2 \mid x \ge 1, y = -\frac{1}{x}\}$. Let $v_1 := (1,1)$ and $w_1 := (1,-1)$. Then choose f_1 tangential to the lower hyperbola. Let v_2 be the point where the tangent of the lower hyperbola intersects the upper hyperbola, that is, $v_2 = (1+\sqrt{2}, \frac{1}{1+\sqrt{2}})$. Choose w_2 to be the point where the tangent of the upper hyperbola in v_2 intersects the lower hyperbola, and proceed in this way. When reaching w_m,

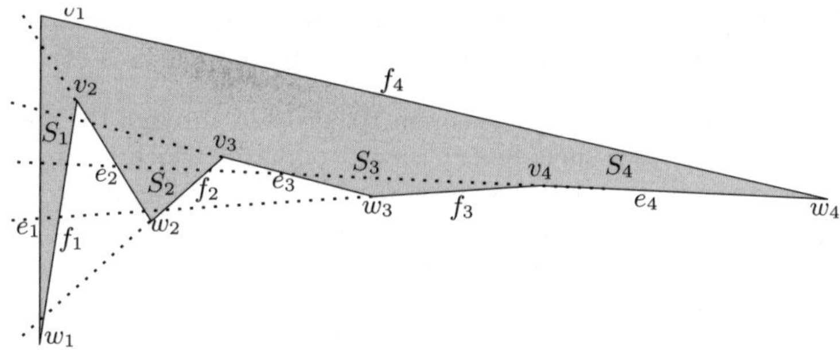

Fig. 3. Example consisting of four spikes

draw the last edge f_m from w_m to v_1 to close the polygon. Due to the convexity of the hyperbolas, P_m has the claimed property.

No two edges of P_m are collinear. Consider the line arrangement defined by the edges of P_m. No two lines intersect outside P_m, unless one of them is the line through f_m. This leads to the following observation.

Observation 3. *In any guarding for P_m every 2-guard that does not cover f_m is a natural vertex guard.*

Theorem 10. *For any even natural number n there exists a simple polygon with n edges which requires at least $n-2$ natural guards.*

We say a guard *belongs* to a spike S_i if it is a natural edge guard on e_i or f_i or if it is a natural vertex guard on v_i or w_i. As only natural guards are allowed, every guard belongs to exactly one spike. The basic idea is that most spikes must have at least two guards. Obviously every spike S_i has at least one guard, since e_i must be covered (Lemma 1).

Lemma 11. *Consider a guarding $\mathcal{G}(P_m)$ using natural guards only, and let $i \in \{1, ..., m-1\}$. If only one guard belongs to S_i, then this guard must be on v_i or on e_i. If neither the guard at w_i nor the guard of f_i appear in $\mathcal{G}(P_m)$, then both the guard at v_{i+1} and the guard of e_{i+1} are in $\mathcal{G}(P_m)$.*

Proof. Assume only one guard from $\mathcal{G}(P_m)$ belongs to S_i. It cannot be the natural edge guard of f_i, because this would leave e_i uncovered (Lemma 1). If we had a guard on w_i only, there would be no guard to distinguish a point near v_i outside P_m from a point near v_{i+1} located inside P_m and below the line through f_i (see the two circles in the figure). Now assume there are no guards at w_i nor on f_i. Then to cover the edge f_i there must be a vertex guard on v_{i+1}. Furthermore, the edge guard on e_{i+1} is the only remaining natural guard to distinguish a point at the apex of S_i near w_i from a point located to the right of the apex of S_{i+1} near w_{i+1} and above the line through e_{i+1} (depicted by two crosses). □

This lemma immediately implies Theorem 10. Proceed through the spikes from left to right. As long as a spike has at least two guards which belong to it, we are fine. Whenever there appears a spike S_i with only one guard, we know that there must be at least two guards in S_{i+1} namely at v_{i+1} and on e_{i+1}. Either there is a third guard that belongs to S_{i+1}, and thus both spikes together have at least four guards; or again we know already two guards in S_{i+2}. In this way, we can go on until we either find a spike which at least three guards belong to or we have gone through the whole polygon. So whenever there is a spike with only one guard either there is a spike with at least three guards that makes up for it, or every spike till the end has two guards. Hence there can be at most one spike guarded by one guard only that is not made up for later. For the last spike S_m the lemma does not hold and we only know that it has at least one guard. So all in all there are at least $2(m-2)+1+1=n-2$ guards.

If we allow general (vertex) guards, it is possible to find guardings for P_m using roughly $2n/3$ guards, which is in accord with the lower bound in [3]. c

Theorem 12. *For any even natural number n there exists a simple polygon with n edges which requires at least $\lceil(3n-4)/5\rceil$ guards.*

Proof. Consider a polygon P_m as defined above, and let $\mathcal{G}(P_m)$ be a guarding for P_m. Define a to be the number of 2-guards in $\mathcal{G}(P_m)$, and let b be the number of other guards. All the n edges of P have to be covered somehow. An edge can be covered completely by a 2-guard, a 1-guard, or a 1′-guard. If no guard covers it completely, then the edge must be covered by at least two guards partly (Lemma 1). Moreover, at least one of these guards, namely the one covering the section towards the right end of the edge, is a 0′-guard, because the orientation can not be correct to cover a second edge. So if an edge e is not covered by a 2-guard, then there is at least one guard that does not cover any edge other than e. Therefore $2a+b\geq n$.

For any $i\in\{1,...,m-2\}$ let h_i be the directed line segment from the intersection of the lines through e_{i+1} and e_{i+2} to v_{i+2} (see Fig. 4). Similarly, let h'_i be the line segment from w_{i+1} to the intersection of the lines through f_i and f_{i+1}. As in Lemma 11, consider pairs $(p_1,q_1),...,(p_{m-2},q_{m-2})$ and $(p'_1,q'_1),...,(p'_{m-2},q'_{m-2})$ of points infinitesimally close to the starting point or the endpoint of the corresponding line segment, located as follows: $p_i,p'_i\in P_m$ for all i, $q_i,q'_i\notin P_m$ for all i, p_i is outside the natural vertex guard at w_{i+1}, whereas q_i is inside the natural vertex guard at w_{i+2}, and similarly, p'_i is outside the natural vertex guard at v_{i+2}, whereas q'_i is inside the natural vertex guard at v_{i+1}. There are $n-4$ such pairs, and they need to be distinguished somehow (Observation 1). Any natural vertex guard can distinguish at most one pair, and the same is true for any (non-natural) 2-guard located along the line through f_m. Thus any 2-guard in $\mathcal{G}(P_m)$ distinguishes at most one of the pairs (Observation 3).

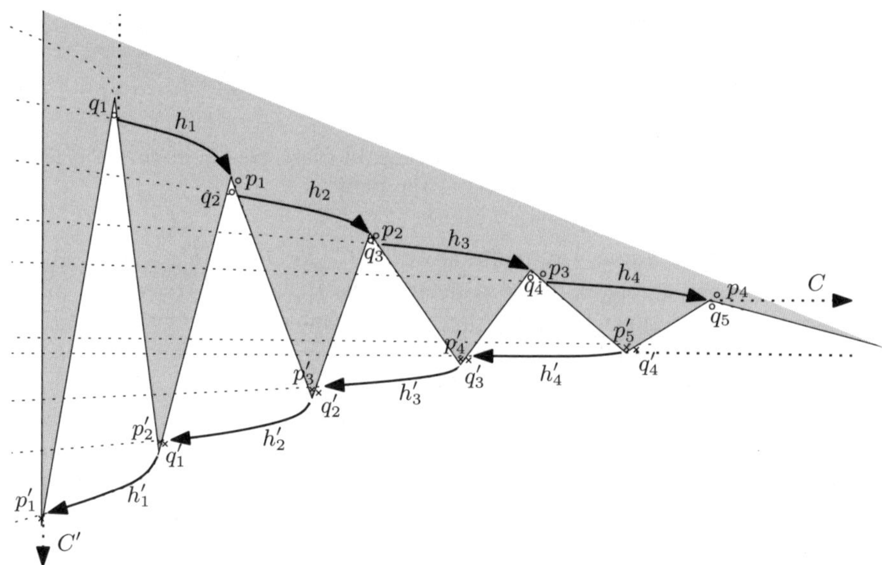

Fig. 4. The pairs (p_i, q_i) and (p'_i, q'_i) must be distinguished

We claim that every guard g in $\mathcal{G}(P_m)$ can distinguish at most three of these pairs. Denote the vertex of g by v_g, and let ℓ_g and r_g denote the left and right ray of g, respectively. Assume g distinguishes p_i from q_i. If v_g is to the left of h_i, then—in order to distinguish p_i from q_i—the ray r_g must intersect h_i. Symmetrically, if v_g is to the right of h_i, then ℓ_g must intersect h_i. Finally, if v_g is on the line through h_i then it must be on the line segment h_i itself. To distinguish p_i from q_i, the endpoint of h_i (i.e. v_{i+2}) must be inside g (possibly on the boundary of g), hence ℓ_g must point to the left side of h_i or in the same direction as h_i, and r_g must point to the right side of h_i or in the same direction. Now assume g distinguishes p'_i and q'_i. If v_g is to the right of h'_i, then ℓ_g must intersect it, if it is to the left r_g must intersect it. If v_g lies on h'_i, ℓ_g leaves to the left and r_g to the right, or either or both rays lie on h'_i. In any case either ℓ_g intersects h_i (h'_i, respectively) coming from the right side of h_i (h'_i) and leaving to the left side, or r_g intersects h_i (h'_i) coming from the left side and leaving to the right, or ℓ_g starts on h_i (h'_i) itself leaving to the left or r_g starts on the line segment itself leaving to the right (see Fig. 5). If r_g leaves an oriented line segment to the right side of the segment or if ℓ_g leaves an oriented line segment to the left side, we say the ray *crosses* the line segment *with correct orientation.* So whenever a pair (p_i, q_i) or (p'_i, q'_i) is distinguished by g, then at least one of the rays ℓ_g or r_g has a correctly oriented crossing with h_i (h'_i, respectively). The line segments $h_1, ..., h_{m-2}$ lie on a oriented convex curve C, which we obtain by prolonging every line segment until reaching the starting point of the next one. Extend the first and last line segment to infinity vertically and horizontally,

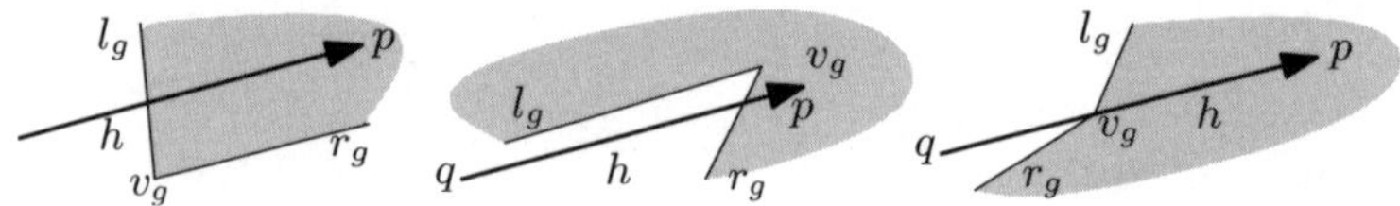

Fig. 5. Different ways g can distinguish p and q. In every case ℓ_g intersects h leaving to the left side or r_g intersects h leaving to the right.

respectively. In the same way define a curve C' for $h'_1, ..., h'_{m-2}$ (see Fig. 4). Any ray can cross a convex curve at most twice. Because of the way C and C' are situated with respect to each other (a line that crosses C twice must have negative slope, to cross C' twice positive slope) a ray can intersect $C \cup C'$ at most three times. But we are only interested in crossings with correct orientation. If a ray crosses a curve twice, exactly one of the crossings has the correct orientation. If a ray crosses both C and C' once, exactly one of the crossings has the correct orientation. Therefore any ray can have at most two correctly oriented crossings. If one of the rays has two correctly oriented crossings, the other ray has at most one. Thus both rays together can have at most three correctly oriented crossings, and therefore g can distinguish at most three pairs. This leads to the second inequality $a + 3b \geq n - 4$. Both inequalities together imply $a + b \geq \frac{3n-4}{5}$. □

References

1. Aigner, M., Ziegler, G.M.: Proofs from THE BOOK, 3rd edn. Springer, Berlin (2003)
2. Chvátal, V.: A Combinatorial Theorem in Plane Geometry. J. Combin. Theory Ser. B 18, 39–41 (1975)
3. Damian, M., Flatland, R., O'Rourke, J., Ramaswami, S.: A New Lower Bound on Guard Placement for Wireless Localization. In: 17th Annual Fall Workshop on Computational Geometry (2007)
4. Dobkin, D.P., Guibas, L., Hershberger, J., Snoeyink, J.: An Efficient Algorithm for Finding the CSG Representation of a Simple Polygon. Algorithmica 10, 1–23 (1993)
5. Eppstein, D., Goodrich, M.T., Sitchinava, N.: Guard Placement for Efficient Point-in-Polygon Proofs. In: Proc. 23rd Annu. Sympos. Comput. Geom. pp. 27–36 (2007)
6. Fisk, S.: A Short Proof of Chvátal's Watchman Theorem. J. Combin. Theory Ser. B 24, 374 (1978)
7. Fragoudakis, C., Markou, E., Zachos, S.: Maximizing the Guarded Boundary of an Art Gallery is APX-complete. Comput. Geom. Theory Appl. 38(3), 170–180 (2007)
8. Kahn, J., Klawe, M.M., Kleitman, D.J.: Traditional Galleries Require Fewer Watchmen. SIAM J. Algebraic Discrete Methods 4, 194–206 (1983)
9. Karavelas, M.I., Tsigaridas, E.P.: Guarding Curvilinear Art Galleries with Vertex or Point Guards. Rapport de recherche 6132, INRIA (2007)
10. Lee, D.T., Lin, A.K.: Computational Complexity of Art Gallery Problems. IEEE Trans. Inform. Theory 32(2), 276–282 (1986)
11. O'Rourke, J.: Galleries Need Fewer Mobile Guards: A Variation on Chvátal's Theorem. Geom. Dedicata 14, 273–283 (1983)

12. O'Rourke, J.: Visibility. In: Goodman, J.E., O'Rourke, J. (eds.) Handbook of Discrete and Computational Geometry, ch. 28, pp. 643–663. CRC Press LLC, Boca Raton (2004)
13. O'Rourke, J.: Computational Geometry Column 48. ACM SIGACT News 37(3), 55–57 (2006)
14. Shermer, T.C.: Recent Results in Art Galleries. Proc. IEEE 80(9), 1384–1399 (1992)
15. Speckmann, B., Tóth, C.D.: Allocating Vertex Pi-guards in Simple Polygons via Pseudo-triangulations. Discrete Comput. Geom. 33(2), 345–364 (2005)
16. Urrutia, J.: Art Gallery and Illumination Problems. In: Sack, J.-R., Urrutia, J. (eds.) Handbook of Computational Geometry, pp. 973–1027. Elsevier Science Publishers B.V, Amsterdam (2000)

Bicriteria Approximation Tradeoff for the Node-Cost Budget Problem

Yuval Rabani[1,⋆] and Gabriel Scalosub[2,⋆⋆]

[1] Computer Science Dept., Technion - Israel Institute of Technology, Haifa 32000, Israel
rabani@cs.technion.ac.il
[2] School of Electrical Engineering, Tel Aviv University, Ramat Aviv 69978, Israel
gabriels@eng.tau.ac.il

Abstract. We consider an optimization problem consisting of an undirected graph, with cost and profit functions defined on all vertices. The goal is to find a connected subset of vertices with maximum total profit, whose total cost does not exceed a given budget. The best result known prior to this work guaranteed a $(2, O(\log n))$ bicriteria approximation, i.e. the solution's profit is at least a fraction of $\frac{1}{O(\log n)}$ of an optimum solution respecting the budget, while its cost is at most twice the given budget. We improve these results and present a bicriteria tradeoff that, given any $\varepsilon \in (0, 1]$, guarantees a $(1 + \varepsilon, O(\frac{1}{\varepsilon} \log n))$-approximation.

1 Introduction

We consider the following problem: Given an undirected graph $G = (V, E)$, a non negative cost function c defined on V, a non negative profit function π defined on V, and a budget B, our aim is to find a connected set of vertices $T \subseteq V$ such that their overall cost does not exceed B, while maximizing the overall profit obtained from T. We refer to this problem as the *Node-Cost Budget Problem* (or the *Budget Problem (BP)* for short). This problem can also be cast as a rooted problem, where we specify a special vertex $r \in V$ as a root, and impose the restriction that $r \in T$. An algorithm for solving the rooted problem can easily be transformed into one which solves the unrooted problem by simply enumerating over all vertices in V as candidates for r, and picking the best among the various solutions. We shall hereafter focus on the rooted version of the problem. In what follows we denote the size of V by n. We note that previous results for this problem all give bicriteria approximation guarantees, where an (α, β)-approximate solution [1] is one which guarantees at least a β fraction of the optimal profit possible using a given budget, while violating the budget restriction by a factor of at most α.

⋆ Work at the Technion supported by BSF grant number 99-00217, by ISF grant number 386/99, by IST contract number 32007 (APPOL), and by the Fund for the Promotion of Research at the Technion.

⋆⋆ Supported in part by BSF grant number 99-00217. This work was done while the author was with the Computer Science Dept. at the Technion, Israel.

J. Gudmundsson (Ed.): SWAT 2008, LNCS 5124, pp. 90–101, 2008.

1.1 Our Results

We present, for every $\varepsilon \in (0,1]$, a $(1+\varepsilon, O(\frac{1}{\varepsilon}\log n))$ bicriteria approximation algorithm for BP. This improves upon previous results discussed below. Our algorithm consists of meticulously defining a collection of instances, and finding approximate solutions to these instances using extensions of the methods described in [2]. Upon finding a collection of approximate solutions to these instances, we show that these can be used to find a collection of feasible solutions to our initial problem. We prove that all these solutions violate the budget restriction by a factor of at most $1+\varepsilon$, while at least one has sufficient profit.

1.2 Related Work

The budget problem was introduced by Guha, Moss, Naor, and Schieber [3]. They motivated the problem by its application to power-outage recovery. Guha et al. presented a $(2, O(\log^2 n))$ bicriteria approximation algorithm for the problem. Their algorithm is based on an $O(\log n)$-approximation algorithm for the node weighted Steiner tree problem, devised by Klein and Ravi [4]. The node weighted Steiner tree problem consists of an undirected graph with costs assigned to vertices, where given a subset T of V, one seeks a connected subset of V which contains all the nodes in T such that its overall cost is minimal.

Moss and Rabani [2] improved these results by presenting a $(2, O(\log n))$ bicriteria approximation. Their algorithm is based on a tree packing devised by using a primal-dual algorithm for approximating the prize collecting problem. This problem consists of an undirected graph with costs and profits defined over its set of vertices. The aim is to find a connected subset S of V so as to minimize the sum of the costs of the vertices in S (the cost) and the profits of the vertices not in S (the penalty). Given the tree packing, an averaging argument then enables them to pick out a tree from the packing with good features.

A closely related problem to BP is the node-cost quota problem. In this problem we are given a graph with costs and profits defined over its set of vertices, and a quota Q. The aim is to find a connected subset S of V whose profit is at least Q, and whose cost is minimal. The best upper bound for this problem guarantees an $O(\log n)$-approximation (see [2]). Furthermore, assuming $P \neq NP$ this result is tight, up to a constant factor because the problem is at least as hard as set cover [5].

There has also been a considerable amount of work concerning edge-costs versions of similar problems, such as variants of the prize-collecting Steiner tree problem [6,7,8], the k-MST problem [9], and the constrained minimum spanning tree problem [10].

The best approximation lower bound for BP is based on the tight lower bound for the budgeted maximum coverage problem, where due to an approximation preserving reduction from this problem to BP, one can obtain a lower bound of $1 - \frac{1}{e}$ on the approximation ratio of BP [11,5].

2 Notation and Preliminaries

Given a connected set of vertices $T \subseteq V$, we will speak in terms of any spanning tree induced by T. We may assume without loss of generality that the cost and the profit of the root of an instance is 0. Otherwise, we may solve an altered instance where we assign cost 0 to the root, and strengthen the budget restriction to be at most the original budget from which we subtract the cost of the root in the original problem. Any solution to the altered problem induces a solution with the same value to the original problem and vice-versa, up to re-adding the profit of the root. We begin by introducing the concept of *distance* and *reachability* among vertices.

Given a graph $G = (V, E)$ and a non-negative cost function c defined on V, we define the *cost-distance* $d(u, v)$ between vertices u and v to be the minimum total cost of the inner vertices on any path connecting u and v, with respect to the cost function c. Formally:

$$d(u,v) = \min \left\{ \sum_{i=2}^{\ell-1} c(v_i) \,\middle|\, \begin{array}{l} \{v_i\}_{i=1}^{\ell} \text{ is a } u\text{–}v \text{ path} \\ \text{such that } v_1 = u \text{ and } v_\ell = v \end{array} \right\}.$$

If $u = v$ or $(u, v) \in E$ we define $d(u, v) = 0$. If there is no path connecting u and v we define $d(u, v) = \infty$. We further denote by $\mathrm{path}(u, v)$ the set of inner vertices of a path achieving minimal cost. Formally:

$$\mathrm{path}(u,v) = \arg\min \left\{ \sum_{i=2}^{\ell-1} c(v_i) \,\middle|\, \begin{array}{l} \{v_i\}_{i=1}^{\ell} \text{ is a } u\text{–}v \text{ path} \\ \text{such that } v_1 = u \text{ and } v_\ell = v \end{array} \right\} \setminus \{u, v\}.$$

In addition, we say that a vertex v is *reachable with cost at most* p from $u \neq v$ if $d(u, v) + c(v) \leq p$, and any vertex v is reachable from itself with cost 0.

By the above definitions, for any instance of BP, we may assume without loss of generality that all vertices in G are reachable from the root r with cost at most B, since any vertex not reachable from r with cost at most B will not be part of any feasible solution. In particular we may assume G is connected.

For any $S \subseteq V$ we let $c(S) = \sum_{v \in S} c(v)$ denote the cost of S and $\pi(S) = \sum_{v \in S} \pi(v)$ denote the profit obtained by S. We further denote the *density* of S by $\gamma(S) = \frac{\pi(S)}{c(S)}$. For any subtree T of G and any $v \in T$, denote by $CH_T(v)$ the set of children of v in T. We further denote by T_u the subtree of T rooted at vertex $u \in T$, i.e. T_u consists of all vertices in T such that the path connecting them to the root r, contains u.

3 Finding Good Candidate Solutions

In finding a good approximate solution we make use of the notion of tree packing. This enables us to find a connected set of vertices with some desirable properties.

Definition 1 (Tree Packing). *Given a graph $G = (V, E)$, a* tree packing *in G relative to a function $d : V \mapsto \mathbb{Q}^+$ is an assignment of weights λ to a set $\mathcal{T}$ of trees in G, which satisfies $\sum_{T \in \mathcal{T} | v \in T} \lambda_T \leq d(v)$, where λ_T is the weight of tree T in the packing.*

In what follows, for every $S \subseteq V$ we let $\partial S = \{v \in V | \exists u \in S \text{ s.t. } (u,v) \in E\}$. Consider the integer program for BP, denoted IP:

$$\begin{aligned}
&\text{maximize} \quad \textstyle\sum_{i \in V} \pi(i) d_i \\
&\text{subject to} \\
&\qquad \textstyle\sum_{i \in V} c(i) d_i \leq B && (1)\\
&\qquad d_i \leq \textstyle\sum_{j \in \partial S} d_j \quad \forall S \subseteq V \setminus \{r\}, \forall i \in S && (2)\\
&\qquad d_r = 1 && (3)\\
&\qquad d_i \in \{0,1\} \qquad \forall i \in V \setminus \{r\}. && (4)
\end{aligned}$$

In the above program, vertex i is part of the solution if and only if its associated variable d_i satisfies $d_i = 1$. Constraint (2) ensures the connectivity of the output, i.e. that the vertices for which $d_i = 1$ make up one connected component. Notice that assuming G is not trivial (i.e. $n > 1$), considering constraint (2) for the case where $S = V \setminus \{r\}$ we have $\partial S = \{r\}$ which implies $d_i \leq 1$ for all $i \in V \setminus \{r\}$. Let us denote by OPT_B the value of an optimal solution for IP. If we replace constraints (4) by

$$d_i \geq 0 \qquad \forall i \in V \setminus \{r\} \tag{5}$$

we obtain a linear programming relaxation for BP which we denote by LP. This linear program can be solved in polynomial time using the ellipsoid algorithm [12]. Let d denote an optimal solution to LP. As shown in the following theorem, we can use such a solution to find a tree packing such that every vertex is covered sufficiently by the packing:

Theorem 1 ([2]). *Let $G = (V, E)$ be an undirected graph with non-negative node weights $d : V \mapsto \mathbb{Q}^+$, considered rooted at $r \in V$. Assume d satisfies inequalities (2) and (3). Then, there exists a polynomial time algorithm that computes a tree packing in G of trees containing r such that for every node $v \in V$*

$$\frac{d(v)}{c \log n} \leq \sum_{T \in \mathcal{T} | v \in T} \lambda_T \leq d(v)$$

for some constant c independent of n.

Let $\mathcal{T}$ be the support of the tree packing guaranteed by Theorem 1, let $\mathcal{L} = \{T \in \mathcal{T} | c(T) \leq B\}$ and $\mathcal{H} = \{T \in \mathcal{T} | c(T) > B\}$. The following lemma will serve as a starting point in our quest for finding a good approximate solution.

Lemma 1 (A good tree exists in the packing [2]). *Given the support $\mathcal{T}$ of the packing guaranteed by Theorem 1, at least one of the following conditions holds:*

1. *$\exists T \in \mathcal{L}$ such that $\pi(T) \geq \frac{1}{2c \log n} OPT_B$;*
2. *$\exists T \in \mathcal{H}$ such that $\gamma(T) = \frac{\pi(T)}{c(T)} \geq \frac{1}{2c \log n} \frac{OPT_B}{B}$.*

Furthermore a tree T satisfying one of the above conditions can be found in time polynomial in the size of the original input.

In what follows we focus our attention on the case where we have a tree T which satisfies condition 2 in Lemma 1. Note that the cost of such a tree may very well exceed the available budget. In the following section we show that under some conditions on the underlying instance, one can trim such a high-density tree while guaranteeing that the resulting tree is not too costly.

3.1 Candidate Solutions with High Density

Consider an instance of BP with graph $G = (V, E)$ such that all vertices are reachable from the root $r \in V$ with cost at most d, and a budget restriction B, where $d \leq B$. In what follows we show that for any $\alpha > 0$ and any subtree T of G rooted at r such that $\gamma(T) \geq \alpha \cdot \frac{\mathrm{OPT}_B}{B}$ and $c(T) > B$, T can be trimmed into a tree T^H satisfying $\pi(T^H) \geq \frac{\alpha}{4} \cdot \mathrm{OPT}_B$ and $c(T^H) \leq B + d$.

Let T be any subtree of G rooted at r such that $\gamma(T) \geq \alpha \cdot \frac{\mathrm{OPT}_B}{B}$ and $c(T) > B$. In order to obtain T^H, we will accumulate subtrees of T, while making sure the overall cost remains within a certain range. The following Lemma gives a sufficient condition for the resulting tree having sufficient profit:

Lemma 2. *Let $T \subseteq G$ be a tree such that $\gamma(T) \geq \alpha \cdot \frac{\mathrm{OPT}_B}{B}$ and $c(T) > B$. For any set of vertices $U \subseteq T$ such that for every $u \in U$, $\gamma(T_u) \geq \gamma(T)$, if $\sum_{u \in U} c(T_u) \geq \frac{B}{2}$ then $\sum_{u \in U} \pi(T_u) \geq \frac{\alpha}{2} \cdot OPT_B$.*

Proof. Since for every $u \in U$ we have $\gamma(T_u) \geq \gamma(T)$, we are guaranteed that for every $u \in U$ we have $\pi(T_u) \geq c(T_u) \cdot \gamma(T)$. By summing over all $u \in U$ and using the assumptions that $\gamma(T) \geq \alpha \cdot \frac{\mathrm{OPT}_B}{B}$ and $\sum_{u \in U} c(T_u) \geq \frac{B}{2}$ we obtain

$$\begin{aligned} \textstyle\sum_{u \in U} \pi(T_u) &\geq \textstyle\sum_{u \in U} c(T_u) \cdot \gamma(T) \\ &\geq \alpha \cdot \frac{OPT_B \cdot \sum_{u \in U} c(T_u)}{B} \\ &\geq \frac{\alpha}{2} \cdot OPT_B \end{aligned}$$

□

The following corollary is an immediate consequence of Lemma 2:

Corollary 1. *Let $T \subseteq G$ be a tree such that $\gamma(T) \geq \alpha \cdot \frac{\mathrm{OPT}_B}{B}$ and $c(T) > B$. For any $v \in T$, and any set of vertices $U \subseteq CH_T(v)$, if U satisfies $\frac{B}{2} \leq \sum_{u \in U} c(T_u) \leq B$ and for every $u \in U$, $\gamma(T_u) \geq \gamma(T)$, then the set $T^H = \bigcup_{u \in U} T_u \cup \{v\} \cup \mathrm{path}(r, v) \cup \{r\}$ is a tree which satisfies $\pi(T^H) \geq \frac{\alpha}{2} \cdot OPT_B$ and $c(T^H) \leq B + d$.*

Proof. Note that by the assumption that $U \subseteq CH_T(v)$ for some $v \in T$, we are guaranteed to have for every $u \neq u'$ in U, $T_u \cap T_{u'} = \emptyset$. It therefore follows that $c(\bigcup_{u \in U} T_u) = \sum_{u \in U} c(T_u)$ and $\pi(\bigcup_{u \in U} T_u) = \sum_{u \in U} \pi(T_u)$. Furthermore, since $U \subseteq CH_T(v)$, then clearly T^H is a tree. Since $\bigcup_{u \in U} T_u \subseteq T^H$ then by Lemma 2 we are guaranteed to have $\pi(T^H) \geq \frac{\alpha}{2} \cdot OPT_B$. On the other hand, since every node is reachable from the root $r \in V$ with cost at most d, the cost of the path $\{r\} \cup \mathrm{path}(r, v) \cup \{v\}$ is at most d. It therefore follows that the overall cost of T^H is at most $B + d$, as required. □

The following lemma, whose proof is omitted due to space constraints, guarantees we can trim a high density tree such that the resulting tree carries a sufficiently large profit, while having a limited cost.

Lemma 3. *If* $T = \arg\max\{\gamma(T')|T' \in \mathcal{T} \setminus \mathcal{L}\}$ *satisfies* $\gamma(T) \geq \frac{1}{2c\log n}\frac{OPT_B}{B}$ *for some constant* c*, then there exists a polynomial trimming algorithm* TRIM*, such that* $T^H = \text{TRIM}(T)$ *satisfies* $\pi(T^H) = \Omega\left(\frac{1}{\log n}\right) OPT_B$ *and* $c(T^H) \leq B + d$.

3.2 An Algorithm for Finding a Good Candidate

Clearly any tree T satisfying condition 1 in Lemma 1, would suffice for our purpose, since such a tree does not violate the budget constraint, while carrying at least an $\Omega\left(\frac{1}{\log n}\right)$ fraction of the optimal profit. On the other hand, by Lemma 3, we can trim any tree T satisfying condition 2 in Lemma 1, so as to obtain a tree with cost at most $B + d$, carrying at least an $\Omega(\frac{1}{\log n})$ fraction of the optimal profit. Algorithm EXTRACT described in Algorithm 1 therefore summarizes the method for finding a subtree $T \subseteq G$ such that $c(T) \leq B + d$ and $\pi(T) = \Omega\left(\frac{1}{\log n}\right) \text{OPT}_B$.

Algorithm 1. EXTRACT (tree packing $\mathcal{T}$, budget B)

set $\mathcal{L} = \{T \in \mathcal{T} | c(T) \leq B\}$
set $\mathcal{H} = \mathcal{T} \setminus \mathcal{L}$
set $T^L = \arg\max\{\pi(T)|T \in \mathcal{L}\}$.
set $T = \arg\max\{\pi(T)/c(T)|T \in \mathcal{H}\}$
set $T^H = \text{TRIM}(T)$
return $\arg\max\{\pi(T^L), \pi(T^H)\}$

The above proves the following lemma:

Lemma 4. *Given any instance of BP with graph* $G = (V, E)$ *such that all vertices are reachable from the root* $r \in V$ *with cost at most* d*, and a budget restriction* B*, where* $d \leq B$*, algorithm* EXTRACT *produces a solution* T *such that* $c(T) \leq B + d$ *and* $\pi(T) = \Omega\left(\frac{1}{\log n}\right) \text{OPT}_B$.

4 Structure of an Optimal Solution

In this section, we study the structure of an optimal solution. We show that assuming all optimal solutions have sufficiently large cost, there exists a decomposition of an optimal solution into disjoint subtrees, such that at least one of them has sufficiently small cost, while carrying at least an $O\left(\frac{1}{\log n}\right)$ fraction of the profit attained by the optimal solution. Our study later motivates the definition of a sequence of instances of BP, each corresponding to one of these subtrees, such that at least one of these instances can be transformed into a $(1 + \varepsilon, O(\frac{1}{\varepsilon}\log n))$-approximate solution.

First note that if there exists an optimal solution T^*, such that $c(T^*) \leq \frac{B}{2}$, then by applying the algorithm of [2] over the same instance, with a budget restriction of $B' = \frac{B}{2}$, we are guaranteed to obtain a $(1, O(\log n))$-approximate solution. We can therefore assume that for every optimal solution T^*, $c(T^*) > \frac{B}{2}$.

Let T^* be any such optimal solution, and let $\varepsilon \in (0, 1]$. Define $k = \lceil \frac{1+\varepsilon}{2\varepsilon} \rceil = \Theta\left(\frac{1}{\varepsilon}\right)$, and let $\overline{\sigma} = \overline{\sigma}(k) = (\sigma_1, \ldots, \sigma_{k-1})$ be a sequence of proportions, $\sigma_i \leq 1$ for all $i = 1, \ldots, k-1$, such that $\sigma_1 \leq \frac{1}{2}$. In what follows we let $\rho_i = \prod_{j=1}^{i} \sigma_j$, $i = 1, \ldots, k-1$ and define $\rho_0 = 1$. We now describe a recursive partition of T^* into k disjoint connected components $\{T_i^*\}_{i=0}^{k-1}$ according to $\overline{\sigma}$. Let $r_0 = r$ and let $G_0^* = T^*$. Given G_{i-1}^*, let $r_i \in G_{i-1}^*$ be such that

$$c(T_{r_i}^*) \geq \rho_i B \tag{6}$$

$$c(T_u^*) \leq \rho_i B \qquad \forall u \in CH_{T^*}(r_i), \tag{7}$$

and define $G_i^* = T_{r_i}^*$.

First note that for every $i = 1, \ldots, k-1$, there exists a node $r_i \in G_{i-1}^*$ which satisfies conditions (6) and (7). To see this, note that by our assumption that $\sigma_1 \leq \frac{1}{2}$ we have $\rho_1 B \leq \frac{B}{2}$. On the other hand, we have assumed that $c(T^*) > \frac{B}{2}$, hence $c(T_{r_0}^*) = c(T^*) \geq \rho_0 B$. Since $\rho_i \geq \rho_{i+1}$, assuming we have found a node $r_i \in G_{i-1}^*$ satisfying (6) and (7), we are guaranteed that r_i also satisfies $c(T_{r_i}^*) \geq \rho_{i+1} B$. Consider any maximal path of nodes $r_i = v_0, \ldots, v_\ell$ in G_i^*, such that for every $j = 0, \ldots, \ell$, $c(T_{v_j}^*) \geq \rho_{i+1} B$. Such a path necessarily exists since the tree is finite, and $c(T_{v_0}^*) \geq \rho_{i+1} B$. By maximality if follows that we can pick $r_{i+1} = v_\ell$, which would satisfy both condition (6) and condition (7).

After having defined subtrees $G_0^*, \ldots, G_{k-1}^*$ as described above, we can assume without loss of generality that for all $i = 0, \ldots, k-2$, $G_i^* \neq G_{i+1}^*$, and define $T_i^* = G_i^* \setminus G_{i+1}^*$ for all $i = 0, \ldots, k-2$ and $T_{k-1}^* = G_{k-1}^*$. We call such a partition a *$\overline{\sigma}$-partition* of T^*. Note that such a partition can be identified by its corresponding sequence of roots $r_0, \ldots, r_{k-1}$. See Figure 1 for the schematics of a $\overline{\sigma}$-partition for $k = 4$.

We will first bound the reachability of vertices in a component T_i^* from r_i.

Lemma 5. *For any k, given a $\overline{\sigma}$-partition of T^*, for every $i = 1, \ldots, k-1$, all vertices in G_i^* are reachable from r_i with cost at most $\rho_i B$.*

Proof. Let v be a vertex in G_i^*. If $v = r_i$ we are done. Otherwise, there exists a vertex $u \in CH_{T^*}(r_i)$ such that $v \in T_u^*$. In particular, $\text{path}(r_i, v) \cup \{v\} \subseteq T_u^*$, which by the definition of reachability and the choice of r_i yields

$$d(r_i, v) + c(v) = c(\text{path}(r_i, v) \cup \{v\}) \leq c(T_u^*) \leq \rho_i B$$

□

Since T_i^* is a subset of G_i^* we have the following corollary:

Corollary 2. *Given a $\overline{\sigma}$-partition of T^*, for every $i = 1, \ldots, k-1$, all vertices in T_i^* are reachable from r_i with cost at most $\rho_i B$.*

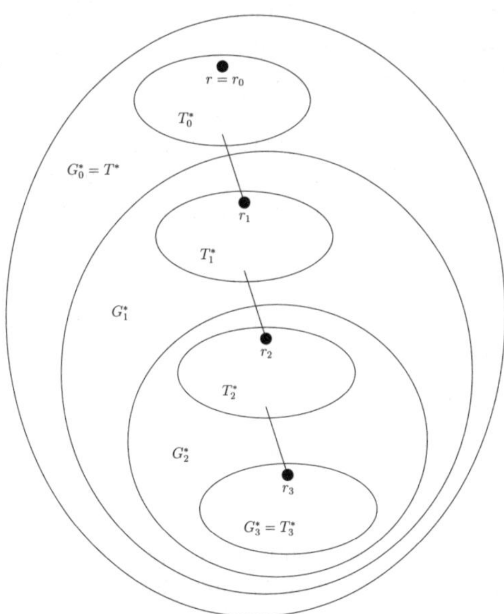

Fig. 1. Schematic $\overline{\sigma}$-partition for $k = 4$

We now turn to bound the cost of every component T_i^*.

Lemma 6. *For any k, given a $\overline{\sigma}$-partition of T^*, the following holds:*

1. $c(T_0^*) \leq (1 - \rho_1)B$
2. $c(T_i^*) \leq B - d(r, r_i) - \rho_{i+1}B$ *for all $i = 1, \ldots, k - 2$.*
3. $c(T_{k-1}^*) \leq B - d(r, r_{k-1})$

Proof. To prove part 1 simply note that by the feasibility of T^*, its cost is upper bounded by B, and since by the choice of r_1, $c(G_1^*) \geq \rho_1 B$, the result follows from the definition of T_0^*. In an even simpler manner, part 3 also follows from the feasibility of T^* and the definition of T_{k-1}^* as its subtree. As for part 2, since $\rho_i B \leq c(G_i^*) \leq B - d(r, r_i)$ then by the definition of T_i^* we have

$$\begin{aligned} c(T_i^*) &= c(G_i^* \setminus G_{i+1}^*) \leq \\ &\leq B - d(r, r_i) - \rho_{i+1}B \end{aligned}$$

which completes the proof. □

5 The Algorithm

5.1 Road Map

The structural analysis of an optimal solution presented in the previous section motivates the definition of a sequence of instances of BP, each corresponding to a different component in the decomposition.

Let T^* be an optimal solution to an instance I of BP, with budget restriction B, and let $k = \Theta\left(\frac{1}{\varepsilon}\right)$, as defined in section 4. For every $i = 0, \ldots, k-1$ and $v \in V$ we define a new instance of BP, $I_{i,v}$. In instance $I_{i,v}$ we actually "guess" that v is the root of T_i^* (i.e. r_i). We then consider all vertices with distance d_i from v, and a budget restriction B_i. Using Lemma 4, we can conclude that we can find a tree $T'_{i,v}$ with cost at most $B_i + d_i$. Our choices of B_i and d_i will be such that two conditions are met; First, for every i and every v, $T'_{i,v}$ can be extended to a solution $T_{i,v}$ to the original instance, with overall cost at most $(1+\varepsilon)B$. Second, for every i, T_i^* is a feasible solution to at least one of the new instances. It follows that at least one of solutions $T_{i,v}$ carries a profit of at least an $\Omega\left(\frac{\varepsilon}{\log n}\right)$ fraction of the optimal profit.

5.2 Detailed Description

Let I be any instance of BP over a graph $G = (V, E)$, with a budget restriction B. Given any $v \in V$ and $i \in \{0, \ldots, k-1\}$, every instance $I_{i,v}$ will be determined by two parameters, whose values are motivated by Corollary 2 and Lemma 6; d_i representing a reachability radius around v, and a budget B_i. Specifically, we consider the subset $V_{i,v} \subset V$ consisting of all vertices with distance at most d_i from v, and let instance $I_{i,v}$ be the instance over the subgraph of G induced by $V_{i,v}$, with budget restriction B_i. Table 1 shows the different values of d_i and B_i, depending on i.

Table 1. Values of reachability distance and budget of the new instances $I_{i,v}$

i	d_i	B_i	$B_i + d_i$
0	$(1-\rho_1)B$	$(1-\rho_1)B$	$2(1-\rho_1)B$
$1, \ldots, k-2$	$\rho_i B$	$B - d(r,v) - \rho_{i+1}B$	$(1+(1-\sigma_{i+1})\rho_i)B - d(r,v)$
$k-1$	$\rho_{k-1}B$	$B - d(r,v)$	$(1+\rho_{k-1})B - d(r,v)$

For every instance $I_{i,v}$, we solve the corresponding LP, compute the tree packing $\mathcal{T}_{i,v}$ which corresponds to $I_{i,v}$ implied by Theorem 1, and let $T'_{i,v} = \textsc{Extract}(\mathcal{T}_{i,v}, B_i)$. Note that by Corollary 2 and Lemma 6, for every $i = 0, \ldots, k-1$, T_i^* is a feasible solution for I_{i,r_i}. The following lemma is an immediate consequence:

Lemma 7. *Given the above notation, for every $v \in V$ and $i = 0, \ldots, k-1$, $c(T'_{i,v}) \le B_i + d_i$, as specified in Table 1. Furthermore, for every $i = 1, \ldots, k-1$, $\pi(T'_{i,r_i}) = \Omega\left(\frac{1}{\log n}\right)\pi(T_i^*)$.*

For every $v \in V$ and $i = 0, \ldots, k-1$, let $T_{i,v} = T'_{i,v} \cup \mathrm{path}(r,v) \cup \{r\}$. Note that every $T_{i,v}$ is a tree rooted at r. The following corollary follows immediately from Lemma 7:

Corollary 3. *Given the above notation*

1. $c(T_{0,v}) \leq 2(1-\rho_1)B$, *for all* $v \in V$
2. $c(T_{i,v}) \leq (1+(1-\sigma_{i+1})\rho_i)B$, *for all* $i = 1,\ldots,k-2$ *and* $v \in V$
3. $c(T_{k-1,v}) \leq (1+\rho_{k-1})B$, *for all* $v \in V$
4. $\pi(T_{i,r_i}) = \Omega\left(\frac{1}{\log n}\right)\pi(T_i^*)$ *for all* $i = 0,\ldots,k-1$

The following lemma ensures we can bound the cost of every $T_{i,v}$ constructed above by $(1+\varepsilon)B$:

Lemma 8. *For* $\overline{\sigma}$ *defined by*

$$\sigma_i = \begin{cases} \dfrac{k-i}{k-i+1} & i = 2,\ldots,k-1 \\ \dfrac{k-1}{2k-1} & i = 1, \end{cases} \tag{8}$$

we have $c(T_{i,v}) \leq (1+\varepsilon)B$ *for all* $i = 0,\ldots,k-1$ *and* $v \in V$.[1]

Proof. Note that the upper bounds given on the cost of each tree $T_{i,v}$ are independent of v. We can therefore refer to a tree T_i regardless of $v \in V$, instead of $T_{i,v}$. We show that for every i, $c(T_i) \leq (1+\frac{1}{2k-1})B$, which by the choice of k implies $c(T_i) \leq (1+\varepsilon)B$. First note that by the choice of σ_i we have

$$\rho_i = \prod_{j=1}^{i}\sigma_j = \frac{k-1}{2k-1}\cdot\frac{k-2}{k-1}\cdot\frac{k-3}{k-2}\cdot\ldots\cdot\frac{k-i}{k-i+1} = \frac{k-i}{2k-1}.$$

We distinguish between several cases. For $i = 0$, we have

$$c(T_i) \leq 2(1-\rho_1)B = \left(1+\frac{1}{2k-1}\right)B,$$

as required. For the case where $i \in \{1,\ldots,k-2\}$, we have

$$(1-\sigma_{i+1})\rho_i = \left(1-\frac{k-i-1}{k-i}\right)\frac{k-i}{2k-1} = \frac{1}{2k-1},$$

hence

$$c(T_i) \leq (1+(1-\sigma_{i+1})\rho_i)B = \left(1+\frac{1}{2k-1}\right)B,$$

as required. The last case to consider is the case where $i = k-1$, in which

$$c(T_i) \leq (1+\rho_{k-1})B = \left(1+\frac{1}{2k-1}\right)B,$$

which completes the proof. □

[1] Note that by our choice, we have $\sigma_1 \leq \frac{1}{2}$, which guarantees our solutions correspond to the decomposition described in Section 4.

Algorithm 2. BPA ($G = (V, E)$, costs c, profits π, budget B)

1: $S_1 \leftarrow \mathrm{MR}(G, c, \pi, \frac{B}{2})$ ▷ Apply the algorithm appearing in [2] with half the budget
2: $S_2 \leftarrow \emptyset$
3: **for** every $i \in \{0, \ldots, k-1\}$ and $v \in V$ **do**
4: let $I_{i,v}$ be the appropriate instance
5: solve the LP corresponding to $I_{i,v}$
6: let $\mathcal{T}_{i,v}$ be its corresponding tree packing
7: $T'_{i,v} \leftarrow \textsc{Extract}(\mathcal{T}_{i,v}, B_i)$
8: $T_{i,v} \leftarrow T'_{i,v} \cup \mathrm{path}(r, v) \cup \{r\}$
9: **if** $\pi(T_{i,v}) > \pi(S_2)$ **then**
10: $S_2 \leftarrow T_{i,v}$
11: **end if**
12: **end for**
13: $T \leftarrow \arg\max \{\pi(S_1), \pi(S_2)\}$
14: return T

We can now prove our main result, as to the performance of algorithm BPA described in Algorithm 2.

Theorem 2. *For any $\varepsilon \in (0, 1]$, algorithm* BPA *produces a $(1 + \varepsilon, O(\frac{1}{\varepsilon} \log n))$-approximate solution to BP.*

Proof. The first candidate solution considered by the algorithm (line 1) is the solution produced by applying the algorithm proposed by Moss and Rabani [2]. If there exists an optimal solution with cost at most $\frac{B}{2}$, then this candidate is guaranteed to be a $(1, O(\log n))$-approximate solution. The remainder of the algorithm is designed to deal with the case where every optimal solution has cost greater than $\frac{B}{2}$.

By the analysis presented in Section 4 and the pigeonhole principle we are guaranteed to have at least one subtree T_i^* contributing at least a $\frac{1}{k}$ fraction of the optimum's profit. Let m be the index of such a subtree. Consider the instance corresponding to I_{m,r_m}, and let T'_{m,r_m} be the candidate solution produced by the algorithm in the appropriate iteration.

By Lemma 4 and the fact that $k = \Theta(\frac{1}{\varepsilon})$, the subtree T'_{m,r_m} produced in line 7 satisfies

$$\pi(T'_{m,r_m}) = \Omega\left(\frac{1}{\log n}\right)\pi(T_m^*) = \Omega\left(\frac{\varepsilon}{\log n}\right)\mathrm{OPT}_B\,.$$

Since $T'_{m,r_m} \subseteq T_{m,r_m}$, we are guaranteed to have $\pi(T_{m,r_m}) = \Omega\left(\frac{\varepsilon}{\log n}\right)\mathrm{OPT}_B$.

Since by Lemma 8 the cost of every candidate solution produced by the algorithm in line 8 is at most $(1+\varepsilon)B$, and as shown above at least one of them has profit at least $\Omega\left(\frac{\varepsilon}{\log n}\right) OPT_B$, the result follows. □

6 Concluding Remarks

In this paper we have shown a $(1 + \varepsilon, O(\frac{1}{\varepsilon} \log n))$ bicriteria approximation tradeoff for the node-cost budget problem. One may consider several directions when aiming at improving our results. The first would be eliminating the budget violation entirely, and ensuring an approximate solution in the usual sense. While still under the bicriteria formulation, one might hope for improving upon the approximation ratio with regard to the profit guaranteed while still settling for budget violation of a certain extent. A substantial improvement in this direction however would necessitate a different approach than the one introduced in [2]. This is due to the fact that their tree-packing makes use of an approximation algorithm to the prize-collecting problem, to which an asymptotically tight $\Omega(\log n)$-approximation lower bound is known under reasonable NP-hardness assumptions. An additional point of great interest remains to try and close the gap between the constant factor lower bound for the problem, and the current $O(\log n)$ approximation techniques and their variants.

References

1. Marathe, M., Ravi, R., Sundaram, R., Ravi, S., Rosenkrantz, D. (III), H.H.: Bicriteria network design problems. Journal of Algorithms 28(1), 142–171 (1998)
2. Moss, A., Rabani, Y.: Approximation algorithms for constrained node weighted steiner tree problems. SIAM Journal on Computing 37(2), 460–481 (2007)
3. Guha, S., Moss, A., Naor, J., Schieber, B.: Efficient recovery from power outage. In: Proceedings of the 31st ACM Symposium on Theory of Computing, pp. 574–582 (1999)
4. Klein, P., Ravi, R.: A nearly best-possible approximation algorithm for node-weighted steiner trees. Journal of Algorithms 19(1), 104–114 (1995)
5. Feige, U.: Threshold of $\ln n$ for approximating set cover. Journal of the ACM 45(4), 634–652 (1998)
6. Goemans, M.X., Williamson, D.P.: A general approximation technique for constrained forest problems. SIAM Journal on Computing 24(2), 296–317 (1995)
7. Johnson, D., Minkoff, M., Phillips, S.: The prize collecting steiner tree problem: theory and practice. In: Proceedings of the 11th annual ACM-SIAM symposium on Discrete algorithms, pp. 760–769 (2000)
8. Jain, K., Hajiaghayi, M.: The prize-collecting generalized steiner tree problem via a new approach of primal-dual schema. In: Proceedings of the 17th Annual ACM-SIAM Symposium on Discrete Algorithms, pp. 631–640 (2006)
9. Garg, N.: Saving an epsilon: a 2-approximation for the k-mst problem in graphs. In: Proceedings of the 37th Annual ACM Symposium on Theory of Computing, pp. 396–402 (2005)
10. Ravi, R., Goemans, M.: The constrained minimum spanning tree problem. In: Proceedings of the 5th Scandinavian Workshop on Algorithmic Theory, pp. 66–75 (1996)
11. Khuller, S., Moss, A., Naor, S.: The budgeted maximum coverage problem. Information Processing Letters 70(1), 39–45 (1999)
12. Grötschel, M., Lovász, L., Schrijver, A.: Geometric Algorithms and Combinatorial Optimization, 2nd edn. Springer, Heidelberg (1993)

Integer Maximum Flow in Wireless Sensor Networks with Energy Constraint*

Hans L. Bodlaender[1,**], Richard B. Tan[1,2], Thomas C. van Dijk[1], and Jan van Leeuwen[1]

[1] Department of Information and Computing Sciences
Utrecht University, Utrecht, The Netherlands
{hansb,rbtan,thomasd,jan}@cs.uu.nl
[2] Department of Computer Science
University of Sciences & Arts of Oklahoma, Chickasha, OK, USA

Abstract. We study the integer maximum flow problem on wireless sensor networks with energy constraint. In this problem, sensor nodes gather data and then relay them to a base station, before they run out of battery power. Packets are considered as integral units and not splittable. The problem is to find the maximum data flow in the sensor network subject to the energy constraint of the sensors. We show that this *integral* version of the problem is *strongly* NP-complete and in fact APX-hard. It follows that the problem is unlikely to have a polynomial time approximation scheme. Even when restricted to graphs with concrete geometrically defined connectivity and transmission costs, the problem is still strongly NP-complete. We provide some interesting polynomial time algorithms that give good approximations for the general case nonetheless. For networks of bounded treewidth greater than two, we show that the problem is *weakly* NP-complete and provide pseudo-polynomial time algorithms. For a special case of graphs with treewidth two, we give a polynomial time algorithm.

1 Introduction

A wireless sensor (or smart dust) is a small physical device that contains a microchip with a miniature battery, and a transmit/receive capability of limited range. Sensors have been deployed in different environments to gather data, perform surveillances and monitor situations in diverse areas such as military, medical, traffic and natural environments. (See e.g. Zhao and Guibas [16].)

As the battery power of a sensor is limited and non-replaceable, it is crucial to maximize the lifetime of the wireless sensor network to ensure the continuing function of the whole network. We study the situation of a data-gathering sensor network, where sensors are deployed in the field to gather data and then relay the data packets via other sensors back to a base station. It is desirable to get

* This work was supported by BSIK grant 03018 (BRICKS: Basic Research in Informatics for Creating the Knowledge Society).

** Corresponding author.

J. Gudmundsson (Ed.): SWAT 2008, LNCS 5124, pp. 102–113, 2008.

as many data packets as possible from the source sensors to the base station, before some of the sensor batteries are depleted. This then becomes an instance of the maximum flow problem, subject to the energy constraint of the lasting battery power of each sensor.

Most of the research papers for the maximum flow problem with energy constraint on wireless sensor networks (e.g. [5,6,8,12,13,14,15]) cast the problem into a Linear Programming (LP) form and assume *fractional* flows, i.e., splitting of packets into fractional portions is allowed. The corresponding LP-formulations then have polynomial time algorithms. These papers then present several heuristics that speed up the algorithms and compare various simulation results. A few papers [5,8,12,13] modified the Polynomial Time Approximation Scheme (PTAS) of Garg and Könemann [10] to obtain fast approximation algorithms.

As data packets are usually quite small, there are situations where splitting of packets into fractional ones is not desirable nor practical. We consider a model where data packets are considered as units that cannot be split, i.e. the packet flows are of *integral* values only. We call this the *Integer* Maximum Flow problem for Wireless Sensor Network with Energy Constraint: the Integer Max-Flow WSNC problem. The corresponding LP formulation becomes Integer Programming (IP) and may no longer have a polynomial time solution.

We show that the problem is in fact *strongly* NP-complete, and thus unlikely to have a Fully Polynomial Time Approximation Scheme (FPTAS) or a pseudo-polynomial time algorithm, unless P=NP. This result also holds for a class of graphs with geometrically defined connectivity and transmission costs, even when the nodes lie on a line. Furthermore, we show that even for a special fixed range model, the problem is APX-hard, thus unlikely to have even a PTAS (unless P=NP). We also provide some approximation algorithms for the problem that do give good approximations nonetheless.

Many hard problems have polynomial time solutions when restricted to networks with bounded treewidth (see e.g. [3]). However, we show that for networks with bounded treewidth greater than two, the Integer Max-Flow WSNC problem is *weakly* NP-complete. We provide pseudo-polynomial time algorithms to compute integer maximum flows in this case. For a special case of graphs that have treewidth two, namely those graphs that have treewidth two when we add an edge from the single source to the sink, we provide a polynomial time algorithm.

The paper is organized as follows. In Section 2, we describe the model and the problem in detail. Section 3 covers the complexity issues. We show here the various NP-completeness results and describe some approximation algorithms. Section 4 contains the results for networks with bounded treewidth.

2 Preliminaries

In this section, we discuss the model we use and the precise formulation of the Integer Max-Flow WSNC problem. We also discuss some variants of the problem.

2.1 The Model

Our model of a sensor is based on the *first order radio model* of Heinzelman et al. [11]. A sensor node has limited battery power that is not replenishable. It consumes an amount of energy $\epsilon_{elec} = 50nJ/bit$ to run the receiving and transmitting circuitry and $\epsilon_{amp} = 100pJ/bit/m^2$ for the transmitter amplifier. In order to receive a k-bit data packet, a sensor has to expend $\epsilon_{elec} \cdot k$ energy, while to transmit the same packet from sensor i to sensor j will cost $\epsilon_{elec} \cdot k + \epsilon_{amp} \cdot k \cdot d_{ij}^2$ energy, where d_{ij} is the distance between sensors i and j.

We model a wireless sensor network as a *directed* graph $G = (N, A)$, where $N = \{1, 2, \ldots, n\} \cup \{t\}$ are the n sensor nodes along with a special non-sensor sink node t, and A is the set of directed arcs ij connecting node i to node j, $i, j \in N$. A sensor node i has energy capacity E_i and each arc ij has cost e_{ij}, the energy cost of receiving (possibly from some node) a packet and then transmitting it from node i to node j. We assume that no data is held back in intermediate nodes $\neq t$, i.e., data that flows in will flow out again, subject to the battery constraint of these nodes. All E_i's and e_{ij}'s are non-negative *integer* values.

The general model assumes that each sensor can adjust its power range for each transmission. We also consider in the next section the fixed range model, where each sensor has only a few *fixed* power settings. All our graphs are assumed to be *connected*. For each arc $ij \in A$, there is a directed path from a source node to the sink node that uses this arc.

2.2 The Problem

Given a wireless sensor network G, there is a set S of *source* sensor nodes, used for gathering data. The sink node t is a *base station* and is equipped with electricity and thus has unlimited energy to receive all packets. The remaining nodes are just *relaying* nodes, used to transfer data packets from the source nodes to the sink node. One would like to transmit as many packets as possible from the source nodes to the sink node. This is feasible as long as the battery power in the network suffices to do so. The transmission process can be viewed as a *flow* of packets from the sources to the sink. The problem is then to find the *maximum flow* of data packets in the network subject to the battery power constraint.

We assume that the data packets are quite small, thus it is neither reasonable nor practical to split them further into fractional portions. A *flow* f_{ij} is a function that assigns to each arc ij a non-negative integer value. This corresponds to the number of packets being sent via the arc ij. A flow is a *feasible* flow if $\sum_j f_{ij} \cdot e_{ij} \leq E_i$ for all nodes $i \in N$, where the sum is taken over all j with $ij \in A$; i.e., the flow through a node cannot exceed the battery capacity of the node.

We can now formulate the maximum flow problem for wireless sensor networks as the problem of determining the maximum number of packets that can be received by the sink node. We call this problem the Integer Maximum Flow problem for Wireless Sensor Networks with energy Constraints or *Integer Max-Flow WSNC problem* for short. The problem has the following Integer Linear Programming formulation.

The Integer Max-Flow WSNC problem:
Objective: maximize $F = \sum_{j \in N} f_{jt}$, t is the sink node,
subject to the following constraints:

$$f_{ij} \ \textit{integer}, \qquad \forall ij \in A \qquad (1)$$

$$f_{ij} \geq 0, \qquad \forall ij \in A \qquad (2)$$

$$\sum_{j \in N} f_{ij} = \sum_{j \in N} f_{ji}, \qquad \forall i \in N - S - \{t\} \qquad (3)$$

$$\sum_{j \in N} f_{ij} \cdot e_{ij} \leq E_i, \qquad \forall i \in N \qquad (4)$$

Condition (3) is the conservation of flow constraint. It simply states that with the exception of the source and sink nodes, every node must send along the packets that it has received. Condition (4) is the energy constraint for the feasible flow: the energy needed to (receive and) transmit packets must be within the capacity of the battery power of each node. This condition also distinguishes the (integer) max-flow WSNC problem from the standard max-flow problem: there, the constraint condition is just $f_{ij} \leq c_{ij}$, where c_{ij} is the flow capacity of arc ij.

We note that without loss of generality, we can augment the network with a *super source* node s with unlimited energy to send and connect it to all the source nodes with some fixed cost. We can then view the network as having a single source and a single sink with a single commodity, subject to the battery energy constraint. However, note this may affect the treewidth of the network; the results for networks of bounded treewidth in Section 4 assume a single source.

We do not include a 'fairness' constraint: we maximize the number of packets reaching the sink, but do not balance the load of the source nodes. Note that the hardness proofs in Section 3 do exhibit fair flows, where all source nodes send an equal number of packets.

2.3 Other Variants

Other variants of the problem formulation exist for wireless sensor networks with energy constraint. See for example Floréen et al. [8] and Chang and Tassiulas [6].

3 Complexity

We will first look at the complexity of the problem on general graphs with arbitrary costs at the arcs. We show the problem is NP-complete. Next we show this proof carries over to a restriction of the problem to a class of graphs with geometrically defined connectivity and energy consumption. Then we look at another restriction of the problem, on general graphs again, but with only a fixed amount of distinct energy costs. The problem is polynomial time solvable for one energy level, but with more distinct power levels the problem is shown to be APX-hard. Lastly, we demonstrate a polynomial time approximation algorithm for the general case. Note however that this algorithm does not guarantee a constant approximation ratio.

3.1 General Graphs

Since each data packet is a self-contained unit and cannot be split, the corresponding LP formulation is an Integer Programming (IP) formulation and may no longer have a polynomial time solution. In fact, we prove that the problem is *strongly* NP-complete.

Theorem 1. *The decision variant of the Integer Max-Flow WSNC problem is strongly NP-complete.*

Proof. We reduce the 3-Partition problem to the decision version of the Integer Max-Flow WSNC problem.

> 3-PARTITION
> INSTANCE: Given a multiset $\mathcal{S}$ of $n = 3m$ positive integers, where each $x_i \in \mathcal{S}$ is of size $B/4 < x_i < B/2$, for a positive integer B.
> QUESTION: Can $\mathcal{S}$ be partitioned into m subsets (each necessarily containing exactly three elements) such that the sum of each subset is equal to B?

The 3-Partition Problem is strongly NP-complete [9].

For any instance I of the 3-Partition problem we create an instance I' of a wireless sensor network as follows. Each number $x_i \in \mathcal{S}$ corresponds to a sensor relay node r_i. Additionally, we have m source nodes $s_1, \ldots, s_m$ each having exactly B energy. The source nodes play the role of the subsets. Now connect each of the source nodes s_j with all the relay nodes r_i with arc cost $e_{ji} = x_i$. The intention is that it will cost each source node exactly x_i energy to send one packet to relay node r_i. We further connect all the relay nodes to a sink node t. Each relay node r_i has energy $E_i = B$ and arc cost $e_{it} = B$, just sufficient energy to send only one packet to the sink node.

Then our instance of the 3-Partition problem has a partition into m subsets $\mathcal{S}_1, \ldots, \mathcal{S}_m$, each with sum equals B if and only if for each subset $\mathcal{S}_i = \{x_{i1}, x_{i2}, x_{i3}\}$ the source node s_i sends three packets, one each to relay nodes r_{i1}, r_{i2}, r_{i3} consuming the energies x_{i1}, x_{i2}, x_{i3}, thus draining all of its battery power of $B = \sum_{j=1}^{3} x_{ij}$. This will give a maximum flow of $n = 3m$ packets for the whole network. Thus, 3-PARTITION reduces to the question whether the WSNC network can transmit at least $3m$ packets to the sink.

We have now given a pseudopolynomial reduction (see [9]), and thus we have shown that the Integer Max-Flow WSNC problem is strongly NP-complete. □

Corollary 2. *The Integer Max-Flow WSNC problem has no fully polynomial time approximation scheme (FPTAS) and no pseudo-polynomial time algorithm, unless P=NP.*

Proof. This follows from the fact that a strongly NP-complete problem has no FPTAS and no pseudo-polynomial time algorithm, unless P=NP. (See [9]). □

We show later on that even in a restricted case, the problem is APX-hard, i.e. the problem does not even have a PTAS unless P=NP.

3.2 The Geometric Model

We will now look at the geometric version of the problem in which the nodes are concretely embedded in space. In this version, each node has a position, and transmitting to a node at distance d costs d^2 energy. This quadratic cost is a typical model of radio transmitters.

Definition 3. *A* geometric configuration *is a complete graph where each node $i \in N$ has a location $p(i)$ and initial battery capacity E_i. The cost of the edge between vertices i and j is $e_{ij} = |p(i) - p(j)|^2$.*

In this section we show that the WSNC-problem remains NP-complete when restricted to geometric configurations where all nodes lie on a line. We will prove this for the case where we allow variable battery capacities E_i (Theorem 4). This can be extended to the case where all nodes have equal battery capacity (Theorem 9), but we omit the more elaborate proof [4] for space reasons.

Theorem 4. *The Integer Max-Flow WSNC-problem is strongly NP-complete on geometric configurations on the real line.*

Proof. Again, we give a pseudopolynomial reduction from 3-Partition. First we describe how to construct a geometric configuration $\mathbf{C}$ on the real line, where the WSNC-problem is equivalent to a given instance of 3-Partition. We then construct an equivalent configuration $\mathbf{C_P}$ that can be described in polynomial size. These steps together show that the WSNC-problem is strongly NP-complete on geometric configurations.

Like before, we have m source nodes $s_1, \dots, s_m$. Each starts with $B = \sum x_i/m$ energy. These source nodes again play the role of the subsets and we place them all at the origin of our geometric configuration, i.e., $p(s_i) = 0$ for all i.

Corresponding to each $x_i \in S$ we have a 'relay' node r_i that serves the same purpose as before: to receive one packet from a source node, costing x_i energy for this source node, and relay the packet to the sink. By setting $p(r_i) = \sqrt{x_i}$ we achieve that a source node must use x_i energy to send a packet to r_i.

Finally, we place a sink node t with $p(t) = \sqrt{B}$. We want each relay node to have just enough energy to send exactly one packet to the sink, so we set $E_{r_i} = (p(t) - p(r_i))^2 = B + x - 2\sqrt{B}\sqrt{x}$.

This concludes the construction of our geometric configuration $\mathbf{C}$. The construction is illustrated in Figure 1.

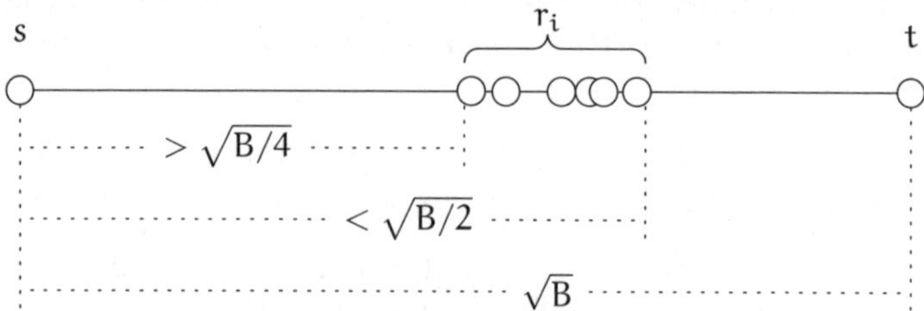

Fig. 1. Configuration **C**

Lemma 5. *Suppose we have a flow of value n in* **C**. *Then every relay node receives exactly one packet from a source and sends it to the sink.*

Proof. If n packets reach the sink, then n packets must have left the sources. By the restriction on the values x_i of the 3-Partition instance, each edge leaving the sources costs strictly more than $B/4$ energy. Since the source nodes start with B energy, no source node can send more than 3 packets. There are only $m = n/3$ source nodes, so every source node must send exactly 3 packets. In particular, no packets are sent directly from a source node to the sink, as this would use up all energy of the source node. Therefore, the sink node only receives packets from relay nodes. No relay node can afford to send more than one packet to the sink, so in fact every relay node sends exactly one packet to the sink. □

Using Lemma 5, the following can now be shown in the same way as Theorem 1 for the case on arbitrary graphs.

Proposition 6. *The configuration* **C** *has a solution of the WSNC-problem with n packets if and only if the corresponding 3-Partition instance is* **Yes**.

Note that this does not yet give an NP-hardness proof for the WSNC-problem on geometric configurations on the line, as configuration **C** has nodes at real-valued coordinates: the specification of the location of the points in **C** contains square roots. We shall now construct a geometric configuration $\mathbf{C_P}$ which is equivalent to **C**, but whose positions are all polynomially representable rational numbers.

We do this by choosing the locations as integer multiples of some ε (value to be determined later), rounding down. The initial power of the batteries also needs to be quantized. We give the source nodes exactly B energy, which is already integer. We give the relay nodes precisely enough energy to send one packet to the sink; this amount can be calculated from the actual distance in $\mathbf{C_P}$.

Lemma 7. *The value for ε can be chosen such that $\mathbf{C_P}$ is equivalent to* **C** *and can be represented in polynomial size.*

We omit the proof that e.g. $\varepsilon = (5\lceil\sqrt{B}\rceil)^{-1}$ satisfies the lemma [4]. The proof of Theorem 4 now follows from Lemmata 5, 7 and Proposition 6: the WSNC-problem is strongly NP-complete on geometric configurations, even when restricted to a line. □

The nodes in $\mathbf{C_P}$ have non-integer positions, but this not essential: scaling can achieve the following.

Corollary 8. *The Integer Max-Flow WSNC-problem is strongly NP-complete on geometric configurations on a line, where each node has an integer coordinate.*

The given results make essential use of the fact that battery capacities can vary. A natural next question is whether the results still hold in case all sensors have equal battery capacity. By a different construction and a non-trivial combinatorial argument one can show that this is indeed the case [4].

Theorem 9. *The Integer Max-Flow WSNC-problem is strongly NP-complete on geometric configurations on a line, even when all battery capacities are equal.*

3.3 Fixed Range Model

Now we return to the case for arbitrary graphs. Suppose that every sensor node has only a *fixed* number of power settings, i.e., only a fixed number of different energy cost values at its outgoing edges. For example, there may be only *one* setting, so that every node within the range is considered a neighbor; or perhaps there are only *two* settings: *short* and *long* range power settings.

It turns out that for the case when there is only one power setting, there is an easy solution. Since now the energy cost e_{ij} is the same for all neighbors j, the maximum flow capacity $f_{ij} = \lfloor E_i/e_{ij} \rfloor$ is also fixed for all outgoing arcs of node i. We can then transform the sensor network into a regular flow network by using the splitting technique in flow networks as follows. (See e.g. the book by Ahuja, et al. [1].) Split each node i into two nodes i and i' and connect them with an arc with capacity $c_{ij} = \lfloor E_i/e_{ij} \rfloor$. The capacity of the original arcs ij will also all be set to c_{ij}, for all $ij \in A$. We then have a new graph that is a flow network with twice as many nodes and n additional arcs. Then it is easy to see that this variant of the Max-Flow WSNC problem is just the standard Max-Flow Min-Cut problem and has a polynomial time algorithm of $O(n^3)$, even in the integer case. This fact has also been noted by Chang and Tassiulas [5]. For the sake of completeness, we record this fact below.

Theorem 10. *If there is only one power setting at each sensor node, then there is a polynomial time algorithm to solve the Integer Max-Flow WSNC problem.*

The situation changes when the number of fixed power settings is increased to two.

Theorem 11. *If there are two power settings at each sensor node, then there is no PTAS for the Integer Max-Flow WSNC problem, unless P=NP.*

Proof. We reduce a restricted version of the Generalized Assignment Problem (GAP) by Chekuri and Khanna [7] to the Integer Max-Flow WSNC problem with two power settings.

2-SIZE 3-CAPACITY GENERALIZED ASSIGNMENT PROBLEM (2GAP-3)
INSTANCE: A set $\mathcal{B}$ of m bins and a set $\mathcal{S}$ of n items. Each bin j has capacity $c(j) = 3$ and for each item $i \in \mathcal{S}$ and bin $j \in \mathcal{B}$, we are given a size $s(i,j) = 1$ or $s(i,j) = 1 + \delta$ (for some $\delta > 0$) and a profit $p(i,j) = 1$.
OBJECTIVE: Find a subset $\mathcal{U} \subseteq \mathcal{S}$ of maximum profit such that $\mathcal{U}$ has a feasible packing in $\mathcal{B}$.

Chekuri and Khanna [7] show that the 2GAP-3 problem is APX-hard, hence it does not have a PTAS (unless P=NP).

Given an instance I of 2GAP-3 we create an instance I' of the Integer Max-Flow WSNC as follows. For each bin $j \in \mathcal{B}$ we have a source node j' with energy capacity $E_{j'} = 3$. Corresponding to each item $i \in \mathcal{S}$ we have a relay node i'. Each source node j' is connected to each of the relay nodes i' by an arc $j'i'$ with energy cost $e_{j'i'} = 1$ if $s(i,j) = 1$ and $e_{j'i'} = 1+\delta$ if $s(i,j) = 1+\delta$. We also have one sink node t. Each of the sensor relay node i' is further connected to the sink node t and provided with just sufficient battery power to have the arc energy cost to send only one packet to the sink node, i.e., we set $E_{i'} = 1$ and $e_{i't} = 1$.

As each node i' that represents an item can forward only one packet, each arc of the form $j'i'$, $j \in \mathcal{B}$, $i \in \mathcal{S}$ can also carry at most one packet. Thus, there is a one-to-one correspondence between integer flows in I fulfilling energy constraints, and feasible packings of sets of items $\mathcal{U} \subseteq \mathcal{S}$: an item i that is placed in bin j corresponds to a unit of flow that is transmitted from j' to i' and then from i' to t. The value of the flow equals the total profit of the packed items.

Thus, we can observe that our reduction is an AP-reduction. As AP-reductions preserve APX-hardness (see e.g., [2]), we can conclude the theorem. □

3.4 Approximation Algorithms

As the Integer Max-Flow WSNC problem has no PTAS (unless P=NP), our hope is to find some approximation algorithms. We first give a very simple approximation algorithm. Then we give a slightly more involved algorithm with a better approximation performance.

Theorem 12. *There is a ρ-approximation algorithm for the Integer Max-Flow WSNC problem, where $\rho = \max_{i \in N} \frac{\max_{ij \in A} e_{ij}}{\min_{ij \in A} e_{ij}}$.*

Proof. Convert the sensor network into a flow network as follows. We give the edges unbounded capacity and we give each node i the capacity $c_i = \lfloor \frac{E_i}{\max e_{ij}} \rfloor$. Then a standard Max-Flow Min-Cut algorithm with node capacities will yield a polynomial time algorithm that is at worst a ρ-factor from the optimum. □

Unfortunately the above approximation is not of constant ratio. Note that for the fixed range model where each sensor has a constant number of fixed power settings, the above algorithm does give a constant ratio approximation.

A better approximation algorithm is the following. The key idea is to first solve the fractional LP-formulation in polynomial time, and then try to find a large integer flow that is close to the optimum value.

Theorem 13. *There is a polynomial time approximation algorithm for the Integer Max-Flow WSNC problem, which computes an integer maximum flow with value $F_{approx} \geq F_{optimum} - m$, where m is the number of arcs of the graph.*

Proof. We give the proof for the case when there is only one source. It is a simple exercise to generalize the proof to the case with multiple sources.

First, we solve the relaxation of the problem optimally, i.e., we allow flows to be of real value. As this is an LP, the ellipsoid method gives us in polynomial time an optimal solution F^*, that can be realized within the energy constraints.

We now find a large integer flow inside F^* in the following way. We use an integer flow function F, which invariantly will map each arc ij to a non-negative integer f_{ij} with $f_{ij} \leq f^*_{ij}$, and which has conservation of flows; i.e., F will invariantly be a flow that fulfills the energy constraints. Initially, set F to be 0 on all arcs.

Now, repeat the following step while possible. Find a path P from s to t, such that for each arc ij on the path, $f^*_{ij} - f_{ij} \geq 1$. Let $f_P = min_{ij \in P} \lfloor f^*_{ij} - f_{ij} \rfloor$ be the minimum over all arcs ij on the path P. Note that $f_P \geq 1$. Now add f_P to each f_{ij} for all arcs ij on the path P. Observe that the updated function F fulfills the energy constraint conditions.

The process ends when each path from s to t contains an arc with $f^*_{ij} - f_{ij} < 1$. We note that $F^* - F$ is a flow, and standard flow techniques show that its value is at most m, the number of arcs. (For let S be the set of nodes, reachable from s by a path with all arcs fulfilling $f^*_{ij} - f_{ij} \geq 1$. Since $t \notin S$, $(S, V - S)$ is a cut and its capacity with respect to the flow $F^* - F$ is at most the number of arcs across the cut.) Since the value of F^* is at least $F_{optimum}$, and hence the value of F is at least $F_{optimum} - m$.

In each step of the procedure given above, we obtain at least one new arc ij with $f^*_{ij} - f_{ij} < 1$; this arc will no longer be chosen in a path in a later step. Thus, we perform at most m steps. Each step can be done easily in linear time. Hence, the algorithm is polynomial, using the time of solving one linear program plus $O(m(n + m))$ time for computing the approximate flow. □

This algorithm is still not of constant ratio but we conjecture that it is a step toward a 2-approximation algorithm.

4 Graphs with Bounded Treewidth

Many hard graph problems have polynomial (sometimes even linear) time algorithms when restricted to graphs of bounded treewidth. This however is not the case for the Integer Max-Flow WSNC problem: the problem remains hard. The treewidth parameter nicely delineates the classes of graphs for which the Integer Max-Flow WSNC is of apparently increasing complexity in the following manner.

Theorem 14. *The Integer Max-Flow WSNC problem can be solved in linear time for graphs of treewidth 1.*

Proof. These are just *forests* and have a very simple linear time algorithm: remove all nodes that do not have a path to the sink t; compute in the resulting tree in post-order for each node the number of packets it receives from its children and then the number of packets it can send to its parent. □

We show that if there is a single source, and the graph with an edge added between this single source and the sink node has treewidth two, then there is a polynomial time algorithm for the Integer Max-Flow WSNC problem. The general case for treewidth two remains open.

Theorem 15. *The Integer Max-Flow WSNC problem, with parallel arcs and arc capacities and with a single source s and sink t, can be solved in $O(m \log \Delta)$ time on directed graphs $G = (N, A)$, such that the graph $(N, A \cup \{st\})$ has treewidth at most two, where Δ is the maximum outdegree of a node in G, and $m = |A|$.*

For larger treewidth we show that the problem is *weakly* NP-complete. We give a pseudo-polynomial time algorithm for this class.

Theorem 16. *The Integer Max-Flow WSNC problem is (weakly) NP-hard for graphs of treewidth three.*

The omitted proof of Theorems 15 and 16 can be found in [4].

As shown in the previous section, the case of unbounded treewidth is *strongly* NP-complete and even APX-hard.

5 Conclusion

The Maximum Flow WSNC problem is an interesting and relevant problem, with practical implications in the context of wireless ad hoc networks. In this paper, we studied the integer variant of the problem. We obtained a good polynomial time approximation algorithm for the problem, which is a step toward an algorithm of constant performance ratio. We also studied how the complexity of the problem depends on the treewidth of the network. We found that except for the case where each sensor has one fixed power setting or when the underlying graph is of treewidth two with an edge joining the source and sink nodes, the problem is weakly NP-complete for bounded treewidth greater than two. It is strongly NP-complete for networks of unbounded treewidth and in fact even APX-hard. It is also strongly NP-complete on geometric configurations on a line.

Acknowledgements. We thank Alex Grigoriev, Han Hoogeveen, Arie Koster, Erik van Ommeren and Gerhard Woeginger for fruitful discussions and helpful comments.

References

1. Ahuja, R.K., Magnanti, T.L., Orlin, J.B.: Network Flows: Theory, Algorithms, and Applications, Upper Saddle River. Prentice-Hall, Englewood Cliffs (1993)
2. Ausiello, G., Crescenzi, P., Gambosi, G., Kann, V., Marchetti-Spaccamela, A., Protasi, M.: Complexity and Approximation: Combinatorial Optimization Problems and Their Approximability Properties. Springer, Berlin (1999)

3. Bodlaender, H.L.: A tourist guide through treewidth. Acta Cybernetica 11, 1–23 (1993)
4. Bodlaender, H.L., Tan, R.B., van Dijk, T.C., van Leeuwen, J.: Integer maximum flow in wireless sensor networks with energy constraint. Technical Report UU-CS-2008-005, Department of Information and Computing Sciences, Utrecht University (2008)
5. Chang, J., Tassiulas, L.: Fast approximate algorithms for maximum lifetime routing in wireless ad-hoc networks. In: Pujolle, G., Perros, H.G., Fdida, S., Körner, U., Stavrakakis, I. (eds.) NETWORKING 2000. LNCS, vol. 1815, pp. 702–713. Springer, Heidelberg (2000)
6. Chang, J., Tassiulas, L.: Maximum lifetime routing in wireless sensor networks. IEEE/ACM Transaction on Networking 12(4), 609–619 (2004)
7. Chekuri, C., Khanna, S.: A polynomial time approximation scheme for the multiple knapsack problem. SIAM J. Comput. 35(3), 713–728 (2005)
8. Floréen, P., Kaski, P., Kohonen, J., Orponen, P.: Exact and approximate balanced data gathering in energy-constrained sensor networks. Theor. Comp. Sc. 344(1), 30–46 (2005)
9. Garey, M.R., Johnson, D.S.: Computers and Intractability, A Guide to the Theory of NP-Completeness, W.H. Freeman and Company, New York (1979)
10. Garg, N., Könemann, J.: Faster and simpler algorithms for multi-commodity flow and other fractional packing problems. In: Proceedings of the 39th Annual Symposium on Foundations of Computer Science, pp. 300–309 (1998)
11. Heinzelman, W.R., Chandrakasan, A., Balakrishnan, H.: Energy-efficient communication protocol for wireless microsensor networks. In: Proceedings 33rd Hawaii International Conference on System Sciences HICSS-33 (2000)
12. Hong, B., Prasanna, V.K.: Maximum lifetime data sensing and extraction in energy constrained networked sensor systems. J. Parallel and Distributed Computing 66(4), 566–577 (2006)
13. Kalpakis, K., Dasgupta, K., Namjoshi, P.: Efficient algorithms for maximum lifetime data gathering and aggregation in wireless sensor networks. Computer Networks 42(6), 697–716 (2003)
14. Ordonez, F., Krishnamachari, B.: Optimal information extraction in energy-limited wireless sensor networks. IEEE Journal on Selected Areas in Communications 22(6), 1121–1129 (2004)
15. Xue, Y., Cui, Y., Nahrstedt, K.: Maximizing lifetime for data aggregation in wireless sensor networks. Mobile Networks and Applications 10, 853–864 (2005)
16. Zhao, F., Guibas, L.: Wireless Sensor Networks: An Information Processing Approach. Morgan Kaufmann, San Francisco (2004)

The Maximum Energy-Constrained Dynamic Flow Problem

Sándor P. Fekete[1], Alexander Hall[2], Ekkehard Köhler[3], and Alexander Kröller[1]

[1] Algorithms Group, Braunschweig Institute of Technology, D-38106 Braunschweig, Germany
{s.fekete,a.kroeller}@tu-bs.de
[2] EECS Department, UC Berkeley, CA 94720, USA
alex.hall@gmail.com
[3] Mathematical Institute, Brandenburg University of Technology, D-03013 Cottbus, Germany
ekoehler@math.tu-cottbus.de

Abstract. We study a natural class of flow problems that occur in the context of wireless networks; the objective is to maximize the flow from a set of sources to one sink node within a given time limit, while satisfying a number of constraints. These restrictions include capacities and transit times for edges; in addition, every node has a bound on the amount of transmission it can perform, due to limited battery energy it carries. We show that this *Maximum energy-constrained dynamic flow problem* (ECDF) is difficult in various ways: it is NP-hard for arbitrary transit times; a solution using flow paths can have exponential-size growth; a solution using edge flow values may not exist; and finding an integral solution is NP-hard. On the positive side, we show that the problem can be solved polynomially for uniform transit times for a limited time limit; we give an FPTAS for finding a fractional flow; and, most notably, there is a distributed FPTAS that can be run directly on the network.

1 Introduction

Energy-Constrained Flows and Our Results: Optimizing the flow in a network is one of the classical problems in the theory and practice of algorithms. Even in the early days, more than 50 years ago, it was recognized by Ford and Fulkerson that flow has a temporal dimension, as the flowing objects need time to get to their destination. This is particularly relevant in many current real-world applications, like traffic or communication, where the aspect of time becomes highly important because there is significant variance in the amount of flow over time, so that stationary flows fail to describe the phenomena that are particularly relevant, like congestion. In recent years, a variety of excellent papers has addressed these problems, making tremendous progress in theory, and providing tools for dealing with the challenges of practical adaptive traffic control.

In this paper we describe an additional aspect that becomes important in the context of another highly relevant topic that has attracted a large amount of attention: when considering the flow of information in a distributed and wireless network, one of the essential features is that nodes have limited energy supply. This limits the amount of information they can transmit to their neighbors before dying due to an exhausted battery; as a consequence, routing algorithms do not just have to deal with edge capacities and transit times, but also with limits on the capabilities of nodes to pass on information.

J. Gudmundsson (Ed.): SWAT 2008, LNCS 5124, pp. 114–126, 2008.

We give the first algorithmic study of the resulting *Maximum energy-constrained dynamic flow problem* (ECDF). Our results are as follows.

- With arbitrary transit times, ECDF is NP-hard (Theorem 1). A solution using flow paths can have exponential size growth (Theorem 2). A solution using edge flow values does not seem to exist.
- Finding integral solutions is NP-hard (Theorem 3).
- For uniform transit times and a polynomially bounded time horizon, we show that the problem can be solved in polynomial time (Lemma 1).
- The complexity of finding optimal fractional solutions is an open problem. The problem admits a FPTAS (Theorem 4).
- There is a distributed FPTAS that can be run directly in the network (Theorem 6).

In the remainder of this section, we discuss related work, both on dynamic network flows and on wireless sensor networks. Section 2 provides formal definitions, while Section 3 shows complexity results. Section 4 focuses on the case of uniform transit times; we show that for polynomially bounded time horizon, there is a polynomial-time algorithm, and develop an FPTAS for general time horizon. In Section 5 we finally give a distributed FPTAS, i.e., a class of algorithms that can be run directly on the network.

Flows over time: Already Ford and Fulkerson [11,12] proposed the *dynamic flows* model (also referred to as flows over time), where the individual edges have transit times, determined by the speed at which flow traverses them. Flow rates into edges may vary over time and are bounded by the given capacities. Ford and Fulkerson show that the *maximum s-t-flow over time problem* can be solved in polynomial time.

The *quickest s-t-flow problem* (here a demand is fixed and the goal is to minimize the time horizon T) can be solved by performing a binary search with respect to T, solving a maximum s-t-flow over time problem in each step. For an integral demand Fleischer and Tardos [10] show that the binary search can be stopped after a polynomially bounded number of steps, yielding the optimal time horizon. A much quicker, strongly polynomial algorithm was given by Burkard et al. [3].

In the *quickest transshipment problem* many sources and sinks with supplies and demands are given. Flow can be routed from any source to any sink. This problem appears to be considerably harder than the quickest s-t-flow problem. Nevertheless, Hoppe and Tardos [15,16] manage to give a polynomial time algorithm, which, however, applies submodular function minimization as a subroutine, rendering the algorithm impractical.

Klinz and Woeginger [17] show that computing a *min-cost quickest s-t-flow* in a network with cost coefficients on the edges is already NP-hard for series-parallel graphs. It is even strongly NP-hard to calculate an optimal temporally repeated flow in presence of costs. For the multicommodity version of this problem Fleischer and Skutella [9] propose a $(2 + \varepsilon)$-approximation algorithm.

For two variations of the multicommodity quickest flow over time problem (without costs) Fleischer and Skutella [9] give an FPTAS which is based on condensed time-expanded graphs. Proofs of NP-hardness and strong NP-hardness for these variations are presented in [14]. The NP-hardness proof inspired our proof of Theorem 1.

The survey articles by Aronson [1] and Powell et al. [22] as well as the book published by Ran and Boyce [23] contain examples and references pertinent to the area of flows over time.

Wireless Sensor Networks: In a classical network setting, there is one central authority that knows the full structure of the network, performs all necessary computations, and makes sure that the results are applied throughout the network. It is clear that such an approach is not without its problems; e.g., see [8] for a discussion in the context of traffic. In recent years, there has been growing interest in *distributed algorithms* that lack such a central authority; see [21] for an introduction.

A particular area that has lead to growing interest in distributed algorithms is the field of *wireless sensor networks:* The nodes in the network have limited knowledge of the environment, local communication in a limited neighborhood, no access to a central authority, limited computing and storage capabilities, and limited energy supply, without any chance of getting recharged. The objective is to allow the network to carry out a variety of tasks, using self-organizing and distributed methods. For a current survey of algorithmic models, see [25], and [26] for a discussion of challenges in distributed computing with respect to sensor networks. For an overview of some of our own related algorithmic work, see [6,7,18].

In the WSN context, a problem similar to ECDF has been studied. The *Maximum Lifetime Routing* problem asks for a flow for given demands that maximizes the time until the first node dies. The motivation is that certain nodes collect sensor data and continuously stream them to one or more base stations. All work on this problem models it as following: "maximize T such that there exists a static flow for given demands, where the flow consumes no more than a $1/T$ fraction of each battery."

Madan et al. consider flows with one sink [19,20], using localized subgradient algorithms. Sankar and Liu [24] solve a multi-commodity flow variant using an exponential penalty function. A combinatorial flow augmentation scheme was proposed by Chang and Tassiulas [5]. Zussman and Segall [27] studied a variant with special relay stations between network and data sink. A comparison of practical protocols can be found in Busse et al. [4].

Simply considering static flows that consume just $1/T$ of each battery leaves a gap of up to $\Theta(n)$ to dynamic flows: consider a network of n nodes, connected in a line, with source and sink at the ends, and $T = n - 1$. Our dynamic flow approach exploits the fact that the source-sink-path can be used exactly once, whereas repeating a static flow would just allocate $1/(n-1)$ of the available battery capacities.

Bodlaender et al. [2] consider the maximization of data flowing from the sensors to the data sink, with battery constraints. They neither have an explicit notion of time, nor do they employ fairness constraints that could turn static solutions into dynamic ones.

2 Problem and Definitions

An instance of the ECDF problem comprises the following: The network $G = (V, E)$ with designated source $s \in V$ and sink $t \in V \setminus \{s\}$ (For the multi-source case, see Section 6). We assume that time is sliced into communication rounds. In each round, a node can forward the data it received in the previous round. We believe this is a sufficiently close approximation of the real timing characteristics of store-and-forward wireless mesh networks. Nodes other than the sink can not store flow to be sent later—sensor nodes are usually extremely limited in memory.

Each edge $e \in E$ models a communication link with *bandwidth* $u_e > 0$, which defines the maximum amount of data that can be sent over e in a single communication round. Data can be sent in both directions in a round, the bandwidth applies to the sum.

Each node $v \in V$ has a non-rechargeable battery with capacity $C_v > 0$. There is an *energy consumption model* $c = (c^s, c^r)$, with non-negative functions $c^s \in \mathbb{R}^E$, $c^r \in \mathbb{R}^E$. Sending one unit of flow over edge uv, from u to v, decreases u's battery capacity by c^s_{uv} and v's capacity by c^r_{uv}. We allow for $C_v = \infty$, $c^s_e = 0$, and $c^r_e = 0$; all situations where this leads to divisions by zero or the appearance of ∞ in linear programs are trivial to resolve. The energy model with $c^s_e = 1, c^r_e = 0$ for all $e \in E$ is called the *trivial* model. In the trivial model, the battery capacities become upper bounds for the total flow that can be routed through a node until it dies. The applicability of our results for more realistic models is discussed in Section 6.

Each link has a *transit time* $\tau_e \in \mathbb{N}$. Data sent over e in round θ can be forwarded from the destination in round $\theta + \tau_e$. For the WSN scenario, we assume *uniform* transit times, i.e., $\tau_e = 1$ for all e; we call this problem variant the 1-ECDF problem. We believe the problem with non-uniform transit times is interesting in itself.

There is a given *time horizon* $T \in \mathbb{N}$. A clarification is necessary to avoid "± 1 confusions": We follow the notation from [16], where there are rounds $0, \ldots, T$. Each link e can be used in rounds $0, \ldots, T - \tau_e$. This reflects our wireless network scenario: If you have a time horizon of 1, you can send data over a link once. (There is an opposing model stemming from continuous flows, e.g., water flowing through a tube. There, you need a horizon of 2 for a unit-transit link: One round until the first drop of water reaches the destination, another until all the water is through.)

ECDF problem definition: Putting it together, the problem we want to solve is: *Send as much flow as possible from s to t, such that the edge capacities are obeyed, the flow is delivered within the time horizon T, and no node sends any data after its battery is exhausted.*

To give a concise problem definition, we need a little bit of notation: We denote by $\mathcal{P}^{st}$ the set of all feasible, simple s-t-paths in G. Because we are only interested in s-t-paths, we can safely treat paths as sequences of undirected edges. The length $\tau(P)$ of a path is defined as $\sum_{e \in P} \tau_e$, i.e., the number of edges in P with uniform transit times. A path is feasible if $\tau(P) \leqslant T$. Hence, the source can relay data over P in communication rounds 0 to $T - \tau(P)$, $\varrho(P) := T - \tau(P) + 1 \geqslant 1$ denotes the number of times P can be used. Now let $P = (e_1, e_2, \ldots, e_k)$. We denote by $\tau_{e_i}(P)$ the delay after which data travelling P reaches e_i, i.e., $\tau_{e_i}(P) = \sum_{j=1}^{i-1} \tau_{e_j}$. For a node $v \in V$, $c^*_{v,P}$ denotes the energy drained from v when 1 unit of flow is sent over P, i.e., $c^*_{v,P} = c^r_{e_i} + c^s_{e_{i+1}}$ for inner nodes of the path, where $v \in e_i$ and $v \in e_{i+1}$ for some i. For s resp. t, $c^*_{v,P}$ equals $c^s_{e_1}$ resp. $c^r_{e_k}$.

So, formally we state the ECDF problem as follows:

$$\max \sum_{P \in \mathcal{P}^{st}} \sum_{\theta=0}^{T-\tau(P)} x_P(\theta) \tag{1}$$

$$\text{s.t.} \sum_{\substack{P \ni e: \\ 0 \leqslant \theta - \tau_e(P) \leqslant T - \tau(P)}} x_P(\theta - \tau_e(P)) \leqslant u_e \;\; \forall e \in E, \theta = 0, \ldots, T \tag{2}$$

$$\sum_{P \ni v} \sum_{\theta=0}^{T-\tau(P)} c^*_{v,P} x_P(\theta) \leqslant C_v \forall v \in V \tag{3}$$

$$x_P(\theta) \geqslant 0 \qquad \forall P \in \mathcal{P}^{st}, \theta = 0, \ldots, T - \tau(P) . \tag{4}$$

In this LP, $x_P(\theta)$ describes the amount of flow that starts traveling along P in round θ. Inequalities (2) model edge capacities, and inequalities (3) describe node battery constraints. Note that both T and $|\mathcal{P}^{st}|$ can be exponential in the problem's encoding size, and this LP consists of more than $T|E|$ constraints and up to $T|\mathcal{P}^{st}|$ variables.

3 Variants and Complexities

In this section, we analyze the complexity structure of ECDF and some variants.

Theorem 1. *ECDF with arbitrary transit times $\tau_e \in \mathbb{N}$ is* NP*-hard.*

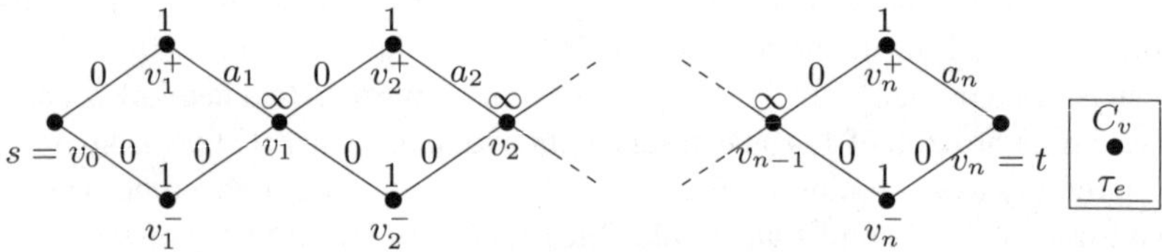

Fig. 1. ECDF instance for the reduction from PARTITION

Proof. Consider an instance of PARTITION: given n positive integers $a_1, \ldots, a_n$ with $\sum_{i=1}^n a_i = 2T$ for some $T \in \mathbb{N}$, partition them into two sets of equal weight, i.e., find a $S \subset \{1, \ldots, n\}$ such that $\sum_{i \in S} a_i = \sum_{i \notin S} a_i = T$. We claim the PARTITION instance is feasible iff the ECDF instance shown in Figure 1 has an optimal solution of value 2, where the horizon is T, and the energy function is the trivial one; it is obvious that more than 2 is impossible.

It is straightforward to see that a feasible solution $S \subset \{1, \ldots, n\}$ for the PARTITION instance induces a feasible pair of paths. For the converse, suppose there is a solution for the ECDF instance that delivers 2 flow units in time. Let $\mathcal{P}$ be the set of flow paths used in the solution, and x_P denote the total flow sent over path $P \in \mathcal{P}$. Because each $\{v_i^+, v_i^-\}$ defines a saturated node cut, $\mathcal{P} \subseteq \{P_S : S \subset \{1, \ldots, n\}, \tau(P_S) \leqslant T\}$. Because each v_i^+ is used by some flow paths with total flow 1, we get

$$\sum_{P \in \mathcal{P}} x_P \tau(P) = \sum_{P \in \mathcal{P}} \sum_{i: v_i^+ \in P} x_P a_i = \sum_{i=1}^n a_i \sum_{P \in \mathcal{P}: v_i^+ \in P} x_P = \sum_{i=1}^n a_i = 2T .$$

Because $\sum_{P \in \mathcal{P}} x_P = 2$ and $\tau(P) \leqslant T$ for all $P \in \mathcal{P}$, this only holds for $\tau(P) = T$, so every $P_S \in \mathcal{P}$ induces a solution S for the PARTITION instance. □

An important property of static network flows is the existence of two popular encoding schemes with polynomial size: First, edge-based, where there is a flow value for every

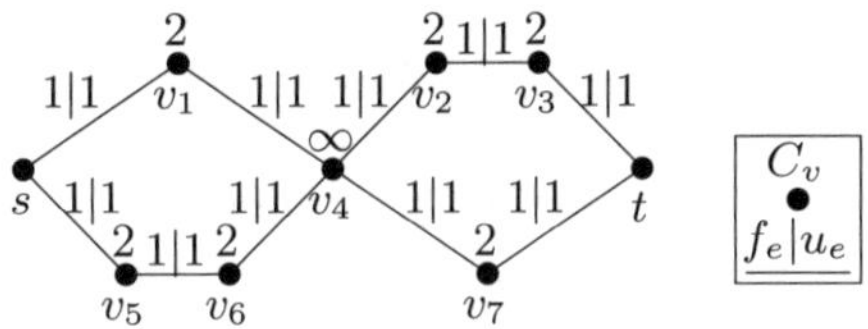

Fig. 2. Network and edge flow values where the path decomposition matters

edge. Second, path-based, where there is a flow value for every path. This even holds for the maximum dynamic flow problem [11]. Now we show that both schemes are not applicable to ECDF with arbitrary transit times. For the edge-based encoding we interpret edge flow values as the maximum flow that is sent over an edge. Consider the network and flow in Figure 2. There are four s-t-paths in it: $P_{\cap\cap} = (s, v_1, v_4, v_2, v_3, t)$, $P_{\cap\cup} = (s, v_1, v_4, v_7, t)$, $P_{\cup\cap} = (s, v_5, v_6, v_4, v_2, v_3, t)$ and $P_{\cup\cup} = (s, v_5, v_6, v_4, v_7, t)$. We show that the given solution—with flow value 1 for every edge—has different values depending on the path decomposition. Let $T = 6$. If we use $P_{\cap\cap}$ and $P_{\cup\cup}$, we can use both paths twice, with a total flow of 4. On the other hand, if we use $P_{\cap\cup}$ and $P_{\cup\cap}$, the paths may be used once resp. twice, giving a total flow of 3. The only edge-based solution encoding that we are aware of assigns time-dependent flow functions $f_e : \{0, \ldots, T\} \to \mathbb{R}$ to the edges, which are not necessarily of polynomial size.

Unfortunately, the usual workaround of using path-based formulation does not help:

Theorem 2. *There are instances for ECDF with arbitrary transit times that allow no optimal solution consisting of a polynomial number of flow paths.*

Proof sketch. The proof employs a sequence of k chained cycles. In each cycle, a flow path can either use a path with transit time 2^{k-1}, or another one with zero transit. Both paths have a flow limit of 2^{k-1}, enforced by batteries. The edge leading to t has a capacity of 1. It can be shown that any optimal flow must deliver one flow unit to t in every time slot from 0 to $T = 2^k - 1$, and each must travel on a different flow path. □

Theorem 3. *Finding an ECDF solution with integral flow values is* NP*-hard.*

Proof. By reduction from the following strongly NP-hard 3-PARTITION variant: given three sets of positive integers $\{a_1, \ldots, a_n\}$, $\{b_1, \ldots, b_n\}$, and $\{c_1, \ldots, c_n\}$ with $L := \sum_{i=1}^n a_i = \sum_{i=1}^n b_i = \sum_{i=1}^n c_i$, find a partition into n triples of equal weight, i.e., find permutations $\alpha, \beta, \gamma \in S_n$ such that $a_{\alpha(i)} + b_{\beta(i)} + c_{\gamma(i)} = 3L/n$ for all i.

We construct an ECDF instance $G = (V, E)$ as follows: For each of the $3n$ numbers, say a_i, there is a chain A_i consisting of a_i nodes connected in line, of which a_i^+ is the first and a_i^- is the last. Additionally, there are the source s and sink t. The source is connected to each entry node of the first set, that is, $sa_i^+ \in E$ for all i. Each exit node of the first set is connected to each entry of the second: $a_i^- b_j^+ \in E$ for all i, j. Analogously, the exits from the second set are linked to the entries of the third, and all exits from the third are linked to t. Each edge $e \in E$ has unit capacity $u_e = 1$. Furthermore, each node $v \neq s, t$ has a battery capacity of $C_v = 1$, where the power consumption model is the trivial one: $c_e^{\mathrm{s}} = 1, c_e^{\mathrm{r}} = 0$ for all $e \in E$. The time horizon is $T = 3L/n + 1$.

We claim that the 3-PARTITION instance is solvable iff the optimal integral ECDF solution value is n. There cannot be a total flow of more than n because $\{a_1^+, \ldots, a_n^+\}$ forms a node cut with total energy n.

Assume there is a feasible, integral ECDF solution $(x_P(\theta))_{P \in \mathcal{P}, \theta \in \theta(P)}$, where $\mathcal{P} \subseteq \mathcal{P}^{st}$, $\theta(P) \subseteq \{0, \ldots, T - \tau(P)\}$, and $x_P(\theta) \in \mathbb{N}^+$ for all $P \in \mathcal{P}, \theta \in \theta(P)$. Assume it has value n. Because each of the sets $\{a_1^+, \ldots, a_n^+\}$, $\{a_1^-, \ldots, a_n^-\}$, $\{b_1^+, \ldots, b_n^+\}$, …, $\{c_1^-, \ldots, c_n^-\}$ is a saturated node cut, it cannot be crossed twice by any path. Hence, each $P \in \mathcal{P}$ is one of the paths $P_{i,j,k} = (s, A_i, B_j, C_k, t)$, $i, j, k = 1, \ldots, n$. Because the flow is integral and all battery capacities are 1, each $x_P(\theta) = 1$. Each of the chains in the graph can be used by just one flow path due to its battery capacity, and because the total flow is n, each chain is used by exactly one path. So $\mathcal{P} = \{P_{\alpha(i),\beta(i),\gamma(i)} : i = 1, \ldots, n\}$ for some $\alpha, \beta, \gamma \in S_n$. We know that

$$\sum_{i=1}^{n} \tau(P_{\alpha(i),\beta(i),\gamma(i)}) = \sum_{i=1}^{n} (a_{\alpha(i)} + b_{\beta(i)} + c_{\gamma(i)} + 1) = \sum_{i=1}^{n} a_i + \sum_{i=1}^{n} b_i + \sum_{i=1}^{n} c_i + n = nT,$$

and, because no path length can exceed T due to the feasibility of the solution, we conclude that each path length must equal $T = 3L/n + 1$. Therefore $a_{\alpha(i)} + b_{\beta(i)} + c_{\gamma(i)} = 3L/n$ for each $i = 1, \ldots, n$, proving that α, β, γ is feasible for the 3-PARTITION instance. It is straighforward to see the converse. □

The proof does not carry over to the fractional ECDF problem: There is a trivial LP formulation for the ECDF instance that uses the n^3 possible flow paths explicitly.

4 Centralized Algorithms for 1-ECDF

In this section, we concentrate on ECDF with uniform transit times.

Lemma 1. *1-ECDF is polynomial-time solvable, if T is polynomially bounded in n.*

Proof. The time-expanded graph $G(T)$ has polynomial size and thus allows for a simple edge-based LP. □

A temporally repeated ("TR") flow is a flow $(x_P(\theta))_{P,\theta}$ where each path carries the same amount of flow at all times, i.e., $x_P(\theta) = x_P(\theta')$ for all $\theta, \theta' \in \{0, \ldots, T - \tau(P)\}$. When $T > 2n$, the problem of finding a temporally repeated 1-ECDF solution can be formulated as follows:

$$\max \sum_{P \in \mathcal{P}^{st}} \varrho(P) x_P \tag{5}$$

$$\text{s.t.} \sum_{P \ni e} x_P \leqslant u_e \qquad \forall e \in E \tag{6}$$

$$\sum_{P \ni v} \varrho(P) c^*_{v,P} x_P \leqslant C_v \quad \forall v \in V \tag{7}$$

$$x_P \geqslant 0\,. \tag{8}$$

The restriction $T > 2n$ comes from inequality (6) which is only valid if all the paths that use some edge e send their flow over e in at least one common point in time.

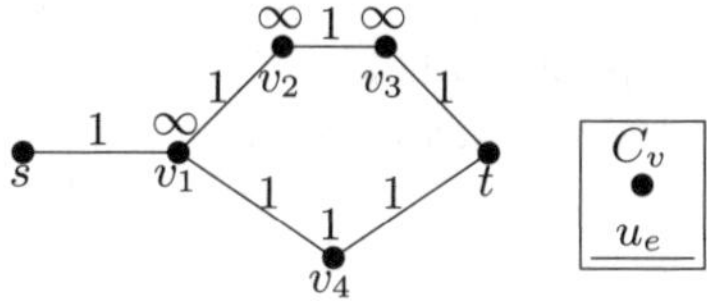

Fig. 3. Network with a gap between optimal and temporally repeated solutions

Lemma 2. *Maximum temporally repeated solutions for 1-ECDF can be found in polynomial time.*

Proof. The dual LP of (5)–(8) is

$$\min \sum_{v\in V} C_v\mu_v + \sum_{e\in E} u_e\pi_e \tag{9}$$

$$\text{s.t.} \sum_{v\in P} \varrho(P)c^*_{v,P}\mu_v + \sum_{e\in P} \pi_e \geqslant \varrho(P) \ \ \forall P \in \mathcal{P} \tag{10}$$

$$\mu_v \geqslant 0 \ \forall v, \qquad \pi_e \geqslant 0 \ \forall e \tag{11}$$

The separation problem for this LP is to find a violated inequality (10), given edge weights $(\pi_e)_{e\in E}$ and node weights $(\mu_v)_{v\in V}$: Find a path $P \in \mathcal{P}^{st}$ satisfying

$$\sum_{v\in P} c^*_{v,P}\mu_v + \sum_{e\in P} \frac{1}{\varrho(P)}\pi_e < 1 \tag{12}$$

or prove that no such path exists. The left-hand-side of (12) can be rewritten as

$$\sum_{v\in P} c^*_{v,P}\mu_v + \sum_{e\in P} \frac{1}{\varrho(P)}\pi_e = \sum_{e=uv\in P} \left(\tfrac{1}{T-\tau(P)+1}\pi_e + c^s_e\mu_u + c^r_e\mu_v\right) \tag{13}$$

which is just the length of P according to some length-dependent edge weights. So the separation problem reduces to the question whether the shortest path in $\mathcal{P}^{st}$ according to this weight function has a length strictly less than 1.

Because $\tau(P) \in \{1, \dots, n\}$ for each $P \in \mathcal{P}^{st}$, there are just n possible values for $\frac{1}{T-\tau(P)+1}$. We can find the shortest path by enumerating over these values. In each step, we seek the shortest path consisting of exactly k edges for some $k \in \{1, \dots, n\}$. This can be done in polynomial time by searching for the shortest path from $s(0)$ to $t(k)$ in the time-expanded graph $G(k)$. □

Lemma 3. *Temporally repeated 1-ECDF solutions are not always optimal.*

Proof. Consider the network shown in Figure 3. The horizon is $T = 4$, communication cost is the trivial one. There are two paths in this network: The "upper" one $P = (s, v_1, v_2, v_3, t)$ that can be used exactly once, and the "lower" one $Q = (s, v_1, v_4, t)$, that can be used twice with a total flow of 1 due to the battery limitation at v_4. An optimal solution sends 1 flow unit along P at time 0 and another unit along Q at time 1, with a total flow value of 2. Optimality holds because edge sv_1 is saturated over time.

A temporally repeated solution sends x_P along P at time 0 and x_Q along Q at times 0 and 1. Because of the capacity of sv_1, $x_P + x_Q \leqslant 1$ holds. Furthermore, $2x_Q \leqslant 1$ due to the battery capacity at v_4. The total flow is $x_P + 2x_Q$, which is maximized by $x_P = x_Q = \frac{1}{2}$ with an objective value of $\frac{3}{2}$. □

Lemma 4. *For $T > \lambda n$, $\lambda \geqslant 2$ the value of a maximum temporally repeated solution* TR *is greater than or equal to* $\frac{\lambda-1}{\lambda}$OPT, *where* OPT *be the value of an optimal 1-ECDF solution.*

Proof. Let $x = (x_P(\theta))_{P,\theta}$ be an optimal solution. We construct a temporally repeated solution $y = (y_P)_P$ from it by averaging over all path flows.

So let $y_P := \frac{1}{\varrho(P)} \sum_{\theta=0}^{T-\tau(P)} x_P(\theta)$ for each $P \in \mathcal{P}^{st}$. This flow satisfies all battery capacity constraints, because each flow path carries the same total flow as in x, and it delivers the same flow within the horizon. It may violate edge capacities though. So let $e \in E$. Then the load on e at time θ is

$$\sum_{\substack{P \ni e: \\ 0 \leqslant \theta - \tau_e(P)}} y_P \leqslant \sum_{P \ni e} y_P \leqslant \frac{1}{T-n} \sum_{P \ni e} \sum_{\theta=0}^{T-\tau(P)} x_P(\theta) \leqslant \frac{1}{T-n} T u_e \leqslant \frac{\lambda}{\lambda-1} u_e \,.$$

Hence $\frac{\lambda-1}{\lambda} y$ is a feasible temporally repeated flow. □

Theorem 4. *1-ECDF admits a FPTAS.*

Proof. Let $\varepsilon > 0$. If $T \leqslant \frac{1}{\varepsilon} n$ or $T \leqslant 2n$, we can solve the problem by Lemma 1. Otherwise, by Lemma 2 we can compute a maximum temporally repeated flow in polynomial time and, by Lemma 4, its value is at least $((\frac{1}{\varepsilon} - 1)/\frac{1}{\varepsilon})\mathrm{OPT} = (1 - \varepsilon)\mathrm{OPT}$. □

5 Distributed Algorithm for 1-ECDF

In this section, we propose a distributed FPTAS for 1-ECDF. The core idea relies on the approximation algorithm [13] for fractional packing problems by Garg and Könemann, so we briefly review their algorithm: Consider a fractional packing LP of the type $\max\{c^\top x | Ax \leqslant b, x \geqslant 0\}$ with a $\tilde{m} \times \tilde{n}$-matrix A, where all coefficients in A, b, and c are nonnegative. Its dual is the covering LP $\min\{y^\top b | y^\top A \geqslant c^\top, y \geqslant 0\}$. Initially, $x = 0$ and $y_j = \delta / b_j$ for all $j = 1, \ldots, \tilde{m}$, where $\delta := (1+\varepsilon)((1+\varepsilon)\tilde{m})^{-1/\varepsilon}$. The algorithm repeats the following iteration until $y^\top b \geqslant 1$: Let i^* be the primal variable (think "maximally violated" dual constraint) that minimizes $(y^T A)_i / c_i$ for $i \in \{1, \ldots, \tilde{n} | c_i > 0\}$. Let j^* be the primal constraint (think "minimum capacity edge/node") that minimizes $b_j / A_{i^* j}$ for $j = 1, \ldots, \tilde{m}$ where $A_{ij} \neq 0$. Now increase x_{i^*} by $b_{j^*} / A_{i^* j^*}$ (corresponding to "sending flow over i^*"), and update the dual variables: $y_j := y_j (1+\varepsilon)(b_{j^*}/A_{i^* j^*})/(b_j / A_{i^* j})$ for all $j = 1, \ldots, \tilde{m}$ with $b_j \neq 0$.

Finally, a feasible primal solution can be obtained by scaling down all variables such that all primal constraints are obeyed; a scaling factor of $1/\log_{1+\varepsilon}((1+\varepsilon)/\delta)$ is sufficient. This can also be done during the increase of primal variables, i.e., during routing of flow, if desired.

Theorem 5 (Garg, Könemann [13]). *Using an oracle that finds a Λ-approximation for the maximally violated constraint, the G&K algorithm computes a $\Lambda(1-\varepsilon)^{-2}$-approximation in $\tilde{m}\lceil \frac{1}{\varepsilon} \log_{1+\varepsilon} \tilde{m} \rceil$ iterations.*

Note that [13] only deals with optimal dual separation, i.e., $\Lambda = 1$. The extension for arbitrary $\Lambda > 1$ is straightforward.

Similar to the previous section, we solve 1-ECDF by distinguishing two cases: $T > \frac{1}{\varepsilon}n$, where the TR gap is small, and $T \leqslant \frac{1}{\varepsilon}n$, where the horizon is polynomially bounded. Actually, the G&K algorithm can easily be distributed, given a fast approximation for the dual separation. For this purpose, we show how to reduce the n shortest path computations we needed in the proof of Lemma 2 to one:

Lemma 5. *Let $T > \lambda n$, $\lambda \geqslant 2$. Let $\pi_e \geqslant 0$ for all $e \in E$ and $\mu_v \geqslant 0$ for all $v \in V$. Then the dual separation problem* (12) *for temporally repeated flows can be $\frac{\lambda}{\lambda-1}$-approximated using a single shortest path computation.*

Proof. The separation problem is to find a shortest path according to the length function

$$z(P) := \sum_{uv \in P} \left(\tfrac{1}{\varrho(P)} \pi_{uv} + c^s_{uv}\mu_u + c^r_{uv}\mu_v\right) . \tag{14}$$

We define another function to approximate z:

$$y(P) := \sum_{uv \in P} \left(\tfrac{1}{T} \pi_{uv} + \tfrac{\lambda-1}{\lambda}(c^s_{uv}\mu_u + c^r_{uv}\mu_v)\right) . \tag{15}$$

Observe that for each $P \in \mathcal{P}^{st}$, $T \geqslant \varrho(P) \geqslant T - n > T - \frac{1}{\lambda}T = \frac{\lambda-1}{\lambda}T$ holds; hence $\frac{1}{T} \leqslant \frac{1}{\varrho(P)} \leqslant \frac{\lambda}{\lambda-1}\frac{1}{T}$. This proves that $y(P) \leqslant z(P) \leqslant \frac{\lambda}{\lambda-1}y(P)$ for every $P \in \mathcal{P}^{st}$. Therefore, the minimum-y path is a $\frac{\lambda}{\lambda-1}$-approximation to the minimum-z path. Now $y(P)$ is just a sum of directed, non-negative edge weights. Then, the minimum-y path can be found by a single run of any shortest path algorithm. □

G&K is turned into distributed algorithm for the large-T case as follows:

Algorithm 1. Distributed algorithm for $T > \frac{1}{\varepsilon}n$

Each node $v \in V$ initializes and stores μ_v and π_e for each $e \in \delta(v)$;
repeat
  s initiates a distributed Bellman-Ford shortest path algorithm;
  The network reports the approximate shortest path's weight and capacity to s;
  The network augments flow along this path, each nodes updates the dual weight, and reports the dual objective increase back to s;
until *s observes that the dual objective is at least* 1 ;
Scale flow for feasibility, unless already done during augmentation.

Lemma 6. *Let $\varepsilon > 0$ and $T > \frac{1}{\varepsilon}n$, $\frac{1}{\varepsilon} \geqslant 2$. Then Algorithm 1 is a $(1-\varepsilon)^{-4}$-approximation for the ECDF problem and runs in time $O(n(n+m)\frac{1}{\varepsilon}\log_{1+\varepsilon}(n+m))$*

Proof. According to Lemma 5, the Bellman-Ford algorithm computes an approximation to the dual separation problem with ratio $\frac{\lambda}{\lambda-1} = (1-\varepsilon)^{-1}$. Hence, Algorithm 1 is a $(1-\varepsilon)^{-3}$-approximation to finding a temporally repeated ECDF solution (Theorem 5). Because a maximum temporally repeated flow is a $(1-\varepsilon)^{-1}$-approximation to the original ECDF problem (Lemma 4), the solution found by Algorithm 1 is a $(1-\varepsilon)^{-4}$-approximation. The runtime results from the iteration bound of Theorem 5 and the $O(n)$ runtime of a distributed Bellman-Ford computation. □

The other case (small T) is mostly analogous, so we just give the result:

Lemma 7. *Let $\varepsilon > 0$ and $T \leqslant \frac{1}{\varepsilon}n$. Then there is a distributed algorithm that finds a $(1-\varepsilon)^{-2}$-approximation in time $O(\frac{1}{\varepsilon^2}mn^2 \log_{1+\varepsilon}(\frac{1}{\varepsilon}mn))$.*

Proof sketch. Run a distributed variant of G&K, similar to Algorithm 1 on the exact LP formulation (1)-(4). □

Together, Lemmas 6 and 7 (and the special case $\varepsilon \geqslant \frac{1}{2}$, $T > \frac{1}{\varepsilon}$, which is trivial to resolve) allow us to state the main theorem of this section:

Theorem 6. *ECDF admits a distributed FPTAS. Each node $v \in V$ needs to store no more than $O(p(v)+1)$ many variables, where $p(v)$ denotes the number of flow paths using v in the solution.*

6 Extensions

Multi-terminal variants: ECDF problems where there is one source and many sinks can be solved by our algorithms, both centralized and distributed, as well. It is sufficient to alter $\mathcal{P}^{st}$ accordingly. The opposite case with one sink and many sources can be solved by exchanging them and reversing time. This just applies to the objective of maximizing the total flow though, as max-min objective variants are no longer fractional packing problems. The situation is similar for multi-commodity settings with many sources and sinks: Maximizing the total flow is possible by adjusting $\mathcal{P}^{st}$—in the distributed setting an additional syncing step between the sources is needed in each iteration. The max-min multi-commodity case can not be solved using our algorithms.

Geometric communication cost functions: In wireless networks, it is a common assumption that the sending cost for transmitting over a link e of length d is $c_e^{\mathrm{s}} = \Theta(d^\alpha)$ for some constant $\alpha \in [2,6]$. The cost for receiving is often modelled as either 0 or $\Theta(c_e^{\mathrm{s}})$. Our constructive results work for any cost function. We have stated the negative results in Section 3 using the trivial cost function for clarity. Note that the problem instances used in the proofs can be embedded such that every link has length 1 (actually, the figures show such embeddings), where the geometric cost function becomes the trivial one. Hence these results apply as well.

7 Conclusion and Open Problems

In this paper, we introduced the novel ECDF problem, which has direct applications in distributed networks, e.g., sensor and ad-hoc network. We proved various negative results and show that an FPTAS exists for the network-motivated 1-ECDF variant, which can be turned into a distributed FPTAS.

There are various related problems of interest. Essentially, we started out with a well-studied problem (maximum dynamic flow) and added two features: firstly, the battery constraints that make the problem much harder (Theorems 1 and 2); and secondly, uniform transit times, making it easier. Because both constraints are important in sensor and ad-hoc networks, studying other dynamic flow problems like quickest flow/transshipment with these extensions poses interesting new challenges.

A tantalizing open problem is the complexity of the fractional ECDF problem with arbitrary transit times. We conjecture that this is in P: consider the path-based LP formulation. Allowing flow changes only in the first and last n steps (i.e., adding $x_P(n) = x_P(n+1) = \ldots = x_P(T-n-1)$ for all P to the formulation) may not change the problem. This new formulation is in P, because the dual separation problem can be easily solved similar to Lemma 2. Alas, we lack a proof.

Another open problem is the existence of an encoding scheme for the solutions with polynomial size (cf. Theorem 2) resp. whether the decision variant of ECDF with arbitrary (or uniform) transit time is in NP.

References

1. Aronson, J.E.: A survey of dynamic network flows. Annals of OR 20, 1–66 (1989)
2. Bodlaender, H., Tan, R., van Dijk, T., van Leeuwen, J.: Integer maximum flow in wireless sensor networks with energy constraint. In: Proc. SWAT (2008)
3. Burkard, R.E., Dlaska, K., Klinz, B.: The quickest flow problem. ZOR — Methods and Models of Operations Research 37, 31–58 (1993)
4. Busse, M., Haenselmann, T., Effelsberg, W.: A comparison of lifetime-efficient forwarding strategies for wireless sensor networks. In: Proc. PE-WASUN, pp. 33–40 (2006)
5. Chang, J.-H., Tassiulas, L.: Maximum lifetime routing in wireless sensor networks. IEEE/ACM Transactions on Networking 12(4), 609–619 (2004)
6. Fekete, S.P., Kröller, A.: Geometry-based reasoning for a large sensor network. In: Proc. SoCG, pp. 475–476 (2006)
7. Fekete, S.P., Kröller, A., Pfisterer, D., Fischer, S.: Algorithmic aspects of large sensor networks. In: Proc MSWSN, pp. 141–152 (2006)
8. Fekete, S.P., Schmidt, C., Wegener, A., Fischer, S.: Hovering data clouds for recognizing traffic jams. In: Proc. IEEE-ISOLA, pp. 213–218 (2006)
9. Fleischer, L., Skutella, M.: Quickest flows over time. SIAM Journal on Computing 36, 1600–1630 (2007)
10. Fleischer, L.K., Tardos, É.: Efficient continuous-time dynamic network flow algorithms. Operations Research Letters 23, 71–80 (1998)
11. Ford, L.R., Fulkerson, D.R.: Constructing maximal dynamic flows from static flows. Operations Research 6, 419–433 (1958)
12. Ford, L.R., Fulkerson, D.R.: Flows in Networks. Princeton University Press, Princeton (1962)
13. Garg, N., Könemann, J.: Faster and simpler algorithms for multicommodity flow and other fractional packing problems. In: Proc. FOCS, p. 300 (1998)
14. Hall, A., Hippler, S., Skutella, M.: Multicommodity flows over time: Efficient algorithms and complexity. Theoretical Computer Science 379, 387–404 (2007)
15. Hoppe, B., Tardos, É.: The quickest transshipment problem. Mathematics of Operations Research 25, 36–62 (2000)
16. Hoppe, B.E.: Efficient dynamic network flow algorithms. PhD thesis, Cornell (1995)

17. Klinz, B., Woeginger, G.J.: Minimum cost dynamic flows: The series-parallel case. In: Balas, E., Clausen, J. (eds.) IPCO 1995. LNCS, vol. 920, pp. 329–343. Springer, Heidelberg (1995)
18. Kröller, A., Fekete, S.P., Pfisterer, D., Fischer, S.: Deterministic boundary recognition and topology extraction for large sensor networks. In: Proc. SODA, pp. 1000–1009 (2006)
19. Madan, R., Lall, S.: Distributed algorithms for maximum lifetime routing in wireless sensor networks. IEEE Transactions on Wireless Communications 5(8), 2185–2193 (2006)
20. Madan, R., Luo, Z.-Q., Lall, S.: A distributed algorithm with linear convergence for maximum lifetime routing in wireless networks. In: Proc. Allerton Conference, pp. 896–905 (2005)
21. Peleg, D.: Distributed computing: a locality-sensitive approach. SIAM, Philadelphia (2000)
22. Powell, W.B., Jaillet, P., Odoni, A.: Stochastic and dynamic networks and routing. In: Network Routing, ch. 3. Handbooks in Operations Research and Management Science, vol. 8, pp. 141–295. North–Holland, Amsterdam, The Netherlands (1995)
23. Ran, B., Boyce, D.E.: Modelling Dynamic Transportation Networks. Springer, Heidelberg (1996)
24. Sankar, A., Liu, Z.: Maximum lifetime routing in wireless ad-hoc networks. In: Proc. INFOCOM, pp. 1089–1097 (2004)
25. Schmid, S., Wattenhofer, R.: Algorithmic models for sensor networks. In: Proc. IPDPS (2006)
26. Wattenhofer, R.: Sensor networks: Distributed algorithms reloaded - or revolutions? In: Flocchini, P., Gąsieniec, L. (eds.) SIROCCO 2006. LNCS, vol. 4056, pp. 24–28. Springer, Heidelberg (2006)
27. Zussman, G., Segall, A.: Energy efficient routing in ad hoc disaster recovery networks. In: Proc. INFOCOM, pp. 682–691 (2003)

Bounded Unpopularity Matchings

Chien-Chung Huang[1,⋆], Telikepalli Kavitha[2,⋆], Dimitrios Michail[3,⋆⋆], and Meghana Nasre[2]

[1] Dartmouth College, USA
villars@cs.dartmouth.edu
[2] Indian Institute of Science, India
{kavitha,meghana}@csa.iisc.ernet.in
[3] INRIA Sophia Antipolis - Méditerranée, France
dmichail@sophia.inria.fr

Abstract. We investigate the following problem: given a set of jobs and a set of people with preferences over the jobs, what is the optimal way of matching people to jobs? Here we consider the notion of *popularity*. A matching M is popular if there is no matching M' such that more people prefer M' to M than the other way around. Determining whether a given instance admits a popular matching and, if so, finding one, was studied in [2]. If there is no popular matching, a reasonable substitute is a matching whose *unpopularity* is bounded. We consider two measures of unpopularity - *unpopularity factor* denoted by $u(M)$ and *unpopularity margin* denoted by $g(M)$. McCutchen recently showed that computing a matching M with the minimum value of $u(M)$ or $g(M)$ is NP-hard, and that if G does not admit a popular matching, then we have $u(M) \geq 2$ for all matchings M in G.

Here we show that a matching M that achieves $u(M) = 2$ can be computed in $O(m\sqrt{n})$ time (where m is the number of edges in G and n is the number of nodes) provided a certain graph H admits a matching that matches all people. We also describe a sequence of graphs: $H = H_2, H_3, \ldots, H_k$ such that if H_k admits a matching that matches all people, then we can compute in $O(km\sqrt{n})$ time a matching M such that $u(M) \leq k-1$ and $g(M) \leq n(1-\frac{2}{k})$. Simulation results suggest that our algorithm finds a matching with low unpopularity.

1 Introduction

The problem of assigning people to positions is a very common problem that arises in many domains. The input here is a bipartite graph $G = (\mathcal{A} \cup \mathcal{P}, \mathcal{E})$, where nodes on one side of the bipartite graph rank edges incident on them in an order of preference, possibly involving ties. That is, the edge set $\mathcal{E}$ is partitioned into $\mathcal{E}_1 \dot\cup \mathcal{E}_2 \ldots \dot\cup \mathcal{E}_r$. We call $\mathcal{A}$ the set of *applicants*, $\mathcal{P}$ the set of *posts*, and $\mathcal{E}_i$ the set of edges with *rank i*. If $(a,p) \in \mathcal{E}_i$ and $(a,p') \in \mathcal{E}_j$ with $i < j$, we say that *a prefers p* to p'. If $i = j$, then a is *indifferent* between p and p'. The ordering of posts adjacent to a is called a's *preference*

⋆ Part of this work was done when the author was visiting Max-Planck-Institut für Informatik, Saarbrücken, Germany.

⋆⋆ Part of this work was carried out during the tenure of an ERCIM "Alain Bensoussan" Fellowship Programme.

J. Gudmundsson (Ed.): SWAT 2008, LNCS 5124, pp. 127–137, 2008.

list. The problem is to assign applicants to posts that is *optimal* with respect to these preference lists.

This problem has been well-studied in economics literature, see for example [3, 15, 17]. It models some important real-world markets, including the allocation of graduates to training positions [9], families to government-owned housing [16], mail-based DVD rental systems such as NetFlix.

Various criteria have been proposed to measure the "goodness" of a matching. Examples include *Pareto optimality* [1, 3, 15], *rank maximality* [10], and *maximum utility*. Most of these criteria use the actual values or numerical ranks expressed by applicants in their preference lists. Such criteria are easily prone to manipulation by people lying about their preferences. Moreover, the preference lists only express the "relative" ranking of the options. Measuring the optimality of a matching as a function of the actual numerical ranks may not be the correct approach. One criterion that does not use numerical ranks is *popularity*. We define it below.

We say that an applicant *a prefers* matching M' to M if (i) a is matched in M' and unmatched in M, or (ii) a is matched in both M' and M, and a prefers $M'(a)$ to $M(a)$ (where $M(a), M'(a)$ are the posts that a is matched to in M and in M', respectively).

Definition 1. *M' is* more popular than *M, denoted by $M' \succ M$, if the number of applicants that prefer M' to M is greater than the number of applicants preferring M to M'. A matching M is* popular *if there is no matching M' that is more popular than M.*

Note that popular matchings may not exist in the given instance. The problem is to determine if a given instance admits a popular matching, and to find such a matching, if one exists. The first polynomial-time algorithms for this problem were given in [2]: when there are no ties in the preference lists, the problem can be solved in $O(n+m)$ time, where $n = |\mathcal{A} \cup \mathcal{P}|$ and $m = |E|$, and more generally, the problem can be solved in $O(m\sqrt{n})$ time. The main drawback of the notion of popular matchings is that such matchings may not exist in the given graph. In this situation, it would be desirable if we can find some good substitutes of a popular matching. This motivates our paper.

1.1 Problem Definition

In this paper, we assume that the input instance G does not admit a popular matching. Our goal is to compute a *least unpopular* matching. We use two criteria given by McCutchen [13] to measure the unpopularity of a matching. We first need the following definitions.

Given any two matchings X and Y in G, define $\phi(X,Y)$ = number of applicants that prefer X to Y. Let us define the following functions to compare two matchings X and Y:

$$\Delta(X,Y) = \begin{cases} \phi(Y,X)/\phi(X,Y) & \text{if } \phi(X,Y) > 0 \\ 1 & \text{if } \phi(X,Y) = 0 \text{ and } \phi(Y,X) = 0 \\ \infty & \text{otherwise.} \end{cases}$$

and $\delta(X,Y) = \phi(Y,X) - \phi(X,Y)$.

Having the above functions, we can define the *unpopularity factor* of a matching M as:

$$u(M) = \max_{M'} \Delta(M, M').$$

The *unpopularity margin* of a matching M is defined as:

$$g(M) = \max_{M'} \delta(M, M').$$

The functions $u(\cdot)$ and $g(\cdot)$ were first introduced by McCutchen, who also gave polynomial time algorithms to compute $u(M)$ and $g(M)$ for any given matching M. A matching M is popular if and only if $u(M) = 1$ and $g(M) = 0$. When G does not admit popular matchings, we are interested in computing a matching M with a low value of $u(M)$. Let us now define a *least unpopular* matching.

Definition 2. *A matching M which achieves the minimum value of $u(M)$ among all the matchings in G is defined as the least unpopularity factor matching in G. Similarly, a matching that achieves the minimum value of $g(M)$ among all matchings in G is defined as the least unpopularity margin matching in G.*

McCutchen recently showed that either computing a least unpopularity factor matching or a least unpopularity margin matching is NP-hard. He also showed that the unpopularity factor of any matching is always an integer. Thus when G does not admit a popular matching, the best matching in terms of the unpopularity factor that one can hope for in G is a matching M that satisfies $u(M) = 2$. Complementing McCutchen's results, we have the following new results here.

- A least unpopularity factor matching can be computed in $O(m\sqrt{n})$ time provided a certain graph H admits an $\mathcal{A}$-*complete* matching. An $\mathcal{A}$-complete matching means all nodes in $\mathcal{A}$ are matched. Such a matching M that we compute in H satisfies $u(M) = 2$.
- We also show a more general result. We construct a sequence of graphs: $H = H_2, H_3, \ldots, H_k, \ldots$ and show that if H_k admits an $\mathcal{A}$-complete matching, then we can compute in $O(km\sqrt{n})$ time a matching M such that $u(M) \leq k-1$ and $g(M) \leq n(1 - \frac{2}{k})$.
- We ran our algorithm on random graphs using a similar setup as in [2]. Our simulation results suggest that when G is a random graph, then for values of $k \leq 4$, we see that H_k admits an $\mathcal{A}$-complete matching. Thus in these graphs our algorithm computes a matching M whose unpopularity factor is a number ≤ 3 and whose unpopularity margin can be upper bounded by $n/2$.

1.2 Background and Related Results

The notion of popular matchings was first introduced by Gardenfors [6] in the context of the stable marriage problem.

When only one side has preferences, Abraham et al. [2] gave polynomial time algorithms to find a popular matching, or to report that none exists. Further variants of popular matchings are studied by Manlove and Sng [12] and by Mestre [14]. As to the existence of popular matchings, Mahdian [11] gave an probabilistic analysis.

Organization of the paper. In Section 2 we describe the popular matching algorithm from [2], which is the starting point of our algorithm. We then describe McCutchen's algorithm to compute the unpopularity factor of a given matching. In Section 3 we describe our algorithm and bound its unpopularity factor and unpopularity margin. In Section 4 we report our experimental results. For a probabilistic analysis of our algorithm, see [4] for details.

2 Preliminaries

In this section, we review the algorithmic characterization of popular matchings given in [2] and the algorithm to compute the unpopularity index of a matching as given by McCutchen.

For exposition purposes, we create a unique strictly-least-preferred post $l(a)$ for each applicant a. In this way, we can assume that every applicant is matched, since any unmatched applicant a can be paired with $l(a)$. From now on, matchings are always $\mathcal{A}$-complete. Also, without loss of generality, we assume that preference lists contain no gaps, i.e., if a is incident to an edge of rank i, then a is incident to an edge of rank $i-1$, for all $i > 1$.

Let $H_1 = (\mathcal{A} \cup \mathcal{P}, \mathcal{E}_1)$ be the graph containing only rank-one edges. Then [2, Lemma 3.1] shows that a matching M is popular in G only if $M \cap \mathcal{E}_1$ is a maximum matching of H_1. Maximum matchings have the following important properties, which we use throughout the rest of the paper.

$M \cap \mathcal{E}_1$ defines a partition of $\mathcal{A} \cup \mathcal{P}$ into three disjoint sets: a node $u \in \mathcal{A} \cup \mathcal{P}$ is *even* (resp. *odd*) if there is an even (resp. odd) length alternating path in H_1 (w.r.t. $M \cap \mathcal{E}_1$) from an unmatched node to u. Similarly, a node u is *unreachable* if there is no alternating path from an unmatched node to u. Denote by $\mathcal{N}$, $\mathcal{O}$ and $\mathcal{U}$ the sets of even, odd, and unreachable nodes, respectively.

Lemma 1 (Gallai-Edmonds Decomposition). *Let $\mathcal{N}$, $\mathcal{O}$ and $\mathcal{U}$ be the sets of nodes defined by H_1 and $M \cap \mathcal{E}_1$ above. Then*

(a) $\mathcal{N}$, $\mathcal{O}$ and $\mathcal{U}$ are pairwise disjoint, and independent of the maximum matching $M \cap \mathcal{E}_1$.
(b) In any maximum matching of H_1, every node in $\mathcal{O}$ is matched with a node in $\mathcal{N}$, and every node in $\mathcal{U}$ is matched with another node in $\mathcal{U}$. The size of a maximum matching is $|\mathcal{O}| + |\mathcal{U}|/2$.
(c) No maximum matching of H_1 contains an edge between a node in $\mathcal{O}$ and a node in $\mathcal{O} \cup \mathcal{U}$. Also, H_1 contains no edge between a node in $\mathcal{N}$ and a node in $\mathcal{N} \cup \mathcal{U}$.

Using this node partition, we make the following definitions: for each applicant a, $f(a)$ is the set odd/unreachable posts amongst a's most-preferred posts. Also, $s(a)$ is the set of a's most-preferred posts amongst all even posts. We refer to posts in $\cup_{a \in \mathcal{A}} f(a)$ as *f-posts* and posts in $\cup_{a \in \mathcal{A}} s(a)$ as *s-posts*. Note that f-posts and s-posts are disjoint, and that $s(a) \neq \emptyset$ for any a, since $l(a)$ is always even. Also note that there may be posts in $\mathcal{P}$ that are neither f-posts nor s-posts. The next theorem characterizes the set of all popular matchings.

Theorem 1 ([2]). *A matching M is popular in G iff (i) $M \cap \mathcal{E}_1$ is a maximum matching of $H_1 = (\mathcal{A} \cup \mathcal{P}, \mathcal{E}_1)$, and (ii) for each applicant a, $M(a) \in f(a) \cup s(a)$.*

Figure 1 contains the algorithm from [2], based on Theorem 1, for solving the popular matching problem.

Popular-Matching($G = (\mathcal{A} \cup \mathcal{P}, \mathcal{E})$)
Construct the graph $G' = (\mathcal{A} \cup \mathcal{P}, \mathcal{E}')$, where $\mathcal{E}' = \{(a,p) : a \in \mathcal{A}$ and $p \in f(a) \cup s(a)\}$.
Construct a maximum matching M of $H_1 = (\mathcal{A} \cup \mathcal{P}, \mathcal{E}_1)$.
//Note that M is also a matching in G'.
Remove any edge in G' between a node in O and a node in $O \cup \mathcal{U}$.
//No maximum matching of H_1 contains such an edge.
Augment M in G' until it is a maximum matching of G'.
Return M if it is $\mathcal{A}$-complete, otherwise return *"no popular matching"*.

Fig. 1. An $O(\sqrt{n}m)$-time algorithm for the popular matching problem (from [2])

2.1 McCutchen's Algorithm

Here we outline the algorithm given by McCutchen for computing the unpopularity factor of a matching. Given a matching M, the idea is to find a series of promotions (of applicants) at the cost of demoting one applicant. The longest such promotion path determines the unpopularity factor of the particular matching. Such a path can be discovered by building a directed weighted graph on the set of posts. We will refer to this graph as the *Posts-Graph* $G_{\mathcal{P}}$. The vertices of $G_{\mathcal{P}}$ represent all the posts $\mathcal{P}$ in the original graph. We add edges into $G_{\mathcal{P}}$ based on the following rules: (let $M(p)$ denote the applicant to which post p is matched to in the matching M)

- an edge with weight -1 is directed from p_i to p_j if $M(p_i)$ prefers p_j to p_i.
- an edge with weight 0 is directed from p_i to p_j if $M(p_i)$ is indifferent between p_i and p_j.

Note that there is no edge from p_i to p_j if $M(p_i)$ prefers p_i to p_j. The series of promotions mentioned above is a negative weight path in this graph. To find the longest negative weight path in this graph, we add a dummy vertex s with 0 weight edges from s to all posts. An algorithm which finds shortest paths from source s to all posts will give the longest negative weight path in $G_{\mathcal{P}}$. Existence of a negative weight cycle implies that there exists a promotion sequence without any demotion and hence the unpopularity factor of the matching is ∞. Let us assume that no negative weight cycles exist. Then all posts have a 0 or negative weight shortest path from the source. The post whose distance from the source is the "most negative" determines the unpopularity index of the matching M. For details of the proof of correctness, refer to [13].

3 Our Algorithm

In this section we describe a *greedy* strategy to compute a matching of G, whose unpopularity can be bounded. Our algorithm is iterative and in every iteration it constructs a graph H_i and a maximum matching M_i in H_i. We show that if M_i is an $\mathcal{A}$-complete matching, then $u(M_i) \leq i-1$ and $g(M_i) \leq n(1-2/i)$.

We will first give some intuition before we formally describe our algorithm. Recall that the popular matching algorithm first finds a maximum cardinality matching M_1 in the graph H_1 (whose edge set is the set of all rank 1 edges). The algorithm then identifies all even applicants/posts using the Gallai-Edmonds decomposition and adds the edges (a,p) where a is even and $p \in s(a)$ to the pruned graph H_1 (all rank 1 edges between an odd node in H_1 and a node that is odd or unreachable in H_1 are removed from H_1). Note that each such edge (a,p) is *new* to H_1, that is, such an edge is not already present in H_1 since by Gallai-Edmonds decomposition (part (iii)), there is no edge between two *even* vertices of H_1, and here both a and p are even in H_1. In this new graph, call it H_2, M_1 is augmented to a maximum cardinality matching M_2. In case M_2 is $\mathcal{A}$-complete, we declare that the instance admits a popular matching. Otherwise no popular matching exists.

The idea of our algorithm here is an extension of the same strategy. Since we are considering instances which do not admit a popular matching, M_2 found above will not be $\mathcal{A}$-complete. In this case, we go further and find the Gallai-Edmonds decomposition of nodes in H_2 and identify nodes that are even in H_1 and in H_2. A node that is odd or unreachable in either H_1 or in H_2 will always be matched by a maximum cardinality matching in H_2 that is obtained by augmenting a maximum cardinality matching in H_1. Hence the nodes that are not guaranteed to be matched by such a matching M_2 are the applicants and posts that are even in both H_1 and H_2.

So let us now add the edges (a,p) to H_2 where a and p are nodes that are even in both H_1 and H_2 and among all posts that are even in both H_1 and H_2, p is a most preferred post of a. We would again like to point out that such an edge (a,p) did not exist in either H_1 or in H_2, since a and p were even in H_1 and in H_2. We also prune H_2 to remove edges that are contained in no maximum cardinality matching of H_2 and call the resulting graph H_3. We then augment M_2 to get M_3 and continue the same procedure till we finally get an $\mathcal{A}$-complete matching M_i.

With the above intuition, we are now ready to formally define the algorithm.

3.1 The Algorithm

We start with $H_1 = (\mathcal{A} \cup \mathcal{P}, \mathcal{E}_1)$ where $\mathcal{E}_1$ is the set of rank 1 edges. Let M_1 be any maximum cardinality matching in H_1. Fig. 2 contains our algorithm.

We note that once a post becomes odd or unreachable in any iteration, it gets marked and hence it cannot get new edges incident upon it in the subsequent iterations. We use this to show that the unpopularity factor of the matching that we produce is bounded by $k-1$ if we find an $\mathcal{A}$-complete matching in H_k. The running time of our algorithm is determined by the least k such that H_k admits an $\mathcal{A}$-complete matching. Since each iteration of our algorithm takes $O(m\sqrt{n})$ time by the Hopcroft-Karp algorithm [8], the overall running time is $O(km\sqrt{n})$, where k is the least number such that H_k admits an $\mathcal{A}$-complete matching.

Initialize $i = 1$ and let all nodes be unmarked.
While M_i is not $\mathcal{A}$-complete do:

1. Partition the nodes of $\mathcal{A} \cup \mathcal{P}$ into three disjoint sets: $\mathcal{N}_i, \mathcal{O}_i, \mathcal{U}_i$.
   – $\mathcal{N}_i$ and $\mathcal{O}_i$ consists of nodes that can be reached in H_i from an unmatched node by an even/odd length alternating path with respect to M_i, respectively.
   – $\mathcal{U}_i$ consists of nodes that are unreachable by an alternating path from any unmatched node in H_i.
2. Mark all unmarked nodes in $\mathcal{O}_i \cup \mathcal{U}_i$.
3. Delete all edges of H_i between a node in $\mathcal{O}_i$ and a node in $\mathcal{O}_i \cup \mathcal{U}_i$
4. Add edges (a, p) to H_i where (i) a in unmarked, (ii) p is unmarked and (iii) p is a's most preferred post among all unmarked posts. Call the resulting graph H_{i+1}.
5. Augment M_i in H_{i+1} to get a new matching M_{i+1} which is a maximum cardinality matching of H_{i+1}.
6. $i = i + 1$.

Fig. 2. An $O(km\sqrt{n})$-time algorithm for finding an $\mathcal{A}$–complete matching

Before we prove our main theorems, we need the following definition that defines a *level* j post for an applicant a. A level 1 post for each applicant is just its rank 1 post. But from levels ≥ 2, a level j post for an applicant need not be its rank j post.

Definition 3. *A level j post for an applicant a is a post p such that (i) p is an even post in $H_1, \ldots, H_{j-1}$ and (ii) p is the most preferred post for a amongst all such posts.*

Theorem 2. *If our algorithm finds an $\mathcal{A}$-complete matching M_k in H_k, then $u(M_k) \leq k-1$.*

Proof. Let M_k be the $\mathcal{A}$-complete matching produced by our algorithm after k iterations. We draw the posts graph $G_{\mathcal{P}}$ corresponding to the matching M_k. The unpopularity index of M_k is the "most negative" distance of a vertex (post) in $G_{\mathcal{P}}$ from the dummy source s as described in Section 2. We now show that the posts in $G_{\mathcal{P}}$ can be partitioned into k layers (corresponding to the k iterations) such that all negative weight edges always go from higher numbered layers to lower numbered layers. If we show this, then it is clear that since there are only k layers and all negative weight edges have weight -1, the longest negative weight path can be of length at most $k-1$.

Let us partition the posts of $G_{\mathcal{P}}$ such that a post belongs to a layer t if it gets marked for the first time in iteration t. Let p be a post that belongs to level i. Recall that in $G_{\mathcal{P}}$ there is a negative weight edge from p to q iff $M_k(p)$ strictly prefers q to p. We now show that any such post q should belong to a layer j such that $j < i$.

First, note that an edge (a, p) is added to the graph at the end of the $(j-1)$-th iteration of our algorithm (for any $j \geq 1$) only if p is a level j post for a. Next, note that since p got marked in the i-th iteration, no new edges are ever added to p in any of the subsequent iterations. Based on these two observations, we can conclude that since the edge $(M_k(p), p)$ exists in $G_{\mathcal{P}}$ it has to be the case that p is a level ℓ post for $M_k(p)$ for some $\ell \leq i$.

That is, at the end of the $(\ell-1)$-th iteration, p was the most preferred *unmarked* post for $M_k(p)$. Hence all the posts that $M_k(p)$ strictly prefers to p were already marked before/during the $(\ell-1)$th iteration. That is, these posts belong to layers j, where $j \leq \ell-1 \leq i-1$. Thus if (p,q) is a negative weight edge out of p, then q belongs to layer j, where $j < i$.

Hence we have shown that all negative weight edges must go from higher numbered layers to lower numbered layers. This implies that the longest negative weight path in the graph $G_{\mathcal{P}}$ corresponding to M_k is at most $k-1$. In other words, $u(M_k) \leq k-1$. □

Theorem 3. *If our algorithm finds an $\mathcal{A}$-complete matching M_k in H_k, then $g(M_k) \leq n(1-\frac{2}{k})$.*

Proof. Let M_k be the $\mathcal{A}$-complete matching produced by our algorithm after k iterations and let M be any other $\mathcal{A}$-complete matching in G. Now let us construct a weighted directed graph $H_{\mathcal{P}}$ similar to the posts graph $G_{\mathcal{P}}$. The vertices of $H_{\mathcal{P}}$ are all posts p such that $M_k(p) \neq M(p)$. For every applicant a we have a directed edge from $M_k(a)$ to $M(a)$ with a weight of $-1,0,+1$ if a considers $M(a)$ better than, the same as or worse than $M_k(a)$. Any post p that does not belong to $H_{\mathcal{P}}$ is matched to the same applicant in M_k as well as in M and hence the corresponding applicant does not contribute to the unpopularity margin. Furthermore, it is clear that the sum of weights of all edges in $H_{\mathcal{P}}$ gives the negative of the unpopularity margin by which M dominates M_k.

First note that $H_{\mathcal{P}}$ is a set of disjoint paths and cycles. This is because, $H_{\mathcal{P}}$ can equivalently be constructed from $S = M_k \oplus M$ by striking off applicants and giving appropriate directions and weights to edges. Thus a path in S continues to be a path in $H_{\mathcal{P}}$ although it may no longer be of even length. The same is true for cycles also. If a path or cycle consists of only 0 weight edges, then we can drop such a cycle/path from the graph, since these edges do not contribute to the unpopularity margin. In addition, note that any cycle or path cannot be composed of only negative and zero weight edges, otherwise the unpopularity factor of M_k is ∞, a contradiction. Hence we can assume that every cycle or path contains at least one positive edge.

Let ρ be any path or cycle in $H_{\mathcal{P}}$. Furthermore, let α and β be the numbers of -1's and $+1$'s in ρ respectively. We define the function:

$$\text{frac-margin}(\rho) = \frac{\alpha - \beta}{\text{number of edges in } \rho}$$

Let us try to bound frac-margin(ρ) for each ρ. For the sake of simplicity, let us first assume that the preference lists are strict. So there are only ± 1 weight edges in $H_{\mathcal{P}}$. Thus frac-margin$(\rho) = (\alpha-\beta)/(\alpha+\beta)$. Since the unpopularity factor of M_k is bounded by $k-1$, it is easy to see that the unpopularity factor of ρ is also bounded by $k-1$ (refer to [13] for a proof), implying $\alpha/\beta \leq k-1$. Thus $\beta/(\alpha+\beta) \geq 1/k$, and $\alpha/(\alpha+\beta) \leq 1-1/k$. Hence frac-margin$(\rho)$ for any path or cycle ρ is at most $1-2/k$. The contribution of ρ towards $\delta(M_k, M)$ is (number of edges in ρ)$\cdot$(frac-margin(ρ)). This is at most $n_\rho(1-2/k)$ where n_ρ is the number of edges in ρ. Since a unique applicant a is associated with each edge $(M_k(a), M(a))$ of $H_{\mathcal{P}}$, it follows that $\sum n_\rho \leq n$. Thus $\delta(M_k, M) \leq n(1-2/k)$ where M is any matching.

The proof for the case with ties also follows from the above argument. Since 0 weight edges in ρ do not affect the numerator of frac-margin(ρ) and only increase the denominator of frac-margin(ρ), it is easy to see that frac-margin(ρ) for a path or cycle ρ with 0 weight edges is dominated by frac-margin(ρ') where ρ' is obtained from ρ by contracting 0 weight edges. Thus frac-margin(ρ) $\leq 1-2/k$ and thus $\delta(M_k, M) \leq n(1-2/k)$ where M is any matching. Thus $max_{M'}\delta(M_k, M) \leq n(1-2/k)$. □

Corollary 1. *Let G be a graph that does not admit a popular matching. If our algorithm produces an $\mathcal{A}$-complete matching M in H_3, then M is a least unpopularity factor matching in G.*

Proof. It follows from Theorem 2 that if our algorithm produces an applicant complete matching M in H_3, then $u(M) \leq 2$. McCutchen [13] showed that the unpopularity factor of any matching is always an integer. Thus if G admits no popular matching, then the lowest value of $u(\cdot)$ we can hope for is 2. Since $u(M) \leq 2$, it follows that this is a least unpopularity factor matching. □

Starting with the above corollary, it is tempting to push the frontier further. Suppose that the algorithm gets an $\mathcal{A}$-complete matching M_4 in the graph H_4 (thus $u(M_4) \leq 3$). Can we also argue that it is impossible to achieve a better matching? Unfortunately, this is not the case. See the full version [4] for an example.

4 Experimental Results

In this section, we present simulation results showing that our algorithm is able to find a matching with small unpopularity in random graphs.

We follow the setting used in [2] so that our experimental results are comparable to those reported in [2]. The number of applicants and posts are equal (denoted by n) and preference lists have the same length k. Existence of ties is characterized by a single parameter t which denotes the probability of an entry in the preference list to be tied with its predecessor.

Table 1 contains simulation results for random graphs with $n = 100$ and $n = 500$ for different values of parameters k and t. The table shows the number of instances (out of 1000 instances) that finish in some particular round of the execution. Round 2 means that the instance has a popular matching. It is easy to observe that the difficult cases are the ones where we only have a few ties (t is small). For a fixed value of k as t decreases the algorithm requires more rounds until it returns a solution. However, the good news are that it never takes more than four rounds.

As Table 1 suggests, the difficult situation is when k is large and t is very small (the preferences have few ties and these ties are of small length). We study this situation further by varying the value of n in order to see whether our observations for $n = 100$ and $n = 500$ are valid for larger values of n. Table 1 shows the number of rounds (again out of a 1000) that is required for different values of n when $t = 0.05$ and $k = n$. The table suggests that as n increases the probability of terminating at the second or third round decreases while the probability of terminating in the fourth round increases. However, this is not accompanied with any increase in rounds larger than 4.

Our experimental results are promising. The algorithm behaves nicely in practice, far away from a possible large approximation.

Table 1. The left and middle tables show the number of instances with $n = 100$ and 500 nodes respectively (out of a 1000 instances) that finish in round number 2 (popular matching), 3 or 4 for different values of the parameters k and t. The table on the right shows the number of instances (out of a 1000 instances) that finish in round number 2 (popular matching), 3 or 4 for fixed $t = 0.05$, $k = n$ and different values of the parameter n.

$n = 100$				
		# rounds		
k	t	2	3	4
	0.05	4	996	
	0.2	28	972	
10	0.5	471	529	
	0.8	729	271	
	1.0	1000		
	0.05		991	9
	0.2	3	991	6
25	0.5	138	861	1
	0.8	773	227	
	1.0	1000		
	0.05		948	52
	0.2	1	978	21
50	0.5	158	832	10
	0.8	793	207	
	1.0	1000		
	0.05		952	48
	0.2	2	973	25
100	0.5	148	836	16
	0.8	783	217	
	1.0	1000		

$n = 500$				
		# rounds		
k	t	2	3	4
	0.05		1000	
	0.2		1000	
10	0.5	176	824	
	0.8	62	938	
	1.0	1000		
	0.05		1000	
	0.2		1000	
25	0.5		999	1
	0.8	93	907	
	1.0	1000		
	0.05		967	33
	0.2		994	6
50	0.5		997	3
	0.8	104	896	
	1.0	1000		
	0.05		828	172
	0.2		942	58
100	0.5		989	11
	0.8	93	907	
	1.0	1000		

n	# rounds		
	2	3	4
10	585	413	2
25	141	844	15
50	6	962	32
100		952	48
250		896	104
500		820	180
1000		667	333
1500		541	459
2000		320	680

References

1. Abraham, D.J., Cechlárová, K., Manlove, D.F., Mehlhorn, K.: Pareto-optimality in house allocation problems. In: Fleischer, R., Trippen, G. (eds.) ISAAC 2004. LNCS, vol. 3341, pp. 3–15. Springer, Heidelberg (2004)
2. Abraham, D.J., Irving, R.W., Kavitha, T., Mehlhorn, K.: Popular matchings. SIAM Journal on Computing 37(4), 1030–1045 (2007), Preliminary version. In: Proc. of 16th SODA, pp. 424-432, (2005)
3. Abdulkadiroğlu, A., Sönmez, T.: Random serial dictatorship and the core from random endowments in house allocation problems. Econometrica 66(3), 689–701 (1998)
4. Huang, C.-C., Kavitha, T., Michail, D., Nasre, M.: Bounded Unpopularity Matchings. Dartmouth Computer Science, Technical Report (2008)-616
5. Gale, D., Shapley, L.S.: College admissions and the stability of marriage. American Mathematical Monthly 69, 9–15 (1962)
6. Gardenfors, P.: Match making: assignments based on bilateral preferences. Behavioural Sciences 20, 166–173 (1975)
7. Gusfield, D., Irving, R.W.: The Stable Marriage Problem: Structure and Algorithms. MIT Press, Cambridge (1989)

8. Hopcroft, J.E., Karp, R.M.: A $n^{5/2}$ Algorithm for Maximum Matchings in Bipartite Graphs. SIAM Journal on Computing 2, 225–231 (1973)
9. Hylland, A., Zeckhauser, R.: The efficient allocation of individuals to positions. Journal of Political Economy 87(2), 293–314 (1979)
10. Irving, R.W., Kavitha, T., Mehlhorn, K., Michail, D., Paluch, K.: Rank-maximal matchings. ACM Transactions on Algorithms 2(4), 602–610 (2006), Preliminary version. In: Proc. of 15th SODA, pp. 68-75 (2004)
11. Mahdian, M.: Random popular matchings. In: Proceedings of the 7th ACM Conference on Electronic-Commerce, pp. 238–242 (2006)
12. Manlove, D.F., Sng, C.: Popular matchings in the capacitated house allocation problem. In: Azar, Y., Erlebach, T. (eds.) ESA 2006. LNCS, vol. 4168, pp. 492–503. Springer, Heidelberg (2006)
13. McCutchen, M.: The least-unpopularity-factor and least-unpopularity-margin criteria for matching problems with one-sided preferences. In: Proceedings of LATIN 2008, the 8th Latin American Theoretical Informatics Symposium, LNCS, vol. 4957, Springer, Heidelberg (2008)
14. Mestre, J.: Weighted popular matchings. In: Bugliesi, M., Preneel, B., Sassone, V., Wegener, I. (eds.) ICALP 2006. LNCS, vol. 4051, pp. 715–726. Springer, Heidelberg (2006)
15. Roth, A.E., Postlewaite, A.: Weak versus strong domination in a market with indivisible goods. Journal of Mathematical Economics 4, 131–137 (1977)
16. Yuan, Y.: Residence exchange wanted: a stable residence exchange problem. European Journal of Operational Research 90, 536–546 (1996)
17. Zhou, L.: On a conjecture by Gale about one-sided matching problems. Journal of Economic Theory 52(1), 123–135 (1990)

Data Structures with Local Update Operations

Yakov Nekrich

Dept. of Computer Science, University of Bonn
yasha@cs.uni-bonn.de

Abstract. In this paper we describe dynamic data structures with restrictions on update operations. In the first part of the paper we consider data structures that support operations insert$_{\Delta}(x,y)$ or insert$_{\Delta}(x)$ instead of general insertions, where insert$_{\Delta}(x,y)$ (insert$_{\Delta}(x)$) inserts a new element x, such that $|x-y| \le \Delta$ for some element y already stored in the data structure. We present a data structure that supports predecessor queries in a universe of size U in $O(\log\log U)$ time, uses $O(n)$ words of space, and supports operations insert$_{\Delta}(x,y)$, and delete(x) in $O(1)$ amortized time, where $\Delta = 2^{2^{O(\sqrt{\log\log U})}}$. We present the dictionary data structure that supports membership queries in $O(\log\log n)$ time and insert$_{\Delta}(x,y)$ and delete(x) in $O(1)$ amortized time, where $\Delta = 2^{2^{O(\sqrt{\log\log n})}}$ We also present a priority queue that supports delete(x), and FindMin in $O(1)$ time and insert$_{\Delta}(x)$ in $O(\log\log n)$ time, where $\Delta = \log^{O(1)} U$. All above data structures also support incrementation and decrementation of element values by the corresponding parameter Δ.

In the second part of this paper, we consider the data structure for dominance emptiness queries in the case when an update changes the relative order of two points or increments/decrements coordinates of a point by a small parameter. We show that in this case dominance emptiness queries can be answered faster than the lower bound for the fully dynamic data structure.

1 Introduction

In this paper we consider data structures that support update operations with locality restriction: an update changes properties of the data structure D only in the vicinity of some point x already stored in D. Examples of such update operations are: incrementing or decrementing an element value by a small parameter Δ, deletions, change of the relative order of two elements (swapping), or insertion of an element that is close to at least one element stored in the data structure. We show that data structures with such restrictions can perform better than the best known fully dynamic data structures. In one case we show that restriction on updates allows us to surpass the lower bound for fully dynamic data structures. Although the repertoire of supported update operations differs, all data structures described in this paper can support incrementation or decrementation of element values (resp. point coordinates) by a small parameter Δ.

J. Gudmundsson (Ed.): SWAT 2008, LNCS 5124, pp. 138–147, 2008.

In the first part of the paper, we consider data structures that instead of insertions support one or more of the following update operations:

- $\text{insert}_{\Delta}(x)$ - insert an element x, if the data structure contains an element y, such that $|x - y| \leq \Delta$. If there is no such y, the operation is undefined.
- $\text{insert}_{\Delta}(x, y)$ - given a location of an element y that is the predecessor/ successor of x in the data structure, such that $|x - y| \leq \Delta$, insert x.
- $\text{increment}_{\Delta}(x, d)$ ($\text{decrement}_{\Delta}(x, d)$) - increase (decrease) the value of element x by $d \leq \Delta$, given a location of x.

Clearly, operations $\text{increment}_{\Delta}(x, d)$ ($\text{decrement}_{\Delta}(x, d)$) can be implemented with operations $\text{insert}_{\Delta}(x, x+\Delta)$ (resp. $\text{insert}_{\Delta}(x, x-\Delta)$) and delete($x$). Hence, a data structure that supports insert_{Δ} and delete also supports operations $\text{increment}_{\Delta}$ and $\text{decrement}_{\Delta}$.

In section 2 we consider a $O(n)$ space deterministic data structure that supports predecessor queries over the universe of size U in $O(\log\log U)$ time and delete(x)[1], $\text{insert}_{\Delta}(x, y)$, $\text{increment}_{\Delta}(x, d)$, and $\text{decrement}_{\Delta}(x, d)$ in $O(1)$ amortized time for $\Delta = 2^{2^{O(\sqrt{\log\log U})}}$. The best known deterministic $O(n)$ space data structure that supports arbitrary updates is the exponential tree of Andersson and Thorup [2] with query time $O(\log\log n \cdot \log\log U)$. A semi-dynamic data structure that supports only deletions can support predecessor queries in $O(\log\log U)$ time [7]. Our data structure supports both delete and insert_{Δ} operations. According to the lower bound of Pătraşcu and Thorup [6] any static data structure that uses linear or pseudo-linear space needs $\Omega(\log\log U)$ time to answer a predecessor query. Thus we demonstrate that the predecessor queries in a universe of size U can be answered in our scenario as efficiently as in the static case.

In section 3 we present a deterministic dictionary data structure that supports membership queries in $O(\log\log n)$ time and operations $\text{insert}_{\Delta}(x, y)$, $\text{increment}_{\Delta}(x, d)$, $\text{decrement}_{\Delta}(x, d)$, and delete($x$) in $O(1)$ amortized time, where $\Delta = 2^{2^{O(\sqrt{\log\log n})}}$. This result is to be compared with the fully-dynamic deterministic dictionary of Pagh [5] that supports membership queries in $O((\log\log n)^2/\log\log\log n)$ time and insertions and deletions in $O((\log n \log\log n)^2)$ time.

In section 4 we describe a priority queue that supports operations delete(x), $\text{increment}_{\Delta}(x)$, $\text{decrement}_{\Delta}(x)$ in $O(1)$ amortized time, FindMin in $O(1)$ time and $\text{insert}_{\Delta}(x)$ in $O(\log\log n)$ time for $\Delta = \log^{O(1)} U$.

In section 5 we consider a different scenario. We describe a data structure for dominance emptiness and one-reporting queries: a set of planar points P is stored in a data structure, so that given a query $Q = (-\infty, a] \times (-\infty, b]$ we can determine whether $P \cap Q \neq \emptyset$ and report one point from $P \cap Q$ if $P \cap Q \neq \emptyset$. We present a data structure that supports such queries in $O(\log\log n \cdot \log\log U)$ time and supports update operations x-move and y-move in $O(1)$ time: If P_x (P_y)

[1] Throughout this paper we assume that the location of the element x in the data structure during operation delete(x) is known.

is the set of x-coordinates (y-coordinates) of all points, then the operation x-move (y-move) increases or decreases the x-coordinate (y-coordinate) of a point p so that the ranks of at most two points in P_x (P_y) are changed. We also describe a data structure that increases or decreases coordinates of a point p by a parameter $d < \Delta$ and supports queries in $O(\log\log n \cdot \log\log U)$ time and updates in $O(\Delta)$ time. As follows from the lower bound for orthogonal range reporting queries proved in [1] and the reduction of orthogonal range reporting queries to dominance queries, a fully dynamic data structure with poly-logarithmic update time needs $O(\log n / \log\log n)$ time to answer a dominance emptiness query.

The model of computation used in this paper is the unit-cost word RAM model. We assume that word size $w = \log U$ where U is the size of the universe. Unless stated otherwise, the space usage of data structures is measured in words.

2 Predecessor Queries in $O(\log\log U)$ Time

The predecessor problem is to store a set A in the data structure, so that for an arbitrary x the predecessor of x in A, $\text{pred}(x, A) = \max\{e \in A \cup \{-\infty\} | e \leq x\}$, can be found. In this section we describe a linear space data structure that supports predecessor queries in $O(\log\log U)$ time and $\text{insert}_{\Delta}(x, y)$, $\text{delete}(x)$ in $O(1)$ time. As mentioned in the Introduction, our data structure also supports increments and decrements of elements by Δ.

We start by describing a data structure that supports updates in $O(\log U)$ time for the case when $\Delta = 1$. Later in this section we will show how our method can be extended to larger values of Δ.

Theorem 1. *There exists a linear space data structure that supports predecessor queries in $O(\log\log U)$ time,* $\text{insert}_{\Delta}(x, y)$ *and* $\text{delete}(x)$ *in $O(\log U)$ amortized time, where U is the size of the universe and $\Delta = O(1)$.*

Proof: The elements of the set A correspond to the leaves of a binary trie with root r. Let b denote the height of the trie; hence, b is the key size – we need b bits to specify an arbitrary element of the universe. We use the standard VEB approach [4] to reduce the key size by a constant factor with one dictionary look-up.

We denote by $h(v)$ the height of a node v; we denote by $rng(v)$ the difference between the maximal and minimal elements that can be stored in v. If A contains more than 2 elements, the predecessor data structure consists of the following components: a dictionary M_r that contains all non-empty trie nodes on level $3h(r)/4$ (node levels are counted from the bottom), a recursively defined data structure D' that contains all non-empty nodes on level $3h(r)/4$. Besides that, for each such non-empty node v all leaf descendants of v are also stored in a recursively defined data structure D_v. To find the predecessor of an integer x, we check whether the node v that must contain x is stored in M_v. If v does not belong to M_v, then v is empty. In this case, we find the largest non-empty node u that precedes v with help of D', and the predecessor of x is the largest leaf descendant of u. Otherwise, if v is not empty, we search for a predecessor of x among all leaf descendants of the node v. Any element stored in D' can be specified with $b/4$ bits. All elements stored in a data structure D_v for some v have the same prefixes of length $b/4$; hence, it suffices to store the last $3b/4$ bits in D_v. Thus, one dictionary look-up reduces the key size at least by a factor 4/3. Hence after $O(\log\log U)$ look-ups the key size is reduced to a constant and the query can be answered in constant time. It remains to show how to implement the dictionary M_v for each node v so that queries are answered in constant time and fast update operations are supported.

Let m be the number of elements stored in a trie node v. We distinguish between two cases. If $m^4 \geq rng(v)$, M_v is a dictionary in the universe of size $O(m)$ and can be implemented as a bit vector. If $m^4 \leq rng(v)$, then we construct a static dictionary S_v and rebuild it after a series of m^3 updates. A static dictionary S_v can be constructed in $O(m^{2+\varepsilon})$ time for any $\varepsilon > 0$ [2] S_v contains all descendants v_i of v on level $3h(v)/4$, such that $v_i \neq \emptyset$, or $v_{i-1} \neq \emptyset$, or $v_{i+1} \neq \emptyset$. The number of nodes in S_v does not exceed $3m$, and S_v can be constructed in $O(m^{2+\varepsilon})$ time for any $\varepsilon > 0$. We also maintain the count of elements stored in v_i, count(v_i), for every $v_i \in S_v$. Obviously, if $v_i \in S_v$ and count$(v_i) > 0$, then $v_i \in M_v$.

Consider consecutive nodes $v_{i-2}, v_{i-1}, v_i, v_{i+1}, v_{i+2}$ on level $3h(v)/4$. Suppose that nodes v_{i-1}, v_i, v_{i+1} are empty when S_v is constructed. For any element x inserted into the data structure during the sequence of m^3 operations, there exists an element y, such that y was stored in the data structure when S_v was constructed and $|x-y| \geq m^3$. Therefore, since $rng(v_{i+1}) > m^3$ and $rng(v_{i-1}) > m^3$, the node v_i will remain empty after m^3 update operations. Hence, only nodes stored in S_v can be non-empty during the sequence of m^3 updates. To determine whether v_i is empty, it suffices to check whether $v_i \in S_v$ and count$(v_i) > 0$.

When some elements appear in a previously empty node v_i, we have to construct a new dictionary M_{v_i}. The dictionary for v_i must contain all non-empty descendants of v_{i+1} on level $3h(v_{i+1})/4$. During the first $2^{(3/4)h(v_i)}$ updates only the leftmost and the rightmost descendants of v_i may be non-empty. If v_i is non-empty after a sequence of $2^{(3/4)h(v_i)}$ updates, we construct the dictionary M_{v_i} (either as a bit vector or with help of S_{v_i}) in $O(2^{(3/4)h(v_i)})$ time.

The total number of trie nodes stored in the data structure for all recursion levels is $O(n \log U)$. Each element $a \in A$ belongs to $O(\log U)$ trie nodes v that are stored in $O(\log U)$ dictionaries. When an element $a \in A$ is updated, those dictionaries may also be updated. Hence, each update takes $O(\log U)$ time.

The space usage can be reduced to $O(n)$ with a standard sub-sampling technique in the same way as in the case of the van Emde Boas data structure: the set A is divided into consecutive groups of $\log U$ elements each, and the set A' contains one representative from each group. Since A' contains $n/\log U$ elements, the above described data structure for A' uses $O(n)$ space. Elements of each group are also stored in a binary tree. We can identify the group in which the predecessor is stored and find the predecessor in the group in $O(\log\log U)$ time. □

We can extend the result of Theorem 1 to support insert$_\Delta(x, y)$ for $\Delta = 2^{\rho(U)}$, where $\rho(U) = 2^{O(\sqrt{\log\log U})}$, and reduce the update time to $O(1)$. Let $E = \{e =$

$a/\delta | a \in A\}$ and $A_e = \{a \in A | a/\delta = e\}$ for $\delta = \Delta \log U$. We combine sets A_e into groups $G_1, G_2, \ldots, G_m$, such that each group G_i contains $\Theta(\log U)$ non-empty sets A_e. To be precise, each G_i contains between $\log U/2$ and $2 \log U$ sets; for any $A_e \in G_i$ and any $A_f \in G_{i+1}$, $e < f$. All elements e, such that $A_e \in G_i$, are stored in data structure $\mathcal{G}_i$. Every set A_e is stored in the exponential tree T_e of Andersson and Thorup [2], so that finger updates are supported in $O(1)$ time and queries are supported in $O((\log \log \delta)^2) = O(\log \log U)$ time. For each G_i, elements $g_i = \max\{\, e \mid A_e \in G_i \,\}$ and $g_i' = \min\{\, e \mid A_e \in G_i \,\}$ are stored in a data structure $\mathcal{E}$ implemented as described above.

To find the predecessor of some x, we identify the group G_i in which the predecessor is stored using $\mathcal{E}$, then identify the set A_e in which the predecessor is stored using $\mathcal{G}_i$. Finally, we search in A_e using T_e. We can search in $\mathcal{G}_i$ in $O(\log \log U)$ time because $\mathcal{G}_i$ contains $O(\log U)$ elements. The search in $\mathcal{E}$ and T_e also takes $O(\log \log U)$ time.

When an element e is inserted, we insert it into the corresponding set A_e. If A_e is empty, we create a new data structure T_e and insert e into the corresponding group $\mathcal{G}_i$. After $\log U/2$ insertions we check the number of sets A_e in G_i; if necessary, we split G_i into two groups and update the $\mathcal{E}$ accordingly. When an element e is deleted, we delete it from T_e and delete e from $\mathcal{G}_i$ if A_e becomes empty. After $\log U/2$ deletions we check the number of sets A_e in G_i; if necessary, we merge G_i with one of its neighbor groups, split the resulting group into two groups if it contains more than $2 \log U$ sets, and update the $\mathcal{E}$ accordingly.

The results can be summed up in the following Theorem

Theorem 2. *There exists a linear space data structure that supports predecessor queries in $O(\log \log U)$ time,* $\text{insert}_\Delta(x, y)$ *and* $\text{delete}(x)$ *in $O(1)$ amortized time, where U is the size of the universe and $\Delta = O(2^{2^{O(\sqrt{\log \log U})}})$.*

As shown in the proof of Theorem 2, we can support queries in time $O(\max(\log \log U, (\log \log \Delta)^2))$ if the value of Δ is known. We can achieve query time $O((\log \log \Delta)^2)$ even if the value of Δ is not known. Let $\Delta_i = 2^{\rho_i}$ and $\rho_i = 2^{2^i \sqrt{\log \log n}}$ for $i = 1, 2, \ldots, (\log \log \log U)/2$. Let $\Delta_e = \max\{\, |e - a| \mid a \in A \,\}$. We write $\Delta_{\max} = \max(\Delta_{x_1}, \Delta_{x_2}, \ldots, \Delta_{x_k})$, where $x_1, \ldots, x_k$ is the sequence of elements inserted into the data structure so far. For each value of Δ_i we construct the data structure D_i that supports queries in $O((\log \log \Delta_i)^2)$ time. We assume that with each element stored in data structure D_i we store a pointer to the same element stored in data structure D_{i+1}. Let $j_{\min}$ be an index, such that $\Delta_i \geq \Delta_{\max}$ for $i \geq j_{\min}$. Predecessor searches are performed with help of data structure $D_{j_{\min}}$ in $O((\log \log(\Delta_{j_{\min}}))^2) = O((\log \log \Delta_{\max})^2)$ time. We maintain only data structures D_i, $i \geq j_{\min}$. When a new element x is inserted, we can identify $\text{pred}(x, A)$ and compute Δ_x in $O((\log \log \Delta_{\max})^2)$ time using data structure $D_{j_{\min}}$. We can find the minimal index j, such that $\Delta_j \geq \Delta_x$, in $O(\log \log \log U)$ time. Then, we set $j_{\min} = \max(j_{\min}, j)$ and insert x into data structures D_i, $i \geq j_{\min}$, in $O(\log \log \log U)$ time. When an element is deleted, we delete it from all data structures D_i, $i \geq j_{\min}$.

We have obtained a data structure $\mathcal{D}$ that supports queries in time $O(\max(\log\log U, (\log\log \Delta_{\max})^2))$, but uses $O(n \log\log\log U)$ space and requires $O(\log\log\log U)$ time for update operations. Again, we can reduce the space usage to $O(n)$ with standard sub-sampling techniques. We can reduce the update time with the same method as in Theorem 2.

Theorem 3. *There exists a linear space data structure that supports predecessor queries in* $O(\max(\log\log U, (\log\log \Delta_{\max})^2))$ *time,* $\text{insert}_{\Delta}(x, y)$ *and* $\text{delete}(x)$ *in* $O(1)$ *amortized time. Here* $\Delta_{\max} = \max(\Delta_{x_1}, \Delta_{x_2}, \ldots, \Delta_{x_k})$*, where* $x_1, \ldots, x_k$ *is the sequence of elements inserted into the data structure so far.*

If the universe size $U = n^{O(1)}$, then the data structure of Theorem 3 supports predecessor queries in $O(\log\log n)$ time as long as $\Delta_{\max} = O(2^{2^{\sqrt{\log\log n}}})$.

2.1 Space-Efficient Implementation

We can implement the predecessor data structure so that is uses $O(n \log \frac{U+n}{n})$ bits without changing the query and update times. This space usage is within the constant factor of the information-theoretic lower bound. We use the following result from [3]

Lemma 1. *Elements of a set* $S \subset [1, U]$, $U = n^{O(1)}$, *can be stored in a list* L *in sorted order, so that* L *uses* $O((|S| \log \frac{U+|S|}{|S|})/\log U)$ *words of* $\log U$ *bits. Each element stored in a word of* L *can be accessed in* $O(1)$ *time. Each word of* L *contains* $O(\log U)$ *elements of* S. *For every word* W *of* L, *we can find the number of elements stored in* W *and extract the* k*-th element from* W *in* $O(1)$ *time. Given a pointer to* $\text{pred}(x, L)$, *we insert* x *into* L *in* $O(1)$ *time. We can delete an element from* L *in* $O(1)$ *time.*

All elements of the set A are stored in L implemented according to Lemma 1. For a word W of L let $\min_w$ denote the minimum element stored in W. The data structure D contains elements $\min_W$ for all words W of L; D is implemented as described in section 2. Since the number of elements in D equals to the number of words in L, D uses the same space as L (up to a constant factor).

To find the predecessor of some e, we first find $m = \mathit{pred}(e, D)$. Then, we search in the word W such that $\min_W = m$ and find $\text{pred}(e, A)$ in $O(\log\log U)$ time using binary search. To delete an element x, we delete it from L; if $\min_W$ for some W is deleted, we delete $\min_w$ from D and insert the new minimal element of W into D. If a word that contained x becomes empty, we delete the corresponding element from D. When a new element x is inserted, we insert it into L. If x is a minimal element in some word W or if W must be split into two words, then D is updated accordingly.

3 Deterministic Dictionaries with $O(\log\log n)$ Query Time

Our approach is very similar to the approach in section 2. Elements of the set A are associated with the leaves of the binary trie. If the size of the universe does not exceed n^3, we can answer predecessor queries for A in $O(\log\log n)$ time with the data structure from section 2. Clearly, the answer to a predecessor query yields an answer to a membership query because $x \in A \Leftrightarrow \text{pred}(x, A) = x$. If the universe size exceeds n^3, the height of the trie exceeds $3\log n$. In this case we store the static dictionary S that contains all nodes v_i on level $3\log n$, such that $v_i \neq \emptyset$, or $v_{i-1} \neq \emptyset$, or $v_{i+1} \neq \emptyset$. In the same way as in section 2, S is re-built after a sequence of n^3 updates. Using S and counts of elements stored in each $v_i \in S$, we can determine whether a node v_k on level $3\log n$ is empty in $O(1)$ time. For every non-empty node v_i, a data structure D_{v_i} that contains all leaf descendants of v_i is maintained. Again, D_{v_i} can be implemented with help of the predecessor data structure from section 2: since all elements of D_{v_i} belong to the universe of size n^3, predecessor queries can be answered in $O(\log\log n)$ time. To determine whether x belongs to the set A, it suffices to check whether $x' \in S$ and $x'' \in D_{x'}$, where x'' is the suffix of x that consists of the last $3\log n$ bits of x, and x' is the prefix of x that is obtained from x by shifting it $3\log n$ bits to the right.

The above data structure supports the operation $\text{insert}_\Delta(x, y)$ for $\Delta = 1$. We can extend our result for the case $\Delta = O(2^{2^{\sqrt{\log\log n}}})$ using the same method as in the previous section.

Theorem 4. *There exists a linear space data structure that supports membership queries in $O(\log\log n)$ time and operations* $\text{insert}_\Delta(x, y)$ *and* $\text{delete}(x)$ *in $O(1)$ amortized time for $\Delta = O(2^{2^{\sqrt{\log\log n}}})$.*

4 Priority Queues

The same approach also leads to a priority queue that supports operation insert_Δ in $O(\log\log n)$ time and operations delete and FindMin in $O(1)$ time for $\Delta = \log^{O(1)} U$. Let $\delta = \Delta \log n$, $A' = \{\, a/\delta \mid a \in A \,\}$ and $A_e = \{\, a \in A \mid a/\delta = e \,\}$. We store the elements of A' in a sorted doubly-linked list $\mathcal{L}$ and in a dictionary $\mathcal{D}$ implemented as described in Theorem 4. Elements of each set A_e are stored in data structure D_e that will be described below.

When a new element x is inserted, we check whether $x' \in \mathcal{D}$ or $(x'-1) \in \mathcal{D}$, or $(x'+1) \in \mathcal{D}$, where $x' = x/\delta$. If $x' \notin \mathcal{D}$, we insert x' into $\mathcal{D}$ in $O(1)$ time, create the data structure $D_{x'}$, put x into $D_{x'}$, and insert x' into $\mathcal{L}$. Observe that when x' is inserted, $x'-1$ or $x'+1$ are already stored in $\mathcal{L}$, therefore insertion of x' into $\mathcal{L}$ takes $O(1)$ time. If $x' \in \mathcal{D}$, we put x into $D_{x'}$. When an element x is deleted, we delete it from $D_{x'}$. If $D_{x'}$ becomes empty, the list $\mathcal{L}$ and the dictionary $\mathcal{D}$ are updated in $O(1)$ time. We can find the minimal element with help of the data structure D_m, where m is the first element in the list $\mathcal{L}$. Given the location of some element x in the data structure, we can insert $x + \Delta$ or $x - \Delta$ in constant time. Hence, operations $\text{increment}_\Delta(x, d)$ ad $\text{decrement}_\Delta(x, d)$ cn be supported in constant time.

Each data structure D_e can be implemented with help of tries with node degree $\log U$. Every trie node can be stored in one word of memory; for every node we maintain the index of the lowest non-zero bit. When an element e stored in the leaf node l is deleted, we set the bit of l corresponding to e to zero. If e was the smallest element in l, we identify the new lowest non-zero bit. If e was the only element stored in l, we delete l and update the parent of node l in the same way. Since the height of the trie is $O(1)$, we can update the trie in $O(1)$ time. Insertions of new elements into D_e can be processed in a similar way.

Theorem 5. *There exists a linear space data structure that supports operations* FindMin *in $O(1)$ time,* delete(x), insert$_\Delta(x, y)$ *in $O(1)$ amortized time, and operation* insert$_\Delta(x)$ *in $O(\log\log n)$ time for $\Delta = \log^{O(1)} U$.*

5 Dominance Queries

In the orthogonal range reporting problem the set of points P must be stored efficiently, so that for an arbitrary axis-parallel query rectangle Q all points from $P \cap Q$ can be reported. Emptiness queries and one-reporting queries are special cases of the range reporting queries. In the case of emptiness queries, we must determine whether $P \cap Q \neq \emptyset$; in the case of one-reporting queries, if $P \cap Q \neq \emptyset$ an *arbitrary* point of $P \cap Q$ must be output. We present the data structure that supports emptiness and one-reporting dominance queries on a $U \times U$ grid: the query range Q is a product of two half-open intervals and all point coordinates are positive integers bounded by U.

In our scenario, the points in the set P can be updated with x-move and y-move operations. Let P_x (P_y) be the set of projections of points from the set P on the x-axis (y-axis). For a set S, we define $\text{prev}(a, S) = \max\{\, e \in S \mid e < a \,\}$ and $\text{fol}(a, S) = \min\{\, e \in S \mid e > a \,\}$. The operation x-move changes the x-coordinate of the given point p, so that the rank of $p.x$ in P_x is incremented or decremented by 1. Consider points q, q', r, and r', such that $q.x = \text{fol}(p.x, P_x)$, $q'.x = \text{fol}(q.x, P_x)$, and $r.x = \text{prev}(p.x, P_x)$, $r'.x = \text{prev}(r.x, P_x)$. The operation x-move increases or decreases the x-coordinates of p, so that $q.x < p.x < q'.x$ or $r'.x < p.x < r.x$. In the former case, we say that p is x-moved behind q; in the latter case, we say that p is x-moved before r. The operation y-move changes the y-coordinate of the given point p, so that the rank of $p.y$ in P_y is incremented or decremented by 1. Consider points u, u', v, and v', such that $u.y = \text{fol}(p.y, P_y)$, $u'.y = \text{fol}(u.y, P_y)$, and $v.y = \text{prev}(p.y, P_y)$, $v'.y = \text{prev}(v.y, P_y)$. The operation y-move increases or decreases the y-coordinates of p, so that $u.y < p < u'.y$ or $v'.y < p.y < v.y$. In the former case, we say that p is y-moved above u; in the latter case, we say that p is y-moved below v.

We demonstrate that in this scenario the dominance one-reporting queries can be answered faster than the lower bound for the fully dynamic scenario [1]. We denote by $\min_y(a) = \min\{\, y \mid (a, y) \in P \,\}$, i.e., $\min_y(a)$ is the minimum y-coordinate of a point $p \in P$ with x-coordinate a. Let $\text{next}(a) = \min\{\, b > a \mid \min_y(b) < \min_y(a) \,\}$, i.e. next$(a)$ is the minimum x-coordinate of a point p whose x-coordinate is greater than a and y-coordinate is smaller than $\min_y(a)$. If $\min_y(a) \leq \min_y(b)$ for all $b > a$, then we assume that $\text{next}(a) = +\infty$. The set L contains elements $x_1 = 1$, $x_i = \text{next}(x_{i-1})$ for all i, such that $\text{next}(x_{i-1}) \neq +\infty$.

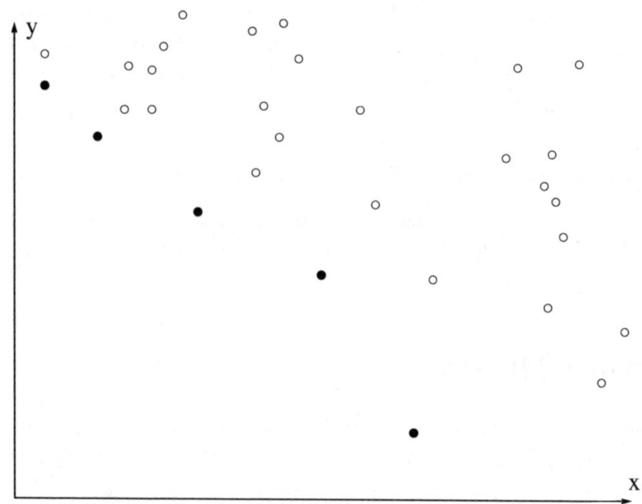

Fig. 1. Example of the data structure: x-coordinates of bold points are stored in array L

The y-coordinates of points with x-coordinates in L decrease monotonously. The minimum y-coordinate of a point whose x-coordinate is at most a equals to $\min_y(m)$ for $m = \text{pred}(a, L)$: for all points p with $p.x < m$, $\min_y(p.x) > \min_y(m)$; for all points p with $p.x \in [m+1, a]$ $p.y \geq \min_y(m)$ because otherwise $p.x$ for at least one such point p would be stored in L. Consider a query $Q = (-\infty, a] \times (-\infty, b]$. $Q \cap P$ is not empty if and only if $\min_y(\text{pred}(b, L)) \leq b$. If $Q \cap P \neq \emptyset$, then $P \cap Q$ contains the point p with $p.x = \text{pred}(b, L)$ and $p.y = \min_y(\text{pred}(b, L))$. Thus a one-reporting query is reduced to answering one predecessor query on L.

Now we show how the set L can be maintained under x-moves and y-moves, so that predecessor queries on L can be answered efficiently. For simplicity we say that a point p belongs to L if $p.x$ belongs to L and $p.y = \min_y(p.x)$. We also say that $p = \text{pred}(a, L)$ if $p.x = \text{pred}(a, L)$ and $p.y = \min_y(p.x)$. When a point p from L is x-moved behind a point r from $P \setminus L$, we check whether $r.y < s.y$, where $s = \text{pred}(r.x, L)$. If $r.y < s.y$, then $r.x$ is inserted into L. When p is x-moved behind a point r from L, we remove p from L. If p is x-moved before some point $q \in (P \setminus L)$, then $p.y < q.y$ and no modifications of L are necessary. If p is moved before $q \in L$, we remove q from L. If a point $u \in (P \setminus L)$ is x-moved before some $v \in L$ and $u.y < w.y$ for $w = \text{pred}(v, L)$, then u is inserted into L. If $p \in L$ is y-moved above some point $r \in (P \setminus L)$ and $p.x > r.x$, then r is inserted into L and p is removed from L. If p is y-moved above some $r \in L$, then p is removed from L. If p is y-moved below some $q \in (P \setminus L)$, then $q.x > p.x$ and no modifications of L are necessary. If p is y-moved below some $q \in L$, then q is removed from L. If $u \in (P \setminus L)$ is moved below some $v \in L$ and $u.x < v.x$, then u is inserted into L and v is removed from L.

Predecessor queries on set L can be supported in $O((\log\log U \cdot \log\log n))$ time and finger updates on L can be supported in $O(1)$ time with help of exponential search trees [2]. When a new element p is inserted into or removed from L, we always know the location of p (resp. the location of the predecessor of p) in L. Hence, all updates on L are supported in $O(1)$ time.

Thus we obtain the following result:

Theorem 6. *There exists a linear space data structure that supports dominance emptiness queries and dominance one-reporting queries on $U \times U$ grid in $O(\log\log U \cdot \log\log n)$ time and operations x-move and y-move in $O(1)$ time.*

We can apply the result of Theorem 6 to the scenario when an update operation may change the values of point coordinates at most by a additive parameter Δ. If coordinates of a point p are changed by some $d < \Delta$, then p is x-moved and y-moved at most 2Δ times. Hence, we can update the data structure described above in $O(\Delta)$ time.

Corollary 1. *There exists a linear space data structure that supports dominance emptiness queries and dominance one-reporting queries on $U \times U$ grid in $O(\log\log U \cdot \log\log n)$ time and operations* $\text{increment}_\Delta(p)$, $\text{decrement}_\Delta(p)$ *in $O(\Delta)$ time, where the operation* $\text{increment}_\Delta(p)$ *(resp.* $\text{decrement}_\Delta(p)$*) increments (decrements) coordinates of a point p by some d, for $0 \leq d \leq \Delta$.*

Acknowledgment

The author would like to thank J. Ian Munro for a stimulating discussion and comments on a preliminary version of this paper.

References

1. Alstrup, S., Husfeldt, T., Rauhe, T.: Marked Ancestor Problems. In: Proc. FOCS 1998, pp. 534–544 (1998)
2. Andersson, A., Thorup, M.: Dynamic Ordered Sets with Exponential Search Trees. J. ACM 54(3), 13 (2007)
3. Blandford, D.K., Blelloch, G.E.: Compact Representations of Ordered Sets. In: Proc. SODA 2004, pp. 11–19 (2004)
4. van Emde Boas, P.: Preserving Order in a Forest in Less Than Logarithmic Time and Linear Space. Inf. Process. Lett. 6(3), 80–82 (1977)
5. Pagh, R.: A Trade-Off for Worst-Case Efficient Dictionaries. Nord. J. Comput. 7(3), 151–163 (2000)
6. Pătraşcu, M., Thorup, M.: Time-Space Trade-Offs for Predecessor Search. In: Proc. STOC 2006, pp. 232–240 (2006)
7. Pătraşcu, M.: Personal Communication
8. Willard, D.E.: Log-Logarithmic Worst-Case Range Queries are Possible in Space Theta(N). Inf. Process. Lett. 17(2), 81–84 (1983)

On the Redundancy of Succinct Data Structures

Alexander Golynski[1], Rajeev Raman[2], and S. Srinivasa Rao[3]

[1] Google Inc.*
[2] Department of Computer Science, University of Leicester, UK
[3] MADALGO**, Department of Computer Science, University of Aarhus, Denmark

Abstract. The *redundancy* of a succinct data structure is the difference between the space it uses and the appropriate information-theoretic lower bound. We consider the problem of representing binary sequences and strings succinctly using small redundancy. We improve the redundancy required to support the important operations of rank and select efficiently for binary sequences and for strings over small alphabets. We also show optimal *density-sensitive* upper and lower bounds on the redundancy for *systematic* encodings of binary sequences.

1 Introduction

Data structures for text indexing [13, 14, 6] and representing semi-structured data [9, 4] often require the very space-efficient representation of a sequence S of length m from the alphabet $\Sigma = \{0, \ldots, \sigma - 1\}$, for some $\sigma \leq m$. There are several measures for the minimum space needed to represent such a sequence:

- The *succinct* bound: since there are σ^m such sequences, a given sequence requires at least $\lceil m \lg \sigma \rceil$ bits[1] in the worst case.
- If the multiplicities of the symbols are known to be $n_0, \ldots, n_{\sigma-1}$, then there are $\binom{m}{n_0, \ldots, n_{\sigma-1}}$ such sequences, so a given sequence requires at least $B(m, n_0, \ldots, n_{\sigma-1}) = \left\lceil \lg \binom{m}{n_0, \ldots, n_{\sigma-1}} \right\rceil$ bits in the worst case.
- The *zeroth-order empirical entropy* of the sequence, $H_0(S) \stackrel{\text{def}}{=} -\sum_{i=0}^{\sigma-1} p_i \lg p_i$, where $p_i = n_i/m$, and $0 \lg 0 = 0$. Fix an integer $k \geq 1$ and a given sequence w of k symbols from Σ. Consider the multiset of symbols x that follow an occurrence of w in S (i.e. the multiset $\{x | wx$ is a contiguous subsequence of $S\}$), and let w_S be this multiset viewed (arbitrarily) as a sequence. Then, $H_k(S) = (\sum_{w \in \Sigma^k} |w_S| H_0(w_S))/m$ is the *k-th order empirical entropy* of S. $mH_k(S)$ is a lower bound on how well S will be compressed by a k-th order compressor.

It can be shown that $H_{k+1}(S) \leq H_k(S)$ and $mH_0(S) \sim B(m, n_0, \ldots, n_{\sigma-1}) \leq m \lg \sigma$. In other words, these space measures get increasingly more stringent.

For the aforementioned applications, it is not enough just to store S. One would like to support the following operations on S, for any $x \in \Sigma$:

---

* Work done when AG was at the Cheriton School of Computer Science, U. Waterloo.

** Center for Massive Data Algorithmics, a Center of the Danish National Research Foundation.

[1] lg denotes the logarithm base 2.

J. Gudmundsson (Ed.): SWAT 2008, LNCS 5124, pp. 148–159, 2008.

- $\mathsf{rank}_x(S, i)$ returns the number of occurrences of x in the prefix $S[1..i]$.
- $\mathsf{select}_x(S, i)$ returns the position of the ith occurrence of x in S.

Our focus will be on data structures that support these operations in *constant* time on the RAM model with word size $O(\lg m)$ bits. In order to support such operations, it appears to be necessary to use additional space, beyond the appropriate bound for storing the sequence itself. This additional space is termed the *redundancy* of the data structure, and this paper focusses on the redundancy required to support the rank and select operations. The redundancy of a data structure is a quantity of both theoretical and practical importance:

- for a string from a given family, the measures B and mH_k can be very small, in which case the redundancy can be the dominant part of the space usage. One often needs to represent sequences that are "highly compressible" by construction, such as binary sequences with few 1s [13, 9].
- if we wish to store S within the succinct bound for S, there are data structures whose redundancy is $o(m \lg \sigma)$, and hence with overall space usage $m \lg \sigma + o(m \lg \sigma)$. While this acceptable from an asymptotic viewpoint, the functions hidden in the $o()$ often grow only slightly slower than $m \lg \sigma$, and can dominate the "higher-order" term for practical values of m and σ.

The recent development of matching upper and lower bounds (particularly for *systematic* data structures [8, 10, 11, 12, 16]) highlights the fundamental nature of the trade-off between redundancy and speed of operations. We discuss three important sub-cases of the problem, *binary sequences*, *binary sequences (systematic encodings)* and *sequences over small alphabets.*

Binary Sequences. If $\Sigma = \{0, 1\}$, then a data structure that supports rank and select is called a *fully indexable dictionary (FID)* [21]. FIDs have found numerous uses in the literature. FIDs are quite powerful: if S is viewed as the characteristic vector of a subset of $\{0, \ldots, m-1\}$, then the predecessor in S of a given integer x can be found in $O(1)$ time as $\mathsf{select}_1(S, \mathsf{rank}_1(S, x))$. For binary sequences, we let $n = n_1$ be the number of 1s and assume, wlog, that $n \leq m/2$. We simplify notation by letting $B(m, n) = B(m, n_0, n_1)$. We also note that $\mathsf{rank}_0(S, i) + \mathsf{rank}_1(S, i) = i$ and refer to both as $\mathsf{rank}(S, i)$ if this is otherwise immaterial.

Our results are summarised in Table 1. The first concerns the redundancy of FIDs that aim for the $B()$ space bound. A redundancy of $O(m \lg \lg m / \lg m)$ was previously achieved by [21], and subsequently improved by a factor of $\Theta(\lg m)$ in [12]. However, this redundancy remains fixed independent of n and is a limiting factor in some applications [12]. If only rank is to be supported, $o(n)$ redundancy

Table 1. Redundancy of $O(1)$-time FIDs

Space for S	Old redundancy	New redundancy
$B(m, n)$	$O(n(\lg \lg n)^2 / \lg n)$ [19, 12]	$O(n \lg \lg n \lg(m/n)/(\lg n)^2)$
$B(m, n)$	$O(m \lg \lg m / (\lg m)^2)$ [12]	
mH_k	$O(m(k + \lg \lg m)/ \lg m)$ [7, 22]	$O(mk/ \lg m)$

was achieved by Pagh [19]. In [12] the same redundancy as Pagh's was achieved while supporting the full FID functionality. Here we improve Pagh's redundancy by a factor of $\Theta(\lg n/\lg(m/n))$. Our new data structure, just like [19, 12], only supports $O(1)$-time operations if $m = n(\lg n)^{O(1)}$ — the predecessor lower bound of [20] rules out $O(1)$-time operations for smaller values of n — so the improvement in redundancy is $\Omega(\lg n/\lg\lg n)$ in the most interesting range of n. Noting that $\lg m = \Theta(\lg n)$ when $m = n(\lg n)^{O(1)}$, the ratio of our redundancy to that of [12] is $O((\lg(m/n))/(m/n))$, which is always $O(1)$. Thus, we subsume all existing results for FIDs that target the $B()$ space measure.

Finally, we improve the redundancy for FIDs where the input sequence is compressed to $mH_k(S)$. Here the improvement in redundancy is smaller — it is, e.g., better by a factor of $\Theta(\lg\lg m)$ if $k = O(1)$. Our ideas also provide a similar improvement to the problem originally considered by [7, 22], namely, that of storing a binary sequence in $mH_k(S)$ space whilst allowing the constant-time retrieval of a contiguous substring of $O(\lg m)$ bits. Both results make use of *informative encoding* [12], and the first result also uses a new, instance-specific, way to partition the input sequence S.

Binary Sequences (Systematic Encodings). For binary sequences S, a *systematic* encoding stores S as given using m bits. This is augmented with an *index*, or a data structure that contains pre-computed information specific to S, that aims to support these operations rapidly; the redundancy is simply the size of the index. A number of recent results give lower bounds on the redundancy of systematic encodings [8, 16, 10]. In addition to making lower bounds a little easier to prove, systematic encodings allow the representation of S to be de-coupled from the set of operations that are being supported, which has many advantages [1].

We now consider the redundancy required for supporting rank and select_1 on binary sequences, but restricted to systematic encodings. It has been shown [16, 10] that $\Omega(m \lg\lg m/\lg m)$ redundancy is needed to support FID operations in $O(1)$ time, matched by upper bounds in [10, 21]. Hence, the redundancy of systematic FIDs appears to be a solved question. The lower and upper bounds, however, are not sensitive to n, the number of 1s in the sequence. For example, when $n = 1$, it is easy to see that redundancy of $O(\lg m)$ bits suffices to support all operations in $O(1)$ time. As already noted, one often has to support rank/select_1 operations on binary sequences that are constructed to have few 1s. Thus, it is interesting to study the redundancy required to support rank/select_1 operations as a function of both m and n. Our new results are shown in Table 2. The lower bounds are shown in the *bit-probe* model, which only counts the number t, of bits of the data structure read by an algorithm to answer the queries. In fact, the lower bound is a complete trade-off that specifies the minimum redundancy required for any values of m, n and t. The matching upper bounds hold in the RAM model as before, and $O(1)$ time is achieved only if, additionally, $m/n = (\lg n)^{O(1)}$. Also note that we only support rank and select_1. The lower bound uses the general *choices tree* framework of [10]. The lower bounds of [10] were optimal only for the case $n = \Theta(m)$, and Miltersen's [16] work implicitly contains an optimal lower bound for the case $n = \Theta(m/\lg m)$.

Table 2. Redundancy of systematic encodings of $O(1)$-time FIDs

Worst case (over all n)	$\Theta(m \lg\lg m/\lg m)$ [16, 10]
Density-sensitive [old]	$\Omega(n \lg\lg m/\lg m)$ [10]
Density-sensitive [new]	$\begin{cases} \Theta\left(\frac{m}{\lg m} \lg\left(\frac{n \lg m}{m}\right)\right), & \text{if } n = \omega(m/\lg m), \text{ and} \\ \Theta\left(n\left(1 + \lg\left(\frac{m}{n \lg m}\right)\right)\right) & \text{if } n = O(m/\lg m). \end{cases}$

Sequences over Small Alphabets. The case $\sigma = (\lg m)^{O(1)}$ is of interest in practice: e.g. XML documents have small alphabets relative to their size [3, 4]. It is also theoretically significant, as rank/select on alphabets of this size (but no larger alphabets) can be supported in $O(1)$ time with reasonable redundancy [5]. The representation of Ferragina *et al.* [5] represents S using $mH_0(S)$ bits, and has a redundancy of $O(m/(\lg m)^{1-\varepsilon})$ bits, for any constant $0 < \varepsilon < 1$. We improve the redundancy to $O(m/(\lg m)^{2-\varepsilon})$, for any constant $0 < \varepsilon < 1$.

2 Preliminaries

We describe some notation and conventions that we will use throughout the paper. For any integer $x \geq 1$, we let $[x]$ refer to the set $\{0, 1, \ldots, x-1\}$. For a given input sequence S, we will often partition S into fixed-length substrings, which we call *blocks*; S may also be divided into variable-length *chunks*. We refer to the length of a chunk (in bits) as its *length*, and the number of 1s it contains as its *weight*. We use the following weak version of [12, Theorem 2]:[2]

Theorem 1. *When $m = n(\lg n)^{O(1)}$, we can store a binary sequence using $O(B(m,n))$ bits and support FID operations in $O(1)$ time.*

We use *informative encodings* of an object in our upper bounds. Informative encodings encode a given object (for our purposes, a block of symbols) using space very close to a desired space measure, but allow some properties of the object to be deduced by looking at a few consecutive bits of the encoding.

Lemma 1 ([12]). *Given a finite universe U, and a partition of U into t sets $C_1, \ldots, C_t$. For any probability distribution $\mathcal{P}$ over U such that for all i, and all $x, y \in C_i$, $\mathcal{P}(x) = \mathcal{P}(y) > 0$, we can encode any $x \in U$ such that the encoding takes at most $\lceil \lg(1/\mathcal{P}(x)) \rceil + 2$ bits, and comprises the concatenation of two parts:*

- *a prefix code of $\lg t + O(1)$ bits encoding the index i such that $x \in C_i$, and*
- *an integer in the range $[|C_i|]$ that specifies x.*

Lemma 2 ([12]). *Let $t > 0$ be an integer and let $\bar{u} = (u_1, \ldots, u_t)$ be a sequence of positive integers. Given t and $\bar{u}$, and a positive integer parameter $z < t$, one can represent any sequence of positive integers $\bar{x} = (x_1, \ldots, x_t)$, where $x_i \in [u_i]$, using $\sum_{i=1}^{t} \lg u_i + O(1 + t/z)$ bits, so that x_i can be accessed in $O(1)$ time, for*

[2] A very simple data structure proves Theorem 1; its practicality is suggested by [18].

any i, on a RAM with word size $O(t \lg z + \lg \max_i\{u_i\})$ bits, using precomputed tables of size $O(t(z^t \lg z + \lg(\sum_{i=1}^t \lg u_i)))$ bits that depend upon $\bar{u}$, t and z, but not upon $\bar{x}$.

3 Density-Sensitive Systematic Encodings

Recall that a systematic encoding of a binary sequence S stores S in raw form together with an *index* – a data structure that contains pre-computed information specific to S – to support the operations rapidly.

3.1 Upper Bounds

Lemma 3. *Provided that $m = n(\lg n)^{O(1)}$, there is an index that supports* rank *and* select_1 *on S, whose size is:*

$$\begin{cases} O\left(\frac{m}{\lg m} \lg\left(\frac{n \lg m}{m}\right)\right) & \text{, if } n = \omega(m/\lg m) \\ O\left(n\left(1 + \lg\left(\frac{m}{n \lg m}\right)\right)\right) & \text{, if } n = O(m/\lg m). \end{cases}$$

Proof. Divide the sequence into blocks of size $(\lg m)/2$ each, and let $b_i \geq 0$ denote the number of 1s in the i-th block. We represent the sequence $\mathrm{OD} = 1^{b_1}01^{b_2}0\ldots$, which has n **1**s and at most $2m/\lg m$ **0**s, as a FID using Theorem 1; the index size is easily verified to be as claimed. Using FID operations on OD, one can reduce rank and select_1 operations on S to rank and select_1 operations on an individual block, which can be performed in $O(1)$ time by table lookup using tables of size $O(m^{2/3})$ bits.

3.2 Density-Sensitive Lower Bounds

In this section, we develop new bounding techniques for binomial coefficients and show the following theorem.

Theorem 2. *The size of the index to support the operations* rank_1 *or* select_1 *on bit vectors of length m and weight n satisfies*

$$r = \begin{cases} \Omega\left(\frac{m}{t} \lg\left(\frac{nt}{m}\right)\right) & \text{, if } \frac{nt}{m} = \omega(1) \\ \Omega(n) & \text{, if } \frac{nt}{m} = \Theta(1) \\ \Omega\left(n \lg\left(\frac{m}{nt}\right)\right) & \text{, if } \frac{nt}{m} = o(1) \end{cases}$$

Golynski [10] showed that $r = \Omega((n/t)\lg t)$ for both rank_1 and select_1. This bound is tight only in the case of constant density bit vectors, i.e. when $n = \Theta(m)$. For sparse bit vectors, e.g. when $n < m/t$, the bound of [10] is smaller than optimal by almost a factor of t.

In this section, we refine the techniques used in [10] and show tight bounds on the index size for rank and select operations in systematic encodings. We prove bounds for the rank problem, and defer the details of select to the full version. Consider γ queries $\mathcal{Q}^* = \{$"$\mathsf{rank}_1(m/\gamma)$", "$\mathsf{rank}_1(2m/\gamma)$", $\ldots\}$, where γ is a parameter which will be chosen later such that γ divides m. Let $I(B)$ denote the index of size r that is used by the rank_1 algorithm on B. Construct the decision tree T for the following procedure: first probe all the r bits stored in I, and then simulate the computation of $\mathcal{Q}^*$ queries one by one. The nodes on the first r levels of this tree are labeled by "$I[p] = ?$" for $1 \le p \le r$, and the rest of the nodes are labeled "$B[p] = ?$" for $1 \le p \le m$. The edges are labeled by 0 or 1. Let x be a leaf of T. For simplicity of presentation, we perform arbitrary extra probes, so that all the leaves of T are at the same depth $r + t\gamma$. Call B *compatible* with x if $I(B)$ corresponds to the first r edges on the path from the root to x, and the probes performed on B by our computation correspond to the rest of the edges on the path. The set of such vectors is denoted by $C(x)$. We note that the bit vectors $B_1, B_2 \in C(x)$ share some common features, e.g. $I(B_1) = I(B_2)$, the locations and the contents of the probed bits by our computation are identical, and the answers to the queries in $\mathcal{Q}^*$ on B_1 and B_2 are also identical.

The idea of the lower bound proof is as follows. Consider the set $\mathcal{H}$ of $\binom{m}{n}$ bit vectors of length m with n 1-bits. These bit vectors are distributed among the leaves in some fashion. Imagine, that we have a bound $|C(x)| \le C^*(x)$, and let C^* be the sum of $C^*(x)$ across all the leaves. Being an upper bound on the number of leaves, C^* is at least $|\mathcal{H}|$. The bounds derived in [10] are such that $C^* = 2^r D^*$, where D^* does not depend on r (intuitively, C^* is proportional to the number of leaves in T). Hence r should be at least $\lg(|\mathcal{H}|/D^*)$.

The bound $C^*(x)$ can be derived as follows. Let us split all the locations in the bit vector into γ blocks, the first block spans positions $1, 2, \ldots, m/\gamma$, the second block spans positions $m/\gamma + 1, m/\gamma + 2, \ldots, 2m/\gamma$ and so on. Let $u_i(x)$ be the number of unprobed locations in the i-th block in the bit vectors that are compatible with x, $y_i(x)$ be the number of 1-probes performed on the block (on the root to leaf path), and $v_i(x) = \mathsf{rank}_1((i+1)m/\gamma) - \mathsf{rank}_1(im/\gamma) - y_i(x)$ be the number of unprobed 1-bits in the block (their locations can be different for different $B \in C(x)$, however the number is fixed for a given leaf). From now on, we omit parameter x and use just u_i, v_i, y_i to denote these quantities, e.g. define $y := \sum_i y_i$. We have,

$$|C(x)| \le C^*(x) = \binom{u_1}{v_1}\binom{u_2}{v_2}\cdots\binom{u_\gamma}{v_\gamma}, \tag{1}$$

where $U := \sum u_i = m - t\gamma$ (since exactly $t\gamma$ positions are probed for each leaf) and $V := \sum v_i = n - y$ (since y is the total number of probed 1-bits). Let L_y be the group of leaves for which there are exactly y 1-probes. Note that $|L_y| = 2^r\binom{t\gamma}{y}$. Let x_y be the leaf in L_y that maximizes the product (1). Hence, we have

$$C^* \le 2^r \sum_{y=0}^{\min\{t\gamma, n\}} \binom{t\gamma}{y} C^*(x_y) \le n2^r X,$$

where X is the maximum of $\binom{t\gamma}{y}\binom{u_1}{v_1}\binom{u_2}{v_2}\cdots\binom{u_\gamma}{v_\gamma}$ over all possible choices of y, u_i's and v_i's, such that $t\gamma + \sum_i u_i = m$, $y + \sum_i v_i = n$, $0 \le u_i \le m/\gamma$, and

$0 \le v_i \le u_i$. The bounding methods of [10] are too crude for our purposes, so we first need to develop a better bounding techniques.

Lemma 4 (Proof omitted). *For values u and v, such that $0 < v \le u/2$, we have*

$$\frac{1}{e} < \frac{\binom{u}{v}}{\frac{1}{\sqrt{v}}\left(\frac{u}{v}\right)^v\left(\frac{u}{u-v}\right)^{u-v}} < \frac{4}{5}.$$

Let us define $u_* = \min u_i$ and $v_* = \min v_i$. In the case where v_i's are of the same order of magnitude, we can use the following lemma.

Lemma 5. *If $u_* \ge 2$ and $v_* \ge 1$, then $\prod_i \binom{u_i}{v_i} \le \binom{U}{V} 2^{-(\gamma/2)\lg v_* - 0.3\gamma + (\lg V)/2}$.*

Proof. (sketch) To bound each individual multiplier, we apply the right part of the inequality of Lemma 4. If $v_i \le u_i/2$, then

$$\prod_i \binom{u_i}{v_i} \le \prod_i \left(\frac{4}{5\sqrt{v_i}}\right)\left(\frac{u_i}{v_i}\right)^{v_i}\left(\frac{u_i}{u_i - v_i}\right)^{u_i - v_i}.$$

The case where $v_i > u_i/2$ can be reduced to the case $v_i = u_i/2$ and is omitted for brevity. Next, we apply the inequality between arithmetic and geometric means for the values

$$\underbrace{\frac{u_1}{v_1}, \ldots, \frac{u_1}{v_1}}_{v_1 \text{ times}}, \underbrace{\frac{u_2}{v_2}, \ldots, \frac{u_2}{v_2}}_{v_2 \text{ times}}, \ldots \underbrace{\frac{u_\gamma}{v_\gamma}, \ldots, \frac{u_\gamma}{v_\gamma}}_{v_\gamma \text{ times}}, \text{ and obtain}$$

$$\prod_i \left(\frac{u_i}{v_i}\right)^{v_i} \le \left(\frac{\sum_i v_i \cdot \frac{u_i}{v_i}}{\sum_i v_i}\right)^{\sum_i v_i} = \left(\frac{U}{V}\right)^V. \text{ Similarly, we obtain}$$

$$\prod_i \left(\frac{u_i}{u_i - v_i}\right)^{u_i - v_i} \le \left(\frac{\sum_i u_i}{\sum_i u_i - v_i}\right)^{\sum_i u_i - v_i} = \left(\frac{U}{U-V}\right)^{U-V}.$$

Finally, we apply the left part of the inequality of Lemma 4,

$$\prod_i \binom{u_i}{v_i} < e\sqrt{V}\binom{U}{V}\prod_i \frac{4}{5\sqrt{v_i}} \le 2^{-\gamma/2 \lg v_* - 0.3\gamma + (\lg V)/2}\binom{U}{V}.$$

Using ideas from [10, Lemma 10 and Corollary 1], we can show that

Lemma 6. *If $u_* V/U \ge 3$, then $\prod_i \binom{u_i}{v_i} \le \binom{U}{V} 2^{-(\gamma/2)\lg(u_* V/U) - 0.3\gamma + (\lg V)/2}$.*

The proof is based on the fact that $\prod_i \binom{u_i}{v_i}$ achieves maximum when v_i/u_i are roughly equal to V/U for all i (we omit the details in this extended abstract).

Density-Sensitive Rank Index. (Theorem 2 for the rank_1 operation).

Proof. (sketch) Let us define $k := m/\gamma$ to be the length of a block. We combine consecutive blocks into larger *superblocks*, such that the number of unprobed bits in the i-th superblock, u_i^*, is between k and $2k$ (except, possibly, for the last superblock). This can be done in a greedy fashion, considering blocks from left to right: we keep adding blocks to a superblock until the number of unprobed bits in it reaches k, at which point we finalize it and start a new one. We will never overshoot the value $2k$, since all u_i's are at most k. It was shown in [10] that the number of superblocks $\gamma_s = \Theta(\gamma)$, and $\prod_i \binom{u_i}{v_i} \leq \prod_i \binom{u_i^*}{v_i^*}$, where v_i^* is the number of unprobed 1-bits in the i-th superblock.

First, consider the case $tn \geq m$. Let us choose γ to be $m/(3t)$. We can apply Lemma 6 to $\binom{t\gamma}{y} \prod_i \binom{u_i^*}{v_i^*}$, since $n \min\{t\gamma, \min\{u_i^*\}\}/m = \min\{n/3, kn/m\} = \min\{n/3, 3tn/m\} \geq 3$. We obtain

$$C^* \leq n2^r \binom{t\gamma}{y} \prod_{i=1}^{\gamma_s} \binom{u_i^*}{v_i^*} \leq n2^r 2^{-(\gamma_s/2)\lg(3tn/m) - 0.3\gamma_s + (\lg n)/2} \binom{m}{n}. \text{ Hence,}$$

$$r \geq (\gamma_s/2)\lg(3tn/m) + 0.3\gamma_s - 3(\lg n)/2 = \Omega((m/t)\lg((nt)/m)).$$

If $cm < tn < m$ for some positive constant c, then pick $\gamma = n/3$. We have, $n \min\{t\gamma, k\}/m \geq \min\{cn/3, 3\} \geq 3$, and obtain

$$C^* \leq n2^r \binom{t\gamma}{y} \prod_{i=1}^{\gamma_s} \binom{u_i^*}{v_i^*} \leq n2^r 2^{-(\gamma_s/2)\lg 3 - 0.3\gamma_s + (\lg n)/2} \binom{m}{n}, \text{ and}$$

$$r \geq (\gamma_s/2)\lg 3 + 0.3\gamma_s - 3(\lg n)/2 = \Omega(n).$$

Finally, if $nt = o(m)$, then we pick $\gamma = \sqrt{nm/t}$ (calculations are omitted).

4 FID with Reduced Redundancy

This section is devoted to proving the following theorem:

Theorem 3. *When $m = n(\lg n)^{O(1)}$, we can support the FID operations in $O(1)$ time using $B(m,n) + O(n \lg\lg n \lg(m/n)/(\lg n)^2)$ bits.*

We first give a procedure that takes two positive parameters, u and b, and partitions the given bit sequence S into chunks:

> PARTITION(u, b) Initialise the current chunk to be the empty string. Append the next unread bit in S to the current chunk, and update the length ℓ and the weight w of the current chunk. End the chunk whenever either $\ell \geq b$ or $\binom{\ell}{w} \geq u$. Output the chunk, ℓ and w.

We now choose $b = m(\lg n)^2/n$, $u = m^{1/4}$, and $t = c \lg m / \lg\lg m$, for some sufficiently small $c > 0$. Scanning S from left to right, we repeatedly call PARTITION(u, b) until all of S has been read and partitioned into chunks (ignoring, for simplicity, the case where S ends inside a call to PARTITION).

Lemma 7. *The above procedure creates $g = O(n \lg(m/n)/\lg n)$ chunks.*

Proof. We call a chunk of length ℓ and weight w *complete* if $\binom{\ell}{w} \geq u$, and *incomplete* otherwise. There are at most $m/b = O(n/(\lg n)^2)$ incomplete chunks. Let k be the number of complete chunks. Using [19, Lemma 4.1]:

$$\sum_{i=1}^{k} \lg \binom{\ell_i}{w_i} \leq \lg \binom{m}{n} + k \implies k \lg m^{1/4} \leq \lg \binom{m}{n} + k \implies k = O(n \frac{\lg(m/n)}{\lg m}).$$

Denote the set of all bit strings of length ℓ and weight w that could be output by PARTITION(u, b) by $\pi_{\ell,w,u,b}$ (we omit u and b if they are clear from the context). Note that $|\pi_{\ell,w}| \leq \binom{\ell}{w}$, but π could be much smaller: e.g., consider $\pi_{6,2,10,10}$. Since $\binom{5}{2} = 10$, the sixth bit in all chunks in $\pi_{6,2,10,10}$ must be 1, and $|\pi_{6,2,10,10}| = 5$. We consider t consecutive chunks produced by the above procedure as a *superchunk*. Let $\mathcal{S}$ denote all strings that could be output as a superchunk; note that any string in $\mathcal{S}$ has length (and hence weight) at most bt. We let $\mathcal{S}_{L,W}$ denote the set of superchunks of total length L and total weight W, for any $0 \leq L, W \leq bt$ (for some L, W, e.g. $W > L$, $\mathcal{S}_{L,W}$ will be empty).

Letting $p = n/m$ and $q = 1 - p$, we define, for any $x \in \mathcal{S}$, $\Pr[x] = p^W q^{L-W}$, if $x \in \mathcal{S}_{L,W}$. The procedure for creating superchunks ensures that $\mathcal{S}$ is prefix-free, and that every infinite binary string has a prefix in $\mathcal{S}$. From this it follows that $\sum_{0 \leq L,W \leq bt} |\mathcal{S}_{L,W}| p^W q^{L-W} = 1$. We can now use Lemma 1 to show:

Lemma 8. *Given a superchunk $x \in \mathcal{S}$, viewed as t contiguous chunks, there is an encoding of this superchunk, which is a concatenation of the following parts:*

1. *a prefix code of at most $O(\lg \lg m)$ bits specifying L and W where $x \in \mathcal{S}_{L,W}$,*
2. *an encoding of x using $\lg |\mathcal{S}_{L,W}| + O(1)$ bits, itself comprising:*
   (a) a prefix code of at most $(\lg m)/2$ bits, that specifies the number of 1s in each chunk that constitutes x, and
   (b) an encoding of the positions of the 1s in each chunk.

The encoding uses at most $\lceil W \lg(1/p) + (L - W) \lg(1/q) \rceil + 4$ bits.

Proof. By Lemma 1, there is an encoding of any $x \in \mathcal{S}_{L,W}$ that takes at most $\lceil W \lg(1/p) + (L - W) \lg(1/q) \rceil + 2$ bits, where (L, W) is encoded by a prefix code of length $2 \lg(bt) + O(1) = O(\lg \lg m)$ bits (since $\mathcal{S}$ is partitioned into at most $(bt)^2$ sets). The remainder of the encoding specifies x using $\lceil \lg |\mathcal{S}_{L,W}| \rceil$ bits.

We now replace the latter by an encoding that uses $\lceil \lg |\mathcal{S}_{L,W}| \rceil + 2$ bits, but rapidly allows the sequence $(\ell_1, w_1), \ldots, (\ell_t, w_t)$ to be read, where (ℓ_i, w_i) is the length/weight of the i-th chunk. For this, we consider the universe $\mathcal{S}_{L,W}$ and postulate a uniform distribution on $\mathcal{S}_{L,W}$. We partition $\mathcal{S}_{L,W}$ into sets $\mathcal{S}_{L,W}^{(\ell_1,w_1),\ldots,(\ell_t,w_t)}$, over all combinations of non-negative integers (ℓ_i, w_i) that sum up to L and W, respectively. A string $x \in \mathcal{S}_{L,W}$ belongs to $\mathcal{S}_{L,W}^{(\ell_1,w_1),\ldots,(\ell_t,w_t)}$ exactly if its t chunks have the specified lengths and weights (note that some such sets may prove to be empty, e.g. those where $w_i > \ell_i$ for some i).

Note that $|\mathcal{S}_{L,W}^{(\ell_1,w_1),\ldots,(\ell_t,w_t)}| = \prod_{i=1}^{t} |\pi_{\ell_i,w_i}|$, and the number of sets into which $\mathcal{S}_{L,W}$ is partitioned is at most $\binom{L+t-1}{t}^2 \leq (L+t)^{2t} \leq (bt+t)^{2t} < \sqrt{m}$, if the constant c used to define t is chosen small enough. □

Finally, we group t consecutive superchunks into a *megachunk* (we ignore the last megachunk, which is immaterial). We define the length/weight of a megachunk to be the sum of the lengths/weights of all the superchunks comprising it. To represent a megachunk, we concatenate the prefix codes of all the superchunks in it to form a *megachunk summary*. The megachunk summary is still at most $(\lg m)/2$ bits for c sufficiently small. Thus, having read the megachunk summary, one can determine the upper bound on the length of the encoding of a component superchunk, as specified by Lemma 8. Assuming that superchunk encodings are always padded out to their maximum length, we can now determine the start and end of each superchunk encoding within the megachunk. A superchunk is encoded as a prefix of at most $(\lg m)/2$ bits, followed by a series of integers in the range $[|\pi_{\ell_i,w_i}|]$, for $i = 1,\ldots,t$, represented using Lemma 2.

The redundancy introduced so far is $O(1)$ bits per superchunk (Lemma 8) plus $O(\lg\lg m/\lg m)$ bits per chunk (Lemma 2), which is also $O(1)$ bits per superchunk. Summed over all $O(g/t)$ superchunks, the total space used is $n\lg(1/p) + (m-n)\lg(1/q) + O(g/t) = B(m,n) + O(n \lg\lg m \lg(m/n)/(\lg m)^2)$ bits, giving the desired redundancy. However, we still need to support rank/select.

Let $k = O(g/t^2)$ be the number of megachunks, and let z_i, o_i and l_i denote the number of 0s, the number of 1s and the length of the ith megachunk. We store the following bitstrings, represented as FIDs using Theorem 1:

1. ZD ('zeros distribution') equals $\mathbf{0}^{z_1}\mathbf{1}\ldots\mathbf{0}^{z_k}\mathbf{1}$. ZD contains k **1**s and $m-n$ **0**s, and occupies $O(k\lg(m/k)) = o(n/(\lg n)^2)$ bits.
2. OD ('ones distribution') equals $\mathbf{0}^{o_1}\mathbf{1}\ldots\mathbf{0}^{o_k}\mathbf{1}$. OD contains k **1**s and n **0**s, and occupies $O(k\lg(n/k)) = o(n/(\lg n)^2)$ bits.
3. MCL ('megachunk lengths') equals ℓ_i. MCL $= \mathbf{0}^{\ell_1}\mathbf{1}\ldots\mathbf{0}^{\ell_k}\mathbf{1}$. MCL contains k **1**s, and m **0**s, and occupies $O(k\lg(m/k)) = o(n/(\lg n)^2)$ bits.

We thereby reduce rank/select on S to rank/select on a megachunk in $O(1)$ time:

rank(i): select$_0$ on MCL finds megachunk j in which position i lies. Note that there are $(j-1)s$ **1**s before megachunk j.

select$_1$(i): select$_0$ on OD finds the megachunk j in which ith **1** lies. select$_1$(j) on MCL then gives the starting position of megachunk j, as well as the number of **1**s in megachunks $1,\ldots,j-1$.

select$_0$(i): select$_0$ on ZD finds the megachunk j in which ith **0** lies. select$_1$(j) on MCL then gives the starting position of megachunk j in X, as well as the number of **0**s in megachunks $1,\ldots,j-1$.

To support rank and select within a megachunk, we first read the megachunk summary, which stores the lengths and weights of all the constituent superchunks (as a prefix code). Using this we can reduce the rank/select on the megachunk to rank/select on a superchunk. This in turn can be reduced to rank/select within a chunk, by reading the superchunk summary. Finally, rank/select within a chunk are supported in constant time using precomputed tables.

5 Further Applications of Informative Encoding

We consider representing a sequence S of length m over $[\sigma]$ to support the $\mathsf{rank}_x(S,i)$ and $\mathsf{select}_x(S,i)$ operations (aiming for the $B()$ space bound), as well as H_k-compressed FIDs for binary strings. We also consider random access to the symbols in S. We first show:

Theorem 4. *There is a representation of a string S of length m over the alphabet $[\sigma]$ that uses $mH_0(S)+O(m/(\lg m)^{2-\varepsilon})$ bits, for any fixed constant $0<\varepsilon<1$, and supports* rank, select *and* access *in $O(\lg\sigma/\lg\lg m)$ time.*

Here, access(i) returns the i-th symbol of S. Theorem 4 uses the *wavelet tree* method [13] to reduce the problem to the case where σ is $(\lg m)^{O(1)}$:

Lemma 9. *Given a string S of length m over an alphabet of size $\sigma = O(\lg^\alpha m)$, for positive constant $\alpha < 1$, one can support the operations* rank, select *and* access *in $O(1)$ time using $mH_0(S) + O(m(\lg\lg m)^3/(\lg m)^{2-2\alpha})$ bits.*

The proof of Lemma 9 follows [12, Theorem 3] and is omitted. Next, we show:

Theorem 5. *A binary sequence S of length m can be stored using $mH_k(S) + O(mk/\lg m)$ bits so that* rank *and* select *over S can be supported in constant time, and any length-ℓ substring can be retrieved in $O(1+\ell/\lg m)$ time.*

Proof. As in [7], we divide S into blocks of length $b = \lfloor(1/2)\lg m\rfloor$, and consider it as a sequence $S_b = x_1x_2\ldots x_{m/b}$ of length m/b over the alphabet $\Sigma = [2^b]$; [7, Theorem 3] shows that $H_0(S_b) \le bH_k(S) + O(k)$, for all $k \le b$.

For any $x \in \Sigma$ that occurs m_x times in S_b, we let $\Pr[x] = n_x/(m/b)$. Using these probabilities, construct a Huffman code $h : \Sigma \to \{0,1\}^+$. We partition Σ into subsets $\Sigma_{i,j}$, where $x \in \Sigma_{i,j}$ iff the weight of x is i and $|h(x)| = j$. Note that $\Pr[x] \ge 1/m$, for all x, so $|h(x)| = O(\lg m)$ for all x, and Σ is partitioned into $(\lg m)^{O(1)}$ classes. Letting $\Pr[x] = 2^{-j} = 2^{-|h(x)|}$ for any $x \in \Sigma_{i,j}$, we get an informative encoding e of x, such that $|e(x)| \le |h(x)| + O(1)$, and such that a prefix of length $O(\lg\lg m)$ of $e(x)$ encodes both $|e(x)|$ and the weight of x. We store the string $S_b^e = e(x_1)e(x_2)\ldots e(x_{m/b})$, together with data structures and precomputed tables of negligible size, enabling us to retrieve $e(x_i)$ for any i in $O(1)$ time, and to support FID operations on S in $O(1)$ time. However:

$$|S_b^e| \le \sum_{i=1}^{m/b} |h(x_i)| + O(m/b) \le (m/b)\cdot H_0(S_b) + O(m/b)$$
$$\le (m/b)(bH_k(S) + O(k)) + O(m/b) \le mH_k(S) + O(mk/b). \qquad \square$$

Remark: If S is drawn from the alphabet $[\sigma]$, the techniques above can be used to show that S can be represented using $mH_k(S) + O(mk\lg\sigma/\lg_\sigma m)$ bits such that any length-ℓ substring can be retrieved in $O(1+\ell/\lg_\sigma m)$ time.

References

1. Barbay, J., He, M., Munro, J.I., Rao, S.S.: Succinct indexes for strings, binary relations and multi-labeled trees. In: Proc. 18th SODA, pp. 680–689 (2007)
2. Benoit, D., Demaine, E.D., Munro, J.I., Raman, R., Raman, V., Rao, S.S.: Representing trees of higher degree. Algorithmica 43, 275–292 (2005)
3. Delpratt, O., Raman, R., Rahman, N.: Engineering succinct DOM. In: Proc. 11th EDBT, pp. 49–60 (2008)
4. Ferragina, P., Luccio, F., Manzini, G., Muthukrishnan, S.: Structuring labeled trees for optimal succinctness, and beyond. In: Proc. 46th FOCS, pp. 184–196 (2005)
5. Ferragina, P., Manzini, G., Mäkinen, V., Navarro, G.: Compressed representations of sequences and full-text indexes. ACM Trans. Algorithms 3, Article 20 (2007)
6. Ferragina, P., Manzini, G.: Indexing compressed text. J. ACM 52, 552–581 (2005)
7. Ferragina, P., Venturini, R.: A simple storage scheme for strings achieving entropy bounds. Theor. Comput. Sci. 372, 115–121 (2007)
8. Gál, A., Bro Miltersen, P.: The cell probe complexity of succinct data structures. Theor. Comput. Sci. 379, 405–417 (2007)
9. Geary, R.F., Raman, R., Raman, V.: Succinct ordinal trees with level-ancestor queries. ACM Transactions on Algorithms 2, 510–534 (2006)
10. Golynski, A.: Optimal lower bounds for rank and select indexes. Theor. Comput. Sci. 387, 348–359 (2007)
11. Golynski, A.: Upper and Lower Bounds for Text Indexing Data Structures. PhD thesis, University of Waterloo (2007)
12. Golynski, A., Grossi, R., Gupta, A., Raman, R., Rao, S.S.: On the size of succinct indices. In: Arge, L., Hoffmann, M., Welzl, E. (eds.) ESA 2007. LNCS, vol. 4698, pp. 371–382. Springer, Heidelberg (2007)
13. Grossi, R., Gupta, A., Vitter, J.S.: High-order entropy-compressed text indexes. In: Proc. 14th SODA, pp. 841–850 (2003)
14. Grossi, R., Vitter, J.S.: Compressed suffix arrays and suffix trees with applications to text indexing and string matching. SIAM J. Comput. 35, 378–407 (2005)
15. Jacobson, G.: Space efficient static trees and graphs. In: Proc. 30th FOCS, pp. 549–554 (1989)
16. Miltersen, P.B.: Lower bounds on the size of selection and rank indexes. In: Proc. 16th SODA, pp. 11–12 (2005)
17. Munro, J.I.: Tables. In: Chandru, V., Vinay, V. (eds.) FSTTCS 1996. LNCS, vol. 1180, pp. 37–42. Springer, Heidelberg (1996)
18. Okanohara, D., Sadakane, K.: Practical entropy-compressed rank/select dictionary. In: Proc. 9th ALENEX, pp. 59–70. SIAM, Philadelphia (2007)
19. Pagh, R.: Low redundancy in static dictionaries with constant query time. SIAM J. Computing 31, 353–363 (2001)
20. Patrascu, M., Thorup, M.: Time-space trade-offs for predecessor search. In: Proc. 38th STOC, pp. 232–240 (2006)
21. Raman, R., Raman, V., Rao, S.S.: Succinct indexable dictionaries, with applications to representing k-ary trees, prefix sums and multisets. ACM Trans. Algorithms 4, 26 (2007), Article 43
22. Sadakane, K., Grossi, R.: Squeezing succinct data structures into entropy-compressed bounds. In: Proc. 17th SODA, pp. 1230–1239 (2006)

Confluently Persistent Tries for Efficient Version Control

Erik D. Demaine[1,⋆], Stefan Langerman[2,⋆⋆], and Eric Price[1]

[1] MIT Computer Science and Artificial Intelligence Laboratory, 32 Vassar Street, Cambridge, MA 02139, USA
{edemaine,ecprice}@mit.edu

[2] Computer Science Department, Université Libre de Bruxelles, CP 212, Bvd. du Triomphe, 1050 Brussels, Belgium
stefan.langerman@ulb.ac.be

Abstract. We consider a data-structural problem motivated by version control of a hierarchical directory structure in a system like Subversion. The model is that directories and files can be moved and copied between two arbitrary versions in addition to being added or removed in an arbitrary version. Equivalently, we wish to maintain a confluently persistent trie (where internal nodes represent directories, leaves represent files, and edge labels represent path names), subject to copying a subtree between two arbitrary versions, adding a new child to an existing node, and deleting an existing subtree in an arbitrary version.

Our first data structure represents an n-node degree-Δ trie with $O(1)$ "fingers" in each version while supporting finger movement (navigation) and modifications near the fingers (including subtree copy) in $O(\lg \Delta)$ time and space per operation. This data structure is essentially a locality-sensitive version of the standard practice—path copying—costing $O(d \lg \Delta)$ time and space for modification of a node at depth d, which is expensive when performing many deep but nearby updates. Our second data structure supporting finger movement in $O(\lg \Delta)$ time and no space, while modifications take $O(\lg n)$ time and space. This data structure is substantially faster for deep updates, i.e., unbalanced tries. Both of these data structures are functional, which is a stronger property than confluent persistence. Without this stronger property, we show how both data structures can be sped up to support movement in $O(\lg \lg \Delta)$, which is essentially optimal. Along the way, we present a general technique for global rebuilding of fully persistent data structures, which is nontrivial because amortization and persistence do not usually mix. In particular, this technique improves the best previous result for fully persistent arrays and obtains the first efficient fully persistent hash table.

⋆ Supported in part by MADALGO — Center for Massive Data Algorithmics — a Center of the Danish National Research Foundation.

⋆⋆ Chercheur qualifié du F.R.S.-FNRS.

J. Gudmundsson (Ed.): SWAT 2008, LNCS 5124, pp. 160–172, 2008.

1 Introduction

This paper is about a problem in persistent data structures motivated by an application in version control. We begin by describing the motivating application, then our model of the underlying theoretical problem, followed by our results.

Version control. Increasingly many creative works on a computer are stored in a *version control system* that maintains the history of all past versions. The many motivations for such systems include the ability to undo any past mistake (in the simplest form, never losing a file), and the ability for one or many people to work on multiple parallel branches (versions) that can later be merged. Source code has been the driving force behind such systems, ranging from old centralized systems like RCS and CVS, to the increasingly popular centralized system Subversion, to recent distributed systems like Bazaar, darcs, GNU arch, Git, Monotone, and Mercurial. By now version control is nearly ubiquitous for source code and its supporting documentation. We also observe a rise in the use of the same systems in academic research for books, papers, figures, classes, etc.[1] In more popular computer use, Microsoft Word supports an optional form of version control ("change tracking"), Adobe Creative Suite (Photoshop, Illustrator, etc.) supports optional version control ("Version Cue"), and most Computer Aided Design software supports version control in what they call Product Data Management (e.g., Autodesk Productstream for AutoCAD and PDMWorks for SolidWorks). Entire modern file systems are also increasingly version controlled, either by taking periodic global snapshots (as in AFS, Plan 9, and WAFL), or by continuous change tracking (as in Apple's new Time Machine in HFS+, and in experimental systems CVFS, VersionFS, Wayback, and PersiFS [PCD05]). As repositories get larger, even to the point of entire file systems, high-performance version control is in increasing demand. For example, the Git system was built simply because no other free system could effectively handle the Linux kernel.

Requirements for version control. Most version control systems mimic the structure of a typical file system: a tree hierarchy of directories, each containing any number of linear files. Changes to an individual file can therefore be handled purely locally to that file. Conceptually these changes form a tree of versions of the file, though all systems represent versions implicitly by storing a delta ("diff") relative to the parent. In this paper, we do not consider such file version tracking, because linear files are relatively easy to handle.

The more interesting data structural challenge is to track changes to the hierarchical directory structure. All such systems support addition and removal of files in a directory, and creation and deletion of empty subdirectories. In addition, every system since Subversion's pioneering innovation supports moving or copying an entire subdirectory from one location to another, possibly spanning two different versions. This operation is particularly important for merging different version branches into a common version.

[1] For example, this paper is maintained using Subversion.

Persistent trie model. Theoretically, we can model version control of a hierarchical directory structure as a confluently persistent trie, which we now define.

A *trie* is a rooted tree with labeled edges. In the version-control application, internal nodes represent directories, leaves represent files, and edge labels represent file or directory names.[2] The natural queries on tries are navigation: placing a finger at the root, moving a finger along the edge with a specified label, and moving a finger from a node to its parent. We assume that there are $O(1)$ fingers in any single version of the trie; in practice, two fingers usually suffice. Each node has some constant amount of information which can be read or written via a finger; for example, each leaf can store a pointer to the corresponding file data structure. The structural changes supported by a trie are insertion and deletion of leaves attached to a finger (corresponding to addition and removal of files), copying the entire subtree rooted at one finger to become a new child subtree of another finger (corresponding to copying subdirectories), and deleting an entire subtree rooted at one finger (enabling moving of subdirectories). Subtree copying propagates any desired subset of the fingers of the old subtree into the new subtree, provided the total number of fingers in the resulting trie remains $O(1)$.

The trie data structure must also be "confluently persistent". In general, persistent data structures preserve old versions of themselves as modifications proceed. A data structure is *partially persistent* if the user can query old versions of the structure, but can modify only the most recent version; in this case, the versions are linearly ordered. A data structure is *fully persistent* if the user can both query and modify past versions, creating new branches in a tree of versions. The strongest form of persistence in *confluent persistence*, which includes full persistence but also supports certain "meld" operations that take multiple versions of the and produce a new version; then the version dependency graph becomes a directed acyclic graph (DAG). The version-control application demands confluent persistence because branch merging requires the ability to copy subdirectories (subtrees) from one version into another.

Related work in persistence. Partial and full persistence were mostly solved in the 1980's. In 1986, Driscoll et al. [DSST89] developed a technique that converts any pointer-based data structure with bounded in-degree into an equivalent fully persistent data structure with only constant-factor overhead in time and space for every operation. In 1989, Dietz [Die89] developed a fully persistent array supporting random access in $O(\lg\lg m)$ time, where m is the number of updates made to any version. This data structure enables simulation of an arbitrary RAM data structure with a log-logarithmic slowdown. Furthermore, this slowdown is essentially optimal, because fully persistent arrays have a lower bound of $\Omega(\lg\lg n)$ time per operation in the powerful cell-probe model.[3]

More relevant is the work on confluent persistence. The idea was first posed as an open problem by [DSST89]. In 1994, Driscoll et al. [DST94] defined confluence

[2] We assume here that edge labels can be compared in constant time; in practice, this property is achieved by hashing the file and directory name strings.

[3] Personal communication with Mihai Pătraşcu, 2008. The proof is based on the predecessor lower bounds of [PT07].

and gave a specific data structure for confluently persistent catenable lists. In 2003, Fiat and Kaplan [FK03] developed the first and only general methods for making a pointer-based data structure confluently persistent, but the slowdown is often suboptimal. In particular, their best deterministic result has a linear worst-case slowdown. Although their randomized result has a polylogarithmic amortized slowdown, it can take a linear factor more space per modification, and furthermore the answers are correct only with high probability; they do not have enough time to check the correctness of their random choices.

Fiat and Kaplan [FK03] also prove a lower bound on confluent persistence. They essentially prove that most interesting confluently persistent data structures require $\Omega(\lg p)$ space per operation in the worst case, even with randomization, where p is the number of paths in the version DAG from the root version to the current version. Note that p can be exponential in the number m of versions, as in Figure 1, resulting in a lower bound of $\Omega(m)$ space per operation. The lower bound follows from the possibility of having around p addressable nodes in the data structure; in particular, it is easy to build an exponential amount of data (albeit with significant repetition) using a linear number of confluent operations. However, their $\Omega(\lg p)$ lower bound requires a crucial and unreasonable assumption: that all nodes of the structure can be addressed at any time. From the perspective of actually using a data structure, it is much more natural for the user to have to locate the data of interest using a sequence of queries. For this reason, our use of trie traversals by a constant number of fingers is both natural and critical to our success.

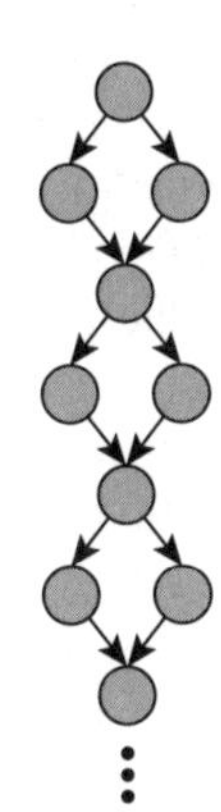

Fig. 1. This version DAG has exponentially many paths from top to bottom, and can result in a structure with an exponential data

Functional data structures. Given the current lack of general transformations into confluently persistent data structures, efficient such structures seem to require exploiting the specific problem. One way to attain confluent persistence is to design a *functional* data structure, that is, a read-only (pointer-based) data structure. Such a data structure can create new cells with new initial contents, but cannot modify the contents of old cells. Each modifying operation requires a pointer to the new version of the data structure, described by a newly created cell. Functional data structures have many useful properties other than confluent persistence; for example, multiple threads can use functional data structures without locking. Pippenger [Pip97] proved a logarithmic-factor separation between the best pointer-based data structure and the best functional data structure for some problems. On the other hand, many common data structures can be implemented functionally with only a constant-factor overhead; see Okasaki [Oka98]. One example we use frequently is a functional *catenable deque*, supporting insertion and deletion at either end and concatenation of two deques in constant time per operation [Oka98].

Path copying. Perhaps the simplest technique for designing functional data structures is *path copying* [Oka98]. This approach applies to any tree data structure where each node modification depends on only the node's subtree. Whenever we would modify a node v in the ephemeral (nonpersistent) structure, we instead create new copies of v and all ancestors of v. Because nodes depend on only their subtrees and the data structure becomes functional (read only), we can safely reuse all other nodes. Figure 2 shows an example of path copying in a binary search tree (which achieves logarithmic worst-case performance).

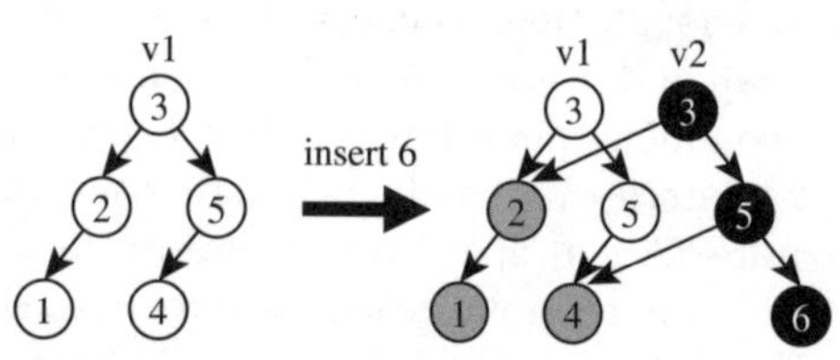

Fig. 2. Path copying in a binary search tree. The old version (v1) consists of white and grey nodes; the new version (v2) consists of black and grey nodes.

Version control systems including Subversion effectively implement path copying. As a result, modifications to the tree have a factor-$\Theta(d)$ overhead, where d is the depth of the modified node. More precisely, for a pointer-based data structure, we must split each node of degree Δ into a binary tree of height $O(\lg \Delta)$, costing $O(d \lg \Delta)$ time and space per update.

Imbalance. The $O(d \lg \Delta)$ cost of path copying is potentially very large because the trie may be extremely unbalanced. For example, if the trie is a path of length n, and we repeatedly insert n leaves at the bottommost node, then path copying requires $\Omega(n^2)$ time and space.

Douceur and Bolosky [DB99] studied over 10,000 file systems from nearly 5,000 Windows PCs in a commercial environment, totaling 140 million files and 10.5 terabytes. They found that d roughly follows a Poisson distribution, with 15% of all directories having depth at least eight. Mitzenmacher [Mit03] studies a variety of theoretical models for file-system creation which all imply that d is usually logarithmic in n.

Our results. We develop four trie data structures, two of which are functional and two of which are efficient but only confluently persistent; see Table 1. All four structures effectively break through the lower bound of Fiat and Kaplan.

Table 1. Time and space complexity of data structures described in this paper. Operations are on an n-node trie at a node of depth d and degree Δ.

Method	Finger movement		Modifications	
	Time	Space	Time	Space
Path copying	$\lg \Delta$	0	d	d
Locality-sensitive (functional)	$\lg \Delta$	$\lg \Delta$	$\lg \Delta$	$\lg \Delta$
Locality-sensitive (fully persistent)	$\lg\lg \Delta$	$\lg\lg \Delta$	$\lg\lg \Delta$	$\lg\lg \Delta$
Globally balanced (functional)	$\lg \Delta$	0	$\lg n$	$\lg n$
Globally balanced (fully persistent)	$\lg\lg \Delta$	0	$\lg n$	$\lg n$

Our first functional trie enables exploiting locality of reference among any constant number of fingers. Both finger movement (navigation) and modifications around the fingers (including subtree copy) cost $O(\lg \Delta)$ time and space per operation, where Δ is the average degree of the nodes directly involved. Note that navigation operations require space as well, though the space is permanent only if the navigation leads to a modification; stated differently, the space cost of a modification is effectively $O(t \lg \Delta)$ where t is the distance of the finger from its last modification. This data structure is always at least as efficient as path copying, and much more efficient in the case of many deep but nearby modifications. In particular, the quadratic example of inserting n leaves at the bottom of a length-n path now costs only $O(n \lg \Delta)$ time and space.

Our second functional trie guarantees $O(\lg n)$ time and space per modification, with no space required by navigation, while preserving $O(\lg \Delta)$ time per navigation. This data structure is substantially more space-efficient than the first data structure whenever modifications are deep and dispersed. For example, if we insert n leaves alternately at the top and at the bottom of a length-n path, then the time cost from navigation is $\Theta(n^2)$, but the space cost is only $O(n \lg n)$. The only disadvantage is that nearby modifications still cost $\Theta(\lg n)$ time and space, whereas the $O(t \lg \Delta)$ cost of the first data structure can be a bit smaller.

Our two confluently persistent trie data structures are based on the functional data structures, replacing each height-$O(\lg \Delta)$ binary tree representation of a degree-Δ trie node with a new log-logarithmic fully persistent hash table. For the first structure, we obtain an exponentially improved bound of $O(\lg \lg \Delta)$ time and space per operation. For the second structure, we improve the movement cost to $O(\lg \lg \Delta)$ time (and no space). These operations have matching $\Omega(\lg \lg \Delta)$ time lower bounds because in particular they implement fully persistent arrays.

To our knowledge, efficient fully persistent hash tables have not been obtained before. The obvious approach is to use standard hash tables while replacing the table with the fully persistent array of Dietz [Die89]. There are two main problems with this approach. First, the time bound for fully persistent arrays is $O(\lg \lg m)$, where m is the number of updates to the array, but this bound can be substantially larger than even the size Δ of the hash table. Second, hash tables need to dynamically resize, and amortization does not mix well with persistence: the hash table could be put in a state where it is about to pay a huge cost, and then the user modifies that version repeatedly, each time paying the huge cost.

The solution to both problems is given by a new general technique for global rebuilding of a fully persistent data structure. The classic global rebuilding technique from ephemeral data structures, where the data structure rebuilds itself from scratch whenever its size changes by a constant factor, does not apply in the persistent context. Like the second problem above, we cannot charge the linear rebuild cost to the elements that changed the size, because such elements might get charged many times, even with de-amortized global rebuilding. Nonetheless, we show that clever choreography of specific global rebuilds works in the fully persistent context. As a result, we improve fully persistent arrays to support operations in $O(\lg \lg \Delta)$ time and space, where Δ is the current size of the array, matching the $\Omega(\lg\lg \Delta)$ lower bound. We also surmount the amortization and randomization issues with this global rebuilding technique.

2 Locality-Sensitive Functional Data Structure

Our first functional data structure represents a trie T with a set F of $O(1)$ fingers $f_1, f_2, \ldots, f_k$ while supporting finger movements, leaf insertion and deletion, and subtree copies and removals in $O(\lg \Delta)$ time and space per operation.

Let T' be the Steiner tree with terminals f_i, that is, the union of shortest paths between all pairs of fingers. Let PF be the set of nodes with degree at least 3 in T' that are not in F. The elements of PF are called *prosthetic fingers* and will be maintained dynamically. Let $F' = F \cup PF$. Note that $|F'| \leq 2k = O(1)$. Let T'' be the compressed Steiner tree obtained from T' by contracting every vertex not in F' (each of degree 2). For any two fingers in F' that are adjacent in T'', the shortest path in T' connecting them is called a *tendon*. A subtree of T that does not contain any finger of F' is called a *knuckle*.

We represent a tendon by a deque, where each element of the deque corresponds to a vertex of T and is represented by a balanced tree of depth $O(\lg \Delta)$ containing the neighbors of that vertex in T other than those in the deque. Each of the nodes in that tree contains a knuckle. The tendon also contains the labels of the two fingers to which it is attached. We represent a knuckle either by a vertex containing a balanced tree of depth $O(\lg \Delta)$, where each node of the tree represents a neighbor of that vertex in T', or by a deque representing a path starting at the root, whose structure is identical to that of the tendon.

The functional data structure stores all fingers in F' in a balanced binary search tree called the *hand*, where all fingers are ordered by the label of the corresponding node. Every finger stores a balanced tree of depth $O(1)$ for the tendons attached to this finger, and another balanced tree of depth $O(\lg \Delta)$ for the knuckles attached to this finger.

To complete this description, it remains to show how to perform update operations and how to move fingers. When performing an update at a finger, we modify the balanced tree attached to that finger: adding a neighbor for a leaf insertion or subtree copy, and deleting a neighbor for a leaf deletion or subtree removal. Then we use the path-copying technique on both that tree and the hand. To move a finger, we essentially transfer vertices between its neighboring nodes, knuckles, and tendons:

1. If a finger enters a neighboring knuckle (stored in a node of its balanced tree), we will move the finger to its new position. This might involve extracting the new finger from a deque and modifying a constant number of neighbors in its balanced tree. We have two cases:
   (a) If the finger has degree 1 in T'', then it is attached to exactly one tendon τ. We insert into τ the vertex at the previous position of the finger.
   (b) If the finger has degree 2 or more in T'', then after the move that vertex has degree at least 3 and we create a new prosthetic finger at that

position. The hand must then be modified so that the tendons that were adjacent to the finger now point to the prosthetic finger. This costs $O(1)$.
2. If a finger moves along a tendon, we proceed similarly, but now, the previous finger is either transferred to a neighboring knuckle or tendon, or becomes a new prosthetic finger. The new vertex for the finger is extracted from the tendon, or if the tendon is empty, two fingers become equal.

The operations change only $O(1)$ nodes from the balanced trees or the deques outside the hand, for a cost of $O(\lg \Delta)$, and modify the hand, which has size $O(1)$.

3 Globally Balanced Functional Data Structure

Our second functional data structure represents the trie as a balanced binary tree, then makes this tree functional via path copying. Specifically, we will use a balanced representation of tries similar to link-cut trees of Sleator and Tarjan [ST83]. This representation is natural because the link and cut operations are essentially subtree copy and delete.[4] Sleator and Tarjan's original formulation of link-cut trees cannot be directly implemented functionally via path copying, and we explain how to address these issues in Section 3.1. In addition to being able to modify the trie, we need to be able to navigate this representation as we would the original trie. We discuss how to navigate in Section 3.2.

A key element of our approach is the *finger*. In addition to the core data structure representing a trie, we also maintain a constant number of locations in the trie called fingers. A finger is a data structure in itself, storing more than just a pointer to a node. Roughly speaking, a finger consists of pointers to all ancestors of that node in the balanced tree, organized to support several operations for computing nearby fingers. These pointers contrast nodes in the balanced tree representation, which do not even store parent pointers. Modifications to our data structure must preserve fingers to point to the same location in the new structure, but fortunately there are only finitely many fingers to maintain. Section 3.4 details the implementation of fingers.

3.1 Functional Link-Cut Trees

In the original link-cut trees [ST83], nodes store pointers to other nodes outside of their subtree, which prevents us from applying path copying. We show how to modify link-cut trees to avoid such pointers and thereby obtain functional link-cut trees via path copying.

There are multiple kinds of link-cut trees; we follow the worst-case logarithmic link-cut trees of [ST83, Section 5]. These link-cut trees decompose the trie into a set of "heavy" paths, and represent each heavy path by a globally biased binary tree [BST85], tied together into one big tree which we call the *representation*

[4] Euler-tour trees are simpler, but linking multiple occurrences of a node in the Eulerian tour makes path copying infeasible.

tree. An edge is *heavy* if more than half of the descendants of the parent are also descendants of the child. A *heavy path* is a contiguous sequence of heavy edges. Because any node has at most one heavy child, we can decompose the trie into a set of heavy paths, connected by *light* (nonheavy) edges. Following a light edge decreases the number of nodes in the subtree by at least a factor of two, so any root-to-leaf path intersects at most $\lg n$ heavy paths. The *weight* w_v of a node v is 1 plus the number of descendants of v through a light child of v. The depth of a node v in its globally biased search tree T is at most $\lg \left[\left(\sum_{u \in T} w_u\right)/w_v\right]$. If the root-to-leaf path to a node v intersects k heavy paths and w_i is the weight of the last ancestor of v in the ith heavy path down to v, then the total depth of v is $\lceil \lg(n/w_1) \rceil + \lceil \lg(w_1/w_2) \rceil + \cdots + \lceil \lg(w_{k-1}/w_k) \rceil$, or $O(\lg n)$.

Link-cut trees augment each node in a globally biased search tree to store substantial extra information. Most of this information depends only on the subtree of the node, which fits the path-copying framework. The one exception is the parent of the node, which we cannot afford to store. Instead, we require a parent operation on fingers, which returns a finger pointing to the parent node in the representation tree. For this operation to suffice, we must always manipulate fingers, not just pointers to nodes. This restriction requires one other change to the augmentation in a link-cut tree. Namely, instead of storing pointers to the minimum (leftmost) and maximum (rightmost) descendants in the subtree, each node stores *relative fingers* to these nodes. Roughly speaking, such a relative finger consists of the nodes along the path from the node to the minimum (maximum). We require the ability to concatenate partial fingers, in this case, the finger of the node with a relative finger to the minimum or maximum, resulting in a (complete) finger to the latter.

To tie together the globally biased search trees for different heavy paths, the link-cut tree stores, for each node, an auxiliary globally biased tree of its light children. More precisely, leaves of the auxiliary tree point to (the root of) the globally biased search tree representing the heavy path starting at such light children. In addition, the top node in a heavy path stores a pointer to its parent in the trie, or equivalently, the root of the tree of light children of that parent. We cannot afford to store this parent pointer, so instead we define the finger to ignore the intermediate nodes of the auxiliary tree, so that the parent of the current finger gives us the desired root. Nodes in an auxiliary tree also contain pointers to their maximum-weight leaf descendents; we replace these pointers with relative fingers. Sleator and Tarjan's link-cut trees order auxiliary trees by decreasing weight; this detail is unnecessary, so we instead order the children by key value to speed up navigation.

With these modifications, the link-cut trees of Sleator and Tarjan can be implemented functionally with path copying. To see that path copying induces no slowdown, note that although the finger makes jumps down, the finger is shortened only one node at a time. Each time we take the parent of a finger, the node at the end can be rebuilt from the old node and the nodes below it in constant time. Furthermore, because the maximum finger length is $O(\lg n)$, at the end of the operation, one can repeatedly take the parent of the finger to get the new root. Another way to see that path copying induces no slowdown is that the link-cut trees in the original paper also involved augmentation, so every ancestor of a modified node was already being rebuilt.

These functional link-cut trees let us support modification operations in $O(\lg n)$ time given fingers to the relevant nodes in our trie. It remains to see how to navigate a finger, the topic of Section 3.2, and how to maintain other fingers as the structure changes, the topic of Section 3.3.

3.2 Finger Movement

In this section, we describe three basic finger-movement operations: finding the root, navigating to the parent, and navigating to the child with a given label. In all cases, we aim simply to determine the operations required of our finger representation.

The root of the trie is simply the minimum element in the topmost heavy path, a finger for which is already stored at the root of the representation tree. Thus we can find the root of the trie in constant time.

The parent of a node in the trie is the parent of the node within its heavy path, unless it is the top of its heavy path, in which case it is the node for which this node is a light child. Equivalently, the parent of a node in the trie is the predecessor leaf in the globally biased search tree, if such a predecessor exists, or else it is the root of the auxiliary tree containing the node. We also defined the parent operation on fingers to solve the latter case. For the former case, we require a predecessor operation on a finger that finds the predecessor of the node among all ancestors of the node. If this operation takes constant time, then we can find the predecessor leaf within the globally biased search tree by taking the maximum descendant of the predecessor ancestor found by the finger.

A child of a node is either the node immediately below in the heavy path or one of its light children. The first case corresponds to finding the successor of the node within its globally biased search tree. Again we can support this operation in constant time by requiring a successor operation on a finger that finds the successor of the node among all ancestors of the node. Thus, if a child stores the label of the edge above it in the trie, we can test whether the desired child is the heavy child. Otherwise, we binary search in the auxiliary search tree of light children in $O(\lg \Delta)$ worst-case time. Because the total depth of a node among all trees of light children is $O(\lg n)$, the total time to walk to a node at depth d can be bounded by $O(d + \lg n)$ as well as $O(d \lg \Delta)$.

3.3 Multiple Fingers

This section describes how to migrate the constant number of fingers to the new version of the data structure created by an update, in $O(\lg n)$ time. We distinguish the finger at which the update takes place as *active*, and call the other fingers *passive*. Just before performing the update, we decompose each passive finger into two relative fingers: the longest common subpath with the active finger, and the remainder of the passive finger. This decomposition can be obtained in $O(\lg n)$ time given constant-time operation to remove the topmost or bottommost node from a relative finger. First, repeatedly remove the topmost nodes from both the active and passive fingers until the removed nodes differ, resulting in the remainder part of the passive finger. Next, repeatedly remove the bottommost node of (say) the active finger until reaching this branching point, resulting in the common part of the passive finger. Now perform the update, and modify the nodes along the active finger according to path copying. We can find the replacement top part of the passive finger again by repeatedly removing the bottommost node of the active finger; the remainder part does not need to change. Then we use the concatenate operation to join these two relative fingers to restore a valid passive finger. Of course, we perform this partition, replacement, and rejoin trick for each relative finger.

3.4 Finger Representation

Recall that each finger to a node must be a list of the ancestors in the representation tree of that node supporting: push and pop (to move up and down); concatenate (to follow relative fingers); inject (to augment relative fingers based on a node's children); eject (to allow for multiple fingers); parent outside path (for faster parent query); and predecessor and successor among the ancestors of the node (to support parent and child in the trie). Our finger must also be functional, because nodes store relative fingers.

As a building block, catenable deques support all the operations we want except for predecessor or successor. Moreover, functional catenable deques have been well researched, with Okasaki giving a fairly simple $O(1)$ method [Oka98] that is amortized, but in a way that permits confluent usage if one allows memoization. Furthermore, Kaplan and Tarjan have shown a complicated $O(1)$ worst case purely functional implementation of catenable deques [KT95].

In order to implement predecessor and successor queries, decompose the path in the representation tree into a sequence of right paths (sequence of nodes where the next element on the path is the right child) and left paths (sequence of nodes where the next element on the path is the left child). Then, the predecessor of a node among its ancestors is the last element of the last right path. The successor of a node among its ancestors is the last element of the last left path.

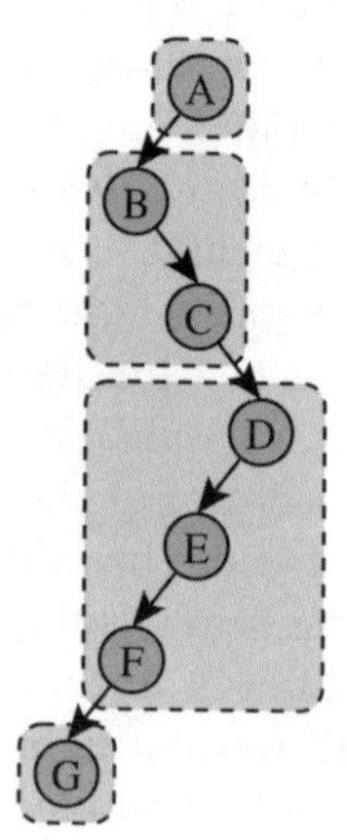

Fig. 3. A finger to G is represented by G and a deque of (outlined) deques of ancestors of G

Instead of maintaining the finger as one catenable deque, we represent the finger as a deque of catenable deques, alternating right and left paths; see Figure 3. Then, because the last right path (or left path, respectively) is either the ultimate or penultimate deque in the sequence, its last element can be retrieved in $O(1)$ time. All other operations of the standard catenable deque can be simulated on the deque of deques with only $O(1)$ overhead. We thus obtain a structure that supports all of the operations of normal catenable deques of nodes, predecessor and successor in $O(1)$ time.

This suffices to describe the basic data structure for functional tries. Their query time is $O(\lg \Delta)$ for navigation downward, $O(1)$ for navigation up, and $O(\lg n)$ for updates.

4 Adding Hash Tables

Our data structures above take $O(\lg \Delta)$ to move a finger to a child node. We can improve this bound to $O(\lg \lg \Delta)$ by introducing fully persistent hash tables. The resulting data structures are confluently persistent, but no longer functional.

For our first structure (Section 2), we show in the full paper how to construct a fully persistent hash table on n elements that performs insertions, deletions, and searches in $O(\lg \lg n)$ expected amortized time. Using this structure instead of the balanced trees of neighboring vertices at every vertex, the time and space cost of updates and finger movements improves to $O(\lg \lg \Delta)$ expected amortized.

We can also use this method to improve our second structure (Section 3.1). Given a set of elements with weights $w_1, \ldots, w_n$, we develop in the full paper a weight-balanced fully persistent hash table with $O(\lg \frac{\sum_j w_j}{w_e})$ expected amortized time modification, $O(\lg \lg n)$ find, and $O(\lg n)$ insert and delete, where n is the number of elements in the version of the hash table being accessed or modified and w_e is the weight of the element being accessed.

To use a hash table to move a finger down the trie, we modify each node of our data structure to include a hash table of relative fingers to light children *in addition* to the binary tree of light children. The binary tree of light children is necessary to support quickly recomputing the heavy child on an update; this would be hard with just a hash table. Except for inserts and deletes, the hash table achieves at least as good time bounds as the weight-balance tree, and each trie operation involves at most $O(1)$ inserts or deletes, so we can maintain both the table and tree in parallel without overhead. As a result, updates still take $O(\lg n)$, moving the finger up still takes $O(1)$, and moving it down now takes $O(\lg \lg \Delta)$, where Δ is the degree of the node being moved from. The hash tables depend on fully persistent arrays, which are expected amortized, so updates and moving a finger down become expected amortized.

5 Open Problems

It would be interesting to combine our two functional data structures into one that achieves the minimum of both performances. In particular, it would be nice to be able to modify a node at depth d in $O(\min\{d \lg \Delta, \lg n\})$ time and space. One approach is to develop modified globally biased binary tree where the depth of the ith smallest node is $O(\min\{i, w_i\})$ and supporting fast splits and joins.

Acknowledgments. We thank Patrick Havelange for initial analysis of and experiments with version control systems such as Subversion.

References

[BST85] Bent, S.W., Sleator, D.D., Tarjan, R.E.: Biased search trees. SIAM J. Comput. 14(3), 545–568 (1985)

[DB99] Douceur, J.R., Bolosky, W.J.: A large-scale study of file-system contents. SIGMETRICS Perform. Eval. Rev. 27(1), 59–70 (1999)

[Die89] Dietz, P.F.: Fully persistent arrays. In: Dehne, F., Santoro, N., Sack, J.-R. (eds.) WADS 1989. LNCS, vol. 382, pp. 67–74. Springer, Heidelberg (1989)

[DSST89] Driscoll, J.R., Sarnak, N., Sleator, D.D., Tarjan, R.E.: Making data structures persistent. J. Comput. Syst. Sci. 38(1), 86–124 (1989)

[DST94] Driscoll, J.R., Sleator, D.D.K., Tarjan, R.E.: Fully persistent lists with catenation. J. ACM 41(5), 943–959 (1994)

[FK03] Fiat, A., Kaplan, H.: Making data structures confluently persistent. J. Algorithms 48(1), 16–58 (2003)

[KT95] Kaplan, H., Tarjan, R.E.: Persistent lists with catenation via recursive slow-down. In: STOC 1995, pp. 93–102 (1995)

[Mit03] Mitzenmacher, M.: Dynamic models for file sizes and double Pareto distributions. Internet Math. 1(3), 305–333 (2003)

[Oka98] Okasaki, C.: Purely Functional Data Structures. Cambridge University Press, Cambridge (1998)

[PCD05] Ports, D.R.K., Clements, A.T., Demaine, E.D.: PersiFS: A versioned file system with an efficient representation. In: SoSP (2005)

[Pip97] Pippenger, N.: Pure versus impure lisp. ACM Trans. Program. Lang. Syst. 19(2), 223–238 (1997)

[PT07] Pătraşcu, M., Thorup, M.: Randomization does not help searching predecessors. In: SODA 2007, pp. 555–564 (2007)

[ST83] Sleator, D.D., Tarjan, R.E.: A data structure for dynamic trees. J. Comput. Syst. Sci. 26(3), 362–391 (1983)

A Uniform Approach Towards Succinct Representation of Trees

Arash Farzan and J. Ian Munro

School of Computer Science,
University of Waterloo,
Waterloo, Ontario, Canada
{afarzan,imunro}@cs.uwaterloo.ca

Abstract. We propose a uniform approach for succinct representation of various families of trees. The method is based on two recursive decomposition of trees into subtrees. Recursive decomposition of a structure into substructures is a common technique in succinct data structures and has been shown fruitful in succinct representation of ordinal trees [7,10] and dynamic binary trees [16,21]. We take an approach that simplifies the existing representation of ordinal trees while allowing the full set of navigational operations. The approach applied to cardinal (*i.e.* k-ary) trees yields a space-optimal succinct representation allowing cardinal-type operations (*e.g.* determining child labeled i) as well as the full set of ordinal-type operations (*e.g.* reporting the number of siblings to the left of a node). Existing space-optimal succinct representations had not been able to support both types of operations [2,19].

We demonstrate how the approach can be applied to obtain a space-optimal succinct representation for the family of free trees where the order of children is insignificant. Furthermore, we show that our approach can be used to obtain entropy-based succinct representations. We show that our approach matches the degree-distribution entropy suggested by Jansson *et al.* [13]. We discuss that our approach can be made adaptive to various other entropy measures.

1 Introduction

With the ever increasing size of data sets, an important aspect in handling information is their storage requirement. A succinct representation of a combinatorial object is an encoding which supports a reasonable set of operations on the object in constant time and has a storage requirement matching the information theoretic lower bound, to within lower order terms. Succinct data structures perform under the uniform-time word RAM-model with $\Theta(\lg n)^1$ word size.

Trees are fundamental data structures in computer science and as a result a great deal of research has been done on their succinct representation. Succinct representation of two major families of trees have been well studied: *ordinal* and *cardinal*. In ordinal trees the order of children of nodes is significant and

[1] $\lg n$ denotes $\log_2 n$.

J. Gudmundsson (Ed.): SWAT 2008, LNCS 5124, pp. 173–184, 2008.

Table 1. Space lower bounds in bits to represent families of trees with n nodes and references to succinct representations

Tree family	Space lower bound (Highest order term)	Succinct representation
Ordinal trees	$2n$ [14]	[11,5,15,2,7,10]
Ordinal trees with a given degree distribution	$\sum_i n_i \lg \frac{n}{n_i}$ [20]	[13]
Cardinal trees	$(k \lg k - (k-1)\lg(k-1))\, n$ [8]	[2,19]
Binary trees	$2n$ [8]	[11,12,4,5,15]
Free trees	$1.56n$ [18]	this paper
Free binary trees	$1.31n$ [6,22]	this paper

preserved. However, in cardinal trees (also known as k-ary trees), each node has k-slots for edges to children which can be independently occupied or not. Binary trees are a subclass of cardinal trees for value $k = 2$. Here we also study the succinct encoding of another family of trees: *free trees*, in which the order of children of a node is not significant.

Certain subfamilies of these major families of trees have been studied in the context of succinct representations. Binary trees ($k = 2$) or DNA trees ($k = 4$) form two of the best known subfamilies of cardinal trees. Ordered trees with a given degree distribution where a list of numbers n_i ($i \geq 0$) is given and the tree is guaranteed to have exactly n_i nodes with i children form a subfamily of ordinal trees studied recently by Jansson *et al.* [13]. We also investigate *free binary trees* which are free trees with maximum two children per node.

Space lower bounds on the required number of bits to represent each class of trees is obtained via information theory by counting the number of trees in the class. Table 1 illustrates the space lower bounds for these classes along with existing references to succinct representations which achieve the optimal space within lower order terms and support a variety of operations.

1.1 Contribution

We propose a uniform approach for representing trees succinctly that encompasses the families of trees in table 1. The method is based on two-level decomposition of a tree into subtrees. The recursive decomposition method is a common technique in succinct representations of various data structures [5,19,1] and has been used to represent trees [16,7,10].

In the case of ordinal trees, our approach supports a wide range of operations proposed by He *et al.* [10] and simplifies implementation of the supported operations. In the case of cardinal trees, there is no known succinct representation that supports a wide range of navigational operations. Raman *et al.* [19] state that their succinct representation for cardinal trees cannot support subtree size. Our succinct representation of cardinal trees can support all ordinal-tree-type operations listed by He *et al.* [10] (such as subtree size) as well as cardinal-tree-type operations suggested by Raman *et al.* [19] (such as following the edge labeled i from a node where $1 \leq i \leq k$).

To show the power of our method, we consider free trees which are trees with no order imposed on children of nodes and show that we can have a succinct representation taking the optimal $(1.56\ldots)n$ bits supporting all navigational operations. Similarly, free binary trees, which are free trees with maximum two children per node, can be represented in the optimal $(1.31\ldots)n$ number of bits.

Existing succinct encodings of trees assume a uniform distribution over the family of trees and therefore give worst case space guarantees. In practice however, there might be many reasons to have trees with certain property that are more likely than others, and therefore an entropy-based succinct representation is necessary. Jansson *et al.* [13] considered this case when the distribution is based on degrees of nodes and gave a representation that matches the degree-distribution entropy. Our succinct tree representation not only can match the degree-distribution entropy, but can be made adaptive to a variety of other entropy measures: *e.g.* trees with a particular probability distribution of number of children (a node has i children with probability p_i).

2 Tree Decomposition

At the heart of our method is the tree decomposition technique. Vaguely speaking, we aim to decompose the tree into subtrees of roughly the same size. Geary *et al.* [7] and He *et al.* [10] use the same decomposition algorithm to match the decomposition algorithm of Munro *et al.* [16] in the binary tree case. Given the subtree size L, the algorithm decomposes a tree into subtrees with size between L and $3L$ (with possible exception of the root subtree). Furthermore, these subtrees are disjoint other than their roots.

The drawback with their algorithms is that the number of child subtrees of a component can grow arbitrarily large (roughly as large as the size of components). With our decomposition technique, the number of child subtrees of a subtree does not exceed the original node degrees. We guarantee this by allowing (a small number of) undersized subtrees.

Theorem 1. *A tree with n nodes can be decomposed into $\Theta(n/L)$ subtrees of size at most $2L$. These are pairwise disjoint aside from the subtree roots. Furthermore, aside from edges stemming from the component root nodes, there is at most one edge per component leaving a node of a component to its child in another component.*

Figure 1 depicts the result of our decomposition algorithm. We start the proof by considering the nodes that have many descendants:

Definition 1. *For a fixed parameter L, a node is* **heavy** *if it has at least L descendants (including itself). Ancestors of a heavy node are heavy by the definition. Therefore, heavy nodes form a subtree on the original tree. We call this tree as the* **heavy-subtree**. *A* **branching node** *is a node which has at least two heavy children.* **Branching edges** *are the edges between a branching node and its heavy children.*

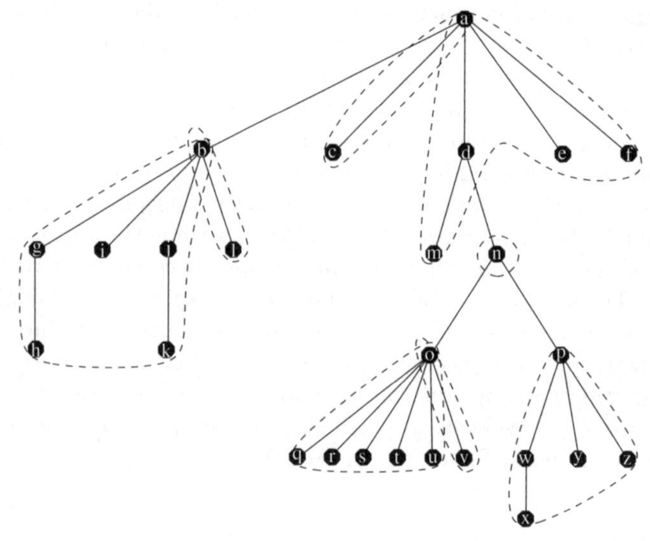

Fig. 1. A tree decomposed into component subtrees for value $L = 5$

For instance, in the tree of figure 1, heavy nodes are a, b, d, n, o, p. Branching nodes are a, n and branching edges are ab, ad, no, np. A crucial observation is that the number of branching edges is bounded (we omit the proof):

Lemma 1. *The number of branching nodes and edges in a tree with n nodes and parameter L is $O(n/L)$.* □

As with previous decomposition methods [5,16,7], we use a recursive bottom-up approach. To decompose a tree rooted at a node v, we first recursively decompose the trees rooted at its children $u_1, \ldots, u_k$. Each recursive call decomposes a tree rooted at a node and returns the component subtrees. Component subtrees that do not contain any of $u_1, \ldots, u_k$ are *permanent* and remain intact. The root components that contain one of $u_1, \ldots u_k$ are exception: they can be declared *temporary*. The temporary components and the parent v can possibly be merged together. The merging of the temporary components depends on the number of heavy children of v:

1. if v has no heavy children (*e.g.* node b in figure 1), entire children subtrees are temporary components to be merged together. We create a new component initially containing only v. We scan the list of children $u_1, \ldots, u_k$ from left to right adding the entire tree rooted at the current child to the component. If the component size exceeds L, we finalize that component and create a new component containing v only and continue in this manner. Since none of the children is heavy, the size of components does not exceed $2L$. The last such component can have size less than L. If we had created at least another component aside from the last component, we finalize the last component (charging its small size to its neighbor component which has the right size).

Otherwise if there is only one component, we have put all descendants of v together in a component which we declare as temporary and send up to the parent of v .

2. if v has only one heavy child u_i (*e.g.* node d in figure 1 as n is heavy), we put children of v into components analogously to the previous case. The only difference occurs where the component containing u_i has been declared permanent as opposed to temporary. In this case, we simply ignore u_i, skipping from u_{i-1} to u_{i+1} during the scan.
3. If v is a branching node–*i.e.* with two or more heavy children (*e.g.* nodes a, n in figure 1)– then among children $u_1, \ldots, u_k$, there are $h \geq 2$ heavy nodes $u_{i_1}, \ldots, u_{i_h}$. We first declare permanent the components containing these heavy nodes. If the component containing u_{i_j} for some j is undersized, we charge it to the branching edge vu_{i_j}.

   If all left children are heavy, v by itself is a permanent single-node component (we charge this undersized component to branching node v itself). Otherwise, the remaining children of v are broken by the heavy nodes into intervals of consecutive non-heavy nodes. We consider the intervals separately, treating each as in the first case. The difference is we charge the possible undersized component at the end of each such interval l to one of the interval's end edges $vu_{i_{l-1}}$ or vu_{i_l} (note that $vu_{i_{l-1}}, vu_{i_l}$ are branching edges) .

One can verify that the manner the components are constructed guarantees that the number of nodes within a subtree does not exceed $2L$ and moreover, aside from the components' root nodes there is at most one edge stemming out of a node of a component to a child in another component. Furthermore, to bound the number of components, one only has to account for the undersized components. One can charge undersized components to branching edges and nodes which we know by lemma 1, there are $\Theta(n/L)$ of them. Therefore, the number of undersized components is $O(n/L)$ and thus the total number of components is $\Theta(n/L)$.

3 Ordinal Trees

In this section, we outline our succinct representation for ordinal trees. The representation is analogous to Munro *et al.* [16] and the simple representation of Geary *et al.* [7] in that it is a two-level recursive decomposition of a given tree. In the first level of recursion, the tree with n nodes is first decomposed into subtrees using value $L = \lceil \lg^2 n \rceil$, and subsequently these subtrees are, in turn, decomposed into yet smaller subtrees using value $L = \lceil \lg n/4 \rceil$ to obtain the subtrees on the second level of recursion. Using the terminology of Geary *et al.* [7], we refer to as the subtrees on the first level by *mini-trees* and the second level by *micro-trees.*

Micro-trees which have size less than $\lceil \lg n/2 \rceil$ are small enough to be represented by a look-up table. The table requires $o(n)$ bits and stores encodings of all trees with sizes up to $\lceil \lg n/2 \rceil$ along with answers to variety of types of queries for each of those trees. The representation of a micro-tree with k nodes consists of two fields: the first field simply is the size of the micro-tree ($O(\log k) = O(\lg\lg n)$ bits) and the second field is an index to the look-up table ($2k$ bits). These indices sum up to $2n$ bits over all micro trees and are the dominant space term; other auxiliary data amounts to $o(n)$ bits.

Mini-trees consist of micro-trees and links between them. Links between different micro-trees can be either in form of a common root node or an edge from a non-root node from a micro-tree to the root of another micro-tree. We represent such an edge vr by introducing a *dummy node* on it. we introduce a dummy node d on the edge between the micro-tree root r and node v and replace vr by vd and dr. Edge vd is accounted for the micro-tree representation, and edge dr is explicitly stored as a $O(\log\log n)$-bit pointer. We refer to edges with a dummy parent such as dr as *dummy edges*. It is easy to see that these pointers require $o(n)$ space overall mini-trees.

To represent the common roots among micro-trees, we use the fully indexable dictionary (FID) of Raman *et al.* [19]. They showed that given a set S a subset of a universe U, there is a FID on S that requires $\lg\binom{|U|}{|S|} + O(|U|\log\log|U|/\log|U|)$ bits and supports rank/select on elements and non-elements of S in constant time. Given a root node v with children $u_1, \ldots, u_k$ in p different micro-trees, if $i_1, \ldots, i_p$ are the indices of children that belong to a different micro-tree than their immediate left siblings, we form set $I = \{i_1, i_2, \ldots, i_p\}$ over the universe of $[k]$. We represent this set as a FID to navigate on children of v. The required space for this FID is $\lg\binom{k}{p} + O(k\log\log k/\log k)$ bits. We omit the details of the proof that the collective space of these structures contributes only to $o(n)$.

The tree consists of mini-trees and links between them. The tree over mini-trees is represented analogously to the manner a mini-tree is represented over micro-trees: *i.e.* explicit pointers for edges coming out of mini-trees from non-root nodes and a FID to represent edges out of a common mini-tree root. One can assess the space analogously to $o(n)$.

Operations. We demonstrate that various operations on ordinal trees can be implemented trivially using our representation. Table 2 defines a comprehensive list of operations suggested by He *et al.* [10] for ordinal trees. We show an implementation for the subtree size operation as an example of how straightforward the support of operations becomes in our representation.

To compute the subtree size of node v, we explicitly store the subtree size at mini-tree roots and we store the subtree size at micro-tree roots within mini-trees, and the look-up table contains the subtree size within a micro-tree for each node of the micro-tree.

If v is a mini-tree root then the value is explicitly stored. Otherwise, if v is a micro-tree root, we have the subtree size within the mini-tree stored. If v is not a micro-tree root then we determine the subtree size within the micro-tree from the look-up table to get the first value. Thus far, we have counted the descendants within the mini-tree; We determine if v is an ancestor of the dummy node of the mini-tree (the ancestor query is easy to support) and if so we add the subtree size value of the mini-tree stemming off the dummy node.

Table 2. Comprehensive list of operations on an ordinal tree suggested by [10]

Operations	Definition
child(v,i), child_rank(v)	i^{th} child of node v, Number of left siblings of node v
degree(v), subtree_size(v)	number of children of v, Number of descendants of v
depth(v), height(v)	the depth/height of node v
leftmost(rightmost)_leaf(v), leaf_size(v)	v's leftmost/rightmost descendant leaf , number of descendant leaves
leaf_rank(v), leaf_select(v)	number of leaves before v in preorder, i^{th} leaf of the tree in preorder
node_rank$_{\text{pre}}$(i), node_select$_{\text{pre}}$(v)	position of v in preorder, i^{th} node in pre order
node_rank$_{\text{post}}$(i), node_select$_{\text{post}}$(v)	position of v in post order, i^{th} node in post order
level_anc(v, i), LCA(x, y), distance(x, y)	ancestor of v at level i, lowest common ancestor and distance of x, y
level_left/rightmost(i), level_succ/pred(v)	left/right most node at level i, successor or predecessor of v on its level

4 Cardinal Trees

In this section, we show the uniform approach can be applied to represent cardinal (k-ary) trees. This representation is the first succinct structure that supports, in constant time, cardinal-type queries such as "find the child labeled j" as well as all ordinal-type queries such as subtree size, degree, or the i-th child.

The number of k-ary trees is $C(n,k) = \frac{1}{kn+1}\binom{kn+1}{n}$ [8] which suggests that a space-optimal representation requires $\lg C(n,k) = (k \lg k - (k-1)\lg(k-1))n - O(\lg(kn))$ bits. We assume a RAM model with word size $w = \max\{\lg n, \lg k\}$.

The representation is a two-level recursive decomposition of the tree analogous to the representation for ordinal trees in section 3. We decompose the tree with value $L = \lg^2 w$ into mini-trees and then recursively decompose each mini-tree into micro-trees with value $L = \max\left\{\frac{\lg w}{4 \lg k}, 1\right\}$. Without loss of generality, we assume $n \geq k$ and thus $w = \lg n$. All the arguments go through analogously where $k > n$ which causes $w = \lg k$ and mini-trees with $L = \lg^2 k$ and micro-trees with $L = 1$. Hence, we can assume $L = \lg^2 n$ for mini-trees and $L = \max\left\{\frac{\lg n}{4 \lg k}, 1\right\}$ for micro-trees.

The representation only differs from that of ordinal trees in how we form the look-up table and represent the roots of mini/micro-trees. We single out nodes that are roots of a micro or a mini tree and represent them separately. The representation we use is the indexable dictionary (ID) of Raman *et al.* [19] (as opposed to their fully indexable dictionary (FID)). They showed that given a set S a subset of a universe U, there is an ID on S that requires $\lg\binom{|U|}{|S|} + o(|S|) + O(\log\log|U|)$ bits and supports rank/select on elements S (In contrast to a FID, we cannot perform rank on non-elements). Here, the universe is k-slots $U = \{1, 2, \ldots, k\}$ and our subset S is the set of present edges. In contrast to ordinal trees, in a root of a micro-tree, we do not confine ourselves within the framework of the containing mini-tree and use the ID on all edges of the root. We note that all ordinal-tree structures are included in our representation such as the FID on roots of micro-trees built over the universe of present edges (confined to the containing mini-tree). The ID and FID will help us answer the cardinal-type queries as well as ordinal-type queries on a root node.

The look-up table must contain all possible micro-trees. Since we keep root nodes' information separately, the trees in the look-up table are such that all nodes are k-ary except for the roots whose children are only ordered. We refer to such trees as *root-relaxed* cardinal trees. We enumerate all root-relaxed tree of size less than $\frac{\lg n}{4\lg k}$ and list them in the lookup table.

The rest of the representation is the same as the ordinal representation: for instance, dummy nodes and edges are introduced and represented in the same manner. Now we argue that the representation is space optimal within lower order terms.

Space optimality: All auxiliary data pertinent to the ordinal tree can be proved to sum to $o(n \lg k)$ bits analogously to the proof in section 3. Thus, we only have to account for the new structures we have introduced: IDs on the root nodes and the sum of indices to the look-up table. An ID on a root node v with d_v children requires $\lg \binom{k}{d_v} + o(d_v) + O(\log\log k)$ from which the first term is dominant. Hence, the contribution of IDs to the space over the entire tree is $\sum_{v:\,root} \lg \binom{k}{d_v}$.

The space required to represent a micro-tree is the size of the index to the look-up table. Consider a root-relaxed tree T with root r and root children $r_1, \ldots, r_d$. We define T_i as the subtree rooted at child r_i and refer to its size as $n_i = |T_i|$. We use enumeration to encode root-relaxed trees and thus we obtain the shortest code. Thus the encoding requires fewer than $\sum_{i=1}^{d} (\lg n_i + \lg C(n_i, k))$ bits which sum as follows:

$$\sum_{i=1}^{d} (\lg n_i + \lg C(n_i, k)) = \lg \prod_{i=1}^{d} n_i C(n_i, k) = \lg \prod_{i=1}^{d} \binom{kn_i}{n_i - 1} \leq \lg \binom{k(|T|-1)}{|T|-1-d_v}.$$

Over all micro-trees these terms together with space for IDs which is $\lg \binom{k}{d_v}$ for each root v sum to:

$$\sum_{T_i} \left(\lg \binom{k(|T_i|-1)}{|T_i|-1-d_{root}} + \lg \binom{k}{d_{root}} \right) = lg \left(\prod_{T_i} \binom{k(|T_i|-1)}{|T_i|-1-d_{root}} \binom{k}{d_{root}} \right),$$

which is less than $\lg \binom{kn}{n}$. Thus, the space requirement of our representation matches the lower bound, to within lower order terms: $\lg C(n,k) + o(n \log k)$.

Operations in constant time: We can support all ordinal-type operations listed in table 2 analogously to ordinal trees. To determine the child labeled i of a node v, if v is not a micro-tree root, then the answer is looked-up from the table. If v is a root node, then we use its ID to see if there is a child at that label. If it exists, we perform rank(i) to know how many siblings to the left there are. Then we can use select on the FID to actually find the mini-tree and then the micro-tree and finally the child labeled i.

5 Free Trees

A *free tree* as a tree with no particular order among children of nodes. We are interested in succinct encodings of such trees allowing navigation in the tree in constant time. A *free binary tree* is a free tree such that nodes have at most two children, or alternatively a binary tree in which ignore the distinction between left and right branches. To show the power of the uniform approach we explain how these families of trees can be encoded succinctly.

Lower bounds. The lower bounds for binary and general free trees come directly from counting by information theory. Define $FB(n)$ and $F(n)$ as the number of free binary trees, and general free trees with n nodes, respectively.

There is no known explicit closed form formula for $FB(n)$ or $F(n)$. Nevertheless, asymptotic behavior of either series is well-studied [9,18,6,22]. The sequence $(FB(n)), n = 1, 2, \ldots$ is known as Etherington-Wedderburn [6,22] sequence and from its asymptotic behavior one can infer that asymptotically $\lg FB(n) = (1.3122\ldots)n+o(n)$. Otter [18] described the asymptotic behavior for $F(n)$ from which one obtains that asymptotically $\lg F(n) = (1.5639\ldots n)+o(n)$. This implies that free trees can potentially be represented more space-efficiently than ordinal trees which require $2n + o(n)$ bits.

Theorem 2. *The information-theoretic lower bound on the number of bits required to represent free binary trees and free general trees with n nodes is $(1.3122\ldots)n + o(n)$ and $(1.5639\ldots)n + o(n)$ respectively.* □

Upper bounds. The representation differs from that of ordinal trees in section 3 in the look-up table. In the case of free binary trees, all free binary trees of size up to $\frac{1}{4}\lg n$ are enumerated modulo isomorphisms and listed in the look-up table in increasing order of their sizes. To represent a micro-tree we use a pair (k, i) index to the table. k is the size of the micro-tree and i is the index to the look-up table which is an offset from the start location of trees with size k. All auxiliary data are carried forward from ordinal trees as they only take $o(n)$ space. One can easily argue that the total bits required by the first fields of pairs (k) is also $o(n)$. Therefore, the dominant field is the sum of the bits of the second fields of pairs (i). Theorem 2 suggest that the size of this field for a tree of size t is $(1.31\ldots)t+o(t)$ bits. The second term $o(t)$ term adds up to $o(n)$ over the entire tree. The first term is the dominant term which adds up to $(1.31\ldots)n + o(n)$ over the entire tree when n is the number of nodes. The $(1.56\ldots)n + o(n)$ bit representation for free general trees is analogous; the look-up table lists free general trees as opposed to free binary trees:

Theorem 3. *The succinct representation for free binary trees and free general trees with n nodes requires $(1.3122\ldots)n + o(n)$ and $(1.5639\ldots)n + o(n)$ bits respectively and supports all navigational operations listed in table 2 in constant time.* □

6 Entropy-Based Succinct Encodings

Thus far, we have assumed a uniform distribution among trees belonging to a certain family of trees. However, there might be many applications so that some trees are biased against other trees within the tree family, and therefore the distribution is non-uniform. Thus, entropy-based succinct encodings are necessary. Jansson *et al.* [13] were the first to give entropy-based succinct encodings for the degree-distribution entropy. In this section, we show how our method can be used to match the degree-distribution entropy as well as a variety of other entropy measures.

6.1 Succinct Encoding Based on Degree-Distribution Entropy

The degree-distribution of an ordinal tree with n nodes is a series of numbers $(n_0, n_1, \ldots)$ such that the tree has n_i nodes with exactly i children ($\sum_i n_i = n$ and $\sum_i i\, n_i = n - 1$). Rote [20] showed that the number of trees with a given degree-distribution is $\frac{1}{n}\binom{n}{n_0,\ldots,n_{n-1}}$, the logarithm of which is $L(T) = \sum_i n_i \lg \frac{n}{n_i}$ to within lower order terms. $L(T)$ is therefore a lower bound on the required number of bits to represent the tree.

Jansson *et al.* [13] gave a representation that requires $L(T) + O\left(\frac{n(\log\log n)^2}{\log n}\right)$ number of bits and supports a variety of operations in constant time. Using our approach we obtain another space-optimal succinct representation with $L(T) + O\left(n \log\log\log n / \log\log n\right)$ number of bits supporting all operations in table 2 in constant time. They did not assume that the degree-distribution is explicitly given. We observe that we can make explicit the assumption that the degree-distribution is given as it takes negligible space to encode the sequence and have it explicitly stored. Thus we can accompany it with the succinct representation.

Our succinct representation sensitive to degree-distribution entropy is the same as that of the ordinal trees with the difference in the look-up table. The look-up table contains all trees with less than $\frac{1}{4}\lg n$ as in ordinal trees; However, the trees are ordered based on their degree-distribution sequence in the lexicographical order and listed in the table accordingly. In order to encode a micro-tree T, we use an index to the table. The index to the table is a pair $(\mathcal{N}, k)$ where $\mathcal{N}$ is the degree-distribution encoding of the tree and k is an offset in the table from where the trees with degree-distribution $\mathcal{N}$ start to the actual position of tree T we reference to.

One can easily verify that the total number of bits required by the first fields of indices (*i.e.* $\mathcal{N}$) is negligible. The second field k is the dominant term. The size of this field for a micro-tree T_t, by a counting argument, is $\lceil L(T_t) \rceil = \left\lceil \lg\left(\frac{1}{|T_t|}\binom{|T_t|}{n_{t,0}\ldots n_{t,|T_t|}}\right)\right\rceil$ where $n_{t,i}$ is the number of nodes with i children in tree T_t. A node with more than $\log n$ children in the original tree is a micro-tree root and its children are split among different micro-trees. Since there are $\Theta(n \log n)$ of these roots and each micro-tree root contributes at most $\log\log n$

bits to the space, the sum of these terms is $O(n \log\log n / \log n)$ and within our space bound. The sum over all micro-trees $T_1, \ldots, T_m$ modulo their roots can be assessed as follows:

$$\sum_{t=1}^{m} \lg\left(\frac{1}{|T_t|}\binom{|T_t|-1}{n_{t,0},\ldots,n_{t,|T_t|}}\right) = \lg\prod_t \frac{1}{|T_t|}\binom{|T_t|-1}{n_{t,0}\ldots n_{t,|T_t|}}$$

$$\leq \lg\left(\frac{1}{n}\binom{n}{n_0,\ldots,n_{n-1}}\right) \approx L(T).$$

Hence, the representation has the optimal space within lower order terms and clearly we can perform all operations listed in table 2 in constant time as in an ordinal tree.

6.2 Other Entropy Measures

Similar to the manner in which we represented trees adaptive to their degree-distribution entropy, we can use the approach to obtain succinct representations adaptive to various other combinatorial properties and entropy measures. For instance, consider the family of ordinal trees such that internal nodes have at least two children. The number of such trees with n nodes is known as Riordan number [3]. The logarithm based two of Riordan numbers is asymptotically $\lg(3)n+o(n) \approx 1.58n+o(n)$. One can use our approach to encode this family of trees. Similarly, The family of ordinal trees with a fixed constant upper bound d on the number children of a node can be represented in the same manner. More generally, where there is a probability distribution for the number of children of a node, our representation can match the entropy bound.

Another interesting family of trees is AVL trees which consists of binary trees such that the height of left and right subtrees differ by at most one. Odlyzko [17] showed that if a_n is the number of AVL trees with n nodes, $\lg a(n) \approx (0.9381\ldots)n$, our representation matches this entropy bound and thus can represent AVL trees in optimal number of bits to within lower order terms.

7 Conclusion

In this paper, we proposed a uniform approach towards succinct representation of trees. We showed that all families of trees with an existing succinct representation can be represented using our framework. Our representation improves on the existing ones on cardinal trees as we are able to answer ordinal-type queries such as subtree size as well as cardinal-type queries. We introduce a new family of trees: *free trees*. We demonstrated how easily our approach can represent these trees succinctly We argued that our approach can represent trees succinctly adaptive to the degree-distribution entropy. We discussed that a variety of other entropy measures can be dealt with similarly.

References

1. Barbay, J., He, M., Munro, J.I., Rao, S.S.: Succinct indexes for strings, binary relations and multi-labeled trees. In: Proceedings of the 18th ACM-SIAM Symposium on Discrete Algorithms (SODA), pp. 680–689. ACM-SIAM, New York (2007)
2. Benoit, D., Demaine, E.D., Munro, J.I., Raman, R., Raman, V., Rao, S.S.: Representing trees of higher degree. Algorithmica 43(4), 275–292 (2005)
3. Bernhart, F.R.: Catalan, motzkin, and riordan numbers. Discrete Mathematics 204, 72–112 (1999)
4. Clark, D.R.: Compact pat trees. PhD thesis, Waterloo, Ontario. Canada (1998)
5. Clark, D.R., Munro, J.I.: Efficient suffix trees on secondary storage (extended abstract). In: SODA, pp. 383–391 (1996)
6. Etherington, I.M.H.: Non-associate powers and a functional equation. The Mathematical Gazette 21(242), 36–39 (1937)
7. Geary, R.F., Raman, R., Raman, V.: Succinct ordinal trees with level-ancestor queries. ACM Transactions on Algorithms 2(4), 510–534 (2006)
8. Graham, R.L., Knuth, D.E., Patashnik, O.: Concrete Mathematics: A Foundation for Computer Science. Addison-Wesley Longman Publishing Co. Inc. Boston (1994)
9. Harary, F., Palmer, E.M.: Graphical Enuemration. Academic Press, New York (1973)
10. He, M., Munro, J.I., Rao, S.S.: Succinct ordinal trees based on tree covering. In: Arge, L., Cachin, C., Jurdziński, T., Tarlecki, A. (eds.) ICALP 2007. LNCS, vol. 4596, pp. 509–520. Springer, Heidelberg (2007)
11. Jacobson, G.: Space-efficient static trees and graphs. In: Foundations of Computer Science. 30th Annual Symposium on (30 October-1 November 1989), pp. 549–554 (1989)
12. Jacobson, G.J. Succinct static data structures. PhD thesis, Pittsburgh, PA, USA (1988)
13. Jansson, J., Sadakane, K., Sung, W.-K.: Ultra-succinct representation of ordered trees. In: Bansal, N., Pruhs, K., Stein, C. (eds.) SODA, pp. 575–584. SIAM, Philadelphia (2007)
14. Knuth, D.E.: The Art of Computer Programming, 3rd edn. vol. 1. Addison-Wesley, Reading (1997)
15. Munro, J.I., Raman, V.: Succinct representation of balanced parentheses, static trees and planar graphs. In: IEEE Symposium on Foundations of Computer Science, pp. 118–126 (1997)
16. Munro, J.I., Raman, V., Storm, A.J.: Representing dynamic binary trees succinctly. In: SODA, pp. 529–536 (2001)
17. Odlyzko, A.M.: Some new methods and results in tree enumeration, (May 04, 1984)
18. Otter, R.: The number of trees. The Annals of Mathematics, 2nd Ser. 49(3), 583–599 (1948)
19. Raman, R., Raman, V., Rao, S.S.: Succinct indexable dictionaries with applications to encoding k-ary trees and multisets. In: SODA, pp. 233–242 (2002)
20. Rote, G.: Binary trees having a given number of nodes with 0,1, and 2 children. Séminaire Lotharingien de Combinatoire 38 (1997)
21. Storm, A.J. Representing dynamic binary trees succinctly. Master's thesis, School of Computer Science, University of Waterloo, Waterloo, Ontario, Canada (2000)
22. Wedderburn, J.H.M.: The Functional Equation $g(x^2) = 2ax + [g(x)]^2$. The Annals of Mathematics, 2nd Ser. 24(2), 121–140 (1922)

An O($n^{1.75}$) Algorithm for $L(2,1)$-Labeling of Trees*

Toru Hasunuma[1], Toshimasa Ishii[2], Hirotaka Ono[3], and Yushi Uno[4]

[1] Department of Mathematical and Natural Sciences, The University of Tokushima, Tokushima 770–8502 Japan
hasunuma@ias.tokushima-u.ac.jp
[2] Department of Information and Management Science, Otaru University of Commerce, Otaru 047-8501, Japan
ishii@res.otaru-uc.ac.jp
[3] Department of Computer Science and Communication Engineering, Kyushu University, Fukuoka 812-8581, Japan
ono@csce.kyushu-u.ac.jp
[4] Department of Mathematics and Information Sciences, Graduate School of Science, Osaka Prefecture University, Sakai 599-8531, Japan
uno@mi.s.osakafu-u.ac.jp

Abstract. An $L(2,1)$-labeling of a graph G is an assignment f from the vertex set $V(G)$ to the set of nonnegative integers such that $|f(x) - f(y)| \geq 2$ if x and y are adjacent and $|f(x) - f(y)| \geq 1$ if x and y are at distance 2 for all x and y in $V(G)$. A k-$L(2,1)$-labeling is an assignment $f : V(G) \to \{0, \ldots, k\}$, and the $L(2,1)$-labeling problem asks the minimum k, which we denote by $\lambda(G)$, among all possible assignments. It is known that this problem is NP-hard even for graphs of treewidth 2. Tree is one of a few classes for which the problem is polynomially solvable, but still only an $O(\Delta^{4.5}n)$ time algorithm for a tree T has been known so far, where Δ is the maximum degree of T and $n = |V(T)|$. In this paper, we first show that an existent necessary condition for $\lambda(T) = \Delta + 1$ is also sufficient for a tree T with $\Delta = \Omega(\sqrt{n})$, which leads a linear time algorithm for computing $\lambda(T)$ under this condition. We then show that $\lambda(T)$ can be computed in $O(\Delta^{1.5}n)$ time for any tree T. Combining these, we finally obtain an $O(n^{1.75})$ time algorithm, which substantially improves upon previously known results.

Keywords: frequency/channel assignment, graph algorithm, $L(2,1)$-labeling, vertex coloring.

1 Introduction

Let G be an undirected graph. An $L(2,1)$-*labeling* of a graph G is an assignment f from the vertex set $V(G)$ to the set of nonnegative integers such that $|f(x) - f(y)| \geq 2$ if x and y are adjacent and $|f(x) - f(y)| \geq 1$ if x and y are at distance 2 for all x and y in $V(G)$. A k-$L(2,1)$-labeling is an assignment $f : V(G) \to \{0, \ldots, k\}$, and the $L(2,1)$-*labeling problem* asks the minimum k among all possible assignments. We call this invariant, the minimum value k, the $L(2,1)$-*labeling number* and is denoted by $\lambda(G)$. Notice that

* This research is partly supported by INAMORI FOUNDATION and Grant-in-Aid for Scientific Research (KAKENHI), No. 18300004, 18700014, 19500016 and 20700002.

J. Gudmundsson (Ed.): SWAT 2008, LNCS 5124, pp. 185–197, 2008.

we can use $k + 1$ different labels when $\lambda(G) = k$ since we can use 0 as a label for conventional reasons.

The original notion of $L(2, 1)$-labeling can be seen in Hale [8] and Roberts [10] in the context of frequency/channel assignment, where 'close' transmitters must receive different frequencies and 'very close' transmitters must receive frequencies that are at least two frequencies apart so that they can avoid interference. Due to its practical importance, the $L(2, 1)$-labeling problem has been widely studied. On the other hand, this problem is also attractive from the graph theoretical point of view since it is a kind of vertex coloring problem. In this context, $L(2, 1)$-labeling is generalized into $L(h, k)$-labeling for arbitrary nonnegative integers h and k, and in fact, we can see that $L(1, 0)$-labeling ($L(h, 0)$-labeling, actually) is equivalent to the classical vertex coloring.

Related Work: There are also a number of studies about the $L(2, 1)$-labeling problem from the algorithmic point of view. It is known to be NP-hard for general graphs [7], and it still remains NP-hard for some restricted classes of graphs, such as planar, bipartite, chordal graphs [1], and recently it turned out to be NP-hard even for graphs of treewidth 2 [5]. In contrast, only a few graph classes are known to have polynomial time algorithms for this problem. Among those, Chang and Kuo [4] established a polynomial time algorithm for the $L(2, 1)$-labeling problem for trees. Their polynomial time algorithm fully exploits the fact that $\lambda(T)$ is either $\Delta + 1$ or $\Delta + 2$ for any tree T. It is based on dynamic programming, and runs in $O(\Delta^{4.5}n)$ time, where Δ is the maximum degree of a tree T and $n = |V(T)|$.

Our Contributions: In this paper, we first show that an existent necessary condition for $\lambda(T) = \Delta + 1$ for a tree T is also sufficient for trees with $\Delta = \Omega(\sqrt{n})$, which leads a linear time algorithm for computing $\lambda(T)$ under this condition. Then we show that the $L(2, 1)$-labeling problem can be solved in $O(\Delta^{1.5}n)$ time for any input tree. Our approach is based on dynamic programming similar to Chang and Kuo's $O(\Delta^{4.5}n)$-time algorithm [4], where its $\Delta^{2.5}$-factor comes from the complexity of solving the bipartite matching problem of a graph with order Δ, and its $\Delta^2 n$-factor from the number of iterations for solving bipartite matchings. In spite that our algorithm is also under the same framework, the running time $O(\Delta^{1.5}n)$ is attained by reducing the number of the matching problems to be solved, together with detailed analyses of the algorithm. As a result, our algorithm achieves $O(n^{1.75})$ running time, and greatly improves the best known result $O(\Delta^{4.5}n)$ time, which could be $O(n^{5.5})$ in its worst case.

Organization of this Paper: The rest of this paper is organized as follows. Section 2 gives basic definitions and related results. Section 3 shows that a necessary condition that $\lambda(T) = \Delta + 1$ for a tree T is also sufficient for trees with $\Delta = \Omega(\sqrt{n})$. In Section 4, after introducing fundamental ideas of dynamic programming for solving this problem, then we show that $\lambda(T)$ can be computed in $O(\Delta^{1.5}n)$ time for a tree T. Combining the results in Sections 3 and 4, Section 5 presents the overall $O(n^{1.75})$ time algorithm for any input tree. Finally, Section 6 gives some concluding remarks.

2 Preliminaries

2.1 Definitions and Notations

A graph G is an ordered set of its vertex set $V(G)$ and edge set $E(G)$ and is denoted by $G = (V(G), E(G))$. We assume throughout this paper that all graphs are undirected, simple and connected, unless otherwise stated. Therefore, an edge $e \in E(G)$ is an unordered pair of vertices u and v, which are *end vertices* of e, and we often denote it by $e = (u, v)$. Two vertices u and v are *adjacent* if $(u, v) \in E(G)$, and two edges are *adjacent* if they share one of their end vertices. A graph $G = (V(G), E(G))$ is called *bipartite* if the vertex set $V(G)$ can be divided into two disjoint sets V_1 and V_2 such that every edge in $E(G)$ connects a vertex in V_1 and one in V_2; such G is denoted by (V_1, V_2, E).

For a graph G, the *(open) neighborhood* of a vertex $v \in V(G)$ is the set $N_G(v) = \{u \in V(G) \mid (u, v) \in E(G)\}$, and the *closed neighborhood* of v is the set $N_G[v] = N_G(v) \cup \{v\}$. The *degree* of a vertex v is $|N_G(v)|$, and is denoted by $d_G(v)$. We use $\Delta(G)$ to denote the maximum degree of a graph G. A vertex whose degree is $\Delta(G)$ is called *major*. We often drop G in these notations if there are no confusions. A vertex whose degree is 1 is called a *leaf vertex*, or simply a *leaf*. A *path* in G is a sequence $v_1, v_2, \ldots, v_\ell$ of vertices such that $(v_i, v_{i+1}) \in E$ for $i = 1, 2, \ldots, \ell - 1$, or equivalently, a sequence $(v_1, v_2), (v_2, v_3), \ldots, (v_{\ell-1}, v_\ell)$ of edges (v_i, v_{i+1}) for $i = 1, 2, \ldots, \ell - 1$. The *length* of a path is the number of edges on it. The *distance* between two vertices u and v is the minimum length of paths connecting u and v. A path $v_1, v_2, \ldots, v_\ell$ is a *cycle* if $v_1 = v_\ell$. A graph is a *tree* if it is connected and has no cycle.

For describing algorithms, it is convenient to regard the input tree to be rooted at an arbitrary vertex r of degree 1. Then we can define the parent-child relationship on vertices in the usual way. For any vertex v, the sets of its children and grandchildren are denoted by $C(v)$ and $C^2(v)$, respectively. For a vertex v, define $d'(v) = |C(v)|$.

2.2 Related Results and Basic Properties

In general, $L(h, k)$-labelings of a graph G are defined for arbitrary nonnegative integers h and k, as an assignment of nonnegative integers to $V(G)$ such that adjacent vertices receive labels at least h apart and vertices connected by a 2-length path receive labels at least k apart. This problem is one of the generalizations of the vertex coloring problem since $L(h, 0)$-labeling problem is equivalent to it. Therefore, we can hardly expect that the $L(h, k)$-labeling problem is tractable, and in fact, $L(0, 1)$- and $L(1, 1)$-labeling problems are known to be NP-hard, for example. We can find a lot of related results on $L(h, k)$-labelings in comprehensive surveys by Calamoneri [2] and Yeh [12].

As for the $L(2, 1)$-labeling problem, it is also known to be NP-hard for general graphs [7]. It remains NP-hard for planar graphs, bipartite graphs, chordal graphs [1], and even for graphs of treewidth 2 [5]. In contrast, very few affirmative results are known, e.g., we can decide the $L(2, 1)$-labeling number of paths, cycles, wheels [7] and trees [4] within polynomial time.

We here review some significant results on $L(2, 1)$-labeling of graphs or trees that will become relevant later in this paper. We can see that $\lambda(G) \geq \Delta + 1$ holds for any graph G. Griggs and Yeh [7] showed a necessary condition for $\lambda(G) = \Delta + 1$ on any graph G, by observing that any major vertex in G must be labeled 0 or $\Delta + 1$ when $\lambda(G) = \Delta + 1$.

Lemma 1. [7] *If $\lambda(G) = \Delta + 1$, then for any $v \in V(G)$, $N_G[v]$ contains at most two major vertices.*

Lemma 2. [7] *For any tree T, $\lambda(T)$ is either $\Delta + 1$ or $\Delta + 2$.*

Concerning the latter lemma, they also conjectured the problem of determining if $\lambda(T)$ is $\Delta+1$ or $\Delta+2$ is NP-hard. Chang and Kuo [4] disproved this by presenting a polynomial time algorithm for computing $\lambda(T)$, whose running time is $O(\Delta^{4.5}n)$. Since tree is one of the most basic graph classes, this yields several affirmative results for more general graph classes, e.g., *p-almost trees*, for which $\lambda(G)$ is computed in $O(\lambda^{2p+4.5}n)$ time [6].

3 A Linear Time Algorithm for Trees with $\Delta = \Omega(\sqrt{n})$

From now on, we focus on the $L(2,1)$-labeling problem on trees. Obviously, Chang and Kuo's algorithm [4] runs in linear time if $\Delta = O(1)$. In this section, we show that the $L(2,1)$-labeling problem for trees can be also solved in linear time if $\Delta > \sqrt{n+3} + 3$. Let T be a tree. As shown in Lemmas 1 and 2, we have a necessary condition for $\lambda(T) = \Delta + 1$ (or a sufficient condition for $\lambda(T) = \Delta + 2$), but no simple necessary and sufficient condition is known although some research such as [11] gave a sufficient condition for $\lambda(T) = \Delta + 1$. Here, we present another sufficient condition for $\lambda(T) = \Delta + 1$, which implies that the necessary condition for $\lambda(T) = \Delta + 1$ of Lemma 1 is also sufficient for large Δ. Let $N^3[v]$ denote the set of vertices whose distance from v is at most three.

Theorem 1. *If for any $v \in V(T)$, $N^3[v]$ contains at most $\Delta - 6$ major vertices and $N[v]$ contains at most two major vertices, then $\lambda(T) = \Delta + 1$.*

Proof. Suppose that for any $v \in V(T)$, $N^3[v]$ contains at most $\Delta - 6$ major vertices and $N[v]$ contains at most two major vertices. At first, label every major vertex with 0 or $\Delta + 1$ so that two major vertices within distance two do not have the same label. Since for any $v \in V(T)$, $N[v]$ contains at most two major vertices, this labeling can be correctly done. Next, regard T as a rooted tree by choosing one vertex as the root. Following the definition of the $L(2,1)$-labeling, we will label each vertex in the rooted tree in the breadth-first-search order. Suppose that a vertex v is labeled b and the parent of v is labeled a, where $|a - b| \geq 2$. Divide the set $C(v)$ of children of v into $C'(v)$, $C''(v)$ and $R(v)$ as follows:

- $C'(v) = \{w \in C(v) \mid w$ is not a major vertex and has a major vertex in $C(w) \cup C^2(w)\}$,
- $C''(v) = \{w \in C(v) \mid w$ is a major vertex $\}$,
- $R(v) = C(v) - C'(v) - C''(v)$.

Note that $|C'(v)| \leq \Delta - 6$, $|C''(v)| \leq 2$, and if $d(v) = \Delta$ then $|C'(v)| \leq \Delta - 7$ and $|C''(v)| \leq 1$.

Case 1: $d(v) < \Delta$. Let $U(a,b) = \{a, b-1, b, b+1\} \cup \{0, 1, \Delta, \Delta + 1\}$, and $\bar{U}(a,b) = \{0, 1, \ldots, \Delta + 1\} - U(a,b)$. Assign injectively labels in $\bar{U}(a,b)$ to vertices in $C'(v)$. Since $|C'(v)| \leq \Delta - 6$ and $|\bar{U}(a,b)| = \Delta + 2 - |U(a,b)| \geq \Delta - 6$, such a labeling is possible. Let $L(v)$ be the set of labels in $\bar{U}(a,b)$ which are not used in the labeling of $C'(v)$.

Case 1-1: $|C''(v)| = 0$. Assign injectively labels in $L(v) \cup (\{0, 1, \Delta, \Delta + 1\} - \{a, b - 1, b, b + 1\})$ to vertices in $R(v)$.

Case 1-2: $|C''(v)| = 1$. Assign injectively labels in $L(v) \cup (\{1, \Delta, \Delta + 1\} - \{a, b - 1, b, b + 1\})$ (resp., $L(v) \cup (\{0, 1, \Delta\} - \{a, b - 1, b, b + 1\})$) to vertices in $R(v)$, if the major vertex in $C(v)$ is labeled 0 (resp., $\Delta + 1$).

Case 1-3: $|C''(v)| = 2$. Assign injectively labels in $L(v) \cup (\{1, \Delta\} - \{b, a - 1, a, a + 1\})$ to vertices in $R(v)$.

Here, $|L(v)| = |\bar{U}(a, b)| - |C'(v)| = \Delta + 2 - |U(a, b)| - |C'(v)|$. Also,

- $|\{0, 1, \Delta, \Delta + 1\} - \{a, b - 1, b, b + 1\}| = |U(a, b)| - 4$,
- $|\{1, \Delta, \Delta + 1\} - \{a, b - 1, b, b + 1\}| \geq |U(a, b)| - 5$,
- $|\{0, 1, \Delta\} - \{a, b - 1, b, b + 1\}| \geq |U(a, b)| - 5$,
- $|\{1, \Delta\} - \{a, b - 1, b, b + 1\}| \geq |U(a, b)| - 6$.

Since $|R(v)| \leq \Delta - 2 - |C'(v)| - |C''(v)|$, each labeling in Case 1 is possible.

Case 2: $d(v) = \Delta$. Let $U(a) = \{a\} \cup \{0, 1, \Delta, \Delta + 1\}$ and $\bar{U}(a) = \{0, 1, \ldots, \Delta + 1\} - U(a)$. Assign injectively labels in $\bar{U}(a)$ to vertices in $C'(v)$. Since $|C'(v)| \leq \Delta - 7$ and $|\bar{U}(a)| = \Delta + 2 - |U(a)| \geq \Delta - 3$, such a labeling is possible. Let $L(v)$ be the set of labels in $\bar{U}(a)$ which are not used in the labeling of $C'(v)$.

Case 2-1: $|C''(v)| = 0$. Assign injectively labels in $L(v) \cup (\{\Delta, \Delta + 1\} - \{a\})$ (resp., $L(v) \cup (\{0, 1\} - \{a\})$) to vertices in $R(v)$, if v is labeled 0 (resp., $\Delta + 1$).

Case 2-2: $|C''(v)| = 1$. Assign injectively labels in $L(v) \cup (\{\Delta\} - \{a\})$ (resp., $L(v) \cup (\{1\} - \{a\})$) to vertices in $R(v)$, if v is labeled 0 (resp., $\Delta + 1$).

Here, $|L(v)| = |\bar{U}(a)| - |C'(v)| = \Delta + 2 - |U(a)| - |C'(v)|$. Also,

- $|\{\Delta, \Delta + 1\} - \{a\}| \geq |U(a)| - 3$,
- $|\{0, 1\} - \{a\}| \geq |U(a)| - 3$,
- $|\{\Delta\} - \{a\}| \geq |U(a)| - 4$,
- $|\{1\} - \{a\}| \geq |U(a)| - 4$.

Since $|R(v)| = \Delta - 1 - |C'(v)| - |C''(v)|$, each labeling in Cases 2-1 and 2-2 is possible.

It can easily be checked that the labeling of $C(v)$ is valid. Therefore, $\lambda(T) = \Delta + 1$. □

From Theorem 1, we can see that the necessary condition for $\lambda(T) = \Delta + 1$ in Lemma 1 is also sufficient if the number of major vertices is at most $\Delta - 6$.

Corollary 1. *If the number of major vertices is at most $\Delta - 6$, then $\lambda(T) = \Delta + 1$ if and only if for any $v \in V(T)$, $N[v]$ contains at most two major vertices.* □

Corollary 2. *If $\Delta > \sqrt{n+3} + 3$, then $\lambda(T) = \Delta + 1$ if and only if for any $v \in V(T)$, $N[v]$ contains at most two major vertices.*[1]

Proof. Suppose that for any $v \in V(T)$, $N[v]$ contains at most two major vertices. Assume that there are $\Delta - 5$ major vertices. They have at least $\Delta - 2$ non-major children for each, and they are all distinct; the number of vertices is bounded by $(\Delta - 5)(\Delta - 2) + (\Delta - 5) + 1$. Thus we have $n \geq (\Delta - 5)(\Delta - 1) + 1$, which yields $\Delta \leq \sqrt{n+3} + 3$. Therefore, if $\Delta > \sqrt{n+3} + 3$, then the number of major vertices is at most $\Delta - 6$. Hence, from Corollary 1, this corollary follows. □

[1] We can also obtain a better bound $\Delta > \sqrt{n + 65/16} + 11/4$ by another analysis.

Clearly, the condition of Corollary 2 can be checked in linear time. Thus, when $\Delta > \sqrt{n+3}+3$, we can decide $\lambda(T)$ in linear time, and if $\lambda(T) = \Delta+1$, then a $(\Delta+1)$-$L(2,1)$-labeling of T can be obtained by an algorithm based on the proof of Theorem 1, which runs in linear time. Otherwise, we can obtain $(\Delta + 2)$-$L(2, 1)$-labeling by Algorithm Greedy: *traverse T in the breadth first order, and if reached vertex v where $f(v) = a$ and $f(u) = b$ for its parent u, label vertices in $C(v)$ from* $\{0, 1, \ldots, \Delta+2\} - \{b, a-1, a, a+1\}$. This is always possible since $|C(v)| \le |\{0, 1, \ldots, \Delta + 2\} - \{b, a - 1, a, a + 1\}|$ for any v, and it gives $(\Delta + 2)$-$L(2, 1)$-labeling.

4 An $O(\Delta^{1.5}n)$-Time Algorithm

4.1 Chang and Kuo's Algorithm

In this subsection, we review a dynamic programming algorithm for the $L(2, 1)$-labeling problem of trees, which is proposed by Chang and Kuo [4], since our algorithm also utilizes the same formula of the principle of optimality. For a tree T with maximum degree Δ, Griggs and Yeh [7] proved that $\lambda(T) = \Delta+1$ or $\Delta+2$. The algorithm determines if $\lambda(T) = \Delta + 1$, and if so, we can easily construct the labeling with $\lambda(T) = \Delta + 1$.

Before describing the algorithm, we introduce some notations. We assume that T is rooted at some leaf vertex r for explanation. Given a vertex v, we denote the subtree of T rooted at v by $T(v)$. Let $T(u, v)$ be a tree rooted at u that forms $T(u, v) = (\{u\} \cup V(T(v)), \{(u, v)\} \cup E(T(v)))$. Note that this u is just a virtual vertex for explanation and $T(u, v)$ is uniquely decided for $T(v)$ in a sense. For a rooted tree, we call the length of the longest path from the root to a leaf its *height*. For $T(u, v)$, we define

$$\delta((u, v), (a, b)) = \begin{cases} 1 & \text{if } \lambda(T(u, v) \mid f(u) = a, f(v) = b) \le \Delta + 1, \\ 0 & \text{otherwise,} \end{cases}$$

where $\lambda(T(u, v) \mid f(u) = a, f(v) = b)$ denotes the $L(2, 1)$-labeling number on $T(u, v)$ under the assumption that $f(u) = a$ and $f(v) = b$, that is, the minimum k of k-$L(2, 1)$-labeling on $T(u, v)$ satisfying $f(u) = a$ and $f(v) = b$. This δ function satisfies the following:

$$\delta((u, v), (a, b)) = \begin{cases} 1 & \text{if there is a distinct assignment } c_1, c_2, \ldots, c_{d'(v)} \text{ on} \\ & w_1, w_2, \ldots, w_{d'(v)}, \text{ where } c_i \text{ is different from } a, b, \\ & b-1, b+1, \text{ and } \delta((v, w_i), (b, c_i)) = 1 \text{ for each } i, \\ 0 & \text{otherwise,} \end{cases} \tag{1}$$

where $w_1, w_2, \ldots, w_{d'(v)}$ are the children of v. The existence of an assignment $c_1, c_2, \ldots, c_{d'(v)}$ on $w_1, w_2, \ldots, w_{d'(v)}$ as above is formalized as the maximum bipartite matching problem; we consider a bipartite graph $G(u, v, a, b) = (V(v), X, E(u, v, a, b))$, where $V(v) = \{w_1, w_2, \ldots, w_{d'(v)} \in C(v)\}$, $X = \{0, 1, \ldots, \Delta, \Delta + 1\}$ and $E(u, v, a, b) = \{(w, c) \mid \delta((v, w), (b, c)) = 1, c \in X - \{a\}, w \in V(v)\}$. (Analogously, we also define $G(u, v, -, b)$ by $E(u, v, -, b) = \{(w, c) \mid \delta((v, w), (b, c)) = 1, c \in X, w \in V(v)\}$, which will be used in Subsection 4.3.) We can see that an assignment $c_1, c_2, \ldots, c_{d'(v)}$ on $w_1, w_2, \ldots, w_{d'(v)}$

is feasible if there exists a matching with size $d'(v)$ of $G(u, v, a, b)$. Namely, for $T(u, v)$ and two labels a and b, we can easily (i.e., in polynomial time) determine the value of $\delta((u, v), (a, b))$ if the values of δ function for $T(v, w_i)$ and any two pairs of labels are given. According to these observations, Chang and Kuo proposed a dynamic programming algorithm as shown in Table 1.

Table 1. Algorithm CK

Step 0. Let $\delta((u, v), (*, *)) := 1$ for all $T(u, v)$ of height 1, where $(*, *)$ means all pairs of labels a and b, where $|a - b| \geq 2$. Let $h := 2$.
Step 1. For all $T(u, v)$ of height h, compute $\delta((u, v), (*, *))$.
Step 2. If $h = h^*$ where h^* is the height of root r of T, then goto Step 3. Otherwise let $h := h + 1$ and goto Step 1.
Step 3. If $\delta((r, v), (a, b)) = 1$ for some (a, b), then output "Yes". Otherwise output "No". Halt.

Since Steps 0, 2 and 3 can be done just by looking up the table of δ, the running time is dominated by Step 1; the total running time of the algorithm is $O(\sum_{v \in V} t(v))$, where $t(v)$ denotes the time for calculating $\delta((u, v), (*, *))$. Each calculation of $\delta((u, v), (a, b))$ in Step 1 can be executed in $O(|V(v) \cup X|^{2.5}) = O(\Delta^{2.5})$ time, because an $O(n^{2.5})$ time algorithm is known for the maximum matching of a bipartite graph with n vertices [9]. Since the number of pairs (a, b) is at most $(\Delta + 2) \times (\Delta + 2)$, we obtain $t(v) \leq (\Delta + 2)^2 \times O(\Delta^{2.5}) = O(\Delta^{4.5})$. Thus the total running time of the algorithm is $\sum_{v \in V} t(v) = O(\Delta^{4.5} n)$ [2].

In Section 5, we propose another algorithm. It is also based on the formula (1) but it computes $\delta((u, v), (*, *))$ more efficiently by techniques shown in the following subsections.

4.2 Preprocessing Operations for Input Trees

In this subsection, we introduce preprocessing operations in our algorithm. Let T be an original input tree. These preprocessing operations are carried out for the purpose that (1) remove inessential vertices from T, where "inessential" means that they do not affect the $L(2, 1)$-labeling number of T, and (2) divide T into several subtrees that preserves the $L(2, 1)$-labeling number of T. Obviously, these operations enable to reduce the input size to solve and we may expect some speedup. However, the effect for reducing the size is not important actually, because the preprocessing operations may do nothing for some instances. Instead, a more important effect is that we can restrict the shape of input trees, which enables the amortized analysis of the running time of our algorithm shown later.

First, we describe how to remove inessential vertices.

1. Check if there is a leaf v whose unique neighbor u has degree less than Δ. If so, remove v and edge (u, v) from T until such a leaf does not exist.

This operation does not affect the $L(2, 1)$-labeling number of T, that is, $\lambda(T) = \lambda(T')$ where T is the original tree and T' is the resulting tree. This is because, in T, such leaf vertex v can be properly labeled by some number in $\{0, 1, \ldots, \Delta + 1\}$ if u and any

[2] By a careful analysis, this running time is reduced to $O(\Delta^{3.5} n)$.

other neighbor vertices of u are properly labeled by numbers among $\{0, 1, \ldots, \Delta + 1\}$. Also, the operation does not change the maximum degree Δ. Since this can be done in linear time, the labeling problem for T is equivalent to the one for T' in terms of linear time computation. Thus, from now on, we assume that an input tree T has the following property.

Property 1. All vertices connected to a leaf vertex are major vertices.

We define V_L as the set of all leaf vertices in T. Also we define V_Q as the set of major vertices whose children are all leaves.

Next, we explain how to divide T into subtrees. We call a sequence of consecutive vertices $v_1, \ldots, v_\ell$ a *path component* if $(v_i, v_{i+1}) \in E$ for all $i = 1, \ldots, \ell - 1$ and $d(v_i) = 2$ for all $i = 1, 2, \ldots, \ell$, and we call ℓ the *size* of the path component. For example, consider vertices v_1, v_2, v_3 and v_4 of T where each v_i is connected to v_{i+1} for $i = 1, 2, 3$. If $d(v_1) = \cdots = d(v_4) = 2$ holds, then $v_1, \ldots, v_4$ is a path component with size 4.

2. Check if there is a path component whose size is at least 4, say $v_1, v_2, \ldots, v_\ell$, and let v_0 and $v_{\ell+1}$ be the unique adjacent vertices of v_1 and v_ℓ other than v_2 and $v_{\ell-1}$, respectively. If it exists, assume T is rooted at v_1, divide T into $T_1 := T(v_1, v_0)$ and $T_2 := T(v_4, v_5)$, and remove v_2 and v_3. Continue this operation until such a path component does not exist.

We assume $\Delta \geq 7$, because otherwise the original algorithm CK is already a linear time algorithm. Here, we show that $\lambda(T) = \Delta + 1$ if and only if $\lambda(T_1) = \lambda(T_2) = \Delta + 1$. The only-if part is obvious, and we show the if part. Suppose that $f(v_1) = a$ and $f(v_0) = b$ in a $(\Delta + 1)$-$L(2, 1)$-labeling of T_1, and $f(v_4) = a'$ and $f(v_5) = b'$ in a $(\Delta + 1)$-$L(2, 1)$-labeling of T_2. Then set $f(v_2) = c$ where $|c - a| \geq 2$ and c is neither b nor a', and set $f(v_3) = c'$ where $|c' - c| \geq 2$, $|c' - a'| \geq 2$ and c' is neither a nor b'. This gives a $(\Delta + 1)$-$L(2, 1)$-labeling of T and is always possible since $\Delta \geq 7$. Namely, we can find an $L(2, 1)$-labeling of T by finding $L(2, 1)$-labelings of T_1 and T_2 independently, which guarantees that this preprocessing preserves $(\Delta + 1)$-$L(2, 1)$-labeling of T if it exists. Clearly, this operation can be done in linear time. Thus, from now on, we assume that an input tree T has the following property.

Property 2. The size of any path component of T is at most 3.

4.3 Efficient Search for Augmenting Paths

As observed in Subsection 4.1, the running time of algorithm CK is dominated by Step 1. Step 1 of algorithm CK computes the maximum bipartite matching $O(\Delta^2)$ times for calculating $\delta((u, v), (*, *))$ for $T(u, v)$, which takes $O(\Delta^{4.5})$ time. In this subsection, we show that for $T(u, v)$, $\delta((u, v), (*, *))$ can be calculated more efficiently; for a fixed label b, $\{\delta((u, v), (i, b)) \mid i \in \{0, 1, ..., \Delta+1\}\}$ can be obtained in $O(\Delta^{1.5} d'(v))$ time by computing a single maximum bipartite matching and a single graph search, where $d'(v)$ is the number of children of v. This shows that $t(v) = O(\Delta^{2.5} d'(v))$.

Let $G(u, v, -, b) = (V(v), X, E(u, v, -, b))$ be the bipartite graph defined in Subsection 4.1, where $V(v) = \{w_1, w_2, \ldots, w_{d'(v)}\}$ and $X = \{0, 1, \ldots, \Delta + 1\}$. In this subsection, we may refer to $i \in X$ as a label i. Then the following property holds.

Lemma 3. *If $G(u,v,-,b)$ has no matching of size $d'(v)$, then $\delta((u,v),(i,b)) = 0$ for any label i.* □

Below, consider the case where $G(u,v,-,b)$ has a matching of size $d'(v)$; without loss of generality, let $M = \{(w_{i+1}, i) \mid i \in \{0,1,\ldots,d'(v)-1\}\}$ be such a matching in $G(u,v,-,b)$ (note that by $d'(v) \le \Delta$, each vertex in $V(v)$ is matched). Recall, as mentioned in Subsection 4.1, that for each label $i \in \{0,1,\ldots,\Delta+1\}$, $\delta((u,v),(i,b)) = 1$ if and only if $G(u,v,i,b)$ has a matching of size $d'(v)$. Clearly, $\delta((u,v),(i,b)) = 1$ for each $i \in \{d'(v), d'(v)+1,\ldots,\Delta+1\}$.

Next consider the value of $\delta((u,v),(i,b))$ for $i \in \{0,1,\ldots,d'(v)-1\}$. Let $i \in \{0,1,\ldots,d'(v)-1\}$. Note that $G(u,v,i,b)$ has the matching $M - \{(w_{i+1}, i)\}$ of size $d'(v)-1$. Given a matching M', a path is called *M'-alternating* if its edges are alternately in and not in M'. In particular, an M'-alternating path is called *M'-augmenting* if the end vertices of the path are both unmatched by M'. It is well-known that M' is a maximum matching if and only if there is no M'-augmenting path.

Hence, $G(u,v,i,b)$ has a matching of size $d'(v)$ if and only if $G(u,v,i,b)$ has an $(M - \{(w_{i+1}, i)\})$-augmenting path; $G(u,v,-,b)$ has an $(M - \{(w_{i+1}, i)\})$-augmenting path not passing through vertex i. It follows that for each label $i \in \{0,\ldots,d'(v)-1\}$, we can decide the value of $\delta((u,v),(i,b))$ by checking whether there exists an $(M - \{(w_{i+1}, i)\})$-augmenting path not passing through vertex i in $G(u,v,-,b)$. Notice that for any label i, if such an augmenting path P exists, then one of two end vertices of P is always included in X', where $X' = \{d'(v), d'(v)+1,\ldots,\Delta+1\} \subseteq X$ (note that the other end vertex is w_{i+1}). Moreover, by the following Lemma 4, we can decide the value of $\delta((u,v),(i,b))$, $i \in \{0,1,\ldots,\Delta+1\}$ simultaneously by traversing all vertices which can be reached by an M-alternating path from some vertex in X' in $G(u,v,-,b)$.

Lemma 4. *$\delta((u,v),(i,b)) = 1$ if and only if vertex i can be reached by an M-alternating path from some vertex in X' in $G(u,v,-,b)$.*

Proof. Assume that $\delta((u,v),(i,b)) = 1$. Then, there exists an $(M-\{(w_{i+1}, i)\})$-augmenting path P not passing through vertex i. Note that two end vertices of P are w_{i+1} and some vertex $u \in X'$. Hence, it follows that vertex i can be reached by the M-alternating path $P \cup \{(w_{i+1}, i)\}$ from $u \in X'$.

Assume that vertex i can be reached by an M-alternating path from some vertex in X' in $G(u,v,-,b)$. Let P be such an M-alternating path in which vertex i appears exactly once. Since P starts from a vertex in X', we can observe that the edge which appears immediately before reaching vertex i in P is $(w_{i+1}, i) \in M$. Hence, the path $P - \{(w_{i+1}, i)\}$ is an $(M - \{(w_{i+1}, i)\})$-augmenting path not passing through vertex i, and it follows that $\delta((u,v),(i,b)) = 1$. □

All vertices that can be reached by an M-alternating path from some vertex in X' in $G(u,v,-,b)$ can be computed in $O(|E(u,v,-,b)| + |X'|) = O(\Delta d'(v))$ time, by using the depth first search from vertex s in G_s, where G_s denotes the graph obtained from $G(u,v,-,b)$ by adding a new vertex s and new edges between s and each vertex in X'.

Consequently, $\{\delta((u,v),(i,b)) \mid i \in \{0,1,\ldots,\Delta+1\}\}$ can be obtained by computing a single bipartite matching and a single depth first search.

4.4 Efficient Computation of δ-Values Near Leaves

In Subsections 4.1 and 4.3, we have observed that algorithm CK runs in $O(\sum_{v \in V} t(v)) = O(\Delta^{2.5} \sum_{v \in V} d'(v))$ time. In this subsection, we show that algorithm CK can be implemented to run in $O(\Delta^{2.5} \sum_{v \in V - V_L - V_Q} d''(v))$ time by avoiding unnecessary bipartite matching computations for vertices incident to leaves, where V_L and V_Q are defined in Subsection 4.2 and $d''(v) = |C(v) - V_L|$.

For a vertex $v \in V_L \cup V_Q$, we can easily obtain $\delta((u,v),(*,*))$ without computing the bipartite matching. Actually, for a leaf $v \in V_L$, $\delta((u,v),(a,b)) = 1$ if and only if $|a-b| \geq 2$. For a vertex $v \in V_Q$, we have $\delta((u,v),(a,b)) = 1$ if and only if $b \in \{0, \Delta+1\}$ and $|a-b| \geq 2$ (notice that each vertex in V_Q is major). Thus, the running time of algorithm CK is dominated by Step 1 for vertices $v \in V - V_L - V_Q$; $O(\sum_{v \in V} t(v)) = O(\sum_{v \in V - V_L - V_Q} t(v))$.

Also for a vertex $v \in V - V_L - V_Q$ incident to some leaf, we can gain some saving of time for computing $\delta((u,v),(*,*))$; for a label b, the calculation of $\delta((u,v),(*,b))$ can be done in $O(\Delta^{1.5} d''(v))$ time, instead of $O(\Delta^{1.5} d'(v))$ time. Let v be a vertex in $V - V_L - V_Q$ incident to some leaf; $C(v) \cap V_L \neq \emptyset$. Note that v is major by Property 1, and that $\delta((u,v),(*,b)) = 0$ for each $b \notin \{0, \Delta+1\}$. Thus, we have only to decide the value of $\delta((u,v),(*,b))$ for $b \in \{0, \Delta+1\}$.

Then, we can observe that for computing $\delta((u,v),(*,b))$, it suffices to check whether there exists a feasible assignment only on $C(v) - V_L$, instead of $C(v)$. Actually, if $b = 0$ and there exists a feasible assignment on $C(v) - V_L$, then the number of the remaining labels is $\Delta + 2 - |C(v) - V_L| - |\{a, 0, 1\}| = |C(v) \cap V_L|$ and we can assign to each leaf in $C(v) \cap V_L$ distinct labels among the remaining labels (note that $|C(v)| = \Delta - 1$ since v is major). The case of $b = \Delta + 1$ can also be treated similarly. Therefore, it follows that the calculation of $\delta((u,v),(*,b))$ is dominated by the maximum matching computation in the subgraph of $G(u,v,-,b)$ induced by $(V(v) - V_L) \cup X$; its time complexity is $O(\Delta^{1.5}|V(v) - V_L|) = O(\Delta^{1.5} d''(v))$.

Consequently, algorithm CK can be implemented to run in $O(\sum_{v \in V - V_L - V_Q} t(v)) = O(\Delta^{2.5} \sum_{v \in V - V_L - V_Q} d''(v))$ (note that $d''(v) = d'(v)$ for each vertex v with $C(v) \cap V_L = \emptyset$).

4.5 Amortized Analysis

In Subsections 4.2–4.4, we have observed that by an efficient implementation of algorithm CK, $\lambda(T)$ can be decided in $O(\sum_{v \in V - V_L - V_Q} t(v)) = O(\Delta^{2.5} \sum_{v \in V - V_L - V_Q} d''(v))$ time. Below, we show that $O(\Delta^{2.5} \sum_{v \in V - V_L - V_Q} d''(v)) = O(\Delta^{1.5} n)$ by amortized analysis; namely, we show the following lemma.

Lemma 5. *Algorithm* CK *can be implemented to run in $O(\Delta^{1.5} n)$ time.* □

Let V_B be the set of vertices $v \in V - V_L - V_Q$ with $d''(v) \geq 2$, V_P be the set of vertices $v \in V - V_L - V_Q$ with $d'(v) = 1$, and $V'_P = V - (V_L \cup V_Q \cup V_B \cup V_P)$. Note that each vertex in V_P belongs to a certain path component. Also each $v \in V'_P$ satisfies $d''(v) = 1$ and $C(v) \cap V_L \neq \emptyset$, and hence by Property 1, it is incident to exactly $\Delta - 2$ leaves.

Now by Property 2, for each vertex $v \in V_P$, there exist the root r or a vertex in $V_B \cup V'_P$ among its ancestors which are at most at distance 3 from v. Hence, we have

$|V_P| \le 3\sum_{v\in V_B\cup V'_P} d''(v)+3$. By $\sum_{v\in V_P} d''(v) = |V_P|$, it follows that

$$\begin{aligned}\Delta^{2.5}\textstyle\sum_{v\in V-V_L-V_Q} d''(v) &= \Delta^{2.5}\textstyle\sum_{v\in V_B\cup V'_P\cup V_P} d''(v)\\ &= \Delta^{2.5}\textstyle\sum_{v\in V_B\cup V'_P} d''(v) + \Delta^{2.5}|V_P|\\ &\le \Delta^{2.5}(\textstyle\sum_{v\in V_B\cup V'_P} 4d''(v)+3)\\ &= O(\Delta^{2.5}(\textstyle\sum_{v\in V_B\cup V'_P} d''(v)+1)).\end{aligned}$$

Thus, in order to prove $O(\Delta^{2.5}\sum_{v\in V-V_L-V_Q} d''(v)) = O(\Delta^{1.5}n)$, it suffices to show that $\sum_{v\in V_B\cup V'_P} d''(v) = O(n/\Delta)$.

Lemma 6. $\sum_{v\in V_B\cup V'_P} d''(v) = O(n/\Delta)$.

Proof. Let E' be the set of all edges connecting a vertex in $V_B\cup V'_P$ and its non-leaf child. Note that $|E'| = \sum_{v\in V_B\cup V'_P} d''(v)$. Let E_L denote the set of all edges incident to a leaf, and E_P denote the set of all edges connecting a vertex in V_P and its unique child. Also note that $|E_L| = |V_L|$, $|E_P| = |V_P|$, $E_L\cap E_P = \emptyset$, and $(E_L\cup E_P)\cap E' = \emptyset$. Hence, we have $|E'| \le |E| - |E_L| - |E_P| = n-1-|V_L|-|V_P|$. Now, by $V = V_L\cup V_Q\cup V_B\cup V_P\cup V'_P$ and that V_L, V_Q, V_B, V_P and V'_P are disjoint each other, we have $n = |V| = |V_L|+|V_Q|+|V_B|+|V_P|+|V'_P|$. Therefore, it follows that $|E'| \le n-1-(n-|V_B|-|V'_P|-|V_Q|) = |V_B|+|V'_P|+|V_Q|-1$.

Now since each vertex $u\in V_B$ has at least two non-leaf children and each leaf not incident to $V_B\cup V'_P$ is incident to a vertex in V_Q, we can observe that $|V_Q| \ge |V_B|+1$ holds. Since each vertex in V'_P (resp., V_Q) is incident to exactly $\Delta-2$ (resp., $\Delta-1$) leaves, we have $|V_L| \ge |V'_P|(\Delta-2)+|V_Q|(\Delta-1)$. Consequently, we have $\sum_{v\in V_B\cup V'_P} d''(v) = |E'| \le |V_B|+|V'_P|+|V_Q|-1 \le 2|V_Q|+|V'_P|-2 \le 2|V_L|/(\Delta-2)-2$. □

5 An $O(n^{1.75})$-Time Algorithm

Summarizing the arguments given in Sections 3 and 4, we give a description of the overall algorithm named LABEL-TREE in Table 2, for determining in $O(n^{1.75})$ time whether $\lambda(T) = \Delta+1$ or not for any input tree T. We show that algorithm LABEL-TREE can be implemented to run in $O(n^{1.75})$ time. Clearly, all of the preprocessing, Steps 0 and 1 can be executed in linear time. As observed in Subsection 4.5, Steps 2–5 can be executed in

Table 2. Algorithm LABEL-TREE

Preprocessing. Execute the preprocessing described in Subsection 4.2.

Step 0. If $N[v]$ contains at least three major vertices for some vertex $v\in V$, output "No". Halt.

Step 1. If the number of major vertices is at most $\Delta-6$, output "Yes". Halt.

Step 2. For $T(u,v)$ with $v\in V_Q$ (its height is 2), let $\delta((u,v),(a,0)) := 1$ for each label $a\ne 0,1$, $\delta((u,v),(a,\Delta+1)) := 1$ for each label $a\ne \Delta,\Delta+1$, and $\delta((u,v),(*,*)) := 0$ for any other pair of labels. Let $h := 3$.

Step 3. For all $T(u,v)$ of height h, compute $\delta((u,v),(*,*))$ by fixing $f(v) := b$ and applying the method described in Subsections 4.3 and 4.4 for each label b.

Step 4. If $h = h^*$ where h^* is the height of root r of T, then goto Step 5.
Otherwise let $h := h+1$ and goto Step 3.

Step 5. If $\delta((r,v),(a,b)) = 1$ for some (a,b), then output "Yes". Otherwise output "No". Halt.

$O(\Delta^{1.5}n)$ time. Moreover, as shown in the proof of Corollary 2, if $N[v]$ contains at most two major vertices for any vertex $v \in V$ and the total number of major vertices is at least $\Delta - 5$, we have $\Delta = O(\sqrt{n})$. Thus, Steps 2–5 take $O(n^{1.75})$ time, and it follows that the running time of algorithm Label-Tree is $O(n^{1.75})$.

Moreover, we remark that in both cases of $\lambda(T) = \Delta+1, \Delta+2$, we can easily construct a $\lambda(T)$-$L(2,1)$-labeling in the same complexity. Actually, if $\lambda(T) = \Delta + 2$, then a $\lambda(T)$-$L(2,1)$-labeling can be obtained by Algorithm Greedy in Section 3. If $\lambda(T) = \Delta + 1$ is determined as a result of Step 1, then according to the proof of Theorem 1, a $\lambda(T)$-$L(2,1)$-labeling can be obtained in linear time. Also if $\lambda(T) = \Delta + 1$ is determined as a result of Step 5, then we can obtain the $\lambda(T)$-$L(2,1)$-labeling in $O(\Delta^{1.5}n)$ time, following the dynamic programming based procedure of Steps 2–5. Namely we have the following result.

Theorem 2. *For trees, the $L(2,1)$-labeling problem can be solved in $O(\min\{n^{1.75}, \Delta^{1.5}n\})$ time.* □

6 Concluding Remarks

Finally, we remark that our results can be extended to apply to some wider variations of labeling problems, as well as the $L(2,1)$-labeling problem on trees.

It is known that Chang and Kuo's algorithm [4] can be extended to solve the $L(h,1)$-labeling problem on trees [3] and *p-almost trees* [6], where a p-almost tree is a connected graph with $n+p-1$ edges. By extending the original Chang and Kuo's algorithm, the $L(h,1)$-labeling problem on trees can be solved in $O((h+\Delta)^{5.5}n) = O(\lambda^{5.5}n)$ time, and $L(2,1)$-labeling on p-almost trees can be solved in $O(\lambda^{2p+4.5}n)$ time for λ given as an input. Our techniques in Subsection 4.3 can also be applied to speed up those algorithms. In fact, it is easy to show that our techniques can solve the $L(h,1)$-labeling problem on trees in $O(\lambda^{3.5}n)$ time, and the $L(2,1)$-labeling problem on p-almost trees in $O(\lambda^{2p+2.5}n)$ time. Moreover, if some properties such as Theorem 1 hold, then we may expect some more improvement on these problems.

References

1. Bodlaender, H.L., Kloks, T., Tan, R.B., van Leeuwen, J.: Approximations for λ-coloring of graphs. The Computer Journal 47, 193–204 (2004)
2. Calamoneri, T.: The $L(h,k)$-labelling problem: A survey and annotated bibliography. The Computer Journal 5, 585–608 (2006)
3. Chang, G.J., Ke, W.-T., Kuo, D., Liu, D.D.-F., Yeh, R.K.: On $L(d,1)$-labeling of graphs. Discrete Mathematics 220, 57–66 (2000)
4. Chang, G.J., Kuo, D.: The $L(2,1)$-labeling problem on graphs. SIAM J. Disc. Math. 9, 309–316 (1996)
5. Fiala, J., Golovach, P.A., Kratochvíl, J.: Distance constrained labelings of graphs of bounded treewidth. In: Caires, L., Italiano, G.F., Monteiro, L., Palamidessi, C., Yung, M. (eds.) ICALP 2005. LNCS, vol. 3580, pp. 360–372. Springer, Heidelberg (2005)
6. Fiala, J., Kloks, T., Kratochvíl, J.: Fixed-parameter complexity of λ-labelings. Discrete Applied Mathematics 113, 59–72 (2001)

7. Griggs, J.R., Yeh, R.K.: Labelling graphs with a condition at distance 2. SIAM J. Disc. Math. 5, 586–595 (1992)
8. Hale, W.K.: Frequency assignment: theory and applications. Proc. IEEE 68, 1497–1514 (1980)
9. Hopcroft, J.E., Karp, R.M.: An $n^{5/2}$ algorithm for maximum matchings in bipartite graphs. SIAM J. Comput. 2(4), 225–231 (1973)
10. Roberts, F.S.: T-colorings of graphs: recent results and open problems. Discrete Mathematics 93, 229–245 (1991)
11. Wang, W.-F.: The $L(2,1)$-labelling of trees. Discrete Applied Mathematics 154, 598–603 (2006)
12. Yeh, R.K.: A survey on labeling graphs with a condition at distance two. Discrete Mathematics 306, 1217–1231 (2006)

Batch Coloring Flat Graphs and Thin

Magnús M. Halldórsson[1] and Hadas Shachnai[2]

[1] School of Computer Science, Reykjavik University, 103 Reykjavik, Iceland
mmh@ru.is
[2] Department of Computer Science, The Technion, Haifa 32000, Israel
hadas@cs.technion.ac.il

Abstract. A batch is a set of jobs that start execution at the same time; only when the last job is completed can the next batch be started. When there are constraints or conflicts between the jobs, we need to ensure that jobs in the same batch be non-conflicting. That is, we seek a coloring of the conflict graph. The two most common objectives of schedules and colorings are the makespan, or the maximum job completion time, and the sum of job completion times. This gives rise to two types of *batch coloring problems*: *max-coloring* and *batch sum coloring*, respectively.

We give the first *polynomial time approximation schemes* for batch sum coloring on several classes of "non-thick" graphs that arise in applications. This includes paths, trees, partial k-trees, and planar graphs. Also, we give an improved $O(n \log n)$ exact algorithm for the max-coloring problem on paths.

1 Introduction

In the classic (unbounded) *p-batch* scheduling problem [6], we are given a set of n jobs where job J_j has processing time, or *length*, p_j. We need to partition the jobs into *batches*. All jobs in a batch start at the same time, and the batch is completed when its last job finishes. The length of a batch is then the length of the longest job in the batch.

Real-life scenarios frequently impose restrictions on the subsets of jobs that can be processed simultaneously. Such conflicts among the jobs are often modeled by an undirected conflict graph $G = (V, E)$, with $V = \{1, 2, \ldots, n\}$ where the length of vertex j is the processing time of J_j, and there is an edge $(i, j) \in E$ if the pair of jobs J_i and J_j cannot be processed in the same batch. Each batch forms an independent set in G, hence a valid schedule is a proper coloring of the graph.

We consider the above problems of batch scheduling with conflicts under three different measures. In the *max-coloring (*MAX-COL*)* problem we minimize the *makespan* of the batch schedule, or the sum of the batch lengths. In the *minimum sum of job completion times problem* (SJC), we minimize the sum of the completion times of the jobs, and in the *minimum sum of batch completion times problem* (SBC), we minimize the sum, over all the jobs J_j, of the completion time of the batch that contains J_j. Let BSC refer to either of the batch scheduling problems under the sum measure: SJC or SBC.

J. Gudmundsson (Ed.): SWAT 2008, LNCS 5124, pp. 198–209, 2008.

In this paper we give the first *polynomial time approximation schemes (PTAS)*, as well as exact algorithms, for BSC on several classes of graphs that arise in applications. We focus on "thin" graphs, namely, graphs with bounded treewidth, and "flat" graphs, i.e., graphs that can (almost) be embedded on flat surface.[1] In particular, we consider paths, trees and partial k-trees (which are "thin"), as well as planar and other "flat" graphs.

We say that an algorithm $\mathcal{A}$ yields an approximation ratio of r, for some $r \geq 1$, if for any instance I of our problems $\mathcal{A}(I) \leq r \cdot OPT(I)$, where $OPT(I)$ is the value of an optimal solution. A problem admits a PTAS if it can be approximated in polynomial time within factor $1 + \epsilon$, for any constant $\epsilon > 0$. When the complexity of the scheme is of the form $f(\epsilon)n^c$, where $c > 0$ is some constant, we get an *efficient PTAS (EPTAS)*.

Our problems naturally arise in the following applications.

Metropolitan Networks: Transmission of real-time messages in metropolitan networks is done by assigning to each sender node a set of fixed length slots, in which the message content is filled and transmitted in sequence to the receiver. To improve transmission times, the same slots can be assigned to messages with non-intersecting paths [15]. A group of messages using the same set of slots can be viewed as a batch. The number of slots in the set is the maximum length of any message transmitted in these slots. The problem of minimizing the number of slots used to transmit all messages then yields an instance of MAX-COL.

Distributed Computation: In distributed operating systems (see, e.g., [21, 16]) the scheduler identifies subsets of non-conflicting, or cooperating processes, that can benefit from running at the same time interval (e.g., since these processes do not use the same resources, or communicate frequently with each other); then, each subset is executed simultaneously on several processors, until *all* the processes in the subset complete. When the objective is to minimize the sum of job completion times, we get an instance of SJC; when we want to minimize the total completion time of the schedule, we get an instance of MAX-COL.

Other applications include, e.g., batch production [14], dynamic storage allocation and memory management in wireless protocol stack (see, e.g., [12, 19].) The graph classes that we study here arise in these applications from tree-like (or planar) network topologies [17] ([18]), conflict graphs of processes generated for computer programs [21], and scheduling jobs with bounded resource requirements (which yield *pebbly* graphs; see in Section 3.4).

1.1 Related Work

There is a wide literature on batch scheduling, namely, our problems with empty conflict graph; see a a comprehensive survey in [6]. In the following we mention known results for our problems on various classes of conflict graphs.

Batch sum coloring: The problem of batch coloring so as to minimize the sum of job completion times was introduced in [3]. Constant factor approximations

[1] Formal definitions can be found, e.g., in [22].

for SJC were given in [2] for, e.g., various classes of perfect graphs, with improved constant factor obtained in [9]. The SBC variant was introduced in [9] and approximated on various graph classes. The common special case where all jobs are of unit length is known as the *sum coloring problem*. This implies that BSC is hard to approximate within factor $n^{1-\varepsilon}$, for any $\varepsilon > 0$ [2].

Max-coloring: The max-coloring problem was first considered in [12], where an $O(n^4)$-time algorithm was given for paths. A quadratic algorithm was presented more recently in [10]. A PTAS is known for trees [19, 10] and for partial k-trees [10]. Constant factor approximation and NP-hardness results are known for various classes of perfect graphs (see the recent work of [8] for details).

Multicoloring problems: Batch coloring problems relate to certain multicoloring problems where vertices must be assigned a given number of contiguous colors, corresponding to a non-preemptive schedule. The difference is that in these problems, jobs are not restricted to start in batches. We shall be using some of the ideas developed for the non-preemptive sum multicoloring problem, NPSMC, defined as follows. Given a graph G with a length $x(v)$ for each vertex v, find an assignment f of starting times for the vertices such that no neighbor of v starts in the interval $[f(v), \ldots, f(v)+x(v)-1]$. For a comprehensive survey of known results for multicoloring problems with minsum objectives see, e.g., [11].

1.2 Our Results

In this paper we focus on batch coloring of several amenable classes of graphs that we could term either "flat" or "thin" (see in Section 3). Given the hardness of these problems on general graphs, it is natural to seek out classes of graphs where effective solutions can be obtained efficiently.

Batch coloring with minsum objective: In Section 2 we describe our main approximation technique. In Section 3 we develop EPTASs for BSC on partial k-trees and later for planar graphs. The complexity of these schemes are of the form $2^{\epsilon^{-O(1)}} n$, when guaranteeing a $(1+\epsilon)$-approximation. By a result of [4, 7], this implies that BSC is fixed parameter tractable on these classes of graphs. Our scheme for planar graphs improves the running time of a PTAS proposed in [13] for the special case of the sum coloring problem to a linear time EPTAS, more precisely $2^{\epsilon^{-1}\log\epsilon^{-1}} n$ time. In Section 3.3 we show how our results for planar graphs can be extended to other "flat" graphs. To the best of our knowledge, we give here the first approximation schemes for batch sum coloring on partial k-trees and planar (or more generally, "flat") graphs.

Max-coloring: In Section 4 we give an $O(n \log n)$ exact algorithm for MAX-COL on paths. This improves upon the $O(n^2)$ algorithm of [10].

Contribution: Our main approximation technique combines technical tools developed in [13] for approximating NPSMC with efficient enumeration of batch length sequences. A key component in our technique is truncation of the number of batches in the schedule, so that all possible batch length sequences can be enumerated in polynomial time. This defines a general framework for solving batch coloring problems with minsum objective, for any graph class for which (a) the number of batches can be truncated with small harm to the objective function, and (b) given a sequence of batch lengths, a proper batch coloring with minimum total cost can be found efficiently.

A framework proposed earlier, for approximately solving the max-coloring problem on certain classes of graphs, shares some similarities with ours (see [12,19]); however, BSC differs from MAX-COL in two ways. (i) For many classes of graphs, we can bound the total number of batches used in a MAX-COL schedule using structural properties of the graph (e.g., the maximum degree of any vertex), while for batch coloring with minsum objective this number may depend on the distinct number of job lengths (see Lemma 1). (ii) Given a batch length sequence, the problem of finding a MAX-COL schedule reduces to finding a *feasible* batch coloring, while solving BSC involves optimizing over the set of feasible schedules. Thus, our BSC problems require the usage of different machinery.

We expect that our framework for solving BSC will find more uses for batch coloring with other (more general) minsum objective functions, as well as for solving BSC on other classes of graphs. In fact, as we show in Section 3, the approximation technique that we use for planar graphs is general enough to be applicable to graph classes that contain planar graphs as a subclass (such as bounded-genus graphs).

Notation: Let $p(G) = \sum_{j \in V} p_j$ be the sum of the job lengths. Let $p_{max} = p_{max}(G) = \max_{j \in V} p_j$ be the maximum job length, $p_{min} = \min_{j \in V} p_j$ be the minimum job length, and $\tau = \tau(G) = p_{max}/p_{min}$ be the maximum ratio between two job lengths. We omit G when clear from context.

The *size* or *weight* of a set of vertices is the sum of the vertex lengths. A *batch* is an independent set that is to be scheduled starting at the same time. The *length* of a batch is the largest length of a vertex in the batch. The *density* of a batch B of length ℓ is $|B|/\ell$, or the number of jobs in B per length unit. A *(batch) schedule* is a partition of the vertices into a sequence of batches.

The completion time of a batch B in a schedule is the sum of lengths of the batches up to and including B. The completion time of a job J_j in batch B is the sum of the lengths of batches prior to B, plus the length of J_j.

The *makespan* of a schedule is the completion time of the last job, or the sum of the lengths of all batches in the schedule. Let $\mu(G)$ be the minimum makespan of a batch schedule of G. Let d denote the number of distinct vertex lengths and Δ the maximum degree of the graph.

Due to space constraints, some of the proofs and implementation details are omitted. The detailed results appear in [14].

2 Techniques

Batch coloring problems introduce new difficulties to the classic *multicoloring* and *batch scheduling* problems, which are known to be hard when solved alone.

The fact that all jobs in the same batch must start at the same time limits the number of different starting times in a schedule, and thus can simplify the search for good schedules. However, as opposed to ordinary multicoloring, batch coloring introduces *non-locality*: the allowable starting times of a job depend not only on this job and its neighbors, but also implicitly on the jobs elsewhere in the graph that have been assigned to earlier batches. This implies, for example, that there are almost no non-trivial results known for the sum of completion times problems, even on paths, nor is there a known exact polynomial time algorithm for the max-coloring problem on trees.

Two difficult features are inherited from related multicoloring problems. One is that a large number of batches may be needed even on low-chromatic number graphs. In fact, as we show in the next result, $\Omega(n)$ batches may be required for optimal SBC coloring of paths.[2]

Proposition 1. *There are* SJC *instances on paths and* SBC *instances on empty graphs for which the only optimal solution uses n batches.*

Proof. Consider a path of n vertices where the length of vertex j is $p_j = n^j$. We claim that the only optimal solution is given by the *shortest processing time first (SPTF)* rule; namely, batch i contains only vertex i, for $i = 1, 2, \ldots, n$.

Suppose there is a non-SPFT optimal SJC schedule S, and let j be the smallest number such that vertex j is not assigned alone in batch j. Consider now the following change to S, where we create a new batch, number it j (increasing the index of later batches) and reassign v_j to batch j. This may increase the completion times of v_s, for $s = j+1, \ldots, n$, by a total of $(n-j) \cdot n^j < n^{j+1}$. It will, however, decrease the completion time of v_j by the length of some later vertex, or at least $p_{j+1} = n^{j+1}$. Thus, this new schedule improves upon S, which is a contradiction.

The same set of lengths yields the same argument for SBC, even if the graph contains no edges. ∎

The way to get around the large number of batches is to analyze the cost of truncating the coloring early, and show that there are approximate colorings that use moderately many colors. Alternatively, we can use standard rounding of job lengths to powers of $1+\epsilon$ and bound the number of batches of each length.

Another difficulty has to do with the large range of job lengths. An essential ingredient in any strategy for these problems is a method to break the instance into sub-instances of similar-length jobs. For max-coloring, it suffices to use a fixed geometric sequence to reduce the problem to the ordinary coloring problem, within a constant factor [19]. For our batch coloring problems with minsum objective, BSC, the situation is not as easy, and it is necessary to partition according to the actual length distribution.

We observe that a result of [13], originally stated for sum multicoloring, holds also for BSC. It shows that we can focus on the case where the ratio between the longest and the shortest job is small, if we also bound the makespan of the algorithm.

[2] The paper [19] gives a similar result for the max coloring problem of bipartite graphs.

Theorem 1. *Let n, $q = q(n)$ and σ be given. Suppose that for any $G = (V, E)$ in a hereditary class $\mathcal{G}$ with n vertices and $\tau(G) \leq q$, we can approximate* Bsc *within a factor $1 + \epsilon(n)$ and with makespan of $\sigma \cdot p_{max}(G)$ in time $t(n)$. Then, we can approximate* Bsc *on any graph in $\mathcal{G}$ within factor $1+\epsilon(n)+\sigma/\sqrt{\ln q}$ with makespan $2\sigma \cdot p_{max}(G)$ in $O(t(n))$ time.*

This is based on finding a partition of the instance into length classes that depends on the actual distribution of the lengths. Then, the solutions produced on each length class by the assumed algorithm are simply concatenated in length order. Thus, if the solution produced on each length class is a batch schedule, so is the combined solution.

The techniques that we use for Bsc build on a technique developed in [13] for non-preemptive sum multicoloring (npSMC) of partial k-trees and planar graphs. For the batch problems studied here, we have modified and replaced some of the parts, including the bounds on the number of batches needed. Most crucially, the central subroutines for handling subproblems with a limited number of job lengths were completely changed. The strategy here is to decide first the length sequence of the batches to be used, and then assign the jobs near optimally to those batches. By rounding the job lengths, we limit the number of possible batch lengths and at the same time reduce the number of batches needed.

3 Batch Sum Coloring

In this section we give EPTASs for Bsc on both "thin" graphs (trees, partial k-trees) and "flat" graphs (planar and bounded-genus graphs). First, we argue some general properties of exact and approximate solutions. Recall that $p(G)$ is the sum of vertex length, p_{max} the length of the longest job, $\mu(G)$ the minimum makespan of a batch schedule, and χ the chromatic number of the graph.

3.1 Properties and Tools

Observation 2. *Any optimal* Bsc *schedule satisfies the following properties.*
(i) (Non-increasing density) *Batch densities are monotone non-increasing.*
(ii) (Density reduction) *After $i(2\mu(G) + p_{max})$ steps, the total length of the remaining graph is at most $p(G)/2^i$.*
(iii) (Restricted batch length sequences) *Each batch is preceded by at most Δ longer batches.*

Proof sketch. (i) The claim holds since, otherwise, batches can be swapped to decrease the cost of the schedule (also known as Smith's rule; see in [20,6]). (ii) Density must go down by half each $2\mu(G) + p_{max}$ steps; otherwise, we could schedule the whole remaining graph in the latter $\mu(G)$ steps, at lesser cost. (iii) This is true since, otherwise, the vertices in that batch can all be recolored with earlier batches. ∎

We now give two lemmas regarding batch coloring of χ-colorable graphs.

Lemma 1. *If G is χ-colorable, then* $\text{SBC}(G) \leq \text{SJC}(G) \leq 3\chi \cdot p(G)$.

Proof. Recall that τ is the maximal ratio between the processing times of any two vertices. Use a batch sequence with χ batches of each length $2^i p_{min}$, for $i = 1, 2, \ldots, \lceil \log \tau \rceil$, in non-decreasing order. Order the batches of the same length in non-increasing order of size. The length of a job is at least half the length of its batch. Each job waits for all batches shorter than it; also, averaged over the jobs in batches of the same length, it waits for $(\chi - 1)/2$ batches of length equal to its batch length. Then, on average, the completion time of v is at most $p(v) + \chi \sum_{j=0} p(v)/2^j + (\chi - 1)/2 \cdot 2p(v) = 3\chi p(v)$. Thus, the total schedule cost is at most $3\chi p(G)$. ∎

The following lemma, which extends a result of [13], helps rein in the total length of our approximate schedule.

Lemma 2. *Let G be a χ-colorable graph, and let $\epsilon > 0$. Then, there exists a $(1 + \epsilon)$-approximate* BSC *schedule of G satisfying the following constraints:*

1. *Batch lengths are powers of $1 + \epsilon$,*
2. *There are at most $t = t_{\chi,\epsilon}$ batches of each length, where $t = \chi(1+\epsilon)\epsilon^{-1}$, and*
3. *The makespan of the schedule is at most $(2\mu(G) + p_{max}) \cdot \log(\chi/\epsilon) + 2\chi \cdot p_{max}$*

Proof. First, we consider the effect of rounding all job lengths to powers of $1+\epsilon$. This increases the size of each batch by at most a factor of $1+\epsilon$, which delays the starting time of each job by factor at most $1 + \epsilon$. Thus, the extra cost incurred for the optimal schedule is at most $\epsilon \cdot (OPT(G) - p(G))$.

Next, consider the optimal schedule S^* for the rounded instance. Suppose that more than $\chi \cdot (1 + \epsilon)\epsilon^{-1}$ batches are used for some batch length ℓ. Then, following batch $\chi \cdot \epsilon^{-1}$ of length ℓ, we introduce χ batches of length ℓ, shifting all later batches by $\chi \cdot \ell$. We color all jobs that occurred in later batches of length ℓ using the χ new batches, and delete those later batches. Each batch in the resulting coloring is delayed by $\chi \cdot \ell$, by the new batches of length ℓ, only if it was already preceded by ϵ^{-1} times that many batches. Thus, each batch in the resulting coloring is delay by at most an ϵ-fraction of what previously came before it, or at most $\epsilon \cdot (OPT_J(G) - p(G))$.

Finally, we analyze the effect of cutting a schedule short. By the density reduction property, the total size remaining is at most $\epsilon p(G)/\chi$ after $i(2\mu(G) + p_{max})$ rounds with $i = \log(\chi/\epsilon)$. By Lemma 1, there is a schedule of the remainder of cost at most $3\chi \cdot \epsilon p(G)/\chi \leq 3\epsilon p(G)$. The makespan of that coloring is at most $2\chi \cdot p_{max}$. ∎

3.2 Thin graphs

We give here algorithms for the class of partial k-trees. For formal definition of this class, and that of the related tree decompositions, please refer to, e.g., [5].

We round the job lengths, obtaining an instance with only a limited number of distinct lengths. There are d possible lengths for each of the b batches, for a total of d^b distinct batch length sequences. We solve the problem optimally for each such length sequence, using standard dynamic programming. We sketch it briefly.

Lemma 3. *Given a partial k-tree G and a sequence of b batch lengths, there is an $O(b^{k+1}n)$-time algorithm to find an optimal* Bsc *coloring of G into those batches, or determine that no such coloring exists.*

Proof sketch. For each bag in the tree decomposition of G, form a table of b^k k-tuples, where a given k-tuple represents a particular batch assignment of the jobs in the given bag, and the entry corresponds to the minimum cost schedule constrained to assign the k-tuple in this given way. By having the children update the entries of the parents, and using that adjacent bags need only differ in only a single element, each entry is needed for a constant-time update of at most b entries in its parent bag. Hence, the complexity is $O(b^{k+1}n)$. ∎

We first argue a PTAS for instances of restricted job lengths.

Proposition 2. *There is a $(1+\epsilon)$-approximate algorithm for* Bsc *on partial k-trees, with time complexity $2^{\Theta((\log\chi+\log\epsilon^{-1})\cdot(k+\chi\cdot\epsilon^{-2}\cdot\log\tau))}n$ and makespan of $O(\chi\log\epsilon^{-1}p_{max})$.*

Proof. We use Lemma 3 to search for restricted $(1+\epsilon)$-colorings guaranteed by Lemma 2. The time complexity of the algorithm of Lemma 3 is $O(b^{b+k}n)$. The number d of different batch lengths is $\log_{1+\epsilon}\tau=\Theta(\epsilon^{-1}\log\tau)$. The number t of batches of each length is $\Theta(\chi\epsilon^{-1})$. Thus, the number b of batches is at most $t\cdot d=\Theta(\chi\cdot\epsilon^{-2}\log\tau)$. The time complexity is therefore bounded by $O(2^{\log b\cdot b}n)=2^{\Theta((\log\chi+\log\epsilon^{-1})\cdot\chi\cdot\epsilon^{-2}\cdot\log\tau)}n$. The makespan is as promised by Lemma 2, using that $\mu(G)\le\chi\cdot p_{max}$. ∎

We now combine Proposition 2 with Theorem 1.

Theorem 3. *There is an EPTAS for* Bsc *on partial k-trees, for any fixed k.*

Proof. We set $\sigma=\chi\cdot(\log(\chi/\epsilon)+2)$ and $q=e^{(2\sigma\cdot\epsilon^{-1})^2}$, giving $\sigma/\sqrt{\ln q}=\epsilon/2$. We use Lemma 3 to search for restricted $(1+\epsilon)$-colorings guaranteed by Lemma 2 with these parameters. Theorem 1 then yields a $(1+\epsilon)$-approximate schedule of G.

Let us evaluate the parameters according to this scheme. Since we may assume $\epsilon^{-1}\ge\chi$, we have $\sigma=\Theta(\chi\cdot\log\epsilon^{-1})$. Thus, $\log\tau=\log q=(2\sigma\cdot\epsilon^{-1})^2=\Theta(\chi^2\cdot\epsilon^{-2}\cdot\log^2\epsilon^{-1})$. Hence, the time complexity of the algorithm of Proposition 2 is bounded by $2^{O((\log\chi+\log\epsilon^{-1})\cdot(k+\chi^3\cdot\epsilon^{-4}\cdot\log^2\epsilon^{-1}))}n$. Since the chromatic number of a partial k-tree is at most $k+1$, the time complexity is singly exponential in $1/\epsilon$ and k. The makespan is at most $2\sigma p_{max}=O(k\cdot\log\epsilon^{-1}p_{max})$. ∎

Parametrization: While the existence of an exact polynomial time algorithm for Bsc on trees remains open, it appears unlikely. We consider instead parameterizations that lead to efficient exact solutions. We treat the parameters d, the number of distinct job lengths, and Δ, the maximum degree.

Consider graphs of maximum degree Δ. Since there can be at most Δ batches of the same length in an optimal schedule, there are at most $d^{\Delta d}$ different batch length sequences. However, this does not take the property of restricted batch length sequences (property *(iii)* in Observation 2) into account. Thus, for example, in a batch length sequence for a path, each batch can be preceded by at most two longer batches. We can capitalize on this to obtain improved bounds. The proof of the following lemma is given in [14].

Lemma 4. *The number of possible batch length sequences with d distinct lengths is at most* $2^{(\Delta+1)d}$.

By applying standard dynamic programming, we can therefore solve Bsc efficiently on thin graphs of bounded-degree when the number of different lengths is bounded.

Theorem 4. *There is a* $2^{O(\Delta \cdot k \cdot d)}n$*-time algorithm for* Bsc *in partial k-trees of maximum degree* Δ *and d different weights.*

3.3 Flat graphs

We first treat planar graphs, and then indicate how the approach can be generalized to other "flat" graphs.

We use a classical partitioning technique due to Baker [1]. See [13] for details on the following specific version. A t-outerplanar graph has treewidth at most $3t-1$ [5].

Lemma 5. *Let G be a planar graph and f be a positive integer. Then, there is a vertex-disjoint partitioning of G into graphs G_1, which is f-outerplanar, and G_2, which is outerplanar with at most $2n/f$ vertices and a total vertex length of $2p(G)/t$.*

In view of Theorem 1, we focus on giving an efficient approximation in the case when the vertex lengths are within a limited range. Our method is similar to that of [13], but somewhat simpler. We can particularly take advantage of the feature of batch schedules that it is easy to insert color classes in between batches of a given schedule, which is not so easily done in non-preemptive multicoloring. The result is an EPTAS, as opposed to a PTAS of [13] for npSMC, whose running time is of the form $(\log n)^{\epsilon^{O(1)}} n$.

Lemma 6. *There is a* $(1+\epsilon)$*-approximation algorithm for* Bsc *on planar graphs, that runs in time* $2^{O(\log \epsilon^{-1} \cdot \epsilon^{-2} \cdot \log \tau)}n$. *The makespan of the schedule found is* $O(\epsilon^{-1} \cdot p_{max})$.

Proof. The algorithm proceeds as follows. Let f be to be determined. We first apply Lemma 5 to partition G into a partial $3f$-tree G_1 and a partial 2-tree G_2, where G_2 has at most $2n/f$ vertices and $2p(G)/f$ weight. We then find a $(1+\epsilon)$-approximate schedule S of G_1 using Proposition 2, and find a schedule

S_2 of G_2 using Lemma 1. The issue that remains is how to insert the batches of S_2 into S so as to limit the cost of the resulting schedule.

Let $z = d \cdot \epsilon^{-1}$, where $d = \lg \tau$ is the number of distinct batch lengths in S_2. Recall that there are at most 4 batches of each length in S_2. We shall fit the batches of each length ℓ in S_2 into the schedule S as follows: Before the batch that starts execution at or right before step $i \cdot z \cdot \ell$, insert a batch of length ℓ, for $i = 1, 2, 3, 4$. Then, each job in S is delayed at most $i\ell$ by batches of length ℓ if its completion time in S is at least $i \cdot z \cdot \ell$, or at most a $1/z$-fraction. Thus, summing over the different lengths, each job is delayed by at most a $d/z = \epsilon$-fraction by jobs from S. Now, each job in S_2 is delayed by at most a z factor. So, the cost of scheduling the jobs of G_2 within the new schedule is at most $z \cdot p(G_2) \leq z \cdot p(G)/f$, which is at most $\epsilon p(G)$ if we choose f to be $z/\epsilon = \epsilon^{-2}d$.

The combined cost of the coloring is then at most $1+3\epsilon$ times optimal. We scale ϵ to suit the claim. The makespan of the schedule is the sum of $O(\epsilon^{-1} \cdot p_{max})$ for scheduling G_1 and $O(p_{max})$ for G_2, as argued in Lemma 1, for a total of $O(\epsilon^{-1} p_{max})$.

The time complexity is dominated by the time used by the algorithm of Proposition 2 for G_1. Here, $\chi \leq 4$, but $k = O(f) = O(\epsilon^{-2} \log \tau)$, which is asymptotically equivalent to the number b of batches. Hence, the combined time complexity is $2^{O(\log \epsilon^{-1} \cdot \epsilon^{-2} \cdot \log \tau)} n$. ∎

The following theorem now follows straightforwardly by combining Lemma 6 with Theorem 1. The time complexity is $2^{\Theta(\epsilon^{-4} \cdot \log^3 \epsilon^{-1})} n$.

Theorem 5. *There is an EPTAS for* BSC *on planar graphs.*

Recall that sum coloring (SC) is the common special case of SBC and SJC when all vertices have the same length. It was shown in [13] that SC is strongly NP-hard for planar graphs. This implies that our results are best possible, in that no FPTAS is possible.

We can obtain a more efficient scheme in the case of small values of p_{max}, and in particular for the sum coloring problem. This improves on the algorithm of [13] that has time complexity $exp(O(\ln \ln n \cdot \epsilon^{-1} \log \epsilon^{-1})) \cdot n$.

Theorem 6. *There is an EPTAS for* SC *on planar graphs with running time* $(\log \epsilon^{-1})^{O(\epsilon^{-1})} n$.

In the full version [14], we indicate how the result of Theorem 5 can be extended, e.g., to the larger class of bounded-genus graphs.

3.4 Pebbly graphs

As implied by Proposition 1, SBC is non-trivial for arbitrary job lengths, even when the jobs are independent (i.e., when the graph contains no edges). We give here a strongly polynomial time algorithm for a meaningful simple class of graphs: those consisting of disjoint cliques of bounded size. This class is neither thin nor flat; instead, we can say it is *pebbly*. Disjoint cliques correspond naturally to the case of single-resource constraints, namely, each job needs to use a single resource to be processed; thus, all jobs that compete for the same resource form a clique.

Proposition 3. *There is an algorithm for* Bsc *on disjoint collection of cliques, that runs in time* $\min(h^{O(d)}, 2^{O(h)})n$*, where* h *is the size of the largest clique and* d *is the number of distinct lengths.*

Proof sketch. We give a dynamic programming formulation. For each (multi)set S of possible job lengths, we compute the minimum cost of coloring a certain subgraph. Namely, we assume that a part of the graph has already been colored, and that S contains the lengths of the longest batches already colored. We therefore compute, for each such S, the minimum cost of batch coloring the best possible remaining graph. Given such a set S, it is easy to tell which nodes must have been already colored to leave the best possible remainder; the greedy choice of assigning to each batch the longest vertex that fits in the batch is an optimal strategy.

Hence, we have a table indexed by h-bounded multisets (sets of size at most h). We can fill into this table, bottom up, by trying all possible values for extending the given set S, namely values whose addition (with the possible deletion of the smallest value) yields an h-bounded sequence dominating S.

The number of h-bounded length sets is bounded by the number of subsets of the cliques in the graphs, or $\sum_{C\in G} 2^{|C|} \leq 2^h n/h$. When the number of lengths d is fixed, we can also bound this number by h^d, by counting how many times each length ℓ occurs in the set. To compute each entry of the table we need to consult at most all dominated entries, for a total time complexity at most square of the table size.

4 Max Coloring Paths

We sketch below an improved exact algorithm for the max-coloring problem on paths. It uses the observation that any non-trivial solution for Max-Col on paths uses at most three batches (see, e.g., [6]).

Theorem 7. *There is an* $O(n \log n)$ *algorithm for* Max-Col *on paths.*

Let x_s be the height of the s-th batch in some optimal solution, $1 \leq s \leq 3$ (note that $x_1 = p_{max}$). For each $s, t \in \{1, 2, 3\}$ and $1 \leq i \leq j \leq n$, where $j - i$ is a power of 2 dividing both i and j, we compute a vector $h(i, s, j, t)$ of length $(j - i + 1)$ which gives the minimum value of the third length (x_3), for each possible value of the second length (x_2), for vertices on the subpath between i and j, when i is scheduled in batch s and j in batch t. The possible values of x_2 are the lengths of the vertices on this subpath, that is, $x_2 \in \{p_i, p_{i+1}, \ldots, p_j\}$; the entries of the vector h are sorted in non-increasing order by the value of x_2. The vector $h(i, s, j, t)$ can be calculated recursively.

We find an optimal solution for max-coloring by calculating $h(1, s, n, t)$ for all $1 \leq s, t \leq 3$, and selecting (within these 9 vectors) the entry which gives the minimum makespan.

References

1. Baker, B.S.: Approximation algorithms for NP-hard problems on Planar Graphs. J. of the ACM 41, 153–180 (1994)
2. Bar-Noy, A., Bellare, M., Halldórsson, M.M., Shachnai, H., Tamir, T.: On chromatic sums and distributed resource allocation. Inf. Comput. 140(2), 183–202 (1998)
3. Bar-Noy, A., Halldórsson, M.M., Kortsarz, G., Salman, R., Shachnai, H.: Sum multicoloring of graphs. J. of Algorithms 37(2), 422–450 (2000)
4. Bazgan, C.: Approximation schemes and parameterized complexity. PhD thesis, INRIA, Orsay, France (1995)
5. Bodlaender, H.L., Koster, A.M.: Combinatorial Optimization on Graphs of Bounded Treewidth. The Computer Journal (2007), doi:10.1093/comjnl/bxm037
6. Brucker, P.: Scheduling Algorithms, 4th edn. Springer, Heidelberg (2004)
7. Cesati, M., Trevisan, L.: On the Efficiency of Polynomial Time Approximation Schemes. Information Processing Letters 64, 165–171 (1997)
8. Epstein, L., Levin, A.: On the max coloring problem. In: Proc. of WAOA (2007)
9. Epstein, L., Halldórsson, M.M., Levin, A., Shachnai, H.: Weighted Sum Coloring in Batch Scheduling of Conflicting Jobs. Algorithmica (to appear)
10. Escoffier, B., Monnot, J., Paschos, V.T.: Weighted Coloring: Further complexity and approximability results. Inf. Process. Lett. 97(3), 98–103 (2006)
11. Gandhi, R., Halldórsson, M.M., Kortsarz, G., Shachnai, H.: Improved Bounds for Sum Multicoloring and Scheduling Dependent Jobs with Minsum Criteria. ACM Transactions on Algorithms (to appear)
12. Guan, D.J., Zhu, X.: A Coloring Problem for Weighted Graphs. Inf. Process. Lett. 61(2), 77–81 (1997)
13. Halldórsson, M.M., Kortsarz, G.: Tools for Multicoloring with Applications to Planar Graphs and Partial k-Trees. J. Algorithms 42(2), 334–366 (2002)
14. Halldórsson, M. M., Shachnai, H.: Batch Coloring Flat Graphs and Thin. full version. `http://www.cs.technion.ac.il/~hadas/PUB/batch_col.pdf`
15. Han, C.-C., Hou, C.-J., Shin, K.J.: On slot reuse for isochronous services in DQDB networks. In: Proc. of 16th IEEE Real-Time Systems Symposium, pp. 222–231 (1995)
16. Liu, H., Beck, M., Huang, J.: Dynamic Co-Scheduling of Distributed Computation and Replication. In: Proc. of CCGRID, pp. 592–600 (2006)
17. Mihail, M., Kaklamanis, C., Rao, S.: Efficient Access to Optical Bandwidth. In: Proc. of FOCS 1995, pp. 548–557 (1995)
18. Peek, B.R.: High performance optical network architecture. In: All-Optical Networking: Existing and Emerging Architecture and Applications (2002)
19. Pemmaraju, S.V., Raman, R.: Approximation Algorithms for the Max-coloring Problem. In: Caires, L., Italiano, G.F., Monteiro, L., Palamidessi, C., Yung, M. (eds.) ICALP 2005. LNCS, vol. 3580, pp. 1064–1075. Springer, Heidelberg (2005)
20. Smith, W.E.: Various optimizers for single-stage production. Naval Research Logistics Quarterly 3, 59–66 (1956)
21. Tanenbaum, A.S.: Distributed Operating Systems. Prentice-Hall, Englewood Cliffs (1995)
22. West, D.B.: Graph Theory, 2nd edn. Prentice-Hall, Englewood Cliffs (2001)

Approximating the Interval Constrained Coloring Problem

Ernst Althaus, Stefan Canzar, Khaled Elbassioni, Andreas Karrenbauer, and Julián Mestre*

Max-Planck-Institute for Informatics, Saarbrücken, Germany
{althaus,scanzar,elbassio,karrenba,jmestre}@mpi-inf.mpg.de

Abstract. We consider the *interval constrained coloring* problem, which appears in the interpretation of experimental data in biochemistry. Monitoring hydrogen-deuterium exchange rates via mass spectroscopy experiments is a method used to obtain information about protein tertiary structure. The output of these experiments provides data about the exchange rate of residues in overlapping segments of the protein backbone. These segments must be re-assembled in order to obtain a global picture of the protein structure. The *interval constrained coloring* problem is the mathematical abstraction of this re-assembly process.

The objective of the interval constrained coloring problem is to assign a color (exchange rate) to a set of integers (protein residues) such that a set of constraints is satisfied. Each constraint is made up of a closed interval (protein segment) and requirements on the number of elements that belong to each color class (exchange rates observed in the experiments).

We show that the problem is NP-complete for arbitrary number of colors and we provide algorithms that given a feasible instance find a coloring that satisfies all the coloring requirements within ± 1 of the prescribed value. In light of our first result, this is essentially the best one can hope for. Our approach is based on polyhedral theory and randomized rounding techniques. Furthermore, we develop a quasi-polynomial-time approximation scheme for a variant of our problem where we are asked to find a coloring satisfying as many fragments as possible.

1 Introduction

Our motivation for the *interval constrained coloring* problem comes from an application in biochemistry. The problem has been introduced recently by Althaus *et al.* [1]. To be self-contained, we restrict ourselves to a very brief and informal description in this paper and refer the interested reader to the publication mentioned above.

A challenging and important problem in biochemistry is to determine the tertiary structure of a protein, i.e. the spatial arrangement, which is indispensable for its function. There are various approaches each with advantages and drawbacks. One method for this task is the so-called *hydrogen-deuterium exchange*,

* Research supported by an Alexander von Humboldt fellowship.

J. Gudmundsson (Ed.): SWAT 2008, LNCS 5124, pp. 210–221, 2008.

abbreviated by HDX. This is a chemical reaction where a hydrogen atom of the protein is replaced by a deuterium atom, or vice versa. To this end, the protein solution is diluted by D_2O. Intuitively, the exchange process happens at a higher rate at amino acids, or residues, that are more exposed to the solvent. Put differently, the exchange rates for residues at the outside of the complex are higher than inside. Note that though deuterium is heavier than hydrogen, they are almost identical from a chemical point of view. Hence, the exchange rate may be monitored by mass spectroscopy while the tertiary structure remains unaffected by the process. However, this method does not deliver that fine grained information such that the exchange rate for each residue can be determined directly. Rather, we get bulk information for fragments of the protein. For example, we get the number of slow, medium, and fast residues for each of several overlapping fragments covering the whole protein. That is, the experimental data only tells us how many residues of a fragment react at low, medium, and high exchange rate, respectively. Moreover, we know the exact location and size of each fragment in the protein. It remains to find a valid assignment of all residues to exchange rates that matches the experimentally found bulk information. If the solution is not unique, we want to enumerate all feasible of them or a representative subset thereof as a basis for further chemical considerations.

The problem can be rephrased in mathematical terms as follows. We are given a protein of n residues and a set of fragments, which correspond to intervals of $[n]$. The fragments cover the whole protein and may overlap. Furthermore, there are k possible exchange rates to which we refer as colors in the following. The goal is to produce a coloring of the set $[n]$ using k colors such that a given set of requirements is satisfied. Each requirement is made up of a closed interval $I \subseteq [n]$ and a complete specification of how many elements in I should be colored with each color class. We refer to this problem as the *interval constrained coloring* problem.

More formally, let $\mathcal{I}$ be a set of intervals defined on the set $V = [n]$, let $[k]$ be a set of color classes, and let $r : \mathcal{I} \times [k] \to Z^+$ be a requirement function such that $\sum_{c \in [k]} r(I, c) = |I|$ for all $I \in \mathcal{I}$. A coloring $\chi : V \to [k]$ is said to be *feasible* if for every $I \in \mathcal{I}$ we have

$$|\{i \in I \mid \chi(i) = c\}| = r(I, c) \text{ for all } c \in [k] \tag{1}$$

Given this information, we would like to determine whether or not a feasible coloring exists, and if so, to produce one.

The problem is captured by the integer program given below. The binary variable $x_{i,c}$ indicates whether i is colored c or not. Constraint (2) enforces that each residue gets exactly one color and constraint (3) enforces that every requirement is satisfied.

$$\sum_{c \in [k]} x_{i,c} = 1 \qquad \forall\, i \in [n] \tag{2}$$

$$\sum_{i \in I} x_{i,c} = r(I, c) \qquad \forall\, I \in \mathcal{I}, c \in [k] \tag{3}$$

$$x_{i,c} \in \{0, 1\} \qquad \forall\, i \in [n], c \in [k] \tag{4}$$

Let $\mathcal{P}$ be the polytope obtained by relaxing the integrality constraint (4) in the above integral problem. That is $\mathcal{P}$ is the set of values of x obeying (2), (3) and $0 \leq x_{i,c} \leq 1$ for all i and c.

1.1 Previous and Related Work

The polyhedral description has already been introduced in [1] and has served there as a basis to attack the problem by integer programming methods and tools, which perform well in practice. Moreover, the authors established the polynomial-time solvability of the two-color case by the integrality of the polytope $\mathcal{P}$ and provided also a combinatorial algorithm for this case. However, the complexity of the general problem has been left open.

A closely related problem is *broadcast scheduling*, where a server must decide which data item to broadcast at each time step in order to satisfy client requests. The literature in broadcast scheduling is vast and many variations of the problem have been studied (see [2,4] and references therein). In the variant we are concerned with here, a client request is specified by a time window I and a data type A. The request is satisfied if A is broadcast *at least once* in I. The similarities between the two problems should be clear with time steps, time windows and data types in broadcast scheduling playing the respective roles of positions, intervals and colors in interval constrained coloring. There are, however, important differences. First, whereas in broadcast scheduling it does not hurt to broadcast an item more times than the prescribed number, in our problem it does. Second, an interval is satisfied only if *all* the requirements for that interval are satisfied *exactly*, which, undoubtedly, makes our problem significantly harder.

1.2 Contributions of This Paper

As mentioned above, the complexity status for the *interval constrained coloring* problem has been open. In Section 4 we partly settle this by showing that deciding whether a feasible coloring exists is NP-complete when k is part of the input.

Although the polytope $\mathcal{P}$ is integral for $k = 2$, it need not be for $k > 3$. Nevertheless, we can check in polynomial time whether $\mathcal{P} = \emptyset$. If that is the case then we know that there is no feasible coloring. Otherwise we can find a feasible fractional solution. In Section 2 we will show how to round this fractional solution to produce a coloring where *all* the requirements are satisfied within a mere additive error of one.

In practice, the data emanating from the experiments is noisy, which normally causes the instance to be infeasible and in some case even forces $\mathcal{P}$ to be empty. To deal with this problem in Section 3 we study a variant of the problem in which we want to maximize the number of requirements that are satisfied. Another way to deal with noisy data is to model the noise in the linear programming relaxation to get a new set of requirements on which to run the algorithm from Section 2. The latter approach was explored by Althaus *et al.* [1]; the reader is referred to their paper for details.

2 A ±1 Guarantee

Let x be a fractional solution in $\mathcal{P}$. We use the scheme of Gandhi *et al.* [4] to round x to an integral solution $\hat{x}$ with the following properties:

Theorem 1. *Given a fractional solution $x \in \mathcal{P}$ we can construct in polynomial time an integral solution $\hat{x}$ with the following properties*

(P1) For every $i \in [n]$ there exists $c \in [k]$ such that $\hat{x}_{i,c} = 1$ and $\hat{x}_{i,d} = 0$ for all $d \neq c$.
(P2) For every $I \in \mathcal{I}$ and $c \in [k]$ we have $|\sum_{i \in I} \hat{x}_{i,c} - r(I,c)| \leq 1$.
(P3) Every $I \in \mathcal{I}$ is satisfied with probability at least $\gamma_k = \frac{k(k+1-H_{k-1})}{(k+1)!}$.

In other words, each position gets exactly one color (P1), every coloring requirement is off by at most one from the prescribed number (P2), and all the requirements for a given interval I are satisfied *exactly* ($\sum_{i \in I} \hat{x}_{i,c} = r(I,c)$ for all $c \in [k]$) with probability at least γ_k. An interesting corollary of this theorem is that if $\mathcal{P}$ is non-empty then there exists always a coloring satisfying at least $\gamma_k |\mathcal{I}|$ intervals, and such coloring can be found in polynomial time.

The high level idea is to simplify the polytope $\mathcal{P}$ into another integral polytope with basic solutions satisfying (P1) and (P2). Then we show how to select a basic solution satisfying (P3). This is done by defining a set of *blocks* and then setting up an assignment problem instance between $[n]$ and the set of blocks, whose polytope is integral.

For each color class $c \in [k]$ we choose a real number $\alpha_c \in [0,1]$, to be specified shortly. Let us define blocks $B_1^c, B_2^c, \ldots, B_{b_c}^c$: For color c and $j = 2, \ldots, b_c - 1$

$$B_j^c = \left[\min\{t \mid \textstyle\sum_{i \leq t} x_{i,c} > j - 2 + \alpha_c\},\ \min\{t \mid \textstyle\sum_{i \leq t} x_{i,c} \geq j - 1 + \alpha_c\}\right]. \quad (5)$$

The first and last blocks, B_1^c and $B_{b_c}^c$, are defined similarly, but starting at 1 and ending at n respectively.

For each $i \in B_j^c$ we define a variable $y_{i,(c,j)}$. If i belongs to a single block B_j^c of color c then we set $y_{i,(c,j)} = x_{i,c}$. Otherwise, i belongs to two adjacent blocks B_{j+1}^c and B_j^c, in which case we set $y_{i,(c,j+1)} = \sum_{t \leq i} x_{t,c} - (j - 1 + \alpha_c)$ and $y_{i,(c,j)} = x_{i,c} - y_{i,(c,j+1)}$. See Figure 1 for an example of how the blocks and the solution y are constructed. Another, equivalent, way to define y is to ask that $x_{i,c} = \sum_j y_{i,(c,j)}$, $\sum_{i \in B_1^c} y_{i,(c,1)} = \alpha_c$ and $\sum_{i \in B_j^c} y_{i,(c,j)} = 1$ for every $1 < j < b_c$. Thus y defines a feasible fractional assignment between $[n]$ and the set of blocks. Let $\mathcal{Q}$ be the polytope of this assignment problem, namely,

$$\textstyle\sum_{B_j^c \ni i} y_{i,(c,j)} = 1 \qquad \forall\, i \in [n] \qquad (6)$$

$$\textstyle\sum_{i \in B_j^c} y_{i,(c,j)} = 1 \qquad \forall\, c \in [k] \text{ and } 1 < j < b_c \qquad (7)$$

$$\textstyle\sum_{i \in B_j^c} y_{i,(c,j)} \leq 1 \qquad \forall\, c \in [k] \text{ and } j \in \{1, b_c\} \qquad (8)$$

$$y_{i,(c,t)} \geq 0 \qquad \forall\, i \in [n], c \in [k], t \in [b_c] \qquad (9)$$

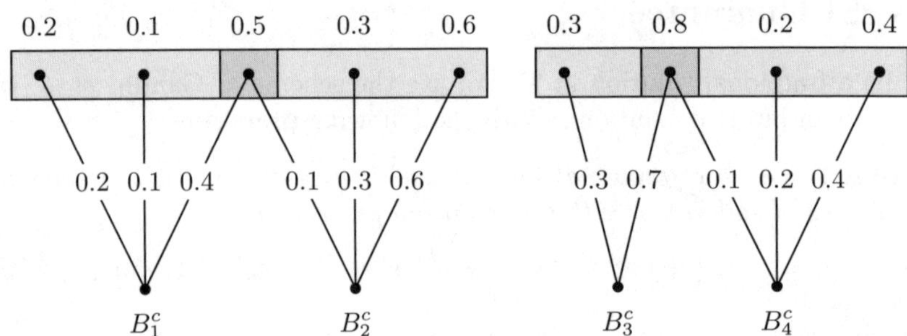

Fig. 1. How the blocks B_j^c are constructed. The $x_{i,c}$ values appear on top and the $y_{i,(c,j)}$ values appear on the edges. Note that a block can only overlap with its predecessor or successor. In this case $\alpha_c = 0.7$.

Because $\mathcal{Q}$ is integral, any fractional solution $y \in \mathcal{Q}$ can be turned into an integral solution $\hat{y} \in \mathcal{Q}$; this can even be done in polynomial time. Notice that an integral solution $\hat{y}$ to $\mathcal{Q}$ induces an integral solution $\hat{x}$ by setting $\hat{x}_{i,c} = 1$ if and only if $y_{i,(c,j)} = 1$. Constraint (6) implies that $\hat{x}$ satisfies (P1). Furthermore, $\hat{x}$ also satisfies (P2).

Lemma 1. *Let $\hat{y}$ be an integral solution for $\mathcal{Q}$ and let $\hat{x}$ be the coloring induced by $\hat{y}$. Then $|\sum_{i\in I} \hat{x}_{i,c} - r(I,c)| \leq 1$ for all $I \in \mathcal{I}$ and $c \in [k]$.*

Proof. Since $\sum_{i\in I} x_{i,c} = r(I,c)$, the number of blocks of color c that intersect I is either $r(I,c)$ or $r(I,c)+1$. Furthermore, at least $r(I,c)-1$ of these blocks lie entirely within I and at most two blocks intersecting I partially. Due to constraints (6) and (7), each internal block will force a different position in I to be colored c. One the other hand, the fringe blocks, if any, can force at most two additional positions in I to be colored c. Hence, the lemma follows. □

It only remains to prove that $\hat{x}$ obeys (P3). To do so, we need to introduce some randomization in our construction. First, we will choose the offset α_c of each color $c \in [k]$ independently and uniformly at random. Second, instead of choosing any extreme point of $\mathcal{Q}$, we choose one using a randomized rounding procedure.

Gandhi *et al.* [4] showed that any fractional solution $y \in \mathcal{Q}$ can be rounded to an integral solution $\hat{y} \in \mathcal{Q}$ s.t. the probability that $\hat{y}_{i,(c,j)} = 1$ is exactly $y_{i,(c,j)}$. It is important to note that these events *are not independent* of each other.

Lemma 2. *Let $\hat{y}$ be the solution output by the randomized rounding procedure and $\hat{x}$ the coloring induced by it. For any interval $I \in \mathcal{I}$, the probability that $\sum_{i\in I} \hat{x}_{i,c} = r(I,c)$ for all $c \in [k]$ is at least $\frac{k(k+1-H_{k-1})}{(k+1)!}$.*

Proof. Let I be an arbitrary, but fixed, interval throughout the proof and for time being let us concentrate on a fixed, but arbitrary, color $c \in [k]$. Let f and l be the indices of the first and last blocks of color class c that intersect I and define $\beta_c = \sum_{i\in I\cap B_f^c} y_{i,(c,f)}$, or, equivalently, $\sum_{i\in I\cap B_l^c} y_{i,(c,l)} = 1 - \beta_c$.

Intuitively, the probability that $\sum_{i \in I} \hat{x}_{i,c} = r(I,c)$ should be greater when the blocks of c are aligned with I (when β_c is close to 0 or 1) and it should be low when they are not (when β_c is around 0.5). By choosing α_c uniformly at random, β_c also becomes a random variable uniformly distributed in $[0,1]$. Thus, we have a decent chance of getting a "good value" of β_c.

Let us formalize and make more precise the above idea. Denote with ξ_f and ξ_l the events $\sum_{i \in I \cap B_f^c} \hat{y}_{i,(c,f)} = 1$ and $\sum_{i \in I \cap B_l^c} \hat{y}_{i,(c,l)} = 1$ respectively. Let $\beta = (\beta_1, \ldots, \beta_k)$ be the vector offset for the color classes. For brevity's sake we denote $\Pr[\xi \mid \beta]$ with $\Pr_\beta[\xi]$.

$$\begin{aligned}\Pr_\beta\left[\sum_{i \in I} \hat{x}_{i,c} \neq r(I,c)\right] &= \Pr_\beta\left[\xi_f \xi_l \vee \overline{\xi_f}\,\overline{\xi_l}\right] = \Pr_\beta[\xi_f \xi_l] + \Pr_\beta\left[\overline{\xi_f}\,\overline{\xi_l}\right] \\ &\leq \min\{\Pr_\beta[\xi_f], \Pr_\beta[\xi_l]\} + \min\{\Pr_\beta[\overline{\xi_f}], \Pr_\beta[\overline{\xi_l}]\}\end{aligned}$$

Since $\Pr_\beta[\xi_f] = \beta_c$ and $\Pr_\beta[\xi_l] = 1 - \beta_c$, it follows that

$$\Pr_\beta\left[\sum_{i \in I} \hat{x}_{i,c} \neq r(I,c)\right] \leq 2 \min\{\beta_c, 1-\beta_c\} \tag{10}$$

As a warm-up we first show that the probability that all requirements for I are fulfilled is at least $\frac{1}{(k+1)!}$. Denote with τ the event $\forall c : \sum_{i \in I} \hat{x}_{i,c} = r(I,c)$. Recall that the vector β is distributed uniformly over the domain $D = [0,1]^k$. Conditioning on β and averaging over D gives the desired result.

$$\begin{aligned}\Pr[\tau] &= \int_D \Pr_\beta\left[\forall c : \sum_{i \in I} \hat{x}_{i,c} = r(I,c)\right] \, \mathrm{d}\beta_1 \cdots \mathrm{d}\beta_k \\ &\geq \int_D 1 - \sum_{c \in [k]} \Pr_\beta\left[\sum_{i \in I} \hat{x}_{i,c} \neq r(I,c)\right] \, \mathrm{d}\beta_1 \cdots \mathrm{d}\beta_k \\ &\geq \int_D \max\left\{0, 1 - 2\sum_{c \in [k]} \min\{\beta_c, 1-\beta_c\}\right\} \, \mathrm{d}\beta_1 \cdots \mathrm{d}\beta_k \\ &= 2\int_D \max\left\{0, \tfrac{1}{2} - \sum_{c \in [k]} \min\{\beta_c, 1-\beta_c\}\right\} \, \mathrm{d}\beta_1 \cdots \mathrm{d}\beta_k\end{aligned}$$

The second inequality follows from the union bound and the third from (10). A moment's thought reveals that the function inside the integral is symmetrical in the 2^k orthants around the point $(\frac{1}{2}, \ldots, \frac{1}{2}) \in D$. Therefore, setting $D' = [0, \frac{1}{2}]^k$ we get

$$\Pr[\tau] \geq 2^{k+1} \int_{D'} \max\left\{0, \tfrac{1}{2} - \sum_{c \in [k]} \beta_c\right\} \, \mathrm{d}\beta_1 \cdots \mathrm{d}\beta_k.$$

The right hand side of the above inequality can be interpreted as the volume of a $(k+1)$-dimensional simplex.

$$\Pr[\tau] \geq 2^{k+1} \operatorname{Vol}\left(\lambda \in R_+^{k+1} \,\middle|\, \sum_{i \in [k+1]} \lambda_i \leq \tfrac{1}{2}\right) = 2^{k+1} \frac{(\frac{1}{2})^{k+1}}{(k+1)!} = \frac{1}{(k+1)!}$$

In order to get the stronger bound in the statement of the lemma we need two more ideas. First, we claim that we only need to condition on fulfilling $k-1$ requirements: Because $\sum_{c \in [k]} r(I,c) = |I|$, once we get $k-1$ colors right, the

kth requirement must be satisfied as well. Second, since we can condition on any $k - 1$ colors, we had better condition on the ones with smallest offset, that is, those that are close to 0 or 1.

$$\begin{aligned}
\Pr[\tau] &= \int_D \Pr_\beta\left[\forall c : \textstyle\sum_{i\in I}\hat{x}_{i,c} = r(I,c)\right] \mathrm{d}\beta_1\cdots\mathrm{d}\beta_k \\
&\geq \int_D \max_{d\in[k]}\left\{1 - \textstyle\sum_{c\neq d}\Pr_\beta\left[\sum_{i\in I}\hat{x}_{i,c} \neq r(I,c)\right]\right\} \mathrm{d}\beta_1\cdots\mathrm{d}\beta_k \\
&\geq \int_D \max_{d\in[k]}\left\{\max\left\{0, 1 - 2\textstyle\sum_{c\neq d}\min\left\{\beta_c, 1-\beta_c\right\}\right\}\right\} \mathrm{d}\beta_1\cdots\mathrm{d}\beta_k \\
&= 2^k \int_{D'} \max_{d\in[k]}\left\{\max\left\{0, 1 - 2\textstyle\sum_{c\neq d}\beta_c\right\}\right\} \mathrm{d}\beta_1\cdots\mathrm{d}\beta_k \\
&= 2^{k+1} \int_{D'} \max\left\{0, \tfrac{1}{2} - \textstyle\sum_{c\in[k]}\beta_c + \max_{d\in[k]}\beta_d\right\} \mathrm{d}\beta_1\cdots\mathrm{d}\beta_k
\end{aligned}$$

The last integral can be simplified by assuming that the maximum β_d is attained by the last variable. Of course, the maximum can be any of the k variables, thus these two quantities are related by a factor of k.

$$\Pr[\tau] \geq k\,2^{k+1}\int_0^{\frac{1}{2}}\left[\int_{[0,z]^{k-1}} \max\left\{0, \tfrac{1}{2} - \textstyle\sum_{c\in[k-1]}\beta_c\right\} \mathrm{d}\beta_1\cdots\mathrm{d}\beta_{k-1}\right] \mathrm{d}z$$

Let $T(z)$ denote $\mathrm{Vol}\left(\lambda \in R_+^k \;\middle|\; \sum_{i=1}^k \lambda_i \leq \frac{1}{2} \text{ and } \lambda_1,\ldots,\lambda_{k-1} \leq z\right)$. Then we can rewrite the above integral as

$$\Pr[\tau] \geq k\,2^{k+1}\int_0^{\frac{1}{2}} T(z)\,\mathrm{d}z \tag{11}$$

The volume computed by $T(z)$ is not a simplex, but it can be reduced to a summation involving only the volume of simplices using the principle of inclusion/exclusion.

Let $V(\rho)$ denote the volume $\mathrm{Vol}\left(\lambda \in R_+^k \;\middle|\; \sum_{i=1}^k \lambda_i \leq \rho\right)$ and recall that $V(\rho) = \frac{\rho^k}{k!}$. Consider what happens when $z \in \left[\frac{1}{4}, \frac{1}{2}\right)$; clearly $T(z) < V(\frac{1}{2})$ since $V(\frac{1}{2})$ includes points $\lambda \in R_+^k$ such that $\lambda_i > z$ for exactly one coordinate $i \in [k-1]$ (since $z \geq \frac{1}{4}$). Notice that

$$\mathrm{Vol}\left(\lambda \in R_+^k \;\middle|\; \textstyle\sum_{i=1}^k \lambda_i \leq \frac{1}{2} \text{ and } \lambda_i > z\right) = V(\tfrac{1}{2} - z)$$

Thus $T(z) = V(\frac{1}{2}) - (k-1)V(\frac{1}{2} - z)$ for $z \in [\frac{1}{4}, \frac{1}{2}]$, but $T(z) > V(\frac{1}{2}) - (k-1)V(\frac{1}{2} - z)$ for $z \in [0, \frac{1}{4})$ since the volume of points y such the constraint $\lambda_i \leq z$ is violated for two coordinates is subtracted twice. To avoid cumbersome notation, assume $V(\rho) = 0$ if $\rho \leq 0$. A simple inclusion/exclusion argument yields

$$T(z) = \sum_{i=0}^{k-1}\binom{k-1}{i}(-1)^i\,V(\tfrac{1}{2} - iz) \tag{12}$$

Plugging (12) into (11) we get

$$\begin{aligned}\Pr[\tau] &\geq 2^{k+1}k\left(\int_0^{\frac{1}{2}} V(\tfrac{1}{2})\,\mathrm{d}z + \sum_{i=1}^{k-1}\binom{k-1}{i}(-1)^i\int_0^{\frac{1}{2i}} V(\tfrac{1}{2}-iz)\,\mathrm{d}z\right)\\ &= \frac{k}{(k+1)!}\left(k+1+\sum_{i=1}^{k-1}\binom{k-1}{i}\frac{(-1)^i}{i}\right)\end{aligned}$$

The intermediate steps of this last derivation can be found in the full version. Using induction on k, it is straightforward to show that the sum in the last line adds up exactly to $-H_{k-1}$, which gives us the desired bound of γ_k. □

Remark. In our application domain the goal usually is not to find a single solution, but to generate a number of candidate solutions and let the user choose the one that he finds most interesting or relevant for the specific application. Our framework is amenable to this task since there are very efficient algorithms to enumerate all the integral solutions of $\mathcal{Q}$ [6].

3 Maximum Coloring

In this section we study a variant of the interval constrained coloring to deal with instances that do not admit a feasible coloring. For these instances we consider the problem of finding a coloring that maximizes the number of intervals satisfying (1). More generally, we assume a non-negative weight $w(I)$, associated with each interval $I \in \mathcal{I}$, and seek a subset $\mathcal{I}' \subseteq \mathcal{I}$, maximizing $w(\mathcal{I}') \overset{\text{def}}{=} \sum_{I\in\mathcal{I}'} w(I)$, such that there exists a coloring of V satisfying (1) for each $I \in \mathcal{I}'$. We call this problem MAXCOLORING. Let OPT $\subseteq \mathcal{I}$ be a subset achieving this maximum. For $\alpha \in (0,1]$ and $\beta \geq 1$, an (α,β)-approximation of the problem is given by a pair $(\chi,\mathcal{I}')$ of a subset $\mathcal{I}' \subseteq \mathcal{I}$, and a coloring $\chi : V \mapsto [k]$, such that $\sum_{I\in\mathcal{I}'} w(I) \geq \alpha \cdot w(\text{OPT})$, and $\frac{1}{\beta}r(I,c) \leq N_\chi(I,c) \leq \beta r(I,c)$, where $N_\chi(I,c)$ is the number of positions in I colored c by χ.

Theorem 2. *Consider an instance $(V,\mathcal{I})$ of* MAXCOLORING *with $|V| = n$ and $|\mathcal{I}| = m$. Then we can find a $(1, 1+\epsilon)$-approximation in quasi-polynomial time $n^{O(\frac{k^2}{\epsilon}\log n \log m)}$, for any $\epsilon > 0$.*

Note that the above bound is quasi-polynomial for $k = \text{polylog}(n,m)$. To prove Theorem 2 we use a similar technique as in [3]. Our approach can be divided into two parts: (i) Reducing the search space, and (ii) developing a dynamic program. We explain these two steps in more details in the next subsections.

3.1 Reducing the Search Space

Let $\epsilon > 0$ be a given constant. For a vertex $u \in V$ and a set of intervals $\mathcal{I}$ on V, denote respectively by $\mathcal{I}_L(u)$, $\mathcal{I}_R(u)$, and $\mathcal{I}[u]$ the subsets of intervals of $\mathcal{I}$ that lie to the left of u, lie to the right of u, and span u, that is

$$\mathcal{I}_L(u) = \{[s,t] \in \mathcal{I} : \; t \leq u-1\}, \; \mathcal{I}_R(u) = \{[s,t] \in \mathcal{I} : \; s \geq u+1\} \text{ and } \mathcal{I}[u] = \{[s,t] \in \mathcal{I} : \; s \leq u \leq t\}.$$

Denote by $V_L(u)$ and $V_R(u)$ the sets of vertices that lie to the left and right of $u \in V$, respectively: $V_L(u) = \{i \in V : \; i < u\}$ and $V_R(u) = \{i \in V : \; i \geq u\}$.

Definition 1. (Assignments) *Let $V = \{p, p+1, \ldots, q\}$. An assignment on V is a pair $A = (\mathcal{I}_A, r_A)$ of intervals $\mathcal{I}_A$ on V and a function $r_A : \mathcal{I}_A \times [k] \mapsto \{0, 1, \ldots, |V|\}$ such that $r_A(I) \leq r_A(I')$ for all $I, I' \in \mathcal{I}_A$, with $I \subseteq I'$. A is called a left-assignment (respectively, right-assignment) if all intervals in $\mathcal{I}_A$ start at p (respectively, end at q).*

Definition 2. (ϵ-Partial assignments) *Let $u^* \in V$ be a given vertex of $V = \{p, p+1, \ldots, q\}$. A set of $h_1 + h_2 + 4$ intervals $\mathcal{I}_P = \mathcal{I}_{P_l} \cup \mathcal{I}_{P_r}$, $\mathcal{I}_{P_l} = \{I_0, I_1, \ldots, I_{h_1}, I_{h_1+1}\}$ and $\mathcal{I}_{P_r} = \{I'_0, I'_1, \ldots, I'_{h_2}, I'_{h_2+1}\}$, and an $r_P : \mathcal{I}_P \times [k] \mapsto \{0, 1, \ldots, |V|\}$, such that*

(R1) all intervals start or end at u^*: $I_j = [u_j, u^*]$ for $j \in \{0, 1, \ldots, h_1\}$, $I_{h_1+1} = [p, u^*]$, $I'_j = [u^*, u'_j]$ for $j \in \{0, 1, \ldots, h_2\}$, and $I'_{h_2+1} = [u^*, q]$, where $u_{h_1} < u_{h_1-1} < \cdots < u_1 < u_0 < u^* < u'_0 < u'_1 < u'_2 < \cdots < u'_{h_2}$
(R2) $r_P(I, c) \leq r_P(J, c)$ for every $I, J \in \mathcal{I}_P$, with $I \subseteq J$, and every $c \in [k]$,
(R3) $\sum_{c \in [k]} r_P(I, c) = |I|$ for every $I \in \mathcal{I}_P$,
(R4) for every $I \in \mathcal{I}_P$, there exist $c \in [k]$ and $i \in \mathbb{Z}_+$ such that $r_P(I, c) = \lceil (1+\epsilon)^i \rceil$, and
(R5) for every $c \in [k]$ and $i \in \mathbb{Z}_+$ with $i \leq \lfloor (\log r_P(I_{h_1+1}, c) / \log(1+\epsilon) \rfloor$, there exists $I \in \mathcal{I}_{P_l}$ such that $r_P(I, c) = \lceil (1+\epsilon)^i \rceil$; similarly, for every $c \in [k]$ and $i \in \mathbb{Z}_+$ with $i \leq \lfloor (\log r_P(I'_{h_2+1}, c) / \log(1+\epsilon) \rfloor$, there exists $I' \in \mathcal{I}_{P_r}$ such that $r_P(I', c) = \lceil (1+\epsilon)^i \rceil$.

will be called an ϵ-partial assignment *w.r.t. u^*, denoted by $P = (u^*, \mathcal{I}_P, r_P)$.*

The total number $\mu(n)$ of possible ϵ-partial assignments with respect to a given vertex $u^* \in V$ with $|V| = n$ can be bounded as follows:

$$\mu(n) \leq \left(\frac{\ln k + 1}{k-1} \cdot n \right)^{2k^2 \frac{\log n}{\log(1+\epsilon)} + 6k}, \tag{13}$$

which is $n^{\mathrm{polylog}(n)}$ for every fixed $\epsilon > 0$ and $k = \mathrm{polylog}(n)$.

Let $\chi : V \mapsto [k]$ be a coloring of V and $u^* \in V$ be an arbitrary vertex. We say that an assignment $A = (\mathcal{I}_A, r_A)$ is *consistent* with χ if $N_\chi(I, c) = r_A(I, c)$ for all $c \in [k]$ and $I \in \mathcal{I}_A$. Two assignments P_1 and P_2 are said to be consistent if there exists a coloring χ with which both are consistent. The next lemma follows immediately from the definition of ϵ-partial assignments.

Lemma 3. *Let χ be a coloring of V and $u^* \in V$ be an arbitrary vertex. Then there exists an ϵ-partial assignment P on V w.r.t. u^*, that is consistent with χ.*

3.2 The Dynamic Program

The algorithm shown below is parameterized with two assignments P_L and P_R, both initially empty. It is based on a divide-and-conquer approach where a point u^* in the middle of V is picked and all intervals containing u^* are evaluated to see if they should be taken into the solution. To do this evaluation conservatively, the procedure iterates over all ϵ-partial assignments P, consistent with P_L and P_R, w.r.t. to the middle vertex u^*, then recurses on the subsets of intervals to the left and right of u^*.

Algorithm MaxColoringApx$(\mathcal{I}, V, P_L, P_R)$

Input: An instance $(\mathcal{I}, V)$ of MAXCOLORING, and consistent left- and right-assignments P_L and P_R on V

Output: A $(1, 1+\epsilon)$ approximation $(\chi, \mathcal{J})$

if $|\mathcal{I}| = 0$, **then**
    $\chi \leftarrow$ MAXCOLORINGSPECIAL(P_L, P_R)
    return $(\chi, \emptyset)$
let $u^* \in V$ be such that $|\mathcal{I}_L(u^*)| \leq m/2$ and $|\mathcal{I}_R(u^*)| \leq m/2$
for every ϵ-partial assignment P w.r.t. u^* **do**
    if P is consistent with P_L and P_R **then**
        $(\chi_1, \mathcal{J}_1) \leftarrow$ MAXCOLORINGAPX$(\mathcal{I}_L(u^*), V_L(u^*),$ REDUCE$(V_L(u^*), P, P_L, P_R))$
        $(\chi_2, \mathcal{J}_2) \leftarrow$ MAXCOLORINGAPX$(\mathcal{I}_R(u^*), V_R(u^*),$ REDUCE$(V_R(u^*), P, P_L, P_R))$
        let $\chi \leftarrow \chi_1 \cup \chi_2$
        $\mathcal{K} \leftarrow \{I \in \mathcal{I}[u^*] : \frac{r(I,c)}{(1+\epsilon)} \leq r_P(I_{\ell(I,P)}, c) + r_P(I_{j(I,P)}, c) \leq r(I, c)\}$
        $\mathcal{J} \leftarrow \mathcal{K} \cup \mathcal{J}_1 \cup \mathcal{J}_2$
        store $(\chi, \mathcal{J})$
return the recorded solution with largest $w(\mathcal{J})$ value

Algorithm MAXCOLORINGAPX uses two subroutines: MAXCOLORINGSPECIAL checks if a pair of a left- and right-assignments are consistent, and if so, returns a feasible coloring; REDUCE$(V_L(u^*), P, P_L, P_R)$ (REDUCE$(V_R(u^*), P, P_L, P_R)$) combines the assignments P, P_L, P_R into a left- and right assignments P'_L, P'_R on $V_L(u^*)$ (respectively, on $V_R(u^*)$).

When the procedure returns, we get two independent colorings $\chi_1 : V_L(u^*) \mapsto [k]$ and $\chi_2 : V_R(u^*) \mapsto [k]$, which are combined into a coloring $\chi = \chi_1 \cup \chi_2$ defined in the obvious way: $\chi(u) = \chi_1(u)$ if $u \in V_L(u^*)$ and $\chi(u) = \chi_2(u)$ if $u \in V_R(u^*)$.

Lemma 4. *Let $\omega = (n, \mathcal{I}, k, r)$ be an instance of* MAXCOLORING. *If set $\mathcal{I}$ can be partitioned into two sets $\mathcal{I}_1$ and $\mathcal{I}_2$, such that for $x \in \{1, 2\}$ it holds*

(a) $I_i \cap I_j = \emptyset$, $\forall I_i, I_j \in \mathcal{I}_x$, i.e. intervals are disjoint
(b) $\bigcup_{I_j \in \mathcal{I}_x} I_j = [s, t]$, i.e. the union of intervals is an interval again

then the feasibility problem for ω can be solved in time $\mathcal{O}\left(n^k |\mathcal{I}|\right)$.

Proof (sketch). We construct an instance $\omega' = (n, \mathcal{I}', k, r')$, where set $\mathcal{I}'$ itself satisfies conditions (a) and (b) from Lemma 4. In particular, intervals in $\mathcal{I}'$ are disjoint (condition (a)) and therefore feasibility of instance ω' can be determined by verifying for every interval $[a,b] \in \mathcal{I}'$ that $\sum_{c\in[k]} r'([a,b],c) = b-a+1$.

We define $\mathcal{I}'$ to be the partition of $\{1,\ldots,n\}$ into a minimal number of intervals, such that for each interval $I' \in \mathcal{I}'$ and each element $I \in \mathcal{I}$ either $I' \subseteq I$ or $I' \cap I = \emptyset$. If we represent I' by sequence $([a'_1,b'_1],[a'_2,b'_2],\ldots,[a'_p,b'_p])$ it can be shown by induction, that the definition of $r'([a'_1,b'_1],c)$ uniquely determines $r'([a'_i,b'_i],c)$, for $2 \le i \le p$. □

Corollary 1. *The feasibility problem for given left assignment $P_L = (\mathcal{I}_{P_L}, r_{P_L})$ and right assignment $P_R = (\mathcal{I}_{P_R}, r_{P_R})$ on a set of vertices $V = \{1,2,\ldots,n\}$ can be solved in time $\mathcal{O}\left(n^k(|\mathcal{I}_{P_L}| + |\mathcal{I}_{P_R}|)\right)$.*

Let $P = (u^*, \mathcal{I}_P, r_P)$ be an ϵ-partial assignment w.r.t. u^*. Given an interval $I = [s,t] \in \mathcal{I}$, with $u^* \in I$, we let $j(I,P)$, $\ell(I,P)$ be respectively the smallest and largest indices such that $[u_{j(I,P)}, u'_{\ell(I,P)}] \subseteq I$, i.e. $j(I,P) = \min\{i : \ u_i \ge s\}$ and $\ell(I,P) = \max\{i : \ u'_i \le t\}$. If either of these indices does not exist, we set the corresponding $r_P(I_{\ell(I,P)},c)$ or $r_P(I_{j(I,P)},c)$ to 0. Note that by $(R5)$

$$r_P(I_{\ell(I,P)},c) + r_P(I_{j(I,P)},c) \le N_{\chi'}(I,c) \le (1+\epsilon)(r_P(I_{\ell(I,P)},c) + r_P(I_{j(I,P)},c)), \tag{14}$$

holds for any $\chi' : V \mapsto [k]$ and ϵ-partial assignment P consistent with χ'.

Lemma 5. *For $|V| = n$ and $|\mathcal{I}| = m$, algorithm* MaxColoringApx *runs in time $T(n,m) = n^{O(\frac{k^2}{\epsilon}\log n \log m)}$.*

Lemma 6. *Algorithm* MaxColoringApx *returns a coloring $\chi : V \mapsto [k]$ and a subset of intervals $\mathcal{J} \subseteq \mathcal{I}$ such that $w(\mathcal{J}) \ge w(\mathrm{OPT})$ and $r(I,c)/(1+\epsilon) \le N_\chi(I,c) \le (1+\epsilon)r(I,c)$ for all $I \in \mathcal{J}$ and $c \in [k]$.*

Proof. Let (χ^*, OPT) be an optimal solution. By Lemma 3, there is an ϵ-partial assignment P consistent with χ^*, which will be eventually considered by the algorithm in Step 5. If $I \in \mathrm{OPT}[u^*]$, then $N_{\chi^*}(I,c) = r(I,c)$ and thus (14) implies, for $\chi' = \chi^*$ that I belongs to the set $\mathcal{K}$ selected by the algorithm in Step 10, i.e., $\mathrm{OPT}[u^*] \subseteq \mathcal{K}$, and hence $w(\mathcal{K}) \ge w(\mathrm{OPT}[u^*])$. Since the returned coloring χ is consistent with P, we also know by using $\chi' = \chi$ in (14) that $r(I,c)/(1+\epsilon) \le N_\chi(I,c) \le (1+\epsilon)r(I,c)$ for $I \in \mathcal{K}$. By induction, we have $w(\mathcal{K}_1) \ge w(\mathrm{OPT}_L(u^*))$, $w(\mathcal{K}_2) \ge w(\mathrm{OPT}_R(u^*))$, $r(I,c)/(1+\epsilon) \le N_{\chi_1}(I,c) \le (1+\epsilon)r(I,c)$ for $I \in \mathcal{J}_1$, and $r(I,c)/(1+\epsilon) \le N_{\chi_2}(I,c) \le (1+\epsilon)r(I,c)$ for $I \in \mathcal{J}_2$. The lemma follows. □

4 Hardness

In this section we show that, in general, deciding whether a feasible coloring exists is NP-hard.

Theorem 3. *The problem of testing the feasibility of an instance of the interval constrained coloring problem is NP-complete when the number of colors is part of the input.*

Proof. Clearly, the problem belongs to NP. To prove the problem is NP-hard we reduce a known NP-hard problem to it using the approach of Chang *et al.* [2]. In the *exact coverage* problem we are given a ground set $\mathcal{U}$ and a collection $\mathcal{S}$ of subsets of $\mathcal{U}$ and we want to know whether there exists a sub-collection $\mathcal{C} \subseteq \mathcal{S}$ of size t, which forms a partition of $\mathcal{U}$; that is, $\cup_{S \in \mathcal{C}} S = \mathcal{U}$ and for any $R, S \in \mathcal{C}$ if $R \neq S$ then $R \cap S = \emptyset$. It is well known that exact coverage is NP-complete [5] even when the cardinality of sets in $\mathcal{S}$ is 3.

Let $u = |\mathcal{U}|$ and $s = |\mathcal{S}|$. For the instance of the coloring problem we divide $V = [n]$ into u blocks $B_1, \ldots, B_u$ each of length s; thus, $n = us$ and $B_i = [(i-1)s+1, \ldots, i\,s]$. Each color $c \in [k]$ is associated with a specific set S_c in $\mathcal{S}$; thus, $k = s$. Let $\mathcal{U} = \{x_1, \ldots, x_u\}$ and suppose that x_i is contained r_i in sets. For every $i \in [u]$ we have

$$
\begin{aligned}
I_i &= [s\,(i-1)+1, \ldots, s\,i] && \text{and} && r(I_i, c) = 1 \text{ for all } c \in [k] \\
I'_i &= [s\,i - t + 1, \ldots, s\,(i+1) - t] && \text{and} && r(I'_i, c) = 1 \text{ for all } c \in [k] \\
I''_i &= [s\,i - t - r_i, \ldots, s\,i - t + 1] && \text{and} && r(I''_i, c) = 1 \text{ if and only if } x_i \in S_c
\end{aligned}
$$

Realize that any coloring satisfying all the I_i and I'_i intervals must use the same set of t colors for the last t positions of every block and the remaining $s-t$ colors for the first $s-t$ position of every block. We therefore encode the cover $\mathcal{C}$ with the last t colors of each block. To enforce that $\mathcal{C}$ is a partition, we ask that for every element $x \in \mathcal{U}$ *exactly* one set in $\mathcal{C}$ contains x in $\mathcal{S}$, then we include the interval $I''_i = [s\,i - t - r_i, s\,i - t + 1]$ and require $r(I''_i, c) = 1$ if and only if $x_i \in S_c$. Clearly, a feasible coloring encodes a solution for the exact coverage and vice-versa. It follows that the testing feasibility is NP-hard. □

Acknowledgments. Thanks to Hubert Chan for useful discussions.

References

1. Althaus, E., Canzar, S., Emmett, M.R., Karrenbauer, A., Marshall, A.G., Meyer-Basese, A., Zhang, H.: Computing H/D-exchange speeds of single residues from data of peptic fragments. In: 23rd Annual ACM Symposium on Applied Computing (2008)
2. Chang, J., Erlebach, T., Gailis, R., Khuller, S.: Broadcast scheduling: Algorithms and complexity. In: Proceedings of the 19th Annual ACM-SIAM Symposium on Discrete Algorithms (2008)
3. Elbassioni, K.M., Sitters, R., Zhang, Y.: A quasi-PTAS for profit-maximizing pricing on line graphs. In: Proceedings of the 15th Annual European Symposium on Algorithms, pp. 451–462 (2007)
4. Gandhi, R., Khuller, S., Parthasarathy, S., Srinivasan, A.: Dependent rounding and its applications to approximation algorithms. J. ACM 53(3), 324–360 (2006)
5. Garey, M.R., Johnson, D.S.: Computers and Intractability, A Guide to the Theory of NP-Completeness, W.H. Freeman and Company, New York (1979)
6. Uno, T.: A fast algorithm for enumerating bipartite perfect matchings. In: Eades, P., Takaoka, T. (eds.) ISAAC 2001. LNCS, vol. 2223, pp. 367–379. Springer, Heidelberg (2001)

A Path Cover Technique for LCAs in Dags

Mirosław Kowaluk[1,⋆], Andrzej Lingas[2,⋆⋆], and Johannes Nowak[3,⋆⋆⋆]

[1] Institute of Informatics, Warsaw University, Warsaw, Poland
kowaluk@mimuw.edu.pl

[2] Department of Computer Science, Lund University, 22100 Lund, Sweden
Andrzej.Lingas@cs.lth.se

[3] Fakultät für Informatik, Technische Universität München, München, Germany
nowakj@in.tum.de

Abstract. We develop a path cover technique to solve lowest common ancestor (LCA for short) problems in a directed acyclic graph (dag).

Our method yields improved upper bounds for two recently studied problem variants, computing one (representative) LCA for all pairs of vertices and computing all LCAs for all pairs of vertices. The bounds are expressed in terms of the number n of vertices and the so called width $w(G)$ of the input dag G. For the first problem we achieve $\widetilde{O}(n^2 w(G))$ time which improves the upper bound of [18] for dags with $w(G) = O(n^{0.376-\delta})$ for a constant $\delta > 0$. For the second problem our $\widetilde{O}(n^2 w(G)^2)$ upper time bound subsumes the $O(n^{3.334})$ bound established in [11] for $w(G) = O(n^{0.667-\delta})$.

As a second major result we show how to combine the path cover technique with LCA solutions for dags with small depth [9]. Our algorithm attains the best known upper time bound for this problem of $O(n^{2.575})$. However, most notably, the algorithm performs better on a vast amount of input dags, i.e., dags that do not have an almost linear-sized subdag of extremely regular structure.

Finally, we apply our technique to improve the general upper time bounds on the worst case time complexity for the problem of reporting LCAs for each triple of vertices recently established by Yuster[26].

1 Introduction

A *lowest common ancestor* (LCA) of vertices u and v in a *directed acyclic graph* (dag) is an ancestor of both u and v that has no descendant which is an ancestor of u and v. Fast algorithms for finding lowest common ancestors (LCAs) in trees and – more generally – directed acyclic graphs (dags) are indispensable computational primitives. Whereas LCA computations in trees are well studied, see, e.g., [14,22,5], the case of dags has been found an independent subject of research only recently, initiated by the paper of Bender et al. [5]. Due to the limited expressive power of trees they are often applicable only in restrictive or over-simplified settings. There are numerous applications for LCA

⋆ Research supported by the grant of the Polish Ministry of Science and Higher Education N20600432/0806.

⋆⋆ Research supported in part by VR grant 621-2005-4806.

⋆⋆⋆ Research supported in part by DFG (Deutsche Forschungsgemeinschaft), grant MA 870/8-1 (SPP 1307 Algorithm Engineering).

J. Gudmundsson (Ed.): SWAT 2008, LNCS 5124, pp. 222–233, 2008.

queries in dags, e.g., object inheritance in programming languages, lattice operations for complex systems, lowest common ancestor queries in phylogenetic networks, or queries concerning customer-provider relationships in the *Internet*. For a more detailed description of possible applications, we refer to [5,4].

Known results on LCAs in dags. LCA algorithms have been extensively studied in the context of trees with most of the research rooted in [2,25]. The first asymptotically optimal algorithm for the all-pairs LCA problem in trees, with linear preprocessing time and constant query time, was given in [14]. The same asymptotics was reached using a simpler and parallelizable algorithm in [22]. Recently, a reduction to range minimum queries has been used to obtain a further simplification [5].

In the more general case of dags, a pair of nodes may have more than one LCA, which leads to the distinction of *representative* versus *all* LCA problems. In early research both versions still coincide by considering dags with each pair having at most one LCA. Extending the work on LCAs in trees, in [21], an algorithm was described with linear preprocessing and constant query time for the LCA problem on arbitrarily directed trees (or, causal polytrees). Another solution was given in [1], where the representative problem in the context of object inheritance lattices was studied. The approach based on poset embeddings into Boolean lattices [1] yielded $O(n^3)$ preprocessing and $O(\log n)$ query time on lower semi-lattices.

The ALL-PAIRS REPRESENTATIVE LCA problem on general dags has been studied recently in [5,17,9,4]. The works rely on fast matrix multiplications (currently the fastest known algorithm needs $O(n^{\omega})$ operations, with $\omega < 2.376$ [7]) to achieve $\widetilde{O}(n^{\frac{\omega+3}{2}})$ [5] [1] and $\widetilde{O}(n^{2+\frac{1}{4-\omega}})$ [17] upper bounds on the running time. The technique developed in [17] was slightly improved in [9] by applying rectangular matrix multiplication[2] [6] to achieve an upper bound for the representative LCA problem of $O(n^{2+\mu})$.

More efficient solutions have been developed for special classes of dags. For sparse dags, algorithms with running time $O(nm)$ given in [17,4], where m is the number of edges in the input dag, improve the general upper bound. In [11] it was shown that this can be improved to $O(n\,m_{\text{red}})$, where m_{red} is the number of edges in the the transitive reduction of the input dag. In [9] a more efficient solution is described for the ALL-PAIRS REPRESENTATIVE LCA problem in dags of small depth h. More specifically, the algorithm is shown to improve upon the general upper bound whenever $h \le n^{0.42}$. Lately, in [18], it is shown that the problem can also be solved in time $\widetilde{O}(n^2 w(G) + n^{\omega})$, where $w(G)$ is the width of the dag G, see Definition 2.

The authors of [4] study variants of the representative LCA problem, namely (L)CA computations in weighted dags and the ALL-PAIRS ALL LCA problem. For the latter problem, an upper bound of $O(w(G)n^{2+\mu})$ is established which is $O(n^{3+\mu})$ in the worst case. The solution makes use of a minimum path cover of the vertices. The general upper bound for ALL-PAIRS ALL LCA was improved to $O(n^{\omega(2,1,1)})$ in [11]. In the same work it is shown that ALL-PAIRS REPRESENTATIVE LCA and ALL-PAIRS ALL

[1] Throughout this work, we use $\widetilde{O}(f(n))$ for $O(f(n) \cdot \text{polylog}(n))$

[2] Throughout this work, $\omega(x,y,z)$ is the exponent of the algebraic matrix multiplication of a $n^x \times n^y$ with a $n^y \times n^z$ matrix. Let μ be such that $\omega(1,\mu,1) = 1 + 2\mu$ is satisfied. The fastest known algorithms for rectangular matrix multiplication imply $\mu < 0.575$ and $\omega(2,1,1) < 3.334$.

LCA can be solved with running times of $O(n^2 \log n)$ and $O(n^3 \log\log n)$ respectively in the average case. For the average case analysis it is assumed that the input space is distributed according to the $G_{n,p}$ model for random dags which was introduced by Barak and Erdős [3]. In [18] the problem of finding unique lowest common ancestors is shown to be solvable in time $O(n^{\omega} \log n)$.

Our contributions.[3] We elaborate in-depth on using path cover techniques for the solution of LCA problems in dags. To this end, we present a multi-purpose decomposition technique that can be applied to a variety of LCA problems.

We apply our technique to the ALL-PAIRS REPRESENTATIVE LCA problem improving recently developed solutions for dags of small width [18]. Our result implies an upper bound of $\widetilde{O}(n^2 w(G))$ and improves the result in [18] for dags with width $w(G)$ bounded by $O(n^{\omega-2-\delta})$ for a constant $\delta > 0$. Similarly, the application of our approach to the ALL-PAIRS ALL LCA problem yields an upper time bound of $\widetilde{O}(n^2 w(G)^2)$. This improves the general upper bound of $O(n^{\omega(2,1,1)})$ ([11]) for $w(G) = O(n^{\frac{\omega(2,1,1)-2}{2}-\delta})$ as well as the upper bound of $O(n\, m_{\text{red}} \min\{\kappa^2, n\})$ [4], where κ is the maximum size of an LCA set, if the size of the transitive reduction cannot be bounded by $O(n \cdot polylog(n))$.

We show that it is possible to combine the path cover approach with the efficient method for low depth dags given in [9]. Our algorithm has the same asymptotic worst-case time complexity as the fastest up to now algorithm for this problem, i.e., $O(n^{2+\mu})$ [9]. However, the class of dags for which the algorithm needs $\Omega(n^{2+\mu})$ time is considerably limited.

Finally, we describe how our approach can be used to improve the general upper bound for the ALL-K-SUBSETS REPRESENTATIVE LCA problem (i.e., compute representative LCAs for each vertex subset of size k) for $k = 3$, which was recently established by Yuster [26].

2 Preliminaries

Let $G = (V, E)$ be a directed acyclic graph (dag). Throughout this work we denote by n the number of vertices and by m the number of edges in G. Let G_{clo} denote the reflexive transitive closure of G, i.e., the graph having an edge (u, v) if and only if $u \rightsquigarrow v$, i.e., v is reachable from u over some directed path in G. G_{clo} is in one-to-one correspondence with a partial order (poset) $P = (V, \leq)$ on the ground set V where the relation of the poset corresponds to the edges of G_{clo}. In this sense, by slight abuse of notation, we refer to G_{clo} also as a poset. We consider dags equipped with some topological ordering [8].

For a dag $G = (V, E)$ and $x, y, z \in V$, the vertex z is a *common ancestor* (CA) of a pair $\{x, y\}$ if both x and y are reachable from z. By CA$\{x, y\}$, we denote the set of all CAs of x and y. A vertex z is a *lowest common ancestor* (LCA) of x and y if and only if $z \in$ CA$\{x, y\}$ and no other vertex $z' \in$ CA$\{x, y\}$ is reachable from z, that is, there exists no witness for z and $\{x, y\}$.

[3] Some of the proofs are omitted due to space limitations and can be found in [19].

Definition 1 (Witness). *A* witness *for z and $\{x,y\}$ such that $z \in \mathrm{CA}\{x,y\}$ is a vertex $w \in V$ such that $w \in \mathrm{CA}\{x,y\}$ and $z \rightsquigarrow w$.*

We denote the set of all LCAs of a pair $\{x,y\}$ by $\mathrm{LCA}\{x,y\}$. If it is not clear from the context we use a subscript to indicate the graph under consideration, e.g., $\mathrm{LCA}_G\{x,y\}$.

An arbitrary lowest common ancestor z of $\{x,y\}$ is also called a *representative* LCA of x and y. It is a well known and used fact, first observed by Bender et al. [5], that the vertex with the maximum topological number among all vertices in the set $\mathrm{CA}\{x,y\}$ is an LCA of x and y. We denote this special representative LCA by *maximum* LCA. (L)CA problems on dags come in various favors, i.e., computing one (representative) LCA vs. computing all LCAs or computing LCAs for all pairs vs. computing LCAs only for special pairs.

The width of G, denoted by $w(G)$, is of particular interest in this work.

Definition 2 (Width). *The width $w(G)$ of a dag G is defined as the cardinality of a maximum antichain, i.e., incomparable vertices with respect to G_{clo}, in G.*

Note that $\mathrm{LCA}\{x,y\}$ is an antichain by definition and hence $|\mathrm{LCA}\{x,y\}|$ is upper bounded by $w(G)$.

For a dag $G = (V,E)$, a *path cover* $\mathcal{P}$ of G is a set of directed paths in G such for every $v \in V$ there exists at least one path $P \in \mathcal{P}$ including v. A *minimum path cover* is a path cover $\mathcal{P}$ such that $|\mathcal{P}|$ is minimized. The sizes of a maximal antichain and a minimum path cover of G are related by the famous Dilworth Theorem [10]:

Theorem 3 (Dilworth '50). *Let G be a dag. Then, $w(G)$ equals the size of a minimum path cover of G.*

In the context of posets one considers usually *chain covers* instead of path covers. The paths in a chain cover are required to be vertex disjoint. However, it is easy to see that each chain cover in G_{clo} corresponds to a path cover in G and *vice versa*.

3 Path Cover Technique

In this section we present a natural approach to computing LCAs in dags based on decomposing the dag $G = (V,E)$ into a set of paths covering G. The efficiency of this technique depends mainly on the width ($w(G)$) of the underlying dag.

We start by giving an intuitive description of our approach. Let $\{x,y\}$ be any vertex pair of G and let $z \in \mathrm{LCA}_G\{x,y\}$. Suppose now that we start a depth first search (DFS) in G at vertex z. Let T_z be the corresponding DFS tree [8]. Then, it is not difficult to verify that $\mathrm{LCA}_{T_z}\{x,y\} = z$. Observe that the partial order induced by the tree is a suborder of G_{clo}. Moreover, for all vertices $w \in T_z$ it holds that w is reachable from z. Since $z \in \mathrm{LCA}\{x,y\}$, the only common ancestor of x and y in T_z is z. Recall again, that the vertex with the highest topological number among $\mathrm{CA}\{x,y\}$ is a lowest common ancestor. This leads to a first naive decomposition solution for computing representative LCAs: For all $v \in V$, compute DFS trees T_v and preprocess these trees for LCA queries. In order to determine a representative LCA for $\{x,y\}$, compute $Z = \bigcup_{v \in V} \mathrm{LCA}_{T_v}\{x,y\}$ and return the maximum vertex in Z.

The correctness of the above approach follows from the fact that the maximum LCA z is returned at least once by the $O(n)$ LCA queries in the trees, i.e., by the query $\text{LCA}_{T_z}\{x,y\}$. The preprocessing time is $O(nm)$. Subsequent queries for LCA in G can be answered in time $O(n)$ implying the trivial deterministic upper bound of $O(n^3)$ for the ALL-PAIRS REPRESENTATIVE LCA problem. However, we show below that in fact only $w(G)$ DFS trees have to be considered.

Definition 4. *For a directed path $P = \{v_1,\ldots,v_l\}$ in G, T_P is a special DFS tree obtained by starting the DFS at vertex v_1 and first exploring the edges along the path P.*

In the following we denote by $T_P(v_i)$ the subtree of T_P rooted at vertex $v_i \in P$.

Lemma 5. *For a directed path $P = \{v_1,\ldots,v_l\}$ in G, $T_P(v_i)$ corresponds to some DFS tree T_{v_i} for all $1 \le i \le l$.*

The above lemma implies that given a path cover $\mathcal{P}$ of G, only $|\mathcal{P}|$ DFS trees have to be considered.

Lemma 6. *Let $\mathcal{P} = \{P_1,\ldots,P_r\}$ be a path cover of G. Next, let $Z = \bigcup z_i$, where $z_i = \text{LCA}_{T_{P_i}}\{x,y\}$ for $1 \le i \le r$. Then, the following chain of inclusion holds:* $\text{LCA}\{x,y\} \subseteq Z \subseteq \text{CA}\{x,y\}$.

Remark 7. By Lemma 6 both answers to representative LCA queries and all LCA queries can be derived from the set Z. For representative LCAs, we simply have to find the vertex with the maximum number in Z in time $O(|Z|)$. In order to find all LCAs, it is necessary to identify the set of vertices without outgoing edges in the subdag induced by Z. This can be achieved in time $O(|Z|^2)$. To this end, we assume that we have a mapping from V to $\mathcal{P}$ such that for each v a path P_i including v is known. With this mapping reachability queries of two vertices $z_1, z_2 \in Z$ can be answered in constant time by querying the LCA of $\{z_1,z_2\}$ in the corresponding trees. Observe that $z_1 \rightsquigarrow z_2$ implies $\text{LCA}_{P_i}\{z_1,z_2\} = z_1$ if $z_1 \in P_i$.

The following lemma follows immediately from the previously established facts.

Lemma 8. *Let $T_{\text{PC}}(G,r)$ be the time needed to compute a path cover $\mathcal{P} = \{P_1,\ldots,P_r\}$ of G. Then,* ALL-PAIRS REPRESENTATIVE LCA *can be solved in time $O(T_{\text{PC}}(G,r) + rm + rn^2)$.*

In [18] Kowaluk and Lingas show that the ALL-PAIRS REPRESENTATIVE LCA problem can be solved in time $\widetilde{O}(n^{\omega} + w(G)n^2)$. Furthermore, this bound is improved to $\widetilde{O}(w(G)n^2)$ [18] by using randomization. In the following we show that our path cover technique can be applied to obtain a simple algorithm for ALL-PAIRS REPRESENTATIVE LCA running in time $\widetilde{O}(w(G)n^2)$.

For a given minimum path cover of size $r = w(G)$, the above lemma immediately yields the claimed time bounds. However, it is not known how to compute a minimum path cover within time $O(n^2 w(G))$. Nonetheless, we show how to compute a path cover

$\mathcal{P}$ of G_{clo} of size $O(w(G)\log n)$ in time $\widetilde{O}(n^2 w(G))$, i.e., $\mathcal{P}$ is minimal up to a logarithmic factor. We apply the technique to improve the upper bound of the ALL-PAIRS REPRESENTATIVE LCA and ALL-PAIRS ALL LCA problems on dags of small width.

The standard solution for computing a minimum chain cover of a dag reduces the problem to finding a maximum matching in a bipartite graph [13]. The number of edges in the bipartite graph corresponds to the number of edges in G_{clo}. In the deterministic setting, the best algorithm for solving the bipartite matching problem is still the $O(m\sqrt{n})$ solution given by Hopcroft and Karp [15]. Recently, Mucha et al. [20] have shown that a maximum matching in a bipartite graph can be found in $O(n^{\omega})$ using randomization. On the other hand, Felsner et al. [12] showed that it is possible to recognize a poset P of width at most k in time $O(n^2 k)$. In their approach, the parameter k is given beforehand and – in the case that the width of P is bounded by k – their algorithm can be extended to output a chain cover of size k.

Definition 9 (Greedy Path Cover). *Let $G = (V,E)$ be a dag. A greedy path cover of G is obtained by recursively finding paths $P_1,\ldots,P_r$ such that for all P_i, the value $|P_i \setminus \bigcup_{k\leq i-1} P_k|$, i.e., the number of vertices on P_i that are not covered by any of the paths $P_1,\ldots,P_{i-1}$, is maximized.*

That is, we reduce the dag into paths by recursively finding paths containing as much uncovered vertices as possible. In Felsner et al. [12], a similar approach is used to decompose a partial order P in a greedy manner. In their approach, all covered vertices are removed from the partial order. Since we do not want to construct the poset G_{clo}, we cannot discard vertices. However, it is possible to establish a one-to-one correspondence between decomposing G and G_{clo}. Extending the ideas in [12] yields the following lemma.

Lemma 10. *A greedy path cover $P_1,\ldots,P_r$ of G with $r \leq w(G)\log n$ can be computed in time $O(mr)$.*

Proof. In Felsner et al. [12], it is shown that the size of a greedy chain cover of a partial order P is at most $w(P)\log n$. It is not difficult to see that a greedy path cover of a dag is in one-to-one correspondence with a greedy chain cover of the respective induced partial order.

The (weighted) single source longest path problem can be solved in time $O(n+m)$ by using standard algorithms for solving the single source shortest path problem in dags, see, e.g., [8]. Observe that arbitrary edge weights are possible. We assume without loss of generality that G is equipped with a single source s, otherwise we simply add a super source. We initialize each edge in G with weight 1. Then, we recursively solve the single source longest path problem for the source s. After the ith iteration we set the weight of each edge $e = (v,w)$ such that $w \in P_i$ to 0. It is easy to check that the weight of P_i is equal to the number of uncovered vertices on P_i. Hence, we obtain a greedy path cover. It is also easy to see that each step can be implemented in time $O(m)$. □

Lemma 10 implies the following theorem.

Theorem 11. *For a dag G with n vertices and width $w(G)$,* ALL-PAIRS REPRESENTATIVE LCA *can be solved in time $\widetilde{O}(w(G)n^2)$ and* ALL-PAIRS ALL LCA *can be solved in time $\widetilde{O}(w(G)^2 n^2)$.*

This result improves the upper bound of ALL-PAIRS REPRESENTATIVE LCA ([18]) for dags with $w(G) = O(n^{\omega-2-\delta})$ and the general upper time bound for ALL-PAIRS ALL LCA ([11]) for dags with $w(G) = O(n^{\frac{\omega(2,1,1)-2}{2}-\delta})$ for a constant $\delta > 0$. Using results on the expected value of the width of $G_{n,p}$ random dags [3] we get the following corollary. The respective average case complexities match results previously established in [11].

Corollary 12. *Using the path cover technique* ALL-PAIRS REPRESENTATIVE LCA *and* ALL-PAIRS ALL LCA *can be solved in time $\widetilde{O}(n^2)$ on a dag with n vertices in the average case under the assumption that the input space is distributed according to the $G_{n,p}$ model for random dags with edge probability $p = O(1)$.*

4 Combining Small Width and Low Depth

The path cover technique described in the previous chapter can be naturally used together with the solution for dags of low depth given in [9]. The *depth* of a vertex v denoted by $\mathrm{dp(v)}$ is defined as the length of longest path to v in G. The depth of G, $\mathrm{dp(G)}$ is given by $\mathrm{dp(G)} = \max_{\mathrm{v}\in\mathrm{V}}\{\mathrm{dp(v)}\}$.

Theorem 13. *For a dag G with depth* $\mathrm{dp(G)} = \mathrm{n^q}$, ALL-PAIRS REPRESENTATIVE LCA *can be solved in* $\widetilde{O}(n^{q+\omega(1,1-q,1)})$.

This result is achieved by exploiting the fact that common ancestor of maximum depth in a dag G is a lowest common ancestor. Observe that a possible witness would have greater depth by definition. The following lemma can be derived from the above proposition.

Lemma 14. *For a dag $G = (V,E)$ with n vertices and depth* $\mathrm{dp(G)} = \mathrm{n^q}$, *the* ALL-PAIRS REPRESENTATIVE LCA *problem can be solved in time $\widetilde{O}(n^{2+\mu-\delta})$ if $q \le 1 - \mu - \delta$ for an arbitrary small constant $\delta > 0$.*

Proof. ALL-PAIRS REPRESENTATIVE LCA in G can be solved in time $\widetilde{O}(n^{q+\omega(1,1-q,1)})$ (Thm. 13). We make use of Proposition 16. Let $\beta = \frac{\omega-2}{1-\alpha}$. Then we want to solve the following inequality $q + \omega(1, 1-q, 1) \le 2 + \mu - \delta$ (i) for q. Recall that μ satisfies $\omega(1,\mu,1) = 1 + 2\mu$. By Proposition 16 we get: $\mu = \frac{1-\beta\alpha}{2-\beta}$ (ii). Now we plug (ii) into (i) and solve for q to get $q \le 1 - \mu - \delta$. □

Assume first that we have already computed G_{clo}. On a high level, the combination of the above result with our path cover technique works as follows:

1. Construct a partial chain cover $C_1, \ldots, C_r$ of G_{clo} greedily. That is, search for chains of maximum size until a termination criterion (to be specified below) is satisfied. Note that we do not require that the chains cover all vertices of G_{clo}. Note further that in general $r \ne w(G)$.
2. Construct the special DFS trees associated with the chains as described in the previous chapter and prepare them for constant time LCA queries.
3. Reduce G_{clo} along the chains. That is, let $V_C = \bigcup_{1\le i\le r} C_i$. Then we remove all edges (v, w) such that $v \in V_C$. Observe that all edges (v, w) where $v \notin V_C$ are retained. The resulting graph is called the *reduced dag* denoted by $G^R = (V, E^R)$.

4. For all $x, y \in V$, compute maximum depth common ancestors in G^R. The maximum depth CAs are either lowest common ancestors in G or all of their witnesses are in the set V_C. Thus, in a second step we search for possible witnesses by querying the special DFS trees for vertices in V_C. If witnesses exist, i.e., common ancestors that are successors of the maximum depth CAs, we output the witness with highest label which is a lowest common ancestor.

The reasoning behind this approach is as follows. By decomposing the graph along the longest chains we reduce successively the depth of G_{clo}. As soon as we reach a certain threshold depth we can apply the algorithm given in [9] to efficiently compute maximum depth CAs in the reduced dag. Moreover, if a maximum depth CA in the reduced dag G^R is not a lowest common ancestor in the original dag, we know that all its witnesses are covered by the chains. Observe that outgoing edges of non-covered vertices are not removed. Hence, if z' is a witness of z and $\{x,y\}$ and we suppose $z' \notin V_C$, we have $z \rightsquigarrow z'$ and $z' \rightsquigarrow x, y$ in the reduced dag contradicting the fact that z is a maximum depth CA in G^R.

In the following we refer to the algorithm described in this section as the *combined algorithm*. Before proving the key properties of the approach we give a formal description of the decomposition:

Lemma 15. *Let z_h be a maximum depth common ancestor in G^R of a pair $\{x,y\}$. Then either $z_h \in \text{LCA}_G\{x,y\}$ or all witnesses of z_h and $\{x,y\}$ in G are in the set V_C.*

Proof. Suppose that z_h is not an LCA of $\{x,y\}$ in G. In that case there exists a witness z' such that $z' \in \text{LCA}_G\{x,y\}$ and z' is a successor of z_h. This immediately implies $z' \in V_C$. Indeed, suppose that $z' \notin V_C$. This implies (i) $\text{dp}(z') > \text{dp}(z_h)$ since z' is a successor of z and (z_h, z') is not removed by the decomposition and (ii) $z' \in \text{CA}_{G^R}\{x,y\}$. But, (i) and (ii) contradict the fact that z_h is the maximum depth CA of $\{x,y\}$ in G^R. □

Observe that we have constructed special DFS trees for the chains in G_{clo}. However, in general it is also possible to use an approach that allows constructing the trees in G. This involves computing a greedy path cover of G instead of the greedy chain cover of G_{clo}. In any case, the reduced dag G^R results from G_{clo}.

We specify the full algorithm for finding representative LCAs for each vertex pair with respect to G.

The correctness of Algorithm 1 is a consequence of Lemma 15 and the results of Section 3. We turn our attention to the complexity of this approach. Obviously, the running time depends on (i) the cardinality r of the partial chain cover and (ii) $\text{dp}(\text{G}^\text{R})$. Our goal is an algorithm that does not exceed the worst case complexity for the general problem of $O(n^{2+\mu})$ but performs better in as many cases as possible. To this end, one can specify an implementation of the termination criterion for the preprocessing algorithm as follows: Pick threshold parameters W and H and terminate whenever $r \geq W$ or $\text{dp}(\text{G}^\text{R}) \leq \text{H}$. Any reasonable choice of H should satisfy $H \leq n^{1-\mu}$ according to Lemma 14. Observe that any choice of W such that $W \leq n^{\mu}$ is sufficient to guarantee a worst case upper bound of $O(n^{2+\mu})$. Simply modify the combined algorithm such that it uses the general solution whenever $r \geq W$ (which implies $\text{dp}(\text{G}^\text{R}) > \text{H}$). However, there is a more elegant way to find the optimal decomposition. The idea is to look for the

Algorithm 1. Algorithm for representative LCA

Input: The reduced dag $G^R = (V, E^R)$ and special DFS trees $T_{C_1}, \ldots, T_{C_r}$ that are prepared for constant LCA queries (output of the preprocessing algorithm)
Output: All-pairs representative LCA matrix

```
begin
    Compute maximum depth CAs on for all vertex pairs in G^R using the algorithm
    described in [9].
    foreach Vertex pair {x,y} do
        Let z_h be the maximum depth CA of {x,y} with respect to G^R.
        Initialize an empty result set Z.
        foreach chain C_i, 1 ≤ i ≤ r do
            Query LCA{x,y} in T_{C_i} and add the result to Z.
        end
        Remove all vertices from Z that are not successors of z_h.
        Return the maximum vertex in Z and z_h if Z is empty.
    end
end
```

intersection point of the functions $n^{q+\omega(1,1-q,1)}$ and rn^2, i.e., the functions that describe the asymptotic behavior of the two approaches. The respective termination criterion becomes (neglecting polylogarithmic factors)

$$rn^2 > n^{\log_n(\mathrm{dp}(\mathrm{G^R}))+\omega(1,1-\log_n(\mathrm{dp}(\mathrm{G^R})),1)}. \tag{1}$$

In order to prove the following theorem, we make use of the following proposition which is due to Huang and Pan [16]. Recall that $\omega(a,b,c)$ denotes the exponent of the multiplication of an $n^a \times n^b$ by an $n^b \times n^c$ matrix.

Proposition 16 (Rectangular Matrix Multiplication). *Let* $\omega = \omega(1,1,1) < 2.376$ *and let* $\alpha = \sup\{0 \le r \le 1 : \omega(1,r,1) = 2 + o(1)\}$. *Then*

$$\omega(1,r,1) \le \begin{cases} 2+o(1) & \text{if } 0 \le r \le \alpha \\ 2+\frac{\omega-2}{1-\alpha}(r-\alpha)+o(1) & \text{if } \alpha \le r \le 1. \end{cases} \tag{2}$$

The current best bound for α *is* $\alpha \le 0.294$ *and is due to Coppersmith [6].*

Theorem 17. *The time complexity of the combined algorithm is bounded by* $O(n^{2+\mu})$ *where* μ *satisfies* $\omega(1,\mu,1) = 1 + 2\mu$.

Proof. Let in the following $\beta = \frac{\omega-2}{1-\alpha}$. Let r be the cardinality of the path cover produced by the preprocessing algorithm and let $\mathrm{dp}(\mathrm{G^R}) = \mathrm{n^q}$ be the depth of the reduced graph G^R. By Equation (1) we have $n^{q+\omega(1,1-q,1)} < rn^2 < n^{q+\omega(1,1-q,1)} + 1$ and from this we can conclude $r \le n^{q+\omega(1,1-q,1)-2+o(1)}$. Moreover, $\widetilde{O}(n^2 r)$ is obviously an upper time bound for the combined algorithm. We can safely assume that $q \le 1/2$. To see this, suppose first that $q > 1/2$. Since every path covers at least n^q vertices, we have $r \le n^{1-q}$. On the other, $r > n^{q+o(1)}$ by Equation (1) and thus $n^{1-q} > n^{q+o(1)}$, a contradiction for $q > 1/2$. We apply Proposition 16 and get $r \le n^{q+2+\beta(1-q-\alpha)-2+o(1)}$. This simplifies to $r \leq n^{q(1-\beta)+\beta(1-\alpha)+o(1)}$. Since we know that each path in the path cover covers at least h vertices we have $q \leq \log_n \frac{n}{r}$. Combining the above two inequalities yields $r \leq n^{\log_n \frac{n}{r}(1-\beta)+\beta(1-\alpha)+o(1)}$. Since we have $n^{\log_n \frac{n}{r}(1-\beta)+\beta(1-\alpha)} = \left(\frac{n}{r}\right)^{1-\beta} \cdot n^{\beta(1-\alpha)}$, this simplifies to $r \leq n^{\frac{1-\beta\alpha}{2-\beta}+o(1)}$. On the other hand, recall that the parameter μ satisfies $\omega(1,\mu,1) = 1+2\mu$. Again, we use Proposition 16 to obtain $\mu = \frac{1-\beta\alpha}{2-\beta}$, which concludes the proof. □

As an immediate consequence of the above lemma and the current best bounds for ω and α we get the following corollary.

Corollary 18. *The time complexity of the algorithm for solving* ALL-PAIRS REPRESENTATIVE LCA *is* $O(n^{2.575})$.

Observe that the average case bound of $\widetilde{O}(n^2)$ is still valid under the assumption that the input space is distributed according to the $G_{n,p}$ model with constant edge probability p. To see this, observe that the transitive closure of a random dag in the $G_{n,p}$ model for arbitrary values of p can be computed in average case time $\widetilde{O}(n^2)$ using the algorithm given by [24].

The combination of these two techniques narrows down the classes of dags for which no solution faster than $O(n^{2+\mu})$ is known considerably. Indeed, recall that for dags G of depth $\mathrm{dp(G)} \leq \mathrm{n}^{1-\mu-\delta}$ the ALL-PAIRS REPRESENTATIVE LCA problem can be solved in time $\widetilde{O}(n^{2+\mu-\delta})$ by Lemma 14.

Theorem 19. *For a dag G with n vertices and an arbitrary constant* $\mu \geq \delta > 0$ ALL-PAIRS REPRESENTATIVE LCA *can be solved in time* $\widetilde{O}(n^{2+\mu-\delta})$ *if G does not contain a subgraph H that contains at least* $n^{\mu-\delta}$ *(vertex-disjoint) maximum size (w.r.t. the greedy approach) chains of length at least* $n^{1-\mu-\delta}$.

Observe that this is a significant restriction for bad dag classes, namely an almost linear-sized, i.e., $n^{1-2\delta}$ for an arbitrary small constant δ, subdag of extremely regular structure.

The ALL-K-SUBSETS REPRESENTATIVE LCA problem is an extension of the ALL-PAIRS REPRESENTATIVE LCA problem. Given a constant $k \geq 2$, compute a representative LCA for each k-subset of vertices in G. The problem has been considered recently by Yuster [26]. He shows that the ALL-K-SUBSETS REPRESENTATIVE LCA problem can be solved in time $O(n^{3.575})$ for $k = 3$ and $O(n^{k+1/2})$ for $k \geq 4$. We improve slightly upon the bound for $k = 3$ by using our combined approach.

Theorem 20. *The* ALL-K-SUBSETS REPRESENTATIVE LCA *problem can be solved in time* $O(n^{3.5214})$ *for* $k = 3$ *and* $\widetilde{O}(n^{k+\frac{1}{2}})$ *for* $k \geq 4$.

Proof. Let $G = (V,E)$ be a dag of depth $\mathrm{dp(G)} = \mathrm{n^q}$. Combining the ideas in [26] and [9] for computing LCAs in dags of small depth, ALL-K-SUBSETS REPRESENTATIVE LCA can be solved by computing (arbitrary) witnesses for Boolean matrix products MM^T for each level of the dag. More specifically, let l_i be the number of vertices on level L_i, then the matrix M corresponding to level L_i is an $n^{\binom{n}{k/2}} \times n^{l_i}$ matrix. Further, by Jensen's inequality, the time complexity of the algorithm is maximized if the levels of G are of equal size. That is, the running time of this approach can be bounded by $\widetilde{O}(n^{q+\omega(1,\frac{2(1-q)}{k},1)\frac{k}{2}})$. The additional polylogarithmic factor results from the witness computations.

On the other hand, we note that the LCA of a k-subset of vertices in a preprocessed tree can be computed in $O(1)$ since we assume that k is a constant. This implies that ALL-K-SUBSETS REPRESENTATIVE LCA can be solved in time $O(n^k w(G))$ by extending the ideas in Section 3.

Now we combine these two approaches in an analogous way as described above. We get an algorithm with time complexity $\widetilde{O}(rn^k + n^{q+\omega(1,\frac{2(1-q)}{k},1)\frac{k}{2}})$ for $\frac{2(1-q)}{k} > 0.294$ and $\widetilde{O}(rn^k + n^{q+k})$ for $\frac{2(1-q)}{k} \leq 0.294$, where $q \leq \log_n \frac{n}{r}$. Observe first that for $k \geq 4$ the respective function is minimized for $q = \frac{1}{2}$, which implies a time bound of $\widetilde{O}(n^{k+\frac{1}{2}})$.

Let now $k = 3$ and assume first that $\frac{2(1-q)}{3} \leq 0.294$. This implies $q \geq 0.559$ and hence an upper bound of $\widetilde{O}(n^{k+0.559})$. Suppose now that $\frac{2(1-q)}{3} > 0.294$. We solve the equality $rn^k = n^{q+\omega(1,\frac{2(1-q)}{k},1)\frac{k}{2}}$. By using $q \leq \log_n \frac{n}{r}$ and $\beta = \frac{\omega-2}{1-\alpha}$ we find the bound for $r \leq n^{1-\frac{3}{2}\beta\alpha}$ in a similar way as in the proof of Theorem 11. The claim follows now since $0.5214 < 1 - \frac{3}{2}\beta\alpha$. □

Note that Theorem 20 slightly subsumes Yuster's upper time bound for $k = 3$ and matches his remaining upper bounds for $k \geq 4$ [26] ignoring polylogarithmic factors. Again, we observe that if the input dags are distributed according to the $G_{n,p}$ model for random dags such that p is a constant, ALL-K-SUBSETS REPRESENTATIVE LCA can be solved in time $\widetilde{O}(n^k)$ in the average case. Nonetheless, we can even state the following corollary which is valid for all choices of the edge probability p. This result is based on a dynamic programming algorithm analogous to algorithms described in [4].

Corollary 21. ALL-K-SUBSETS REPRESENTATIVE LCA *can be solved in time* $\widetilde{O}(n^k)$ *in the average case.*

We note that our path cover approach can be further applied to develop space-efficient solutions for LCA problems in dags. For details on this application we refer to the full version of this paper [19].

References

1. Aït-Kaci, H., Boyer, R., Lincoln, P., Nasr, R.: Efficient Implementation of Lattice Operations. ACM Transactions on Programming Languages 11(1), 115–146 (1989)
2. Aho, A., Hopcroft, J., Ullman, J.: On Finding Lowest Common Ancestors in Trees. SIAM Journal on Computing 5(1), 115–132 (1976)
3. Barak, A., Erdös, P.: On the maximal number of strongly independent vertices in a random acyclic directed graph. SIAM J. Algebraic Discrete Methods 5, 508–514 (1984)
4. Baumgart, M., Eckhardt, S., Griebsch, J., Kosub, S., Nowak, J.: All-Pairs Common Ancestor Problems in Weighted Directed Acyclic Graphs. In: Chen, B., Paterson, M., Zhang, G. (eds.) ESCAPE 2007. LNCS, vol. 4614, pp. 282–293. Springer, Heidelberg (2007)
5. Bender, M.A., Farach-Colton, M., Pemmasani, G., Skiena, S., Sumazin, P.: Lowest common ancestors in trees and directed acyclic graphs. Journal of Algorithms 57(2), 75–94 (2005); A preliminary version. In: Proc. SODA 2001, pp. 845–853 (2001)

6. Coppersmith, D.: Rectangular matrix multiplication revisited. Journal of Symbolic Computation 13, 42–49 (1997)
7. Coppersmith, D., Winograd, S.: Matrix multiplication via arithmetic progression. Journal of Symbolic Computation 9, 251–290 (1990)
8. Cormen, T.H., Leiserson, C.E., Rivest, R.L., Stein, C.: Introduction to Algorithms, 2nd edn. McGraw-Hill Book Company, Boston (2001)
9. Czumaj, A., Kowaluk, M., Lingas, A.: Faster algorithms for finding lowest common ancestors in directed acyclic graphs. In: The special ICALP 2005, Theoretical Computer Science, vol. 380(1-2), pp. 37–46 (2007)
10. Dilworth, R.: A decomposition theorem for partially ordered sets. Annals of Mathematics 51(1), 161–166 (1950)
11. Eckhardt, S., Mühling, A., Nowak, J.: Fast Lowest Common Ancestor Computations in Dags. In: Arge, L., Hoffmann, M., Welzl, E. (eds.) ESA 2007. LNCS, vol. 4698, pp. 705–716. Springer, Heidelberg (2007)
12. Felsner, S., Raghavan, V., Spinrad, J.: Recognition Algorithms for Orders of Small Width and Graphs of Small Dilworth Number. Order 20(4), 351–364 (2003)
13. Ford, L.R., Fulkerson, D.R.: Flows in Networks. Princeton University Press, Princeton (1962)
14. Harel, D., Tarjan, R.E.: Fast algorithms for finding nearest common ancestors. SIAM Journal on Computing 13(2), 338–355 (1984)
15. Hopcroft, J.E., Karp, R.M.: An $n^{5/2}$ algorithm for maximum matchings in bipartite graphs. SIAM J.Comput. 2(4), 225–231 (1973)
16. Huang, X., Pan, V.Y.: Fast rectangular matrix multiplications and applications. Journal of Complexity 14, 257–299 (1998)
17. Kowaluk, M., Lingas, A.: LCA queries in directed acyclic graphs. In: Caires, L., Italiano, G.F., Monteiro, L., Palamidessi, C., Yung, M. (eds.) ICALP 2005. LNCS, vol. 3580, pp. 241–248. Springer, Heidelberg (2005)
18. Kowaluk, M., Lingas, A.: Unique Lowest Common Ancestors in Dags Are Almost as Easy as Matrix Multiplication. In: Arge, L., Hoffmann, M., Welzl, E. (eds.) ESA 2007. LNCS, vol. 4698, pp. 265–274. Springer, Heidelberg (2007)
19. Kowaluk, M., Lingas, A., Nowak, J.: A Path Cover Technique for LCAs in Dags. Technical Report TUM-I0809, Technische Universität München (2008)
20. Mucha, M., Sankowski, P.: Maximum Matchings via Gaussian Elimination. In: Proc. FOCS 2004, pp. 248–255 (2004)
21. Nykänen, M., Ukkonen, E.: Finding lowest common ancestors in arbitrarily directed trees. Information Processing Letters 50(6), 307–310 (1994)
22. Schieber, B., Vishkin, U.: On Finding Lowest Common Ancestors: Simplification and Parallelization. SIAM Journal on Computing 17(6), 1253–1262 (1988)
23. Shäffer, A.A., Gupta, S.K., Shriram, K., Cottingham Jr., R.W.: Avoiding recomputation in linkage analysis. Human Heredity 44, 225–237 (1994)
24. Simon, K.: An Improved Algorithm for Transitive Closure on Acyclic Digraphs. Theor. Comput. Sci. 58, 325–346 (1988)
25. Tarjan, R.E.: Applications of path compression on balanced trees. Journal of the ACM 26(4), 690–715 (1979)
26. Yuster, R.: All-pairs disjoint paths from a common ancestor in $\widetilde{O}(n^{\omega})$ time. Theoretical Computer Science 396(1-3), 145–150 (2008)

Boundary Labeling with Octilinear Leaders*,**

M.A. Bekos[1], M. Kaufmann[2], M. Nöllenburg[3], and A. Symvonis[1]

[1] School of Applied Mathematical & Physical Sciences, National Technical University of Athens, Greece
[2] Institute for Informatics, University of Tübingen, Germany
[3] Faculty of Informatics, Karlsruhe University, Germany

Abstract. A major factor affecting the readability of an illustration that contains textual labels is the degree to which the labels obscure graphical features of the illustration as a result of spatial overlaps. Boundary labeling addresses this problem by attaching the labels to the boundary of a rectangle that contains all features. Then, each feature should be connected to its associated label through a polygonal line, called *leader*, such that no two leaders intersect.

In this paper we study the boundary labeling problem along a new line of research, according to which different pairs of type leaders (i.e. *do* and *pd*, *od* and *pd*) are combined to produce boundary labelings. Thus, we are able to overcome the problem that there might be no feasible solution when labels are placed on different sides and only one type of leaders is allowed. Our main contribution is a new algorithm for solving the total leader length minimization problem (i.e., the problem of finding a crossing free boundary labeling, such that the total leader length is minimized) assuming labels of uniform size. We also present an NP-completeness result for the case where the labels are of arbitrary size.

1 Introduction

Placing extra information—usually in the form of textual labels—next to features of interest within an illustration, constitutes an important task in the process of information visualization. The interest in algorithms that automate this task has increased, due to the large number of applications that stem from diverse areas such as cartography, geographical information systems etc.

Current research on map labeling has been devoted to labeling point-features, so that each label is placed next to the point that it describes (an extensive bibliography about map labeling is maintained by Strijk and Wolff [14]). In this case, the basic requirement is that the labels should be pairwise disjoint. However, this is not always possible, e.g., in the case where the labels are too

* The work of M. Bekos and A. Symvonis is funded by the project PENED-2003. PENED-2003 is co - funded by the European Social Fund (75%) and Greek National Resources (25%).

** The work of M. Nöllenburg is supported by the German Research Foundation (DFG) under grant WO 758/4-3.

J. Gudmundsson (Ed.): SWAT 2008, LNCS 5124, pp. 234–245, 2008.

large or the feature set is too dense. In practice, large labels are quite usual, e.g., in technical drawings, where it is common to explain certain features of the drawing with blocks of text, arranged on its boundary. As a response to this problem, Bekos et al. [4] proposed *boundary labeling*. In boundary labeling, the labels are attached to the boundary of a rectangle R enclosing all features and each feature is connected with its label by using polygonal lines, called *leaders*.

Several authors have proposed algorithms to produce boundary labelings in different settings [2,3,4,5,6,11]. Recently, Benkert et al. [5,6] studied the boundary labeling problem along a new line of research, according to which the leaders are of type *do*, i.e., polygonal lines consisting of two line segments, where the first one is "diagonal" to the side of R containing the label it leads to, whereas the second one is orthogonal to that side (see Figure 1c). Leaders of type *do* maintain a uniform shape and result in simple and easy-to-read labelings. However, in the work reported in [5] and [6], Benkert et al. study the case where the labels can be attached only to one side of R and they state that the production of a boundary labeling with such leaders is not always feasible. Extending their work, we examine the case of four-sided boundary labeling. We also introduce two new types of leaders and we show that by combining them, the boundary labeling problem is always feasible. To the best of our knowledge, this is the first attempt, where different types of leaders are combined to produce boundary labelings.

2 Problem Definition

The *input* of a boundary labeling problem consists of a set P of n points (referred to as *sites*) $s_i = (x_i, y_i)$, $i = 1, 2, \ldots n$. The site set P is enclosed in an axis-parallel rectangle $R = [0, W] \times [0, H]$, which is called *enclosing rectangle*. Each site s_i is associated with an axis-parallel, $w_i \times h_i$ rectangular label l_i.

The *output* of a boundary labeling problem is a placement of the labels at distinct positions on the boundary of R and a set of leaders connecting each site with its associated label, so that i) the labels do not overlap with each other and ii) the leaders do not intersect or overlap with each other. Such labelings are referred to as *legal boundary labelings* (or simply as *legal labelings*).

Following the naming scheme of Bekos et al. [4], we focus on three different types of leaders, each of which consists of two line segments:

Type-*od* leaders: The first line segment of a leader of type *od* is orthogonal (*o*) to the side of R containing the label it leads to. Its second line segment is "diagonal" (*d*) to that side (see Figure 1a).

Type-*pd* leaders: The first line segment of a leader of type *pd* is parallel (*p*) to the side of R containing the label it leads to. Its second line segment is "diagonal" (*d*) to that side (see Figure 1b).

Type-*do* leaders: The first line segment of a leader of type *do* is "diagonal" (*d*) to the side of R containing the label it leads to. Its second line segment is orthogonal (*o*) to that side (see Figure 1c).

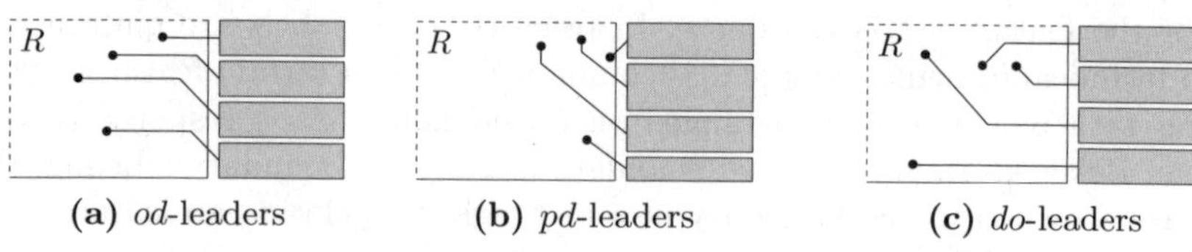

(a) *od*-leaders (b) *pd*-leaders (c) *do*-leaders

Fig. 1. Different types of leaders

In general, the labels are of arbitrary size (*non-uniform labels*; see Figure 1b). We separately consider the case, where the labels are of the same width and height (*uniform labels*; see Figures 1a and 1c). We further assume that the point where each leader touches its associated label (referred to as *port*) is fixed, e.g., the middle point of the label's side that faces the enclosing rectangle R (see Figures 1a, 1b and 1c). Also, the labels are usually attached to one, two or all four sides of the enclosing rectangle and are either placed at predefined locations (*fixed labels*) along the sides or can slide (*sliding labels*).

Keeping in mind that we want to obtain simple and easy-to-read labelings, we consider the *leader length minimization problem*, i.e., the problem of determining a legal labeling, such that the total leader length is minimized.

2.1 Preliminaries

We denote the number of sites (and consequently the number of labels) by n. We also denote by c_i the leader of site s_i. A set of sites is considered to be in *general position* if i) no three sites are collinear, ii) no two sites share the same x- or y-coordinate, iii) no two sites lie on the same diagonal line and iv) the horizontal, vertical and diagonal lines that pass through the ports of the labels do not coincide with the sites. In order to avoid leader overlaps, we usually assume that the input site set P is in general position. We also assume that the sites, the leader bends and the label corners have integer coordinates. Consider a leader c_i which originates from site s_i and is connected to a label l_i on the right side AB of R. The horizontal line which coincides with s_i divides the plane into two half-planes (see the dashed line l of Figure 2). We say that leader c_i is *oriented towards* corner A if both A and the port of label l_i are on the same half-plane, otherwise, we say that leader c_i is *oriented away* from corner A.

Consider a site s_i that has to be connected to a label l_i on the right side AB of R. The lines that pass through the port of label l_i and form 45°, 90° and 135° angles with the left side of label l_i, partition R into four regions $R_{i,1}$, $R_{i,2}$, $R_{i,3}$ and $R_{i,4}$, as in Figure 3. If the site s_i lies within a region incident to A or B (i.e., $R_{i,1}$ or $R_{i,4}$; refer to the light-gray colored regions of Figure 3), then it can only be connected to label l_i using a leader of type *pd*. Otherwise (i.e., site s_i lies within $R_{i,2}$ or $R_{i,3}$; refer to the dark-gray colored regions of Figure 3), it can be connected to l_i using either a leader of type *do* or *od*. Also, observe that connecting a site to its label with a leader of type *do*, requires the same leader length as with a leader of type *od*. So, depending on the location of site s_i, one has to use an appropriate leader to connect it to its label l_i.

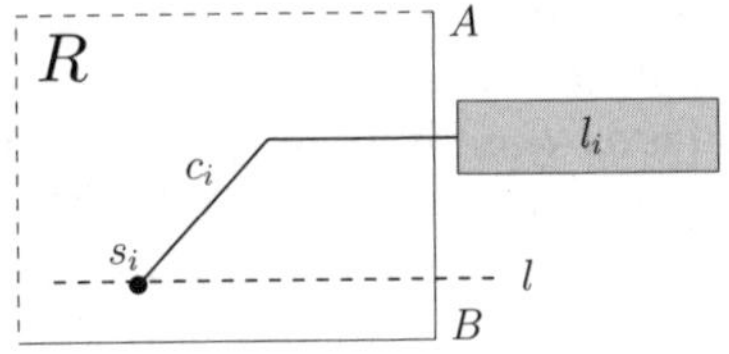

Fig. 2. c_i is oriented towards corner A

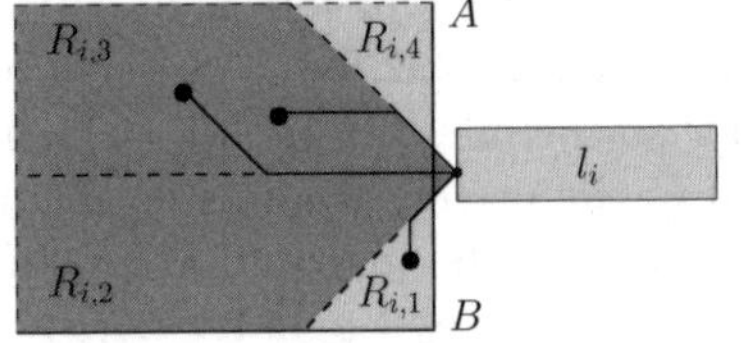

Fig. 3. Connecting site s_i to label l_i

This paper is structured as follows: In Section 3 we prove that the problem of determining a legal boundary labeling of minimum total leader length with leaders of type *do* and *pd* and non-uniform labels is NP-complete. In Sections 4 and 5, we present polynomial time algorithms for obtaining either optimal (in terms of total leader length) or simply legal boundary labelings with labels of uniform size. We conclude in Section 6 with open problems and future work.

3 Boundary Labeling with Non-uniform Labels

In this section, we consider the boundary labeling problem with labels of non-uniform size. We are given a set P of n sites $s_i, i = 1, 2, \ldots n$, each associated with axis-parallel, rectangular label l_i of height h_i. The labels are allowed to be placed on the right side of the enclosing rectangle R. We further assume fixed label ports, i.e., each leader is connected to its corresponding label using the middle point of the label side that faces the enclosing rectangle. For the case where the sites can be placed in arbitrary position, i.e., the general position restriction is relaxed, we can prove:

Theorem 1. *Given a set P of n sites, a label l_i of height h_i for each site s_i and an integer $k \in \mathbb{Z}^+$, it is NP-complete to decide whether there exists a legal boundary labeling of total leader length no more than k assuming type do and pd leaders.*

Proof. Membership in NP follows from the fact that a nondeterministic algorithm needs only guess a positioning of the labels on the boundary of R, a set of leaders connecting each site with its associated label and check in polynomial time that i) the labels do not overlap with each other, ii) the leaders do not intersect with each other and iii) the sum of the lengths of all leaders is no more than k.

We will reduce the following single machine scheduling problem (known as *total discrepancy problem* [8]) to our problem: We are given a set J of $2n+1$ jobs $J_0, J_1, J_2, \ldots, J_{2n}$, which are to be executed on one machine nonpreemptively and a single *preferred midtime* $M \in \mathbb{Z}^+$, which corresponds to the time at which we would like the first half of each job to be completed. Each job J_i is also associated with a known deterministic processing time p_i. Without loss of generality, we assume that M is large (e.g. $M > \sum_{i=0}^{2n} p_i$) and the jobs are ordered so that $p_i < p_j$, $\forall i < j$. Given a schedule σ, we denote the *starting (completion) time* of job J_i in σ by $S_i(\sigma)$ ($C_i(\sigma)$) and we use $M_i(\sigma)$ to denote its *midtime*, i.e., $M_i(\sigma) = S_i(\sigma) + p_i/2$, or equivalently, $M_i(\sigma) = C_i(\sigma) - p_i/2$. Under a schedule σ, a job J_i is considered to be *on-time* if its midtime $M_i(\sigma)$ is equal to the preferred midtime M and in this case, it incurs no penalty. On the other hand, if the midtime $M_i(\sigma)$ of J_i commences prior to M (exceeds M), an earliness (tardiness) penalty $E_i(\sigma) = M - M_i(\sigma)$ ($T_i(\sigma) = M_i(\sigma) - M$) incurs. The objective is to determine a schedule σ, so that the total earliness-tardiness penalty $\sum_{i=0}^{2n}(E_i(\sigma)+T_i(\sigma)) = \sum_{i=0}^{2n}|M - M_i(\sigma)|$ is minimized[1]. Let σ_{opt} be an optimal schedule of the total discrepancy problem. Then, the following hold [8]:

1) σ_{opt} does not have any gaps between the jobs.
2) $M_0(\sigma_{opt}) = M$.
3) If $A(\sigma_{opt}) = \{J_i : M_i(S) < M\}$ and $B(\sigma_{opt}) = \{J_i : M_i(S) > M\}$, then $|A(\sigma_{opt})| = |B(\sigma_{opt})| = n$.
4) $\sigma_{opt} = [A_n, A_{n-1}, \ldots, A_1, J_0, B_1, B_2, \ldots, B_n]$, where $\{A_i, B_i\} = \{J_{2i}, J_{2i-1}\}$, i.e., if $A_i = J_{2i}$ then $B_i = J_{2i-1}$ otherwise $A_i = J_{2i-1}$ and $B_i = J_{2i}$.
5) The minimum total earliness-tardiness penalty is equal to

$$ETP = \sum_{i=1}^{n}(p_{2i} + p_{2i-1})(n - i + 1/2) + np_0.$$

The reduction we propose, can be achieved in linear time. Let I_S be an instance of the total discrepancy problem mentioned above. We proceed to construct an instance I_L of our problem as follows: For each job J_i, we introduce a site s_i placed at point $(2n + 1 - i, M)$, i.e., the sites are collinear, lie on the horizontal line $y = M$ and the horizontal distance between two consecutive sites is one unit. The label l_i associated with site s_i has height h_i equal to the processing time p_i of job J_i. The bottom left corner of the enclosing rectangle R is $(0, 0)$. The height H of R is equal to $2M$, which ensures that all labels can be placed at the right side of R, since $M > \sum_{i=0}^{2n} p_i$. We seek to exclude the case where a site can be connected to its label through a leader of type *pd*. So, the enclosing rectangle should be of appropriate width. We set its width W to be equal to $\frac{\sqrt{2}}{2}H + 2n + 1$ (see Figure 4). This ensures that the gray-colored triangular area contains no sites and therefore, all sites can be connected to their associated labels through leaders of type *do* only.

Then, we can show that we can derive a schedule σ of I_S with total earliness-tardiness penalty ETP if and only if we can determine a legal labeling L of I_L with total leader length at most $(\sqrt{2} - 1)ETP + (2n + 1)(W - n - 1)$. □

Note. The NP-completeness result of Theorem 1 also holds in the case of boundary labelings with *po* leaders. The proof is almost identical. Instead of measuring the length of each leader using the Euclidean metric, we have to use the Manhattan metric.

[1] Surveys on the most important aspects of scheduling research are given at [1,9].

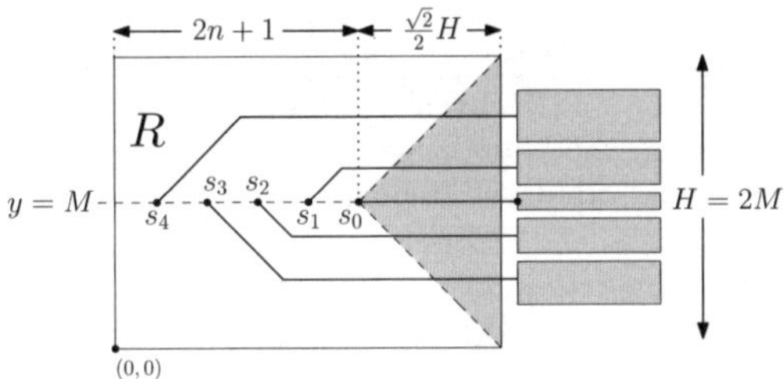

Fig. 4. For each job J_j, we introduce a site s_i placed at $(2n+1-i, M)$

4 Boundary Labeling with Uniform Labels

Theorem 1 implies that, unless $P = NP$, we cannot efficiently determine an optimal solution of the boundary labeling problem with non-uniform labels. Therefore, we proceed to consider the case of uniform labels, which is a reasonable assumption, since in real applications the labels usually contain single line texts (for example a place name or an integer used as a legend).

Let $P = \{s_1, s_2, \ldots s_n\}$ and $L = \{l_1, l_2, \ldots l_n\}$ be the sets of sites and labels, respectively. We assume that the sites are in general position and the labels are placed in fixed positions on the boundary of R. Since the labels are of uniform size, each site s_i can be connected to any label l_j. We seek to connect each site s_i to a label l_j, so that the total leader length is minimized.

Initially, we construct a complete weighted bipartite graph $G = (P \cup L, E, w)$ between all sites $s_i \in P$ and all labels $l_j \in L$, where $E = \{(s_i, l_j); s_i \in P, l_j \in L\}$ and $w : E \rightarrow \mathbb{R}$ is a cost function (see step A of Algorithm 1). Each edge $e_{ij} = (s_i, l_j) \in E$ of G is assigned a weight $w(e_{ij}) = d_{ij}$, where d_{ij} is equal to the length of the leader which connects site s_i with label l_j. Recall that the type of the leader that will be used to connect site s_i to label l_j depends on their relative positions, as stated in Section 2.1. Also, recall that if a site can be connected to its associated label with a leader of type *do*, it can also be connected using an *od* leader. However, in both cases the total length required is the same and, consequently, the edge e_{ij} is assigned the same weight, regardless the type of the leader that will eventually used (i.e., *do* or *od*). Observe that G is regular.

We proceed by computing a minimum-cost bipartite matching on G, i.e., a matching between the sites and the labels that minimizes the total weight of the matched pairs (see step B of Algorithm 1). Since G is regular and bipartite, by Hall's theorem a perfect matching exists [10]. Then, we obtain a labeling M of minimum total leader length as follows: If an edge $e_{ij} = (s_i, l_j) \in E$ is selected in the matching, then we connect site s_i with label l_j using a leader of length $w(e_{ij})$ (see step C of Algorithm 1). However, labeling M may contain crossings, which have to be eliminated while keeping the total leader length unchanged, i.e., equal to that of M (see step D of Algorithm 1). The crossing elimination procedure is described in the remainder of this section and depends on i) the location of the labels and ii) the type of the leaders that are used to produce M.

Algorithm 1. Generic Algorithm

input : A set $P = \{s_1, \ldots, s_n\}$ of n sites and a set $L = \{l_1, \ldots, l_n\}$ of n uniform labels placed on the boundary of R.
output : A crossing free boundary labeling of minimum total leader length.

Step A: *Construct a complete weighted bipartite graph.*
Construct a complete weighted bipartite graph $G = (P \cup L, E, w)$ between all sites $s_i \in P$ and all labels $l_j \in L$. The weight $w(e_{ij})$ of an edge $e_{ij} = (s_i, l_j) \in E$ is the length of the leader, say d_{ij}, which connects s_i with l_j.
Step B: *Compute a Minimum Cost Bipartite Matching.*
Compute a minimum-cost perfect bipartite matching $\mathcal{M}$ of G, i.e., compute a matching between sites and labels that minimizes the total distance of the matched pairs.
Step C: *Obtain an optimal boundary labeling M.*
foreach (edge $e_{ij} = (s_i, l_j) \in E$) **do**
  if $e_{ij} = (s_i, l_j) \in \mathcal{M}$ **then** connect site s_i to label l_j s.t. $length(c_i) = w(e_{ij})$
Step D: *Eliminate crossings.*
Eliminate all crossings among pairs of leaders and obtain a legal boundary labeling M', keeping the total leader length unchanged, i.e., equal to that of M.

4.1 One-Sided Boundary Labeling

We first describe how to eliminate all crossings of labeling M (obtained in Step C of Algorithm 1), assuming that the labels are allowed to be attached to one side of the enclosing rectangle R, say the right side AB. Note that labeling M is of minimum total leader length and the leaders, we have used to produce it in Step C of Algorithm 1, are i) either of type *do* and *pd* or ii) of type *od* and *pd*. Our aim is to eliminate all crossings and obtain a legal labeling M' that keeps the total leader length unchanged.

Lemma 1. *Let M be an optimal one-sided boundary labeling either with type do and pd leaders or with type od and pd leaders (which may contain crossings) obtained in Step C of Algorithm 1. Let c_i and c_j be a pair of intersecting leaders originating from sites s_i and s_j, respectively. Then the following hold:*

i) Leaders c_i and c_j are of the same type.
ii) Leaders c_i and c_j are oriented towards the same corner, say A, of the enclosing rectangle R.
iii) Leaders c_i and c_j can be rerouted so that they do not cross each other, the sum of their leader length remains unchanged, their type remains unchanged and they remain oriented towards corner A of R.

Sketch of proof. Due to space constraints, the detailed proof is omitted. It is based on an exhaustive case analysis on a) the types of the two leaders, b) the orientation of the two leaders (towards the same or different corners) and c) the different regions the two sites may reside. It also makes use of several geometric properties (e.g. triangle inequality, properties of isosceles triangles, orthogonal triangles etc.). Another important property that is heavily used is the assumption that the sites are in general position. □

Lemma 2. *Let M be an optimal one-sided boundary labeling either with type do and pd leaders or with type od and pd leaders (which may contain crossings) obtained in Step C of Algorithm 1. We can always determine a crossing-free labeling M' with total leader length equal to that of M (step D of Algorithm 1). Moreover, labeling M' can be obtained in $O(n^2)$ time.*

Proof. By Lemma 1, it follows that leaders involved in a crossing are of the same type and oriented towards the same corner of R. We show how to eliminate all crossings of labeling M by rerouting the crossing leaders. Our method performs four passes over the sites. In the first and second pass, we eliminate all crossings among the leaders of type *pd*, which are oriented towards the top right and bottom right corner of R, respectively. In the third and fourth pass, we eliminate all crossings among the remaining leaders (i.e. either leaders of type *do* or of type *od*), which are oriented towards the top right and bottom right corner of R, respectively. Due to space constraints, we describe in detail the first pass only.

We examine these sites from right to left. We are interested only in those sites, that have crossing leaders. Let s_i be the first such site and let c_i be the leader that connects it to its corresponding label on the right side AB of R (see the left part of Figure 5). By Lemma 1.(i) and Lemma 1.(ii), all leaders that intersect c_i are also, of type *pd*, oriented towards corner A. Let s_k be the site whose leader c_k intersects c_i and its label is placed bottommost. From Lemma 1.(iii), it follows that we can reroute leaders c_i and c_k so that the total leader length remains unchanged (see the right part of Figure 5). Note that the rerouting possibly eliminates more than one crossing but, in general, it may also introduce new crossings with other type *pd* leaders, oriented towards corner A. However, the crossings are now, located to the left of the vertical line that coincides with s_i (within the gray-colored region of Figure 5). Continuing in the same manner, the line which forms the region containing the crossings in the right-to-left pass is pushed to the left (i.e., the area of this region is reduced at each iteration, in the right-to-left pass), which guarantees that all crossings among leaders of type *pd* that are oriented towards corner A, are eventually eliminated.

When the four independent passes over the site set are completed, we have eliminated all crossings, resulting in a labeling M' without any crossings and of total leader length equal to that of M, i.e., of minimum total leader length. To complete the proof of the lemma, it remains to explain how to obtain in $O(n^2)$ time the new labeling M'. At each pass, we sort appropriately the site set. This can be done in $O(n \log n)$ time. At each iteration over the sorted sets of sites, we are interested in finding a specific site, which crosses the leader of the site that we currently consider. In a straight-forward manner, this can be computed in $O(n)$ time. This results in a total of $O(n^2)$ time for each pass and, consequently, for the elimination of all crossings. □

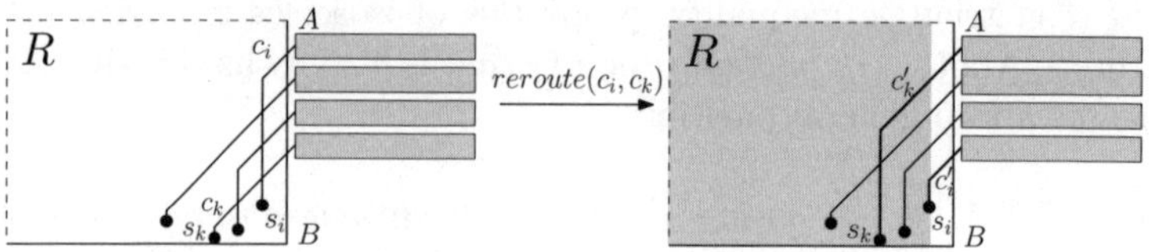

Fig. 5. Rerouting the crossing *pd*-leaders c_i and c_k

Theorem 2. *Given a set P of n sites and a set L of n labels of uniform size placed at fixed positions on one side of the enclosing rectangle R, we can compute in $O(n^3)$ total time a legal boundary labeling of minimum total leader length with type do and pd leaders.*

Proof. In Step A of Algorithm 1, we construct a complete weighted bipartite graph $G = (P \cup L, E, w)$ between all sites $s_i \in P$ and all labels $l_j \in L$, where the weight of an edge $e_{ij} = (s_i, l_j) \in E$ is the length of the leader connecting site s_i to label l_i. The computation of each edge weight requires constant time. Hence, the construction of G can be done in $O(n^2)$ time. In Step B of Algorithm 1, we compute a minimum cost bipartite matching on the graph G, which can be done by means of the Hungarian method in $O(n^3)$ time [12]. Note that we cannot use Vaidya's algorithm [13] to reduce the time complexity of Step B, since the leaders are neither straight lines (Euclidean metric) nor rectilinear (Manhattan metric). The solution obtained in Step C of Algorithm 1 is optimal. However, it may contain crossings. In Step D of Algorithm 1, the crossings are eliminated in $O(n^2)$ time. Thus, the total time complexity of Algorithm 1 for the case of one-sided boundary labeling with type *do* and *pd* leaders is $O(n^3)$. □

4.2 Two-Sided Boundary Labeling

In this subsection, we consider the case where the labels are allowed to be attached to two opposite sides of R. Again, we use Algorithm 1 to obtain a boundary labeling M (not necessarily crossing-free) of minimum total leader length. This can be done in $O(n^3)$ time. We can observe that a possible crossing, between two leaders that lead to labels located at opposite sides of R, cannot occur, since the rerouting of the leaders c_i and c_j results in a solution with smaller total leader length. This result is summarized in the following lemma.

Lemma 3. *In an optimal two-sided boundary labeling, crossings between leaders that connect labels located at opposite sides of the enclosing rectangle, cannot occur.*

From Lemma 3, it follows that we can independently eliminate the crossings along the two opposite sides of R. The following theorem summarizes our result.

Theorem 3. *Given a set P of n sites and a set L of n labels of uniform size, placed at fixed positions, on two opposite sides of the enclosing rectangle R, we can compute in $O(n^3)$ total time a legal boundary labeling of minimum total leader length with either type do and pd leaders or with od and pd leaders.*

4.3 Four-Sided Boundary Labeling

In this subsection, we consider the general case of determining a legal boundary labeling of minimum total leader length with type *od* and *pd* leaders, where the labels are allowed to be attached to all four sides of R. Again, we use Algorithm 1, to obtain a labeling M of minimum total leader length. By Lemma 1, it follows that crossing leaders that connect labels placed at the same side of R, are of the same type, oriented towards the same corner of R. Crossings between leaders that connect labels placed at opposite sides of R cannot occur, because of Lemma 3. For the case, where the leaders which cross each other connect labels placed on two adjacent sides of R, we can show that the following lemma holds.

Lemma 4. *Let M be an optimal four-sided boundary labeling with type od and pd leaders (which may contain crossings) obtained in Step C of Algorithm 1. Let c_i and c_j be a pair of intersecting leaders originating from sites s_i and s_j, respectively. Let also l_i and l_j be their associated labels, which lie on two adjacent sides of the enclosing rectangle R. Then the following hold:*

i) Leaders c_i and c_j are of different type.
ii) Leaders c_i and c_j are oriented towards their incident corner, say corner A.
iii) Leaders c_i and c_j can be rerouted so that they do not cross each other, the sum of their leader length remains unchanged, their type remains unchanged and they remain oriented towards corner A of R.

Lemma 5. *Let M be an optimal four-sided boundary labeling with type od and pd leaders (which may contain crossings) obtained in Step C of Algorithm 1. We can determine a legal labeling M' with total leader length equal to that of M (step D of Algorithm 1). Moreover, labeling M' can be obtained in $O(n^2)$ time.*

Proof. We partition the site set into four disjoint sets S_{TR}, S_{TL}, S_{BR} and S_{BL}, each of those contains the sites whose leaders are oriented towards the top-right, top-left, bottom-right and bottom-left corner of R, respectively (see Figure 6). From Lemma 1 (one side case), Lemma 3 (two opposite sides case) and Lemma 4 (two adjacent sides case), it follows that possible crossings can only occur between leaders that are oriented towards the same corner of R. Thus, we can independently eliminate the crossings at each of the sets S_{TR}, S_{TL}, S_{BR}, S_{BL}.

We describe in detail how to eliminate the crossings of S_{TR}. The remaining are treated similarly. We further partition S_{TR} into two disjoint subsets $S_{1,TR}$ and $S_{2,TR}$ as follows: $S_{1,TR}$ ($S_{2,TR}$) contains the sites of S_{TR} whose leaders are either i) of type *od* leading to a label placed at the right (top) side of R or ii) of type *pd* leading to a label placed at the top (right) side of R. In Figure 6, the sites that constitute $S_{1,TR}$ are the ones whose leaders are drawn as solid lines.

From Lemma 4.(i), it follows that the leaders, which are involved in a crossing and lead to labels placed at two adjacent sides of R, should be of different type. Furthermore, crossing leaders that connect labels placed at the same side of R, should be of the same type. This directly follows from Lemma 1.(i). Hence, we can independently eliminate the crossings at each of the sets $S_{1,TR}$ and $S_{2,TR}$.

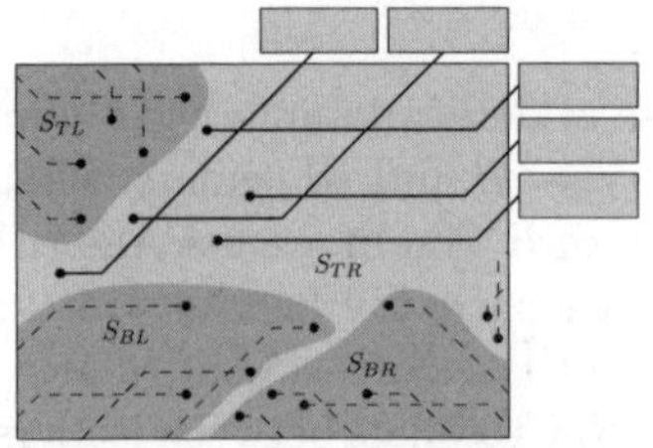

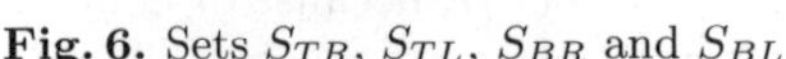

Fig. 6. Sets S_{TR}, S_{TL}, S_{BR} and S_{BL}

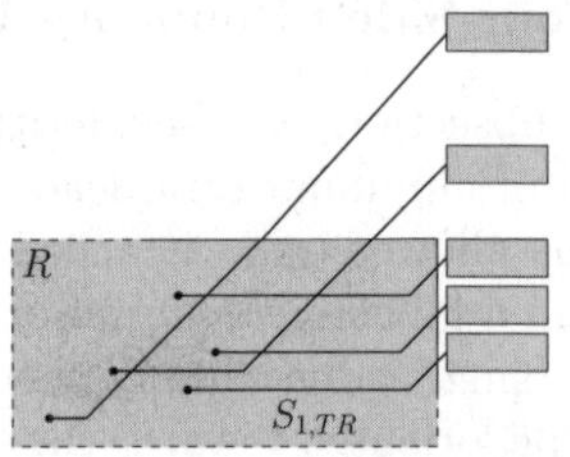

Fig. 7. Extending a *pd* leader

Let $s \in S_{1,TR}$ be a site whose leader c is of type *pd*. Leader c leads to a label at the top side of R. By extending its d-segment, leader c can be viewed as an *od* leader leading to a label at the right side of R (see Figure 7). This implies that we can make use of the algorithm described in the proof of Lemma 2 to eliminate all crossings of $S_{1,TR}$. Since all leaders of the sites of $S_{1,TR}$ have the same orientation (this holds because $S_{1,TR} \subseteq S_{TR}$), their d-segments are parallel to each other, which guarantees that all crossings will occur within R. This ensures that our approach will find a legal labeling. Similarly, we eliminate the crossings of the set $S_{2,TR}$. The total time needed to eliminate the crossings at each of the sets S_{TR}, S_{TL}, S_{BR} and S_{RL}, is $O(n^2)$. Thus, labeling M' can be obtained in $O(n^2)$. □

Theorem 4. *Given a set P of n sites and a set L of n labels of uniform size placed at fixed positions on all four sides of the enclosing rectangle R, we can compute in $O(n^3)$ total time a legal boundary labeling of minimum total leader length with type od and pd leaders.*

5 An Algorithm for Obtaining Legal Boundary Labelings

In this section, we consider the problem of determining a legal boundary labeling with type *od* and *pd* leaders, i.e., we relax the optimality constraint on the resulting labeling. Our aim is to obtain a more efficient algorithm in terms of time complexity.

Theorem 5. *Given a set P of n sites and a set L of n labels of uniform size placed at fixed positions on all four sides of the enclosing rectangle R, we can compute in $O(n^2)$ total time a legal boundary labeling with type od and pd leaders.*

Sketch of proof. Our basic idea is simple: We first develop an algorithm which determines a legal labeling in the case where the labels are attached to one side of R. Its time complexity is $O(n^2)$. Then, using standard plane sweep algorithms [7], we can in $O(n \log n)$ time partition R into four disjoint regions such that the previous algorithm can be applied to each region separately. To achieve this, we have two requirements for a region A in the partition of R: (a) A must be adjacent to a specific side s_A of R and (b) each site in A can be connected to any label attached to s_A either through a leader of type *do* or of type *pd*. □

6 Conclusions

In this paper, we studied boundary labelings with *do*, *od* and *pd* leaders. The focus of our work was on the leader length minimization problem. The $O(n^3)$ time complexity of the proposed algorithms is dominated by the computation of a minimum-cost bipartite matching. Unfortunately, we cannot use Vaidya's algorithm [13] to reduce it, since the leaders are neither straight lines (Euclidean metric) nor rectilinear (Manhattan metric). It is worth trying to derive a more efficient matching algorithm for this metric. The evaluation of different optimization criteria would also be of particular interest.

References

1. Baker, K.R., Scudder, G.D.: Sequencing with earliness and tardiness penalties: a review. Operations Research 38, 22–36 (1989)
2. Bekos, M.A., Kaufmann, M., Potika, K., Symvonis, A.: Polygons labelling of minimum leader length. In: Kazuo, M., Kozo, S., Jiro, T. (eds.) Proc. Asia Pacific Symposium on Information Visualisation, CRPIT 1960, pp. 15–21 (2006)
3. Bekos, M.A., Kaufmann, M., Symvonis, A., Wolff, A.: Boundary labeling: Models and efficient algorithms for rectangular maps. In: Pach, J. (ed.) GD 2004. LNCS, vol. 3383, pp. 49–59. Springer, New York (2005)
4. Bekos, M.A., Kaufmann, M., Symvonis, A., Wolff, A.: Boundary labeling: Models and efficient algorithms for rectangular maps. Computational Geometry: Theory and Applications 36, 215–236 (2007)
5. Benkert, M., Haverkort, H., Kroll, M., Nöllenburg, M.: Algorithms for multi-criteria one-sided boundary labeling. In: Proc. 15th Int. Symposium on Graph Drawing (GD 2007). LNCS, vol. 4875, pp. 243–254. Springer, Heidelberg (2007)
6. Benkert, M., Nöllenburg, M.: Improved algorithms for length-minimal one-sided boundary labeling. In: 23rd European Workshop on Computational Geometry (EWCG 2007), pp. 190–193 (2007)
7. de Berg, M., van Kreveld, M., Overmars, M., Schwarzkopf, O.: Computational Geometry: Algorithms and Applications. Springer, Heidelberg (2000)
8. Garey, M., Tarjan, R., Wilfong, G.: One-processor scheduling with symmetric earliness and tardiness penalties. Mathematics of Operations Research 13, 330–348 (1988)
9. Gordon, V., Proth, J.-M., Chu, C.: A survey of the state-of-the-art of common due date assignment and scheduling research. European Journal of Operational Research 139(1), 1–25 (2002)
10. Hall, P.: On representation of subsets. Journal of the London Mathematical Society 10, 26–30 (1935)
11. Kao, H.-J., Lin, C.-C., Yen, H.-C.: Many-to-one boundary labeling. In: Proc. Asia Pacific Symposium on Information Visualisation (APVIS 2007), pp. 65–72. IEEE, Los Alamitos (2007)
12. Kuhn, H.W.: The Hungarian method for the assignment problem. Naval Research Logistic Quarterly 2, 83–97 (1955)
13. Vaidya, P.M.: Geometry helps in matching. SIAM J. Comput. 18, 1201–1225 (1989)
14. Wolff, A., Strijk, T.: The Map-Labeling Bibliography (1996), `http://i11www.ira.uka.de/map-labeling/bibliography`

Distributed Disaster Disclosure

Bernard Mans[1], Stefan Schmid[2], and Roger Wattenhofer[3]

[1] Department of Computing, Macquarie University, Sydney, NSW 2109, Australia
bmans@ics.mq.edu.au
[2] Institut für Informatik, Technische Universität München, 85748 Garching, Germany
schmiste@in.tum.de
[3] Computer Engineering and Networks Laboratory (TIK), ETH Zurich,
8092 Zurich, Switzerland
wattenhofer@tik.ee.ethz.ch

Abstract. Assume a set of distributed nodes which are equipped with a sensor device. When nodes sense an event, they want to know (the size of) the connected component consisting of nodes which have also sensed the event, in order to raise—if necessary—a disaster alarm. This paper presents distributed algorithms for this problem. Concretely, our algorithms aim at minimizing both the response time as well as the message complexity.

1 Introduction

Governments and organizations around the world provide billions of dollars each year in aid to regions impacted by disasters such as tornadoes, flooding, volcanos, earthquakes, bush-fires, etc. In order to recognize disasters early and in order to limit the damage, endangered environments are often monitored by a large number of distributed sensor devices. The idea is that when these devices sense an event, an alarm should be raised, e.g., to inform helpers in the local community. Unfortunately, in practice, the sensor devices may sometimes wrongfully sense events, and of course false alarms can be quite costly as well. Therefore, nodes sensing an event should make sure that there are other nodes in their vicinity which have sensed the same event. Clearly, as sensor nodes may only be equipped with a limited energy-source (e.g., a small battery), the *number of messages* transmitted by a distributed alarming protocol should be minimized. As a second objective, the algorithm should have a *small latency*: If there is a disaster, it is of prime importance that the alarm is raised as soon as possible.

This paper investigates protocols for distributed disaster detection and alarming. We speak of a *disaster* when more than a given number of nodes is involved, and assume that the more nodes sensing an event the more severe the potential damage. For example, in a sensor network application, an alarm should be raised when more than a given number of sensor nodes detects a certain event, and the alarm message should include the *magnitude* of the disaster.

Apart from wireless systems, the disclosure of disasters is important in wired systems as well, for instance, to respond fast to worm propagations through the

J. Gudmundsson (Ed.): SWAT 2008, LNCS 5124, pp. 246–257, 2008.

Internet and trigger appropriate defense mechanisms when many machines show signs of infection. Disasters with a large impact do not necessarily have to be globally distributed, but are often *local* in nature. For example, a bush-fire, or the emission of toxic chemicals, or even a computer virus, may mostly impact a certain region of the world.

In this paper we tackle the disclosure of such disasters from a viewpoint of *distributed computing*. Our goal is to minimize the communication overhead for computing the disaster's dimension, and the time until detection. Concretely, we consider a network $G = (V, E)$ of n sensor nodes. There may be several events going on simultaneously in the network. However, although our algorithms allow to detect them individually, for ease of presentation we will assume here that there is just *one event* which affects an arbitrary set of nodes $V' \subseteq V$.

When a node senses an event (*event-node*), it seeks to find out how many of the nodes in its vicinity sensed it as well; more concretely: a node aims at aggregating information about the connected component of event-nodes it is in, e.g., at computing the component's size. If the component's size exceeds a certain threshold, at least one node of the component should raise a disaster alarm and report the component's magnitude. In this paper, we assess the quality of a distributed algorithm using the classic quality measures *time and message complexity*, that is, the running time of the algorithm, and the total number of messages transmitted.

There are two major algorithmic challenges. The first challenge we call the *neighborhood problem*: After a node has sensed an event, it has no clue which of its neighbors (if any) are also event-nodes. Distributed algorithms where event-nodes simply ask all their neighbors already leads to a costly solution: If G is the star graph S_n and the star's center node is the only node in V', the message complexity is $\Theta(n)$ while the size of the disaster component is one. Observe that the simple trick to let nodes only ask the neighbors of higher degree does not work either: While it would clearly be a solution for the star graph, it already fails for dense graphs such as the clique graph K_n. Indeed, it may at first sight seem that $\Theta(n)$ is a lower bound for *any* algorithm for K_n, as an event-node has no information about its neighbors! We will show, however, that this (naive) intuition is incorrect.

The second challenge concerns the coordination of the nodes during the exploration of the component. In a distributed algorithm where all nodes start exploring the component independently at the same time, a lot of redundant information is collected, resulting in a too high message complexity. As a lower bound, we know that the time required to compute the disaster component's size is at least linear in the component's diameter d, and the number of messages needed by any distributed algorithm is linear in the component's size s. We are hence striving for distributed algorithms which are output-sensitive and thus *competitive* to these lower bounds.

2 Model

We consider arbitrary undirected graphs $G = (V, E)$ where the nodes V have unique identifiers. We assume that an arbitrary subset of nodes $V' \subseteq V$ senses an event. The nodes V' are called *event-nodes*, and the nodes $V \setminus V'$ are called *non-event nodes.* We are interested in the subgraph induced by the nodes in V', that is, in the subgraph $H = (V', E')$ with $E' := \{\{u, v\} | u, v \in V', \{u, v\} \in E\}$. The subgraph H consists of one or more connected *components* C_i. The total number of nodes in component C_i will be referred to by $size(C_i)$. When the component is clear from the context, we will simply use s for $size(C_i)$. Note that in the following, for ease of presentation, we will often assume that there is only one type of event. However, all our algorithms can also handle concurrent events of different types.

After an event has hit a subset of nodes V', at least one node in each event component C_i is required to determine $size(C_i)$. This paper studies distributed algorithms which try to minimize the message and time complexities. Thereby, we allow the algorithm designer to *preprocess* the graph, e.g., to decompose the network into clusters with desired properties, i.e., to pre-compute *network decompositions* [12] (or, more specifically, sparse neighborhood-covers) of the graph. Note, however, that in this preprocessing phase, it is not clear yet which nodes will be affected by an event, i.e., V' is unknown. Also note that this preprocessing is done *offline* and its resulting structure can be reused for all future events.

During the *runtime phase*, an arbitrary number of events will hit the nodes, and each node $v \in V'$ first has to figure out which of its neighbors also belong to V' (*neighborhood problem*). In Section 3, we will allow non-event nodes to participate in the distributed algorithm as well. We will refer to this model as the *on-duty model.* It is suited for larger sensor nodes which are attached to a constant (infinite) energy supply. For smaller (wireless) nodes which rely on a limited battery, this model may not be appropriate: Typically, in order to save energy, such nodes are in a parsimonious sleeping mode. Only an event will trigger these nodes to wake up and participate in the distributed computation. We will refer to the latter model as the *off-duty model.* It will be discussed quickly in Section 4.

This paper assumes a *synchronous* environment in the sense that events are sensed by all nodes simultaneously and that there is an upper bound (known by all nodes) on the time needed to transmit a message between two nodes. The algorithms are presented in terms of communication *rounds.*

3 The On-Duty Model

In this section, the model is investigated where the non-event nodes are also allowed to participate in the distributed computations during runtime.

3.1 A Simple Solution for the Tree

Before discussing the general problem, we quickly review a simple special graph to acquaint the reader with our problem. Concretely, we look at *undirected trees.*

Consider an event component C_i of (unknown) size in a tree. If we let all s nodes start exploring the component, the message complexity grows quickly and the overhead is large. In contrast, the following ALG_{TREE} algorithm helps to organize the nodes in a simple preprocessing phase, such that component detection at runtime is efficient. Concretely, in the preprocessing phase, ALG_{TREE} makes the entire tree graph directed and rooted, i.e., each node (except the root) is assigned a *parent node*. See Figure 1 (*left*).

During runtime, when a node senses an event, it will *immediately* notify its parent using a dummy packet. This is necessary in order to ensure fast termination. The computation of the component's size then works by an aggregation algorithm on the tree: Leaf nodes—nodes which have not received a notification from their children—inform their parents that they are the only event-node in the corresponding subtrees. Inner nodes wait until the sizes of all their children's subtrees are known, and then propagate this result to their parent node. After $O(d)$ many rounds, the root of the component knows the exact value.

Obviously, Algorithm ALG_{TREE} is asymptotically optimal for trees both in terms of time and message complexity: The time and message complexities for exploring an event component are $O(d)$ and $O(s)$, respectively, where d is the diameter of the (event) component, and s is the component's size.

3.2 The Neighborhood Problem

The neighborhood problem is a first key challenge in distributed disaster disclosure. While for special graphs, e.g., trees, the solution can be straight-forward, the situation for general graphs is less clear. In this section, we present a *network decomposition approach* [1] for the neighborhood problem.

Broadly speaking, the idea of our decomposition is to divide the nodes into different, overlapping sets or *clusters* with corresponding *cluster heads* (e.g., the node with the largest ID in the cluster). These cluster heads provide a local coordination point, where nodes can learn which of their neighbors sensed the event as well.

Before defining our decomposition more formally, we need to introduce the following definition. Two different types of diameters of node sets are distinguished: the *weak* and the *strong* diameters.

Definition 1 (Weak and Strong Diameters). *Given a set S of nodes $S \subseteq V$ of a graph $G = (V, E)$, we call the maximum length of a path between any two nodes $v, u \in S$ the* weak diameter $diam(S) := max_{u,v \in S}(dist_G(u, v))$, *if the path is allowed to include nodes from the entire node set V. On the other hand, for the* strong diameter $Diam(S)$ *of a set S, $Diam(S) := max_{u,v \in S}(dist_S(u, v))$, paths are allowed to use nodes from S only. It thus holds that $diam(S) \leq Diam(S)$. Henceforth, when the set or cluster S is clear from the context, we will just write d and D for $diam(S)$ and $Diam(S)$, respectively.*

We can now define the notion of a *(k,t)-neighborhood cover*—a special form of a network decomposition [12]. In such a cover, each node belongs to at least one, but at most to k sets or *clusters*. The overlap of the clusters guarantees that there is at least one cluster containing the entire t-neighborhood of a node.

Definition 2 (Sparse (k,t)-neighborhood Cover). *[1] A* (k,t)-neighborhood cover *is a collection of sets (or* clusters*) of nodes* $S_1, ..., S_r$ *with the following properties: (1)* $\forall\, v, \exists\, i$ *such that* $N_t(v) \subseteq S_i$*, where* $N_t(v) = \{u|dist_G(u,v) \leq t\}$*, and (2)* $\forall i$, $Diam(S_i) \leq O(kt)$*.*

A (k,t)-neighborhood cover is said to be sparse *if each node is in at most* $kn^{1/k}$ *sets. Finally, we will refer to the node with the largest ID in a given set* S *as the* cluster head *of* S*. In the following, we will sometimes denote a sparse (k,t)-neighborhood cover by* (k,t)-$\mathcal{NC}$*.*

We will propose a solution to the neighborhood problem which—in the preprocessing phase—decomposes the network with such a neighborhood cover. Thereby, we will make use of the following result.

Theorem 1. *[1] Given a graph* $G = (V, E)$, $|V| = n$*, and integers* $k, t \geq 1$*, there is a deterministic (and distributed) algorithm which constructs a t-neighborhood cover in G where each node is in at most* $O(kn^{1/k})$ *clusters and the maximum cluster diameter is* $O(kt)$*.*

The idea for solving the neighborhood problem is to compute a ($\log n$,1)-$\mathcal{NC}$ in the preprocessing phase. At runtime, in the first round, each event-node v sends a message to all cluster heads of the clusters it belongs to. The cluster head of one of those clusters will then reply in the second round with the set of v's neighbors which are also event-nodes. This algorithm has the following properties.

Theorem 2. *The* ($\log n$,1)-$\mathcal{NC}$ *algorithm solves the neighborhood problem for any component in time* $O(\log n)$ *and requires* $O(s \log n)$ *many messages, where* n *is the total number of nodes in the network, and* s *is the event component's size.*

Proof. The time complexity is due to the fact that messages have to be routed to the cluster heads and back, and that—according to Theorem 1—the diameter of clusters in the ($\log n$,1)-$\mathcal{NC}$ is bounded by $O(kt) = O(\log n)$.

As for the message complexity, observe that each of the s nodes in the component sends a message to at most $O(kn^{1/k}) = O(\log n \cdot n^{1/\log n}) = O(\log n)$ cluster heads (Theorem 1). The cluster head's replies add at most a constant factor to the complexity, and hence we have $O(s \log n)$ message transmissions. □

3.3 Hierarchical Network Decomposition

In this section we propose the distributed algorithm ALG_{DC} for exploring the event components. ALG_{DC}'s running time is linear in the diameter of the component, and the message complexity is linear in the component's size (both up to polylogarithmic factors). Obviously, this is asymptotically optimal up to polylogarithmic factors, since the exploration of a graph requires at least d time and requires s messages.

ALG_{DC} makes again use of the sparse (k,t)-$\mathcal{NC}$ of Definition 2. However, instead of using just one decomposition as in the neighborhood problem, we build a *hierarchical* structure for exponentially increasing neighborhood sizes, i.e., for $t = 1, 2, 4, 8$, etc.

The detailed preprocessing and runtime phases are now described in turn (see also Algorithm 1).

ALG_{DC} Preprocessing Phase. In the preprocessing phase, ALG_{DC} constructs a hierarchy of sparse $(\log n,t)$-$\mathcal{NC}$s (Definition 2) for exponentially increasing neighborhood sizes, that is, the decompositions $\mathcal{D}_0 := (\log n, 1)$-$\mathcal{NC}$, $\mathcal{D}_1 := (\log n, 2)$-$\mathcal{NC}$, $\mathcal{D}_2 := (\log n, 4)$-$\mathcal{NC}$, $\mathcal{D}_3 := (\log n, 8)$-$\mathcal{NC}$, ..., $\mathcal{D}_i := (\log n, 2^i)$-$\mathcal{NC}$, ..., $\mathcal{D}_{\log \Delta} := (\log n, \Delta)$-$\mathcal{NC}$, are constructed, where Δ is the diameter of the graph G.[1] Moreover, each node computes the shortest paths to its cluster heads (e.g., using Dijkstra's single-source shortest path algorithm [4]). These paths are allowed to include nodes outside the clusters.

ALG_{DC} Runtime Phase. At runtime, initially, all event-nodes are in the *active* state. The event-nodes then contact their cluster heads to learn about their neighbors which are also event-nodes.

ALG_{DC} then starts with decomposition $\mathcal{D}_0$, switches to the level $\mathcal{D}_1$ afterwards, then to level $\mathcal{D}_2$, and so on, until level $\mathcal{D}_{\log d}$. On a general level i, ALG_{DC} does the following: All event-nodes which are still active inform their cluster heads in the $\mathcal{D}_i$ decomposition about the parts of their component which they already know. Each cluster head h of the clusters C in $\mathcal{D}_i$ then looks at each event component it hears about and performs the checks described next: If a component K is completely contained in C, h computes K's size and informs all nodes in K about s. Thereafter, all corresponding nodes are told to change to the passive state. If, on the other hand, the component K hits the boundary of C, h determines the node v_{max} with the largest ID it sees in the component, and tests whether v_{max}'s entire 2^i-neighborhood is contained in C. If this is the case, h tells v_{max} to remain active and provides it with all the event-nodes in v_{max}'s component which h knows. If not, v_{max} does not need to be notified by this cluster head. All other nodes are told to become passive. Figure 1 (*right*) depicts the situation. This scheme is applied recursively for increasingly larger neighborhood covers.

Theorem 3. *ALG_{DC} always terminates with the correct solution.*

Proof. In the $(\log n,d)$-$\mathcal{NC}$ (the weak diameter is used as clusters may include nodes outside the disaster component), there are definitively no active event-nodes left, and ALG_{DC} terminates. It remains to prove that there will always be at least one active event-node in each component K until a cluster contains the component completely. To see this, consider the (globally) largest ID node v in K. According to Theorem 1, there is always a cluster which completely contains v's neighborhood. This cluster will instruct v to continue, unless K is covered completely. □

---

[1] Note that $\log \Delta$ does not have to be integer. However, in this paper, we simplify the description by omitting corresponding $\lfloor \cdot \rfloor$ and $\lceil \cdot \rceil$ operations.

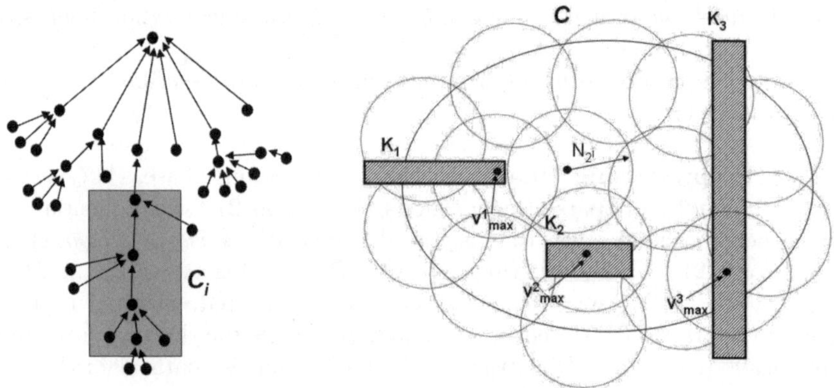

Fig. 1. *Left:* In the preprocessing phase, ALG_{TREE} makes the tree rooted and directed. Information about the event component (shaded) can then efficiently be aggregated at runtime. *Right:* Visualization for ALG_{DC}: Components K_1 and K_3 have nodes which are outside C, while K_2 is completely contained in C. The cluster head of C informs all nodes in K_2 about the component's size and deactivates them. In K_1, the N_{2^i}-neighborhood of the maximal node is completely contained, so v^1_{max} is told to remain active. In K_3, the cluster head instructs all nodes in $K_3 \cap C$ to deactivate, as v^3_{max} is too close to the boundary.

Theorem 4. *ALG_{DC} has a total running time of $O(d \log n)$, and requires at most $O(s \log d \log n)$ many messages, where n is the size of the network, s is the number of nodes in the component and d is the component's weak diameter.*

PROOF. *Time complexity.* The execution of ALG_{DC} proceeds through the hierarchy levels up to level d for exponentially increasing decompositions. For each level, the active event-nodes are involved in a constant number of message exchanges with their cluster heads. On level i, according to Theorem 1, the cluster diameter is $O(2^i \log n)$, and hence the time required is $O(2^i \log n)$ as well. As $\sum_{i=0}^{\log d} 2^i = O(d)$, we have a total execution time of $O(d \cdot \log n)$.

Message complexity. Consider again the $O(\log d)$ many phases through which ALG_{DC} proceeds on the decomposition hierarchy. First, we show that the number of active nodes is at least cut in half after each phase. To see this, recall that according to ALG_{DC}, a node v with maximal identifier can only continue if it its 2^i-neighborhood is completely contained in a cluster, while the entire component v is in is not yet seen by any cluster head (e.g., component K_1 in Figure 1). This implies that for each node which remains active, at least 2^i nodes have to be passive. Consequently, the maximal number of active nodes is divided by two after each phase.

Now observe that in the first phase, all s nodes are active, sending $O(s \log n)$ many messages to their cluster heads. The cluster head's replies are asymptotically of the same order. In the second phase, the diameters of the clusters have doubled, but the number of active nodes is divided by two. Thus, again $O(s \log n)$ many messages are sent by ALG_{DC}. Generally, in phase i, the cluster's diameter

Algorithm 1 ALG_{DC}

```
(* Global Preprocessing *)
for i from 0 to log d:
  D_i := (log n, 2^i)-NC;
(* Runtime *)
i := 1;
∀v ∈ V': v.active := true;
while (∃v : v.active = true)
  ∀ active v: notify v's cluster heads in D_i;
  for all clusters C
    let K := {K_1, ..., K_r} be C's components;
    ∀K ∈ K:
    if (K ⊆ C): output(size(K));
    else
      v_max := max{i|i ∈ (K ∩ C)};
      ∀v ∈ K: v.active := false;
      if (N_{2^i}(v_max) ⊆ C)
        v_max.active := true;
  i++;
```

is $O(2^i \log n)$, but only a fraction of $O(s/2^i)$ many nodes are active. Therefore, the message complexity is bounded by $O(\log d \cdot s \log n)$. □

While ALG_{DC} is asymptotically optimal up to polylogarithmic factors, the main term contains a factor which is a function of n. The subsequent section presents a different approach which aims at being more competitive in this respect. Moreover, ALG_{DC} needs large messages up to the size of the component; the message *sizes* of the algorithm of Section 3.4 are logarithmic in the number of nodes only.

3.4 Forests and Pointer Jumping

This section presents an alternative distributed algorithm ALG_{FOREST} for disaster detection. It is based on the *merging forests paradigm* (e.g., [8,10]), and makes use of *pointer jumping techniques* [2] in order to improve performance—both techniques are known, e.g., from *union-find data structures* [4].

ALG_{FOREST} Preprocessing Phase. ALG_{FOREST} solves the neighborhood problem by a sparse $(\log n,1)$-$\mathcal{NC}$. No additional decompositions are required for ALG_{FOREST}.

ALG_{FOREST} Runtime Phase. First, event-nodes perform a lookup operation at the cluster heads of the $(\log n, 1)$-$\mathcal{NC}$ in order to find out their neighbors which are also event-nodes. Then, each node v selects the node with the largest ID among its neighbors to become its *parent*; in case this ID is smaller than the ID of v itself, no parent is chosen. As cycles are impossible in parental relationships, the relationships define a *forest* among the event-nodes.

The idea of ALG_{FOREST} is to merge these trees efficiently to form one single tree on which all information about the component can be aggregated. However, before merging the trees, each tree is transformed to a *logical star graph*, that is, each node in the tree will learn about the tree's root (i.e., the star's center). This is achieved by the following *randomized pointer jumping technique* (cf. Algorithm 2): First, each node in the tree tosses a fair coin resulting in a bit 0 or 1 with probability 1/2 each. Parents then inform their children about their bit. Let IS

be the set of nodes consisting of all children having a 0-bit and whose parent has a 1-bit. The set IS of nodes forms a random *independent set* on the tree. The nodes in the IS will then establish a (logical) link to their parent's parent. This procedure is repeated until all nodes in the tree have a logical link to the root. Termination follows immediately from the fact that nodes arriving at the root will stop.

By this pointer jumping technique, trees become rooted stars. From now on, the roots then become the coordinators of the tree: First, they perform a *converge cast operation* [12] to learn the size of the tree. Then a root informs its children about its ID and the tree size. Subsequently, the root tells its children to determine in which trees their neighboring nodes are by performing a lookup in the $(\log n, 1)$-$\mathcal{NC}$. Information about the sizes and root IDs of the neighboring trees is then aggregated to the root. A tree seeks to join the largest neighboring tree, where "large" is defined with respect to the number of nodes in the tree, and in case of a tie, with respect to the roots' IDs. If a tree has no larger tree in its neighborhood, it will not send any join requests.

Basically, the rooted stars then become the virtual nodes of the new graph, where the corresponding roots are their coordinators, and the pointer jumping and merging techniques are applied recursively (cf. Algorithm 3; for simplicity, although the algorithm is of course distributed as described in the text, it is here presented in *global* pseudo-code). The algorithm terminates when stars do not have any neighboring stars anymore. Moreover, note that the phases of the trees need not to be synchronized, that is, some trees can be performing pointer jumping operations while other trees are in a converge cast phase.

Algorithm 2 describes the pointer jumping sub-routine for a tree T.

Algorithm 2 ALG_{PJ}

```
while (∃v s.t. v.parent ≠ root)
  ∀v ∈ T:
    with prob = 1/2 v.bit := 0, else v.bit := 1;
  ∀v ∈ T:
    if (v.bit = 0 ∧ v.parent.bit = 1)
      IS := IS ∪ {v};
  ∀v ∈ IS:
    v.parent = v.parent.parent;
```

Algorithm 3 ALG_{FOREST}

```
∀v ∈ V: define v.parent;
let 𝒯 := {T_1, ..., T_f} be set of resulting trees;
while (|𝒯| > 1) do
  ∀T ∈ 𝒯: ALG_PJ(T);
  ∀T ∈ 𝒯:
    T_m := max{X | X ∈ 𝒯, adjacent(X, T)};
    if (T < T_m): merge T ▷ T_m;
  update 𝒯: set of resulting trees;
```

Lemma 1. *Let T be a tree, let h be its height, s its size, and d the weak diameter of the underlying graph. Applying ALG_{PJ} to T requires expected time $O(d \log h)$, and $O(sd \log h)$ many messages on average.*

PROOF. *Time Complexity.* Consider an arbitrary node v, and consider its path to the root. In each round, the length of this path is reduced by a factor $3/4$ in expectation. From this it follows that $O(\log h)$ many iterations are enough to find the root. Moreover, as the virtual links span at most d hops in the underlying graph, the claim follows.

Message Complexity. The message complexity follows immediately from the time complexity, as there are at most $O(d \cdot \log h)$ many rounds and at most s nodes. □

From the description of ALG_{FOREST} it follows that there will never be cycles in the pointer structure, and that all trees of a given component will eventually merge. In the following, the algorithm's performance is analyzed in detail.

Theorem 5. *ALG_{FOREST} has an expected total running time of $O(d \log s + \log s \log n)$, and requires at most $O(s \log s(d + \log n))$ many messages on average, where n is the network's size, s is the component's size, and d is the component's weak diameter.*

PROOF. *Time Complexity.* The time to solve the neighborhood problem using the network decomposition is of course again $O(\log n)$.

By the description of ALG_{FOREST} it follows that a tree always joins a neighboring tree which is of the same size or larger. By a simple induction argument it can be seen that in phase i, the size of the minimal tree is at least 2^i: For $i = 0$, all trees have at least one node, and the claim follows. Now, by the induction hypothesis, assume that in phase i, indeed all trees are of size at least 2^i. Clearly, each tree will either join a neighbor, or will be joined by at least one neighbor, or both. In both cases, the new tree's size at least doubles. Consequently, ALG_{FOREST} will form a single tree after at most $\log s$ many such phases. In each phase, the tree has to be converted to a star by ALG_{PJ}, which—according to Lemma 1—requires expected time $O(d \cdot \log s)$. However, due to the exponentially growing tree sizes, a geometrically declining number of roots performing the pointer jumping operations exists, and hence the overall costs are $O(d \cdot \log s)$ as well. There are two more operations to be taken into account: First, in each phase, a constant number of aggregations or converge cast operations have to be performed in the tree, requiring time at most $O(d)$ per phase. This does not increase the execution time asymptotically. Second, according to ALG_{FOREST}, in each phase the root asks its children about the trees of their neighbors. This is done by a lookup operation in $(\log n, 1)$-$\mathcal{NC}$, which requires time $O(\log n)$ in each of the $\log s$ many phases. This gives the second summand in the formula: $O(\log s \log n)$.

Message Complexity. According to Lemma 1, the pointer jumping algorithm requires $O(s \cdot d \cdot \log d)$ many messages. Since the tree sizes at least double in each phase, the amortized amount of messages for the entire execution is $O(s \cdot d \cdot \log d)$ as well. The total cost for the $(\log n, 1)$-$\mathcal{NC}$ lookups are $O(s \log n)$ for each of the $\log s$ many phases. Finally, the aggregation costs are in $O(s \cdot d \log s)$. Since $s \geq d$, this supersedes the message cost of the pointer jumping. The claim follows. □

4 The Off-Duty Model

So far, we have assumed that both event and non-event nodes can participate in the component's exploration. While this assumption may be justified in certain systems, e.g., in wired networks, it may not be realistic for wireless networks where only the nodes which have sensed an event wake up from energy-saving mode. In the following, we will briefly discuss this *off-duty model*.

Clearly, if the number of messages does not matter, the event component can be explored in optimal time by using a simple flooding algorithm where each event-node floods the entire graph.

If we ignore the time complexity and only seek to minimize the total number of messages, the situation is different: Consider the clique K_n and assume that there are two event-nodes. Clearly, in order to find out about each other, at least $\Omega(n)$ messages need to be sent: Nodes cannot agree on local coordinators in the preprocessing phase, as these coordinators may be sleeping at runtime. On the other hand, for an optimal "offline" algorithm a constant number of messages is sufficient. Consequently, the message complexity of any distributed algorithm must be worse by a factor of at least $\Omega(n)$.

In contrast to the difficulty of the neighborhood detection, the component exploration is well understood. Depending on whether time or communication costs should be optimized, an appropriate distributed *leader election algorithm* can be applied to the resulting graph (e.g., [11] for time-optimality).

5 Related Work

Motivated by the at times tragic consequences of nature's moods, disaster disclosure is subject to a huge body of research, and it is impossible to provide a complete overview of all the proposed approaches. While many systems are based on (or complemented by) satellite techniques, e.g., for damage estimation of landslides in the *Shihmen Reservoir* in Thailand caused by heavy rainfalls, or for post-earthquake damage detection [6], there are also approaches which directly deploy sensor nodes in the region, e.g., for detecting the boundaries of a toxic leach [5]. Distributed event detection also appears in wired environments, e.g., in the defense against Internet worms [9]. Early warning systems are not only useful to react to natural catastrophes, but are also employed in international politics. Techniques to implement such indicators include expected utility models, artificial intelligence methods, or hidden Markov models [13].

This paper assumes an interesting position between local and global distributed computations, as our algorithms aim at being *as local as possible* and *as global as necessary*. While in the active field of local algorithms [12], algorithms are bound to perform their computations based only on the states of their immediate neighbors, many problems are inherently global, e.g., *leader election*. Only recently, there is a trend to look for local solutions for global problems, where the runtime depends on the *concrete problem input* [3,7], rather than considering the worst-case over all possible inputs: if in a special instance of a problem the input behaves well, a solution can be computed quickly. Similar concepts have already been studied outside the field of distributed computing, e.g., for sorting algorithms. Our paper is a new incarnation of this philosophy as performance mostly depends on the output only.

6 Conclusion

This paper has addressed the problem of distributed alarming and efficient disaster detection. We have presented first solutions for this problem by providing competitive distributed algorithms. We believe that there remain many interesting problems for future research. For instance, the question of fault-tolerance has to be addressed: How can our algorithms be adapted for the case that nodes may be faulty, yielding disconnected components? Moreover, it would be interesting to investigate asynchronous environments. Finally, our model could also be extended to incorporate wireless aspects, such as interference.

References

1. Awerbuch, B., Berger, B., Cowen, L., Peleg, D.: Fast Network Decomposition. In: Proc. ACM PODC (1992)
2. Blelloch, G.E., Maggs, B.M.: Parallel Algorithms. In: Atallah, M.J. (ed.) Handbook of Algrorithms and Theory of Comptuation, CRC Press, Boca Raton (1998)
3. Birk, Y., Keidar, I., Liss, L., Schuster, A., Wolff, R.: Veracity Radius: Capturing the Locality of Distributed Computations. In: Proc. ACM PODC (2006)
4. Cormen, T.H., Leiserson, C.E., Rivest, R.L., Stein, C.: Introduction to Algorithms, 2nd edn. MIT Press and McGraw-Hill (2001)
5. Dong, C.J.G., Wang, B.: Detection and Tracking of Region-Based Evolving Targets in Sensor Networks. In: Proc. 14th ICCCN (2005)
6. Eguchi, R.T., Huyck, C.K., Adams, B.J., Mansouri, B., Houhmand, B., Shinozuka, M.: Resilient Disaster Response: Using Remote Sensing Technologies for Post-Earthquake Damage Detection. In: Earthquake Engineering to Extreme Events (MCEER), Research Progress and Accomplishments 2001-2003 (2003)
7. Elkin, M.: A Faster Distributed Protocol for Constructing a Minimum Spanning Tree. In: Proc. 15th SODA (2004)
8. Gallager, R.G., Humblet, P.A., Spira, P.M.: A Distributed Algorithm for Minimum-Weight Spanning Trees. In: ACM TOPLAS (1983)
9. Kim, H.-A., Karp, B.: Autograph: Toward Automated, Distributed Worm Signature Detection. In: Proc. 13th Usenix Security Symposium (2004)
10. Lotker, Z., Patt-Shamir, B., Peleg, D.: Distributed MST for Constant Diameter Graphs. J. of Dist. Comp. (2006)
11. Peleg, D.: Time-optimal Leader Election in General Networks. J. Parallel Distrib. Comput. 8(1), 96–99 (1990)
12. Peleg, D.: Distributed Computing: A Locality-sensitive Approach. SIAM, Philadelphia (2000)
13. Schrodt, P.A.: Early Warning of Conflict in Southern Lebanon Using Hidden Markov Models. In: American Political Science Association (1997)

Reoptimization of Steiner Trees*

Davide Bilò, Hans-Joachim Böckenhauer, Juraj Hromkovič, Richard Královič, Tobias Mömke, Peter Widmayer, and Anna Zych

Department of Computer Science, ETH Zurich, Switzerland
{dbilo,hjb,juraj.hromkovic,richard.kralovic, tobias.moemke,peter.widmayer,anna.zych}@inf.ethz.ch

Abstract. In this paper we study the problem of finding a minimum Steiner Tree given a minimum Steiner Tree for similar problem instance. We consider scenarios of altering an instance by locally changing the terminal set or the weight of an edge. For all modification scenarios we provide approximation algorithms that improve best currently known corresponding approximation ratios.

1 Introduction

Traditional optimization theory focuses on searching for solutions when nothing or very little is known a'priori about the problem instance. In reality, prior knowledge is often at our disposal, because a problem instance can arise from a small modification of a previous problem instance. As an example, imagine that we are given a set of nodes (points in some metric space), and the network graph of shortest length interconnecting them, where the length is the sum of the distances between adjacent nodes. Now imagine one node is excluded from the network. It is intuitively obvious that we should profit from the old network when we try to find a new one. The general idea we pursue is: given a problem instance with an optimal solution, and a variation of the problem instance obtained through a local modification, what can we learn about the new solution? We believe that looking at NP-hard problems from this perspective will allow to explore them deeper and learn more about the nature of their complexity.

The problem we deal with in this paper is reoptimization of the Minimum Steiner Tree (ST) problem. Given an edge-weighted graph and a set of terminal vertices, the ST problem asks for a minimum tree spanning the terminal set. The ST problem is a very prominent optimization problem with many practical applications, especially in network design, see for example [6,7]. The best up to date approximation ratio for non-reoptimization case is ≈ 1.55 [8]. The problem of reoptimizing ST where the local modification consists of adding/deleting one vertex to/from the input graph was considered in [5]. For the local modification of adding/removing a vertex to/from the terminal set a 1.5-approximation algorithms have been provided in [2]. We improve over these results, providing

* This work was partially supported by SBF grant C 06.0108 as part of the COST 293 (GRAAL) project funded by the European Union.

J. Gudmundsson (Ed.): SWAT 2008, LNCS 5124, pp. 258–269, 2008.

1.344-approximation algorithm for adding a terminal vertex, and 1.408 for removing a terminal vertex, while for the scenarios of increasing and decreasing the weight of an edge we obtain 4/3 and 1.302 approximation ratio respectively.

The paper is organized as follows. In Section 2 we provide basic notation, definitions and facts used throughout the paper. In Section 3 we describe the basic techniques used in Section 4, Section 5, Section 6 and Section 7 where we provide our results for the four local modification scenarios we consider.

2 Preliminaries

Let us begin with a formal definition of the Minimum Steiner Tree Problem (ST). We call a complete graph $G = (V, E, c)$ with edge weight function $c : E \to \mathbb{R}^+$ *metric*, if the edge weights satisfy the *triangle inequality*, i.e., $c((u, v)) \leq c((u, w)) + c((w, v))$ for all $u, v, w \in V$.

The ***Minimum Steiner Tree Problem (ST)*** is defined as follows:
Instance: A metric graph $G = (V, E, c)$ and a terminal set $S \subseteq V$
Solution: A Steiner tree, i.e., a subtree T of G containing S
Objective: Minimize the sum of the weights of the edges in the subtree T.

The ST problem is APX-complete even when the edge weights are restricted to the set $\{1, 2\}$ [1]. The best up to date approximation ratio is $\sigma = 1 + \frac{\log 3}{2} \approx 1.55$ for general edge costs and 1.28 for edge costs from $\{1, 2\}$ [8]. We denote the best up to date approximation algorithm for ST as ApprST(G, S) and view it as a function from the instance space to the solution space. The ST problem can be solved optimally in time exponential in the size of S with an algorithm proposed by Dreyfus and Wagner [4]. We denote this algorithm as OptST(G, S).

Below we define the reoptimization problems we investigate in this paper. The objective for all of them is to minimize the cost of Steiner tree T.

1. ***Minimum Steiner Tree Terminal Addition (ST-S+)***
Instance: Metric graph $G = (V, E, c)$, terminal set $S_O \subseteq V$, an optimal Steiner tree T_O for (G, S_O), and terminal set $S_N = S_O \cup \{t\}$ for some $t \in V$
Solution: Steiner tree T for (G, S_N).
2. ***Minimum Steiner Tree Terminal Removal (ST-S−)***
Instance: Metric graph $G = (V, E, c)$, terminal set $S_O \subseteq V$, an optimal Steiner tree T_O for (G, S_O), and terminal set $S_N = S_O \setminus \{t\}$ for some $t \in V$
Solution: Steiner tree T for (G, S_N).
3. ***Minimum Steiner Tree Edge Cost Increase (ST-E+)***
Instance: Metric graph $G_O = (V, E, c_o)$, terminal set $S \subseteq V$, an optimal Steiner tree T_O for (G, S_O), and metric graph $G_N = (V, E, c_n)$, where $c_n = c_o$ on all but one edge $e \in E$ for which $c_n(e) \geq c_o(e)$[1]
Solution: Steiner tree T for (G_N, S).
4. ***Minimum Steiner Tree Edge Cost Decrease (ST-E−)***
Instance: Metric graph $G_O = (V, E, c_o)$, terminal set $S \subseteq V$, an optimal Steiner tree T_O for (G, S_O), and metric graph $G_N = (V, E, c_n)$, where $c_n = c_o$

[1] Whenever c_o and c_n coincide on some edge f, we drop the subscripts and write $c(f)$.

on all but one edge $e \in E$ for which $c_n(e) \leq c_o(e)$[1]
Solution: Steiner tree T for (G_N, S).

Lemma 1 ([2,3]). *The aforementioned problems are strongly* NP*-hard.* □

Let us adopt the following notation. With T_N we denote an optimal Steiner tree for the modified instance (G, S_N) or (G_N, S). Given a simple graph G, we denote its set of vertices with $V(G)$ and its set of edges with $E(G)$. An edge $(u,v) \in E(G)$ can be seen as a subgraph H of G with $V(H) = \{u,v\}$ and $E(H) = \{(u,v)\}$. A vertex $v \in V(G)$ can be seen as a subgraph H of G with $V(H) = \{v\}$ and $E(H) = \emptyset$. The degree of a vertex $v \in V(G)$ is $\deg_G(v)$. If H is a subgraph of G, we write $H \subseteq G$. The notation $c(G)$ denotes the *cost* of G, i.e. the sum of all its edge costs. With $G - v$ we denote graph G after removing node $v \in V(G)$ and incident edges. For two subgraphs of G: $H_1, H_2 \subseteq G$ (H_i can be a single edge) we introduce the following notation. With $H_1 - H_2$ we denote a graph such that $V(H_1 - H_2) = V(H_1)$ and $E(H_1 - H_2) = E(H_1) \setminus E(H_2)$. With $H_1 + H_2$ we denote a graph such that $V(H_1 + H_2) = V(H_1) \cup V(H_2)$ and $E(H_1 + H_2) = E(H_1) \cup E(H_2)$. We denote with $\mathit{CheapestEdge}(H_1, H_2)$ the cheapest edge in G connecting H_1 and H_2. Expression $\min\{H_1, \ldots, H_i\}$ returns a cheaper graph among $H_1, \ldots, H_i$ w.r.t. their cost. A forest F is a graph composed of node-disjoint trees $T_1, \ldots, T_i$. Such tree decomposition of F is denoted as $F = T_1 + \cdots + T_i$. For a connected subgraph $H \subseteq G$ of G, with $\mathit{Contract}(G, H, h)$ we denote a weighted graph $G' = (V', E', c')$ obtained from G by contracting H into node h, where if after contraction multiedges occur between h and any node $v \in V(G) \setminus V(H)$, then we set $c'((v,h)) = c(\mathit{CheapestEdge}(v, H))$. We describe a path in a graph as a sequence of its vertices. The length of a path is its number of edges. The cost of a path is the sum of costs of its edges. In a shortest path, the length of the path is minimized whereas in a cheapest path its cost is minimized.

For a complete graph $G = (V, E, c)$ with an arbitrary edge weight function $c : E \to \mathbb{R}^+$, we define the *metric closure* of G as the graph $\tilde{G} = (V, E, \tilde{c})$ where $\tilde{c}((u,v))$ is defined as the cost of the cheapest path in G from u to v. It is well known (see for example [7]) that a tree T is a minimum Steiner tree for (G, S) if and only if it is also a minimum Steiner tree for $(\tilde{G}, S)$ where $\tilde{G}$ is the metric closure of G. Because of this fact, for ST-S+ and ST-S−, we can assume w.l.o.g. that the given graph G is metric. For problems ST-E+ and ST-E− we assume as well that the local modification preserves metricity, however in this case this assumption is a restriction as changing the weight of one edge in a metric graph G can result in altering the cost of many edges in its metric closure. Finally, we can assume without loss of generality that the given minimum Steiner tree T_O has no non-terminal vertex of degree two; due to the metricity these vertices can be removed using the direct edge between the two adjacent vertices instead.

3 Techniques

In this section we present standard procedures used further in reoptimization algorithms. Algorithm 1 is based on the assumption, that we know a large part

T_s of an optimal solution T_N for (G, S_N) or (G_N, S). Provided that knowledge, we contract T_s to a single node, make it a terminal, and use σ-approximation algorithm ApprST to obtain the solution for the remaining part of the graph.

Algorithm 1. Shrink(G, S, T_s)

Input: A metric graph G, a terminal set $S \subseteq V(G)$, and a tree $T_s \subseteq G$
1: $G' := \mathit{Contract}(G, T_s, t_s)$
2: $T' := \mathit{ApprST}(G', (S \cap V(G')) \cup \{t_s\})$
3: Obtain T: Expand T' by substituting t_s with T_s
Output: T

Lemma 2. *Let T_{opt} be an optimal solution for (G, S). Given that $T_s \subseteq T_{opt}$, and $c(T_s) \geq \alpha c(T_{opt})$, Algorithm 1 applied to (G, S, T_s) returns $(\sigma - \alpha(\sigma - 1))$-approximation of T_{opt}.*

Proof. Let $G' := \mathit{Contract}(G, T_s, t_s)$, $S' = (S \cap V(G')) \cup \{t_s\}$, and T' be as defined in Algorithm 1. Note, that given $T_s \subseteq T_{opt}$, solution $\mathit{Contract}(T_{opt}, T_s, t_s)$ with cost $c(T_{opt}) - c(T_s)$ is optimal for (G', S'). Thus $c(T') \leq \sigma(c(T_{opt}) - c(T_s))$. Since $\sigma \geq 1$, then the cost of solution tree T returned by *Shrink*(G, S, T_s) is:

$$c(T) \leq \sigma(c(T_{opt}) - c(T_s)) + c(T_s) \leq (\sigma - \alpha(\sigma - 1))c(T_{opt}).$$

□

The other technique, shown in Algorithm 2 is used for connecting optimally a given forest F. Provided that the number of trees in $F = T_1 + \cdots + T_i$ is logarithmic in the input size, we can connect F optimally in polynomial time.

Algorithm 2. Connect(G, F)

Input: A metric graph G, a forest $F = T_1 + \cdots + T_i$, where $T_i \subset G$ and $T_1 \ldots T_i$ are pairwise node-disjoint
1: $G_1 := \mathit{Contract}(G, T_1, t_1)$
2: For $j = 2, \ldots, i$ do $G_j := \mathit{Contract}(G_{j-1}, T_j, t_j)$
3: $S := \{t_1, \ldots, t_i\}$
4: $T' := \mathrm{OptST}(G_i, S)$
5: Obtain tree T by substituting each t_i with T_i and keeping the edges of T' for connecting T_i with the rest of graph G
Output: T

Lemma 3. *If $i = O(\log |V(G)|)$, then Algorithm 2 runs in polynomial time.*

Proof. The running time of Dreyfus-Wagner algorithm OptST() is exponential in size of terminal set [4]. Since we apply OptST to S of size $O(\log |V(G)|)$, the overall running time of Algorithm 2 is polynomial. □

Remark 1. Let (G, S) be an instance of ST and T_{opt} be an optimal solution for that instance. If $F = T_1 + \cdots + T_i \subseteq G$ and $T_{j=1\ldots i} \cap S \neq \emptyset$, then solution T returned by Algorithm 2 satisfies $c(T) \leq c(F) + c(T_{opt})$.

4 Removing One Terminal

In this section we present a 1.408-approximation algorithm for ST-S−, thus improving the result in [2]. The algorithm we propose computes several feasible solutions and chooses the one of minimal cost. Let $f_1, \ldots, f_k$ be the edges adjacent in T_O to the terminal t that is supposed to be removed. We distinguish 4 cases depending on the value of k. We remark here, that Algorithm 3 improves the worst case of $k = 2$. The other cases are dealt with in the same manner as in algorithm provided in [2]. For the sake of completeness however, we provide here the full analysis.

Algorithm 3. MinSTP-S-

Input: A metric graph G, a terminal set $S_O \subseteq V(G)$, an optimal Steiner tree $T_O \subseteq G$ for (G, S_O), and $S_N = S_O \setminus \{t\}$ for some terminal $t \in S_O$

Let $f_1, \ldots, f_k$ be the list of edges adjacent to t in T_O
if $k = 3$ **then**
  Compute shortest paths p_1, p_2, p_3 connecting t with S_N, such that $f_i \in E(p_i)$ for $i = 1, \ldots, 3$
  $F := T_O - (p_1 + p_2 + p_3) - t$
  return *Connect*(G, F)
end if
if $k = 2$ **then**
  $T_1 := \min_{f \in E(G)}\{Shrink(G, S_N, f)\}$
  Compute shortest paths p_1, p_2, p_3, p_4 connecting t with S_N, such that $f_1 \in E(p_1), E(p_2)$ and $f_2 \in E(p_2), E(p_3)$
  $F := T_O - (p_1 + p_2 + p_3 + p_4) - t$
  $T_2 :=$ *Connect*(G, F)
  return $\min\{T_1, T_2\}$
end if
if $k = 1$ **then**
  Let $f_1 = (t, v)$. **if** $v \in S_N$ **then return** $T := T_O - t$
  else return MinSTP-S-$(G, S_N \cup \{v\}, T_O - t, S_N)$
end if

Output: T_O

Theorem 1. *Algorithm 3 is a 1.408-approximation algorithm for ST-S−.*

Proof. Let p be a cheapest path connecting t with S_N. Let $\alpha \geq 0$ be a parameter that shall be fixed later. We want to compute the smallest α, for which Algorithm 3 returns $(1+\alpha)$-approximation of T_N. If $c(p) \leq \alpha c(T_N)$, then $T_O \leq c(T_N) + c(p) \leq (1+\alpha)c(T_N)$, therefore we assume from now on, that $c(p) > \alpha c(T_N)$. We distinguish four cases depending on the degree of t in T_O.

Case 1. Assume $\deg_{T_O}(t) > 3$ or equivalently $k > 3$, and note that $c(T_N) \leq c(T_O)$. Since $\deg_{T_O}(t) \geq 4$, there exist four edge-disjoint paths from t to terminals in T_O and thus $c(T_O) \geq 4 \cdot c(p)$. On the other hand, since $T_N + p$ is a solution for (G, S_O), we know $c(T_N) + c(p) \geq c(T_O)$. Therefore, $c(T_N) \geq 3 \cdot c(p)$ and thus $c(T_O) \leq c(T_N) + c(p) \leq \frac{4}{3}c(T_N)$.

Case 2. Assume $\deg_{T_O}(t) = 3$ or equivalently $k = 3$. Let p_1, p_2, p_3 be as in the Algorithm 3 in case $k = 3$ (see Figure 1). Since p_1, p_2, p_3 are paths minimal in the number of nodes ending in a terminal, and each non-terminal is branching ($deg_{T_O}(v) > 2$ for $v \notin S$), thus $|p_i| \leq \log |V(G)|$. Therefore the number of trees in F is bounded by $3 \log |V(G)|$, and thus by Lemma 3 *Connect*(G, F) runs in polynomial time. For connecting the forest F optimally we pay at most $c(T_N)$, because each tree of F contains a terminal from S_N. Therefore for tree T_2 computed by the algorithm we have that $c(T_2) \leq c(T_O) - c(p_1) - c(p_2) - c(p_3) + c(T_N) \leq c(T_N) + c(p) - 3c(p) + c(T_N) \leq 2c(T_N)(1 - \alpha)$. This value is bounded from above by $(1 + \alpha)c(T_N)$ for any $\alpha \geq \frac{1}{3}$.

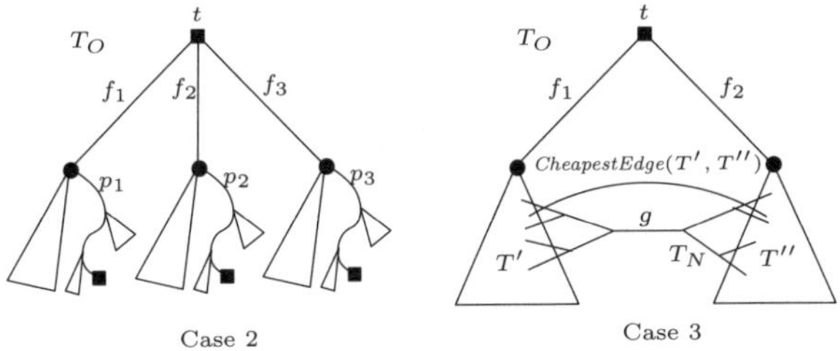

Fig. 1. Illustration for Case 2 and first case of Case 3 (A)

Case 3. Assume $\deg_{T_O}(t) = 2$ or equivalently $k = 2$. Let now β be a parameter that shall be fixed later. We distinguish two cases: $c(f_1) + c(f_2) \leq \beta c(p)$ and $c(f_1) + c(f_2) > \beta c(p)$.

Assume (A) $c(f_1) + c(f_2) > \beta c(p)$. Let μ be a parameter which we will fix later. Let T' and T'' be the trees attached to t in T_O with f_1 and f_2 respectively, as shown in Figure 1. Let p' be the cheapest path in T_N connecting T' with T''. Note, that $T' + T'' + p'$ is a feasible solution in (G, S_N) and that $c(T' + T'' + p') \geq c(T_2)$. If there exists an edge g contained in all paths in T_N from T' to T'', then such an edge is unique. Otherwise let g be an imaginary edge of cost 0. By minimality of p' and due to the fact, that T_N is branching on each endpoint of g into disjoint paths to T' or T'', we have $c(p') \leq \frac{c(T_N) - c(g)}{2} + c(g) = \frac{1}{2}(c(T_N) + c(g))$. If $c(g) > \mu c(T_N)$, then by Lemma 2 *Shrink*(G, S_N, g) returns solution T_1 of cost $c(T_1) \leq (\sigma - \mu(\sigma - 1))c(T_N)$. This value is bounded from above by $(1 + \alpha)c(T_N)$ for $\alpha \geq (\sigma - 1)(1 - \mu)$. If $c(g) \leq \mu c(T_N)$, then $c(T_2) \leq c(T_O) - c(f_1) - c(f_2) + c(p') \leq c(T_N) + c(p) - \beta c(p) + \frac{1}{2}(c(T_N) - c(g))$, therefore $c(T_2) \leq (\frac{3}{2} + \alpha(1 - \beta) + \frac{\mu}{2})c(T_N)$. This in turn is bounded from above by $(1 + \alpha)c(T_N)$ for $\alpha \geq \frac{1+\mu}{2\beta}$. Setting $\frac{1+\mu}{2\beta} = (\sigma - 1)(1 - \mu)$ to obtain later minimal α satisfying both inequalities gives $\mu = \frac{2\beta(\sigma-1)-1}{2\beta(\sigma-1)+1}$ and $\alpha \geq \frac{2(\sigma-1)}{2\beta(\sigma-1)+1}$.

Now assume (B) $c(f_1)+c(f_2) \leq \beta c(p)$. Let p_1, p_2, p_3, p_4 be the shortest paths p_1, p_2, p_3, p_4 connecting t with S_N, such that $p_1 \cap p_2 = f_1$ and $p_2 \cap p_3 = f_2$. If $f_1 = (t, v_1)$ and v_1 is a terminal, then $p_1 = p_2 = f_1$ (same holds for f_2), otherwise there must be two edge disjoint paths from v_1 to S_N not passing trough t. The same argument as in *Case 2* shows that after removing these paths from T_O we can reconnect obtained forest F in polynomial time. The cost of T_2 is bounded from above by $c(T_2) \leq c(T_O)-c(p_1)-c(p_2)-c(p_3)-c(p_4)+c(f_1)+c(f_2)+c(T_N)$, because for reconnecting the obtained forrest optimally we again pay at most $c(T_N)$. Thus $c(T_2) \leq c(T_N) + c(p) - 4c(p) + \beta c(p) + c(T_N)$, what gives $c(T_2) \leq 2c(T_N)-(3-\beta)\alpha c(T_N)$ (assuming $\beta < 3$). This is upper bounded by $(1+\alpha)c(T_N)$ when $\alpha \geq \frac{1}{4-\beta}$. To compute the minimal α for which this inequality and the inequality obtained for Case (A) are satisfied, we set $\frac{1}{4-\beta} = \frac{2(\sigma-1)}{2\beta(\sigma-1)+1}$. That gives $\beta = 2 - \frac{1}{4(\sigma-1)}$ and the minimum value of α is $\frac{4(\sigma-1)}{8(\sigma-1)+1}$. Plugging in $\sigma = 1 + \frac{\log 3}{2}$ guarantees 1.408 approximation ratio for Case 3.

Case 4. Assume $\deg_{T_O}(t) = 1$ or equivalently $k = 1$. In this case, there is exactly one $v \in V(G)$ such that $f_1 = (t, v) \in E(G)$ is incident to t. Tree $T_O - t$ is an optimal Steiner tree for $(G, (S \cup \{v\}) \backslash \{t\})$. Therefore we get a new, smaller problem instance. Since we exclude non-terminals of degree two, either v is a terminal or $\deg_{T_O}(v) \geq 3$. If v is a terminal, the algorithm yields a solution that costs at most $c(T_O - t)$, which is the optimum. Otherwise, $\deg_{T_O - t}(v) \geq 2$ and we have to continue with one of the other three cases. □

5 Adding One Terminal

In this section we present a 1.344-approximation algorithm for the scenario of adding a vertex to the terminal set S, thus improving the result in [5,2].

Algorithm 4. MinSTP-S+

Input: A metric graph G, a terminal set $S_O \subseteq V(G)$, an optimal Steiner tree $T_O \subseteq G$ for (G, S_O) and a new terminal set $S_N := S_O \cup \{t\}$ for some non-terminal t.

$T_1 := T_O + \mathit{CheapestEdge}(T_O, t)$
Let T_2 be any spanning tree in G
for $t' \in V(G) \setminus V(T_O), u \in V(G) \setminus \{t, t'\}, v \in V(G) \setminus \{t, t', u\}$ **do**
  Let T' be a tree on $V(T') = \{t, t', u, v\}$ with edges $E(T') = \{(t, t'), (t', u), (t', v)\}$
  $T_2 := \min\{T_2, \mathit{Shrink}(G, S_N, T')\}$
end for

Output: $\min\{T_1, T_2\}$

Theorem 2. *Algorithm 4 is a 1.344-approximation algorithm for ST-S+.*

Proof. Let $\alpha > \frac{1}{3}$ be a parameter which we fix later. Because T_N is a feasible solution for the old instance, there holds $c(T_O) \leq c(T_N)$. If $t \in T_O$ then $T_1 = T_O$

is optimal. Otherwise, let $f_{min} = (w, t)$ be a cheapest edge connecting T_O with t. We can assume that $c(f_{min}) > \alpha c(T_N)$, otherwise $c(T_1) \leq c(T_O) + c(f_{min}) \leq (1+\alpha)c(T_N)$ gives $(1+\alpha)$-approximation. Let $f_1, \ldots, f_k$ be the edges that are adjacent to t in T_N. There must be k edge disjoint paths in T_N from t to S_O, and by metricity the cost of each of them is greater then $c(f_{min}) > \alpha c(T_N) > \frac{1}{3}c(T_N)$. Therefore $c(T_N) > \frac{k}{3}c(T_N)$ implies $k \leq 2$. We distinguish two cases.

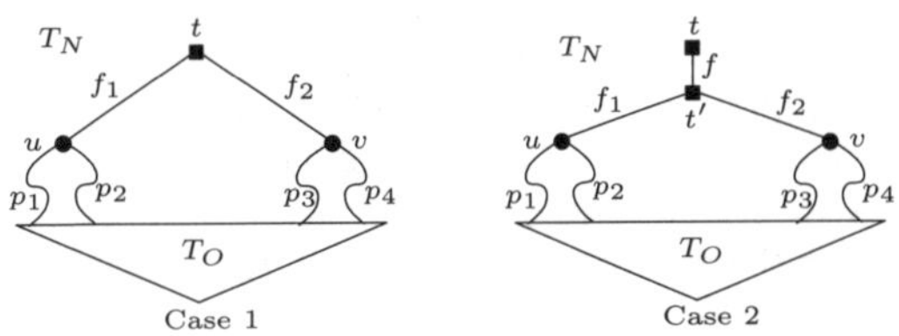

Fig. 2. Pattern for adding a terminal

Case 1: $k = 2$. Let $f_1 = (t, u)$ and $f_2 = (t, v)$. Algorithm 4 exhaustively searches trough all triples of nodes that are candidates for t, u and v. When it hits the triple t, u, v with $t' = t$, it applies $Shrink(G, S_N, T')$ for tree $T' \subseteq T_N$ which contains only edges f_1 and f_2. It looks for minimum T_2 over all triples, thus $c(T_2) \leq Shrink(G, S_N, T')$ for that particular T'. Let p_1 and p_2 be two edge disjoint paths in T_N connecting u with T_O. If $u \notin T_O$, its degree in T_N is at least 2 and therefore such paths exist. If $u \in T_O$, then let $p_1 = p_2 = \{u\}$ and $c(p_1) = c(p_2) = 0$. We define p_3 and p_4 analogically with respect to vertex v. The situation is shown in Figure 2. By metricity the following inequalities hold:

$$\begin{cases} c(f_1) + c(p_1) > \alpha c(T_N) \\ c(f_1) + c(p_2) > \alpha c(T_N) \\ c(f_2) + c(p_3) > \alpha c(T_N) \\ c(f_2) + c(p_4) > \alpha c(T_N) \end{cases} \tag{1}$$

Summing all the inequalities in (1) gives $c(f_1) + c(f_2) + c(T_N) \geq c(f_1) + c(f_2) + c(f_1) + c(f_2) + \sum_{i=1}^{4} c(p_i) > 4\alpha c(T_N)$ and thus $c(f_1) + c(f_2) > (4\alpha - 1)c(T_N)$. Therefore by Lemma 2 we get $c(T_2) \leq c(Shrink(G, S_N, T')) \leq (\sigma - (4\alpha - 1)(\sigma - 1))c(T_N)$. This is bounded by $(1+\alpha)c(T_N)$ for any α satisfying $\alpha \geq \frac{2(\sigma-1)}{4(\sigma-1)+1}$. Plugging in $\sigma = 1 + \frac{\log 3}{2}$ ensures 1.344 approximation ratio in this case.

Case 2: $k = 1$. The situation is shown in Figure 2. Let $f = (t, t')$ be the only edge adjacent to t in T_N. In this case $T_N - t$ is a feasible solution for the old instance, and thus $c(T_O) \leq c(T_N) - c(f)$. Let f_{min} be the cheapest edge connecting t' with T_O. If $c(f_{min}) \leq \alpha c(T_N)$ we have $c(T_1) \leq c(T_O) + c(f_{min}) + c(f) \leq (1+\alpha)c(T_N)$. Assume $c(f_{min}) > \alpha c(T_N)$. If $t' \in T_O$, then T_1 is an optimal solution. Otherwise t' is non terminal and there must be two edges $f_1 = (t', u)$ and $f_2 = (t', v)$ adjacent to t' in T_N. Further analysis is identical as in Case 1, taking t' instead of t, and gives 1.344 approximation ratio. □

6 Increasing the Weight of One Edge

Throughout this and the next section we will assume that a modification of edge weights do not affect the metricity of the graph, as already discussed in Section 2. In this section we consider the local modification where the cost of one edge $e \in E(G)$ increases: $c_n(e) > c_o(e)$. As a consequence $c(T_N) \geq c(T_O)$.

Algorithm 5. MinSTP-E+

Input: A metric graph $G_O = (V, E, c_o)$, a terminal set $S \subseteq V(G)$, an optimal Stainer Tree T_O for (G_O, S), and a new metric graph $G_N = (V, E, c_n)$, where $c_n = c_o$ on all but one edge e for which $c_o(e) \leq c_n(e)$.

Let $T_O{}'$ and $T_O{}''$ be the subtrees obtained from T_O by removing e.
$T_1 := T_O - e + \mathit{CheapestEdge}(T_O{}', T_O{}'')$
Let $e = (u, v)$ and $f_1, \ldots, f_k$ be the edges adjacent to e in T_O
$F := T_O - u - v$
if $k < 3$ **then** $T_3 := \mathit{Connect}(G_N, F)$
$T_4 := \min_{x \in V(G) \setminus \{u,v\}, y \in V(G) \setminus \{x,u\}} \{\mathit{Shrink}(G_N, S, T_{(u,x,y)})\}$
$T_5 := \min_{x \in V(G) \setminus \{u,v\}, y \in V(G) \setminus \{x,v\}} \{\mathit{Shrink}(G_N, S, T_{(v,x,y)})\}$
where $T_{(w,x,y)}$ denotes tree on vertices $\{u, v, x, y\}$ spanning edges $\{(u, x), (x, v)(w, y)\}$

Output: $\min\{T_O, T_1, T_3, T_4, T_5\}$

Theorem 3. *Algorithm 5 is a $\frac{4}{3}$-approximation algorithm for ST-E+.*

Proof. Let $\alpha = \frac{1}{3}$. If $e \notin E(T_O)$ or $e \in E(T_N)$, then T_O is an optimal solution for the new instance. Therefore we consider the only non trivial case when $e = (u, v) \in E(T_O)$ and $e \notin E(T_N)$. Let $f_1, \ldots, f_k$ be the edges adjacent to e in T_O. Assume there is an edge $f_i = (u, w) \in E(T_O)$, such that $c(f_i) \leq \alpha c(T_N)$. Let $g = (w, v)$. Then $c(T_1) \leq c(T_O) - c_O(e) + c(g)$, thus by metricity we have $c(T_1) \leq c(T_N) + c(f_i) \leq (1 + \alpha)c(T_N)$. Therefore we can assume that $c(f_i) > \alpha c(T_N) \geq \frac{1}{3}c(T_N)$. Hence $deg_{T_N}(u) + deg_{T_N}(v) \leq 3$. Moreover, since $c(T_N) \geq c(T_O)$, there are at most two such edges in T_O. We can also assume

$$c_n(e) - c_o(e) > \alpha c(T_N) \tag{2}$$

otherwise $c(T_O) \leq (1 + \alpha)c(T_N)$. Observe that for each f_i adjacent to e holds:[1]

$$c(f_i) \geq \frac{c_n(e) - c_o(e)}{2}. \tag{3}$$

If $k = 2$, then $c(T_3) \leq c(T_O) - c(f_1) - c(f_2) - c_o(e) + c(T_N) \leq 2c(T_N) - \frac{2}{3}c(T_N) = \frac{4}{3}c(T_N)$. The remaining case is when there is only one edge f adjacent to e in T_O. In this particular case both u and v are terminal vertices. We distinguish further cases regarding the number of edges in T_N adjacent to e.

---

[1] Let h be an edge that forms a triangle with e and f_3. By metricity: $c_n(e) + c(f_3) \geq c(h) \geq c_o(e) - c(f_3)$ which gives the inequality we use.

Case 1: $deg_{T_N}(u) + deg_{T_N}(v) = 3$. Let p_{uv} be the path from u to v in T_N. If (A) there are two nodes $x, y \in V(p_{uv})$ other than u and v (see Figure 3, Case 1A), then there must edge disjoint paths from x and y to terminals. Moreover there must be another path edge disjoint with these two from u or v to a terminal, since $deg_{T_N}(u) + deg_{T_N}(v) = 3$. This gives three paths of cost greater than $\frac{1}{3}c(T_N)$ which can not be the case. Assume (B) there is only one node $x \in V(p_{uv})$ other then u and v. This situation is shown in Figure 3, Case 1B. Let $f_1 = (u, x)$, $f_2 = (x, v)$ and w.l.o.g let $f_3 = (v, y)$, $f_1, f_2, f_3 \in E(T_N)$. When computing T_4, the algorithm exhaustively searches trough all candidates for x and y, therefore $c(T_4) \leq c(Shrink(G_N, S, T_{(v,x,y)}))$, where $T_{(v,x,y)}$ spans f_1, f_2, f_3. By metricity and (2) we get $c(f_1) + c(f_2) > \alpha c(T_N)$. By (3) and (2), we get $c(f_3) \geq \frac{c_n(e) - c_o(e)}{2} \geq \frac{\alpha}{2}c(T_N)$. Therefore $c(f_1 + f_2 + f_3) > \frac{3\alpha}{2}c(T_N)$ and Lemma 2 gives the following bound: $c(T_4) \leq \sigma c(T_N) - \frac{3\alpha}{2}(\sigma - 1)c(T_N)$. This guarantees $(1 + \alpha)$-approximation for any $\alpha \geq \frac{2(\sigma-1)}{3(\sigma-1)+2}$. Plugging in best up to date σ ensures in this case 4/3-approximation ratio.

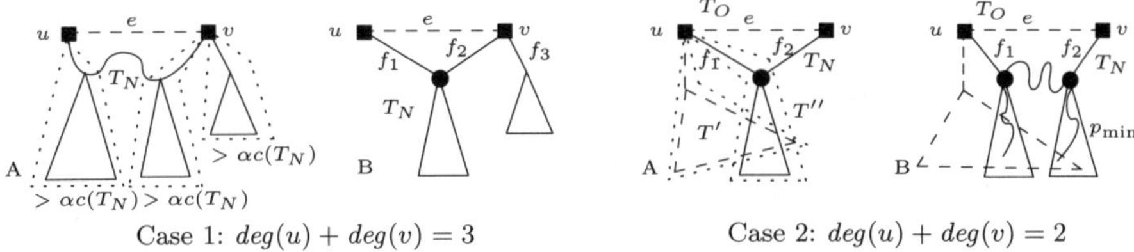

Fig. 3. Pattern for increasing the cost of edge e

Case 2: $\deg_{T_N}(u) + \deg_{T_N}(v) = 2$. In this case we distinguish two subcases (see Figure 3 Case 2), mainly (A) when there is only one node on the path $p_{uv} \subseteq T_N$ from u to v, and when (B) there are at least two nodes on this path. For both cases, let w.l.o.g. $f = (u, w)$ be the only edge in T_O adjacent to e.

Assume that (A) $x \in V(p_{uv})$ is the only node on the path other than u and v. This implies that v is a leaf in T_O. Let $f_1 = (u, x), f_2 = (v, y)$ be the edges adjacent to e in T_N. Since $\deg_{T_N}(u) + \deg_{T_N}(v) = 2$, both u and v are leaves in T_N. Thus, tree $T' = T_O - v$ is an optimal solution for $(G_N, S \setminus \{v\})$ (otherwise we could improve T_O in (G_O, S)). Since v is a leaf in T_N, tree $T'' = T_N - v$ is a feasible solution for $(G_N, S \setminus \{v\})$, and therefore $c(T') \leq c(T'')$. The situation is presented in Figure 3 Case 2A. This gives $c(T_O) - c_o(e) + c(f_2) \leq c(T_N)$. But $c(T_1) \leq c(T_O) - c_o(e) + c(f_1) + c(f_2) \leq c(T_N) + c(f_1)$. It follows immediately, that if $c(f_1) \leq \alpha c(T_N)$, then T_1 is $(1+\alpha)$ - approximation. Otherwise $c(f_1) > \alpha c(T_N)$. From (3) and (2) holds $c(f_2) > \frac{\alpha}{2}c(T_N)$, what together with the above inequality implies $c(f_1) + c(f_2) > \frac{3}{2}\alpha c(T_N)$, and applying $Shrink(G_N, S, T_{(u,x,v)})$ when $y = v$ guarantees by Lemma 2 that $c(T_4) \leq (1 + \alpha)c(T_N)$ for any $\alpha \geq \frac{2(\sigma-1)}{3(\sigma-1)+2}$. Plugging in best up to date σ ensures also in this case 4/3 approximation ratio.

Now assume (B) $f_1 = (u, x), f_2 = (v, y) \in E(p_{uv}) \subseteq T_N$, and $x \neq y$. We may assume $y \notin T_O$, otherwise T_1 is optimal. Since $\alpha \geq \frac{1}{3}$, there must be

$c(T_N) \leq 3\alpha c(T_N)$. By (3) and (2) holds $c(f_1) \geq \frac{\alpha}{2}c(T_N)$ and $c(f_2) \geq \frac{\alpha}{2}c(T_N)$, what implies $c(T_N) - c(f_1) - c(f_2) \leq 2\alpha c(T_N)$. Because y is a non terminal, we have $deg_{T_N}(y) \geq 3$, thus there must be two edge disjoint paths in $T_N - f_1 - f_2$ from y to $S \backslash \{u, v\}$. Minimal such path p_{min} must satisfy $c(p_{min}) \leq \frac{c(T_N)-c(f_1)-c(f_2)}{2} \leq \alpha c(T_N)$. Therefore $c(T_1) \leq c(T_O) - c_o(e) + c(f_2) + c(p_{min}) \leq c(T_N) + c(p_{min}) \leq (1+\alpha)c(T_N)$ gives the desired bound. □

7 Decreasing the Weight of One Edge

In this subsection, we present a 1.302-approximation algorithm for ST-E−. The local modification is the decrease of the cost of one edge: $c_n(e) \leq c_o(e)$.

Algorithm 6. MinSTP-E-

Input: A metric graph $G_O = (V, E, c_o)$, a terminal set $S \subseteq V(G)$, an optimal Stainer Tree T_O for (G_O, S), and a new metric graph $G_N = (V, E, c_n)$, where $c_n = c_o$ on all but one edge $e = (u, v)$ for which $c_o(e) \geq c_n(e)$.

if $u \in S$ **and** $v \in S$ **then**
  Let f_{max} be the most expensive edge on the path from u to v in T_O
  $T_A := T_O - f_{max} + e$
end if
$T_S := \min_{t,w \in \{u,v\}, x,y \in V(G), x \neq y} \{Shrink(G_N, S, T_{(x,t,w,y)})\}$ where $T_{(x,t,w,y)}$ is a tree on $\{x, t, w, y\}$ spanning edges $\{(t, x), (w, y), e\}$

Output: $\min\{T_O, T_A, T_S\}$

Theorem 4. *Algorithm 6 for ST-E− achieves an approximation ratio of* 1.302.

Proof. Let $\alpha = \frac{22}{73}$. Clearly $c(T_N) \leq c(T_O)$. If $e \notin E(T_N)$, then T_N is a feasible solution for (G, S, c_o), and thus T_O is optimal for the new instance. If $e \in E(T_N)$ and $e \in E(T_O)$, then $c_n(T_N) - c_n(e) + c_o(e) \geq c_o(T_O)$ because T_N is feasible in G_O. But that implies $c_n(T_O) \leq c_n(T_N)$ and thus T_O is optimal for (G_N, S). Further we analyze the only non trivial case when $e \notin E(T_O)$ and $e \in E(T_N)$.

Let $f_1, \ldots, f_k \in E(T_N)$ be the edges adjacent to e in T_N. W.l.o.g. let $f_i = (u, x_i)$. Let $g_i = (x_i, v)$. A feasible solution for (G_O, S) is $T_N - e + g_i$, thus $c(T_O) \leq c(T_N) - c_n(e) + c(g_i)$. By metricity $c(g_i) \leq c(f_i) + c_n(e)$, and therefore for each edge f_i adjacent to e in T_N holds

$$c(T_O) \leq c(T_N) + c(f_i). \tag{4}$$

For the remaining part of the proof we can assume that

$$c(f_i) > \alpha c(T_N) \tag{5}$$

$$c_o(e) - c_n(e) > \alpha c(T_N), \tag{6}$$

otherwise, from (4) and from the fact that $c(T_O) \leq c(T_N) - c_n(e) + c_o(e)$ we have that T_O is a $(1+\alpha)$-approximation. We distinguish two cases.

Case 1: $k > 1$ or ($k = 1$ and $c(f_1)+c_n(e) > \frac{3}{2}\alpha c(T_N)$). If there are at least two edges $f_1, f_2 \in E(T_N)$ adjacent to e, $Shrink(G_N, S, T_{(u,v,x,y)})$ must be called at some step for $T_{(u,v,x,y)}$ spanning e, f_1, f_2. Note, that by (5) we get at this step $c(T_{(u,v,x,y)}) \geq c(f_1) + c(f_2) \geq 2\alpha c(T_N) \geq \frac{3}{2}\alpha c(T_N)$. If there is only one edge f_1 adjacent to e in T_N, $Shrink(G_N, S, T_{(u,v,x,y)})$ is called at some step for $T_{(u,v,x,y)}$ spanning e, f_1. Then, $c(T_{(u,v,x,y)}) \geq c(f_1) + c_n(e) > \frac{3}{2}\alpha c(T_N)$. Thus, in both cases from Lemma 2 after calculations and plugging in best up to date σ, we obtain $(1+\alpha)$-approximation for any $\alpha \geq \frac{22}{73}$.

Case 2: $k = 1$ and $c(f_1) + c_n(e) \leq \frac{3}{2}\alpha c(T_N)$. In this case both u and v are terminals, thus Algorithm 6 computes solution T_A. For any edge f adjacent to e in G, by (3), there must hold $c(f) \geq \frac{c_o(e)-c_n(e)}{2}$. Since on the path from u to v in T_O there are two edges adjacent to e, we are guaranteed $c(f_{max}) \geq \frac{c_o(e)-c_n(e)}{2} \geq^{(6)} \frac{\alpha}{2}c(T_N)$. Therefore

$$c(T_A)=c(T_O)-c(f_{max})+c_n(e) \leq^{(4)} c(T_N)+c(f_1)+c_n(e)-c(f_{max}) \leq (1+\alpha)c(T_N).$$

□

References

1. Bern, M.W., Plassmann, P.E.: The Steiner problem with edge lengths 1 and 2. Inf. Process. Lett. 32(4), 171–176 (1989)
2. Böckenhauer, H.-J., Hromkovič, J., Královič, R., Mömke, T., Rossmanith, P.: Reoptimization of steiner trees: Changing the terminal set. Theoretical Computer Science (to appear)
3. Böckenhauer, H.-J., Hromkovič, J., Mömke, T., Widmayer, P.: On the hardness of reoptimization. In: 34th International Conference on Current Trends in Theory and Practice of Computer Science, SOFSEM 2008, pp. 50–65 (2008)
4. Dreyfus, S.E., Wagner, R.A.: The Steiner problem in graphs. Networks 1, 195–207 (1971/1972)
5. Escoffier, B., Milanic, M., Paschos, V.T.: Simple and fast reoptimizations for the Steiner tree problem. Technical Report 2007-01, DIMACS (2007)
6. Hwang, F., Richards, D., Winter, P.: The Steiner Tree Problems. Annals of Discrete Mathematics, vol. 53. North-Holland, Amsterdam (1992)
7. Prömel, H.J., Steger, A.: The Steiner Tree Problem. Advanced Lectures in Mathematics. Friedr. Vieweg & Sohn, Braunschweig (2002)
8. Robins, G., Zelikovsky, A.: Improved Steiner tree approximation in graphs. In: Proceedings of the Eleventh Annual ACM-SIAM Symposium on Discrete Algorithms, SODA 2000, pp. 770–779. ACM Press, New York (2000)

On the Locality of Extracting a 2-Manifold in $\mathbb{R}^3$

Daniel Dumitriu[1], Stefan Funke[2], Martin Kutz[3], and Nikola Milosavljević[3]

[1] Max-Planck-Institut für Informatik
Campus E 1.4, 66123 Saarbrücken, Germany
dumitriu@mpi-inf.mpg.de
[2] Ernst-Moritz-Arndt-Universität
Jahnstr. 15a, 17487 Greifswald, Germany
stefan.funke@uni-greifswald.de
[3] Stanford University, Stanford CA 94305, USA
nikolam@stanford.edu

Abstract. Algorithms for reconstructing a 2-manifold from a point sample in $\mathbb{R}^3$ based on Voronoi-filtering like *CRUST* [1] or *CoCone* [2] still require – after identifying a set of candidate triangles – a so-called *manifold extraction* step which identifies a subset of the candidate triangles to form the final reconstruction surface. Non-locality of the latter step is caused by so-called *slivers* – configurations of four almost cocircular points having an empty circumsphere with center close to the manifold surface.

We prove that under a certain mild condition – local uniformity – which typically holds in practice but can also be enforced theoretically, one can compute a reconstruction using an algorithm whose decisions about the adjacencies of a point only depend on nearby points.

While the theoretical proof requires an extremely high sampling density, our prototype implementation, described in a companion paper [3], preforms well on typical sample sets. Due to its local mode of computation, it might be particularly suited for parallel computing or external memory scenarios.

1 Introduction

Reconstructing a surface Γ in $\mathbb{R}^3$ from a finite point sample V has attracted a lot of attention both in the computer graphics community and in the computational geometry community. While in the former the emphasis is mostly on algorithms that work "well in practice", the latter has focused on algorithms that come with a theoretical guarantee: if the point sample V satisfies a certain *sampling condition*, the output of the respective algorithm is guaranteed to be "close" to the original surface.

In [1], Amenta and Bern proposed a framework for rigorously analyzing algorithms reconstructing smooth closed surfaces. They define for every point $p \in \Gamma$ on the surface the *local feature size* $\text{lfs}(p)$ as the distance of p to the *medial axis*[1] of Γ. A set of points $V \subset \Gamma$ is called a ε-sample of Γ if $\forall p \in \Gamma\ \exists s \in V$:

[1] The medial axis of Γ is defined as the set of points which have at least two closest points on Γ.

J. Gudmundsson (Ed.): SWAT 2008, LNCS 5124, pp. 270–281, 2008.

$|sp| \leq \varepsilon \cdot \text{lfs}(p)$. For sufficiently small ε, Amenta and Bern define a *canonical correct reconstruction* of V with respect to Γ as the set of Delaunay triangles whose dual Voronoi edges in the Voronoi diagram of V intersected the surface Γ. Unfortunately, due to certain point configurations called *slivers* – four (almost) cocircular points that are nearby on the surface and have an empty, (almost) diametral circumsphere – it is not possible to algorithmically determine the canonical correct reconstruction of V with respect to Γ without knowing Γ, no matter how dense the sampling V is. However, there are algorithms that can determine a collection of Delaunay triangles which form a piecewise linear surface that is topologically equivalent to the canonical correct reconstruction, and converges to the latter, both point-wise and in terms of the surface normals, as the sampling density goes to infinity ($\varepsilon \rightarrow 0$). The CoCone algorithm [2] is one example; it proceeds in four stages: 1) The Voronoi diagram of V is computed. 2) For every point $p \in V$ the surface normal $\overrightarrow{n_p}$ at p is estimated as a vector pointing from p to the furthest point in p's Voronoi cell. 3) A set of *candidate triangles* $\mathcal{T}$ is determined by selecting all Delaunay triangles whose dual Voronoi edge intersects the *CoCones*[2] of all three respective sample points. 4) From the set of candidate triangles that form a "thickened" layer near the real surface, the final piecewise-linear surface approximating Γ is extracted.

The last step of the CoCone algorithm first removes triangles with "free" edges, and then determines the final reconstruction as the outside surface of the largest connected component of the remaining triangles. Observe that this is a highly non-local operation. There have been attempts to locally decide for each sample p which of the candidate triangles to keep for the final reconstruction; such local decisions might disagree, though, and hence the selected triangles do not patch up to a closed manifold. Again, the reason why local decisions might disagree is the presence of *slivers* which induce a Voronoi vertex inside the CoCone region of the involved sample points. Each involved sample point has to decide whether in "its opinion" the true surface Γ intersects above or below the Voronoi vertex, and create the respective dual Delaunay triangles. If these decisions are not coordinated contradictions arise. Not only in theory, but even in practice, the manifold extraction step is still quite challenging and requires nontrivial engineering to actually work as desired.

One potential way of obtaining a local manifold extraction step is to decide on triangles/adjacencies in a conservative manner by only creating those triangles/adjacencies which are "safe", i.e. where the respective dual Voronoi edge/face essentially completely pierces the CoCone region and hence are certainly part of the canonical reconstruction as well as any good approximation to it. It is unclear, though, how much connectivity is lost — whether the resulting graph is connected at all and how big potential holes/faces are. The main contribution of this paper is to show that it is actually possible to make local decisions but still guarantee that the resulting graph exhibits topological equivalence to the original surface. That is, it is connected, locally planar, and contains no large holes.

[2] The CoCone region of a point p with estimated surface normal $\overrightarrow{n_p}$ is the part of p's Voronoi cell that makes an angle close to $\pi/2$ with $\overrightarrow{n_p}$ at p.

We want to point out that the CoCone algorithm (like many other algorithms in that area) has an inherent (theoretical) quadratic worst-case running time since it computes a Voronoi diagram/Delaunay triangulation of a point set in $\mathbb{R}^3$ – the algorithm by Funke and Ramos [4] is an exception since it runs in near-linear time by enforcing the relevant Voronoi computations to take place locally; nevertheless, this algorithm also requires a non-local manifold extraction step.

We borrow two ingredients of [4]. In a first step, the algorithm by Funke and Ramos computes a function $\phi(p)\ \forall p \in V$ such that ϕ is Lipschitz[3] and $\phi(p) \leq \varepsilon\,\text{lfs}(p)$. They use this function to "prune" the original point set V to obtain a set S which has certain nice properties. For our theoretical analysis we assume that this function ϕ has been computed in the same manner and use it to construct a local neighborhood graph on which our algorithm operates. For S, the algorithm in [4] then locally computes candidate triangles (which due to the nice properties of S can be done locally in near-linear time), and uses the standard (non-local!) manifold extraction step to obtain a reconstruction of S with respect to Γ. Finally, all samples in $V - S$ are reinserted to produce the final reconstruction; this can be done very elegantly using the reconstruction of S as a "reference surface" with respect to which the restricted Voronoi Diagram/Delaunay triangulation of $V - S$ is considered. The restricted VD/DT can easily be computed locally and efficiently via a 2D weighted Delaunay triangulation on planes supporting the faces of the reconstruction of S. We borrow this last step for our algorithm as we also compute an intermediate reconstruction for a subset of sample points.

In [5] Funke and Milosavljević present an algorithm for computing *virtual coordinates* for the nodes of a wireless sensor network which are themselves unaware of their location. Their approach crucially depends on a subroutine to identify a provably planar subgraph of a communication graph that is a quasi-unit-disk graph. The same subroutine will also be used in our surface reconstruction algorithm presented in this paper.

While we deal with the problem of slivers in some sense by avoiding or ignoring them, another approach called *sliver pumping* has been proposed by Cheng et al. in [6]. Their approach works for smooth k-manifolds in arbitrary dimension, though its practicality seems uncertain. There are, of course, other non-Voronoi-filtering-based algorithms for manifold reconstruction which do not have a manifold extraction step; they are not in the focus of this paper, though.

1.1 Our Contribution

We propose a novel method for extracting a 2-manifold from a point sample in $\mathbb{R}^3$. Our approach fundamentally differs from previous approaches in two respects: first it mainly operates combinatorially, on a graph structure derived from the original geometry; secondly, the created adjacencies/edges are "conservative" in a sense that two samples are only connected if there is a safe, sliver-free

---

[3] More precisely their algorithm computes a δ-approximate ω-Lipschitz function ϕ, that is for x, y we have $\phi(x) \leq (1+\delta)(\phi(y) + \omega|xy|)$.

region around the two samples. Interestingly, we can show that conservative edge creation only leads to small, constant-size faces in the corresponding reconstruction. Hence completion to a triangulated piecewise linear surface can easily be accomplished using known techniques. The most notable advantage compared to previous Voronoi-filtering-based approaches is that the manifold extraction step can be performed locally, i.e. at any point relying only on adjacency information of points that are geometrically close to the part of the manifold being extracted.

While the theoretical analysis requires an absurdly high sampling density – like most of the above mentioned algorithms do – our prototype implementation of the novel local manifold extraction step (see companion paper [3]) suggests that the approach is viable even for practical use. The results are quite promising, and there is potential for considerable speedup e.g. in parallel computing or external memory scenarios due to the local nature of computation in our new method.

From a technical point of view, two insights are novel in this paper (and not a result of the mere combination of the two previous results): first, we show that the neighborhood graph that our algorithms constructs is locally a *quasi-unit-disk graph*; it is this property that allows us to actually make use of the machinery developed in [5]. Second, we provide a more elegant and much stronger result about the density of the extracted planar graph based on a connection between the β-skeleton and a power-spanner property; this insight also improves the overall result in [5].

2 Graph-Based, Conservative Adjacencies

In this section we present an algorithm that, given a ε-sample V from a closed smooth 2-manifold Γ in $\mathbb{R}^3$, computes a faithful reconstruction of V with respect to Γ, as a subcomplex of the Delaunay tetrahedralization of V. The outline of our method is as follows:

1. Determine a Lipschitz function $\phi(v)$ for every $v \in V$ which lower-bounds $\varepsilon\,\mathrm{lfs}(v)$ (as in [4])
2. Construct a local neighborhood graph $G(V)$ by creating an edge from every point v to all other points v' with $|vv'| \leq O(\phi(v))$
3. Compute a subsample S of (V)
4. Identify adjacencies between elements in S based on the connectivity of $G(V)$ (as in [5])
5. Use geometric positions of the points in S to identify faces of the graph induced by certified adjacencies when embedded on the manifold
6. Triangulate all non-triangular faces
7. Reinsert points in $V - S$ by computing the weighted Delaunay triangulations on the respective faces (as in [4])

The core components of the correctness proof of this approach are:

- We show that the local neighborhood graph corresponds locally to a quasi-unit-disk graph for a set of points in the plane.
- The identified adjacencies locally form a planar graph.
- This locally planar graph has faces of bounded size.

Essentially this means that we cover Γ by a mesh with vertex set S consisting of small enough cells that the topology of Γ is faithfully captured. Note that the first and last item from above are original and novel to this paper and do not follow from our previous results in [4] and [5] (the last item makes the theoretical result in [5] much stronger).

We first discuss the 2-dimensional case, where we are given a uniform ε-sampling (i.e. the local feature size is 1 everywhere) of a disk and show that steps 2. to 5. yield a planar graph with "small" faces. Then we show how the same reasoning can be applied to the 3-dimensional case. The main rationale of our approach is the "conservative" creation of adjacencies; that is, we only create an edge between two samples if in any good reconstruction the two points are adjacent, which can be interpreted as creating edges only in the absence of slivers in the vicinity.

2.1 Conservative Adjacencies in $\mathbb{R}^2$

Let V be a set of n points that form a ε-sampling of the disk of radius R around the origin o, that is, $\forall p \in \mathbb{R}^2$ with $|po| \leq R$, $\exists v \in V : |vp| \leq \varepsilon$.

Definition 1. *A graph $G(V, E)$ on V is called a* α-quasi-unit-disk-graph *(α-qUDG) for $\alpha \in [0, 1]$ if for $p, q \in V$*

- *if $|pq| \leq \alpha$ then $(p, q) \in E$*
- *if $|pq| > 1$ then $(p, q) \notin E$*

That is, in G all nodes at distance at most α have to be adjacent, while all nodes at distance more than 1 *cannot be adjacent. For nodes with distances in between, either is possible.*

Within G we consider the distance function d_G defined by the (unweighted) graph distances in $G(V, E)$. Let $k \geq 1$, we call a set $S \subseteq V$ a *tight k-subsample* of V if

- $\forall s_1, s_2 \in S$: $d_G(s_1, s_2) > k$
- $\forall v \in V$: $\exists s \in S$ with $d_G(v, s) \leq k$.

A tight k-subsample of V can easily obtained by a greedy algorithm which iteratively selects a so far unremoved node v into S and removes all nodes at distance at most k from consideration.

The following algorithm determines adjacencies between nodes in S based on a *Graph Voronoi diagram* such that the induced graph on S remains planar.

Graph-Based Conservative Adjacencies. The idea for construction and the planarity property of our construction are largely derived from the geometric intuition. To be specific, the planarity follows from the fact that our constructed graph – we call it *combinatorial Delaunay map* of S, short CDM(S) – is the *dual graph* of a suitably defined partition of the plane into simply connected disjoint regions. In the following we use the method for identifying adjacencies between nodes in S purely based on the graph connectivity as described in [5]. The reasoning relies on the fact that our graph instance is not an arbitrary graph but reflects the geometry of the underlying domain by being a quasi-unit-disk graph.

First we introduce a *labeling of* $G(V, E)$ for a given set $S \subseteq V$ assuming that all elements in V (and hence in S) have unique IDs that are totally ordered.

Definition 2. *Consider a vertex $a \in S$ and a vertex $v \in V - S$. We say that v is an a-vertex (or: labeled with a) if a is one of the elements in S which is closest to v (in graph distance), and a has the smallest ID among such.*

Clearly, this rule assigns unique labels to each vertex due to the uniqueness of nodes' IDs. Also note that any $a \in S$ is an a-vertex. Next we present a criterion for creating adjacencies between vertices in S.

Definition 3. *$a, b \in S$ are adjacent in CDM(S) iff there exists a path from a to b whose 1-hop neighborhood (including the path itself) consists only of a and b vertices, and such that in the ordering of the nodes on the path (starting with a and ending with b) all a-nodes precede all b-nodes.*

We have the following result of [5].

Theorem 1. *If G is an α-qUDG with $\alpha \geq \frac{1}{\sqrt{2}}$ and S is a tight k-subsample of G, then $CDM(S)$ is a planar graph.*

Of course, just planarity as such is not too hard to guarantee – one could simply return a graph with no edges. But we will show in the following that this is not the case; in particular we show that the respective graph is connected and all its faces are bounded by a constant number of edges. The following lemmas and proofs are *not* taken from [5], but also apply there, improving the (weaker) statements about the density of CDM(S) in [5].

CDM(S) is Dense(!). Let us consider the β-skeleton [7] of the points corresponding to the node set S. The β-skeleton of a point set has an edge between two points p, q iff any ball of radius $\beta|pq|/2$ touching p and q is empty of other points. For $\beta = 1$ we obtain the well-known Gabriel graph (there is exactly one ball touching p and q with radius $|pq|/2$). For $\beta > 1$ we get a subgraph thereof (there are always two balls of radius $> |pq|/2$ touching p and q).

First we will show that the graph obtained via the β-skeleton is connected[4] and all internal faces of this graph (when using the obvious straight-line embedding in the plane) have constant complexity. Then we argue that for suitable parameters k and ε, every β-skeleton edge is also present in $CDM(S)$.

β-skeleton is dense. We establish both connectivity as well as bounded face complexity by showing that the β-skeleton for the point set S is a *σ-power-spanner*, more precisely we show that all power distance optimal paths only use edges of the β-skeleton. For the point set S, the *σ-power distance* $d^{\sigma}(p, q)$ between two

[4] In general the β-skeleton need not be connected; in our case it is due to our choice of β and the local uniformity of S.

points $p, q \in S$ is determined by a sequence of points $p = p_0 p_1 \ldots p_l = q$ such that $d^\sigma(p, q) = \sum_{i=0}^{l-1} |p_i p_{i+1}|^\sigma$ is minimal, where $p_i \in S$. Intuitively, the power distance between two points is the minimum amount of energy required to transmit a message between the two points (potentially using intermediate points as relays) assuming that direct communication between two points at distance d has cost d^σ.

Observation 2. *For $\sigma \geq 2$ every $p_i p_{i+1}$ is a Gabriel edge.*

What we want to show is that for a suitable choice of σ dependent on β, all links $p_i p_{i+1}$ in a power-minimal path are even edges of the β-skeleton. We start with a simple observation about the distribution of S.

Lemma 1. *For any pair $s_1, s_2 \in S$ we have $|s_1 s_2| \geq (\alpha - \varepsilon)k$.*

Proof. Follows from the fact that V is an ε-sampling and S being a tight k-sample in a α-qUDG.

The fact that V is an ε-sampling of a disk also implies that all Gabriel (and hence β-skeleton) edges cannot be too long.

Lemma 2. *For any $s_1, s_2 \in S$ such that $|s_1 s_2| > 2(k + \varepsilon)$, $s_1 s_2$ cannot be a Gabriel edge.*

Proof. Assume otherwise. Consider the Gabriel ball of $s_1 s_2$, with center c and radius $r > k + \varepsilon$; due to convexity of a disk and V being a ε-sampling thereof, there must be a point $p \in V$ at distance at most ε from c. But for this p there must exist a $s \in S$ at distance at most k, violating the Gabriel ball property.

Let us now consider one potential edge between nodes p and q which is not part of the β-skeleton and show that this edge cannot be part of any power minimal path.

Lemma 3. *Let $p, q \in S$ but pq not an edge in the β-skeleton. Then if $\sigma \geq \frac{4 \ln 2(k+\varepsilon)}{(\alpha-\varepsilon)k}$ and $\beta \leq \frac{\left(1+\frac{(\alpha-\varepsilon)k}{4(k+\varepsilon)}\right)^2}{2\sqrt{\left(1+\frac{(\alpha-\varepsilon)k}{4(k+\varepsilon)}\right)^2-1}}$, pq cannot be part of any power minimal path.*

Proof. Due to space restrictions we can only provide a very rough sketch. See the authors' homepages for a long version. Due to Observation 2 and Lemma 2, pq cannot be too long. On the other hand, because pq is not part of the β-skeleton, there must be other landmarks inside the two balls of radius $\beta|pq|/2$ touching p and q. Then σ can be chosen such that any power minimal path would rather go via those other landmarks than taking the direct hop pq.

It follows that for the respective choice of σ and β, all power-efficient paths use only edges of the β-skeleton and that the β-skeleton-graph is connected. It remains to show that the graph is somewhat dense, more precisely we want to show that the induced faces (when embedding using the original geometric coordinates) are of constant size.

Lemma 4. *Any bounded face of the β-skeleton is of size* $O\left(\left(\frac{2(k+\varepsilon)}{(\alpha-\varepsilon)k}\right)^{2(\sigma+1)}\right)$.

Proof. Again, we only give a sketch, due to limited space. Fix any bounded face f. Observe that some Delaunay edge $e = (u, v)$ cuts f in a balanced way. Lemmas 1, 2 (a version for Delaunay edges), and 3 upper-bound the hop-distance between u and v in the β-skeleton. A shortest uv-path p need not go along the face boundary, but p and e "enclose" a path p' that does. A packing argument (involving Lemma 1 again) shows that p' cannot be too long either.

At this point we have shown that the β-skeleton of S induces a connected planar graph which has constant-size faces only. It remains to show that all edges of the β-skeleton of S are also identified as adjacencies in $CDM(S)$. To achieve that, observe that any value $\beta > 1$ implies that points $s_1, s_2 \in S$ adjacent in the β-skeleton of S share a "relatively long" Voronoi edge in the Voronoi diagram of S; more precisely s_1 and s_2 share a Voronoi edge of length $|s_1 s_2|\sqrt{\beta^2 - 1}$. Hence it suffices to show that this long Voronoi edge is also reflected in the α-qUDG by respective witness paths. This can be easily achieved by choosing a large enough k (dependent on β). The following corollary follows immediately:

Corollary 1. *The graph induced by S and the adjacencies identified by our algorithm is planar, connected and has (internal) faces of size $O(1)$.*

2.2 Conservative Adjacencies in $\mathbb{R}^3$

All the reasoning so far has been concentrating on a flat, planar setting. Let us now consider the actual setting in $\mathbb{R}^3$. Let V be an ε-sampling of a smooth closed 2-manifold Γ in $\mathbb{R}^3$.

Here the steps of our algorithm are as follows: in step (1) we have to compute a Lipschitz[5] function ϕ with $\phi(p) \leq \varepsilon\,\mathrm{lfs}(p)$ for all $p \in V$. This can be done using the procedure given in [4] in near-linear time. Then in step (2) the graph $G(V, E)$ is constructed by creating edges between samples p_1, p_2 iff $|p_1 p_2| \leq 6 \cdot \phi(p_1)$ or $|p_1 p_2| \leq 6 \cdot \phi(p_2)$ (Note that the constant 6 is somewhat arbitrary and only chosen to make every sample connected to its neighbors in the canonical correct reconstruction). The following steps (3) to (5) are exactly the same as in the 2-dimensional case. We now want to argue that locally around a sample point p the constructed graph looks like an α-quasi-unit-disk graph. To keep the representation simple, we assume that $\phi(p) = \varepsilon\,\mathrm{lfs}(p)$. Although the procedure in [4] yields a ϕ which is a pointwise lower bound for the latter, the argumentation remains the same. The following basic observations are easy to derive (think of $\varepsilon \ll \gamma \ll 1$):

Lemma 5. *Let $p \in \Gamma$ be a point on the surface, T_p the tangent plane at p, $q \in \Gamma$ some point on the surface with $|pq| \leq \gamma\,\mathrm{lfs}(p)$, $\gamma < 1$, q' its orthogonal projection on T_p. Then we have $|qq'| < \gamma^2\,\mathrm{lfs}(p)$.*

[5] In fact one computes a δ-approximate ω-Lipschitz function.

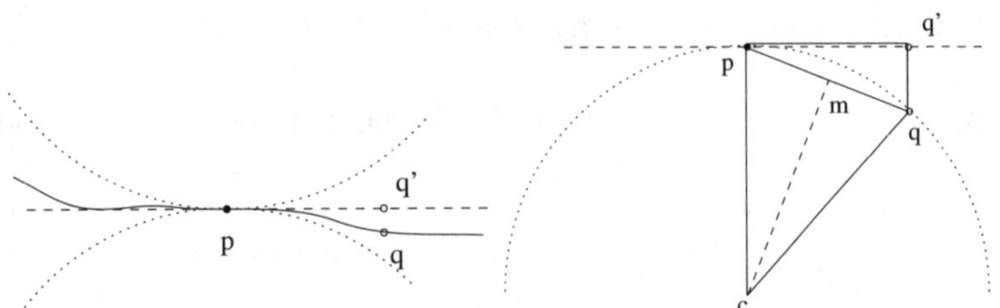

Fig. 1. Sandwiching balls

Proof. Consider the two balls of radius $\mathrm{lfs}(p)$ tangent at p. By definition of the local feature size, no point of Γ is contained in the *interior* of either of the two balls. Let q, q' and T_p be defined as above, see Fig. 1, left. In the worst case, $q \in \Gamma$ lies on one of the two balls tangent at p as in Fig. 1, right. The angle at p in triangle $\Delta pqq'$ is equal to the angle at c in the triangle Δcmp – call this angle θ. Since both these triangles are right-angled, we have that $\sin\theta = |pm|/|pc|| = |qq'|/|pq|$, or in other words $|qq'| = |pm||pq|/|pc|$. But since $|pm| = |pq|/2 \leq \gamma\,\mathrm{lfs}(p)/2$, $|pc| = \mathrm{lfs}(p)$, and $|pq| \leq \gamma\,\mathrm{lfs}(p)$, we obtain $|qq'| \leq \gamma^2\,\mathrm{lfs}(p)/2$.

Observation 3. *Let $p \in \Gamma$ be a point on the surface, T_p the tangent plane at p, $q_1, q_2 \in \Gamma$ two points on the surface with $|pq_i| \leq \gamma\,\mathrm{lfs}(p)$, $\gamma < 1$, q_i' their orthogonal projections on T_p. Then we have $|q_1q_2| - |q_1'q_2'| \leq 2\gamma^2\,\mathrm{lfs}(p)$.*

The latter observation means that when projecting into the tangent plane at p, distances are shortened by at most an additive amount of $2\gamma^2\,\mathrm{lfs}(p)$. We are interested in the largest distance between any two samples s_1, s_2 with $|ps_i| \leq \gamma\,\mathrm{lfs}(p)$ and the smallest (projected!) distance of a *non-adjacent* pair. If the ratio of these two distances is at most $1/\alpha$ we know that locally in a $\gamma\,\mathrm{lfs}(p)$-neighborhood of p the constructed graph is an α-quasi-unit-disk graph.

Lemma 6. *For $\gamma \leq 1/16$ the $\gamma\,\mathrm{lfs}(p)$-neighborhood of p in the constructed graph is an α-quasi-unit-disk graph with $\alpha > 1/\sqrt{2}$.*

Proof. Within a distance of $\gamma\,\mathrm{lfs}(p)$ from p on the surface the local feature size can *increase* by at most $2\gamma^2\,\mathrm{lfs}(p)$ (again a sandwiching argument, as in the previous Lemma). Hence the distance of samples identified as adjacent can increase by a factor of $(1 + 2\gamma^2)$. On the other hand, the local feature size can actually *decrease* by $\gamma\,\mathrm{lfs}(p)$, hence the smallest distance of a non-adjacent pair (taking into account the projection) can be as little as $6\varepsilon\,\mathrm{lfs}(p)(1 - \gamma - 2\gamma^2)$. Therefore the ratio between these two distances is $\frac{1+2\gamma^2}{1-\gamma-2\gamma^2}$. Choosing e.g. $\gamma = 1/16$ makes this ratio less than $\sqrt{2}$.

Now we can invoke Corollary 1 which implies that locally for any $p \in S$ the graph constructed by our algorithm is planar, connected and has internal faces of constant size.

Note: The faces, or more precisely, the local embedding can be simply obtained by reusing the geometry again and locally projecting the adjacent points s_i of a point $s \in S$ into an (almost) tangent plane, then reading off the cyclic order around s. Global connectivity is ensured by choosing γ large enough (compared to ε) such that large faces are completely contained in the $\gamma\,\mathrm{lfs}(p)$-neighborhood of any node bounding the face.

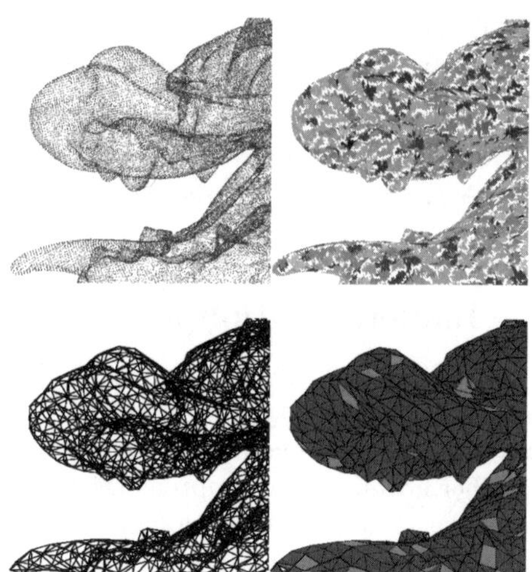

Fig. 2. Main steps of our algorithm: Point cloud, Graph Voronoi Diagram, $CDM(S)$, identified faces

What does this mean? The graph that we constructed on the subsample of points S is a *mesh* that is locally planar and covers the whole 2-manifold. The mesh has the nice property that all its *cells* (aka faces) have constant size (number of bounding vertices). The edge lengths of the created adjacencies between S are proportional to the respective local feature sizes. Therefore its connectivity structure faithfully reflects the topology of the underlying 2-manifold.

Algorithm Epilog. We did not talk about steps (6) and (7) of our approach since they follow exactly the description in [4] and are not novel to this work; we nevertheless give a brief summary here. Essentially, in step (6) we triangulate non-triangular faces by projecting them into a nearby (almost) tangent plane and computing the Delaunay triangulation. The resulting triangulated faces behave nicely since all faces have small size (and hence their vertices are almost coplanar) and because S is a *locally uniform* sampling of the surface. In step (7) the points pruned in step (3) are reinserted by computing a weighted Delaunay triangulation on the supporting planes of the respective faces. The resulting triangulations are guaranteed to patch up.

The proofs for convergence both point-wise as well as with respect to triangle normals can be carried over from [4] since S can be made an arbitrarily good, locally uniform ε'-sampling (the original ε-sampling V has to be accordingly denser, i.e. $\varepsilon \ll \varepsilon'$). Therefore, the same theorem holds for the result of our algorithm:

Theorem 4. *There exists ε^* such that for all $\varepsilon < \varepsilon^*$, smooth surfaces Γ in $\mathbb{R}^3$ and ε-samplings $V \subset \Gamma$, the triangulated surface $\widetilde{\Gamma}$ output by our algorithm satisfies the following conditions:*

1. BIJECTION: *$\mu : \widetilde{\Gamma} \rightarrow \Gamma$, determined by closest point, is a bijection*
2. POINTWISE APPROXIMATION: *For all $x \in \widetilde{\Gamma}$, $d(x, \mu(x)) = O(\varepsilon^2 \operatorname{lfs}(\mu(x)))$*
3. NORMAL APPROXIMATION: *For all $x \in \widetilde{\Gamma}$, $\angle n_{\widetilde{\Gamma}}(x) n_\Gamma(\mu(x)) = O(\varepsilon)$ where $n_F(y)$ denotes the (outside) normal of F at y.*[6]
4. TOPOLOGICAL CORRECTNESS: *Γ and $\widetilde{\Gamma}$ have the same topological type.*

3 Implementation and Experimental Evaluation

We have prototyped the novel steps of our algorithm in C++. This implementation is the topic of the companion paper [3]. This companion paper also highlights another novelty of our approach, namely that most of our computation does *not* use any geometry information: after establishing neighborhood relations between nearby points, the main steps of our algorithm operate combinatorially on a graph structure. As such, they are by far less susceptible to robustness problems due to round-off errors in floating-point arithmetic.

Fig. 3. Output of our implementation for the standard "Dragon" model. Non-triangular faces are denoted in light color.

In Fig. 2 we have visualized the main steps of our algorithms: starting with a point cloud we compute a Graph Voronoi diagram, based on that the $CDM(S)$ and finally we inspect the $CDM(S)$ to identify faces, some of which might be non-triangular (here in light color) due to the conservative edge creation. In Fig. 3 we give a picture of the reconstruction for a complete object (the standard "Dragon" dataset).

[6] For $\widetilde{\Gamma}$ the normal is well-defined in the interior of triangles; at edges and vertices it can be defined as an interpolation from that at the incident triangles.

4 Outlook

Theoretically, our approach has the potential to work for reconstructing 2-manifolds even in higher dimensions. It does not extend to non-2-manifolds, though, as the "local planarity property" of a graph that our algorithm crucially depends upon, has no equivalent for non-2-manifolds.

A more in-depth discussion on the advantages and disadvantages of dropping geometry information early-on can be found in the companion paper to this submission [3]; further studies are required, though, whether this can be extended to a general paradigm when designing geometric algorithms.

In parallel computing or external memory scenarios, it is much easier to obtain efficient algorithms if the performed operations require only *local* access to data. In the former, making non-local data available typically incurs a runtime penalty for the data transfer or more complicated access control mechanisms, in the latter, local data can be cached in internal memory, while non-local data has to be read from external memory again incurring considerable latency. The manifold extraction step as for example employed by the CoCone algorithm is a global, highly non-local operation. It remains to be seen whether the localization property exhibited in this paper leads to practically more efficient algorithms in the parallel computing or external memory scenario.

Acknowledgment

In memory of our dear friend and colleague Martin Kutz.

References

1. Amenta, N., Bern, M.: Surface reconstruction by Voronoi filtering. In: Proc. 14th ACM SoCG. (1998)
2. Amenta, N., Choi, S., Dey, T.K., Leekha, N.: A simple algorithm for homeomorphic surface reconstruction. In: Proc. 16th SoCG (2000)
3. Dumitriu, D., Funke, S., Kutz, M., Milosavljević, N.: How much Geometry it takes to Reconstruct a 2-Manifold in $\mathbb{R}^3$. In: Proc. 10th ACM-SIAM ALENEX, pp. 65–74 (2008)
4. Funke, S., Ramos, E.: Smooth-surface reconstruction in near-linear time. In: Proc. ACM-SIAM SODA (2002)
5. Funke, S., Milosavljević, N.: Network Sketching or: "How Much Geometry Hides in Connectivity? – Part II". In: Proc. ACM-SIAM SODA, pp. 958–967 (2007)
6. Cheng, S.W., Dey, T., Ramos, E.: Manifold reconstruction from point samples. In: ACM-SIAM SODA, pp. 1018–1027 (2005)
7. Kirkpatrick, D.G., Radke, J.D.: A framework for computational morphology. In: Toussaint, G.T. (ed.) Computational Geometry, pp. 217–248. North-Holland, Amsterdam (1985)

On Metric Clustering to Minimize the Sum of Radii*

Matt Gibson, Gaurav Kanade, Erik Krohn, Imran A. Pirwani, and Kasturi Varadarajan**

Department of Computer Science,
University of Iowa, Iowa City, IA 52242-1419, USA
{mrgibson,gkanade,eakrohn,pirwani,kvaradar}@cs.uiowa.edu

Abstract. Given an n-point metric (P, d) and an integer $k > 0$, we consider the problem of covering P by k balls so as to minimize the sum of the radii of the balls. We present a randomized algorithm that runs in $n^{O(\log n \cdot \log \Delta)}$ time and returns with high probability the optimal solution. Here, Δ is the ratio between the maximum and minimum interpoint distances in the metric space. We also show that the problem is NP-hard, even in metrics induced by weighted planar graphs and in metrics of constant doubling dimension.

Keywords: k-clustering, k-cover, clustering, metric clustering, planar metric, doubling metric.

1 Introduction

Given a metric d defined on a set P of n points, we define the ball $\mathrm{B}(v, r)$ centered at $v \in P$ and having radius $r \geq 0$ to be the set $\{q \in P | d(v, q) \leq r\}$. In this work, we consider the problem of computing a minimum cost k-cover for the given point set P, where $k > 0$ is some given integer which is also part of the input. For $\kappa > 0$, a κ-cover for subset $Q \subseteq P$ is a set of at most κ balls, each centered at a point in P, whose union covers (contains) Q. The cost of a set $\mathcal{D}$ of balls, denoted $\mathrm{cost}(\mathcal{D})$, is the sum of the radii of those balls.

This problem and its variants have been well examined, motivated by applications in clustering and base-station coverage [6,4,13,3,1].

Doddi et al. [6] consider the metric min-cost k-cover problem and the closely related problem of partitioning P into a set of k clusters so as to minimize the sum of the cluster diameters. Following their terminology, we will call the latter problem *clustering to minimize the sum of diameters.* They present a bicriteria poly-time algorithm that returns $O(k)$ clusters whose cost is within a multiplicative factor $O(\log(n/k))$ of the optimal. For clustering to minimize the sum of diameters, they also show that the existence of a polynomial time

* Work by the first, second, third, and fifth authors was partially supported by NSF CAREER award CCR 0237431.

** Part of this work was done while the author was visiting the Institute for Mathematical Sciences, Chennai, India.

J. Gudmundsson (Ed.): SWAT 2008, LNCS 5124, pp. 282–293, 2008.

algorithm that returns k clusters whose cost is strictly within 2 of the optimal would imply that $P = NP$. Notice that this hardness result does not imply the NP-hardness of the k-cover problem. Charikar and Panigrahy [4] give a poly-time algorithm based on the primal-dual method that gives a constant factor approximation – around 3.504 – for the k-cover problem, and thus also a constant factor approximation for clustering to minimize the sum of diameters.

The well known k-center problem is a variant of the k-cover problem where the cost of a set of balls is defined to be the maximum radius of any ball in the set. The problem is NP-hard and admits a polynomial time algorithm that yields a 2-approximation [10]. Several other formulations of clustering such as k-median and min-sum k-clustering are NP-hard as well [11,5].

Recently, Gibson et al. [9] consider the geometric version of the k-cover problem where $P \subset \Re^l$ for some constant l. When the L_1 or L_∞ norm is used to define the metric, they obtain a polynomial time algorithm for the k-cover problem. With the L_2 norm, they give an algorithm that runs in time polynomial in n, the number of points, and in $\log(1/\epsilon)$ and returns a k-cover whose cost is within $(1+\epsilon)$ of the optimal, for any $0 < \epsilon < 1$.

1.1 Our Results

Our first result generalizes the algorithmic approach of Gibson et al. [9] to the metric case. For the k-cover problem in the general metric setting, we obtain an exact algorithm whose running time is $n^{O(\log n \cdot \log \Delta)}$, where Δ is the *aspect ratio* of the metric space, the ratio between the maximum interpoint distance and the minimum interpoint distance. The algorithm is randomized and succeeds with high probability. Thus when Δ is bounded by a polynomial in n, the running time of the algorithm is quasi-polynomial. This result for the k-cover problem should be contrasted with the NP-hardness results for problems such as k-center, k-median, and min-sum k-clustering, which hold when the aspect ratio is bounded by a polynomial in n.

The main idea that underlies this result is that if we probabilistically partition the metric into sets with at most half the original diameter [2,7], then with high probability only $O(\log n)$ balls in the optimal k-cover of P are "cut" by the partition. A recursive approach is then used to compute the optimal k-cover.

This algorithmic result raises the question of whether an algorithm whose running time is quasi-polynomial in n is possible even when the aspect ratio is not polynomially bounded. Our second result shows that this is unlikely by establishing the NP-hardness of the k-cover problem. The aspect ratio in the NP-hardness construction is about 2^n. The metrics obtained are induced by weighted planar graphs, thus establishing the NP-hardness of the k-cover problem for this special case.

Our final result is that the k-cover problem is NP-hard in metrics of constant doubling dimension for a large enough constant. This result is somewhat surprising given the positive results of [9] for fixed dimensional geometric spaces.

Before concluding this section, we point out that our algorithmic result for the metric k-cover problem readily yields a randomized approximation algorithm that runs in time $2^{\text{polylog}(n/\epsilon)}$ and returns with high probability a k-cover whose cost is at most $(1+\epsilon)$ times the cost of the optimal k-cover. This approximation algorithm is obtained by applying a simple transformation (involving discretization) that reduces the approximate problem to several instances of the exact metric κ-cover problem with aspect ratio bounded by poly(n/ϵ).

The rest of this article is organized as follows. In Section 2, we present our algorithm for the k-cover problem. In Section 3, we establish the NP-hardness of the k-cover problem for metrics induced by weighted planar graphs. In Section 4, we establish NP-hardness for metrics of constant doubling dimension.

2 Algorithm for General Metrics

We consider the k-cover problem whose input is a metric (P, d), where P is a set of n points and d is a function giving the interpoint distances, and an integer $k > 0$. We assume without loss of generality that the minimum interpoint distance is 1. Let Δ denote diam(P), the maximum interpoint distance. We present a randomized algorithm that runs in $n^{O(\log n \log \Delta)}$ time and with high probability returns the best k-cover for P. We will assume below that $k \leq n$.

The main idea for handling the metric case is that probabilistic partitions [2,7] can play a role analogous to the line separators were used in the geometric case [9]. To formalize this, let Q denote some subset of P such that diam$(Q) \geq 50$, and consider the following randomized algorithm (taken from [7]) that partitions Q into sets of diameter at most diam$(Q)/2$:

Algorithm 1. Partition(Q)

```
Let π denote a random permutation of the points in Q.
Let β denote a random number in the range [diam(Q)/8, diam(Q)/4].
Let R ← Q.
for all i ← 1 to |Q| do
    Let Q_i ← {p ∈ R | d(p, π(i)) ≤ β}.
    Let R ← R \ Q_i.
```

Since each Q_i is contained in a ball of radius at most diam$(Q)/4$, we have that diam$(Q_i) \leq$ diam$(Q)/2$. Clearly, the Q_i also partition Q. Let us say that a ball $B \subseteq P$ is *cut* by this partition of Q if there are two distinct indices i and j such that $(B \cap Q) \cap Q_i \neq \emptyset$ and $(B \cap Q) \cap Q_j \neq \emptyset$. The main property that the probabilistic partition enjoys is encapsulated by the following lemma, whose proof follows via the methods of Fakcharoenphol et al. [7].

Lemma 1. *Let $B \subseteq P$ be some ball of radius r. The probability that B is cut by the partition of Q output by Partition(Q) is at most $\frac{r}{diam(Q)} O(\log |Q|)$.*

Proof. Let $q_1, \ldots q_{|Q|}$ denote the ordering of the points in Q according to increasing order of distance from $B' = B \cap Q$, with ties broken arbitrarily. We may assume that $B' \neq \emptyset$ for otherwise the lemma trivially holds. For each q_j let a_j (resp. b_j) denote the distance to the closest (resp. furthest) point in B'. By the triangle inequality it follows that $b_j - a_j \leq 2r$. We say that $\pi(i)$ *settles* B if i is the first index for which some point in B' belongs to Q_i. Note that exactly one point in Q settles B. We say that $\pi(i)$ *cuts* B if $\pi(i)$ settles B and at least one point in B' is not assigned to Q_i. The probability that B is cut by the partition equals

$$\sum_i \Pr[\pi(i) \text{ cuts } B] = \sum_j \Pr[q_j \text{ cuts } B].$$

The event that q_j cuts B requires the occurrence of two events: E_1, the event that β lands in the interval $[a_j, b_j)$, and E_2, the event that q_j appears before $q_1, \ldots, q_{j-1}$ in the ordering π. Using independence,

$$\begin{aligned} \Pr[q_j \text{ cuts } B] &\leq \Pr[E_1] * \Pr[E_2|E_1] = \Pr[E_1] * \Pr[E_2] \\ &\leq \frac{2r}{\text{diam}(Q)/8} \cdot \frac{1}{j} = \frac{16r}{\text{diam}(Q)} \cdot \frac{1}{j}. \end{aligned}$$

So the probability that B is cut by the partition is bounded above by

$$\frac{16r}{\text{diam}(Q)} \sum_j \frac{1}{j} = \frac{r}{\text{diam}(Q)} O(\log |Q|).$$

□

Let S denote the optimal κ-cover for Q some $\kappa > 0$. The following states the main structural property that S enjoys.

Lemma 2. *The expected number of balls in S that are cut by Partition(Q) is $O(\log |Q|)$. Consequently, the probability is at least $1/2$ that the number of balls in S that are cut by Partition(Q) is at most $c \log n$, where $c > 0$ is some constant.*

Proof. The expected number of balls in S cut is equal to

$$\sum_{B \in S} \Pr[\text{B is cut}] = O(\log |Q|) \sum_{B \in S} \frac{\text{radius}(B)}{\text{diam}(Q)} = O(\log |Q|) \frac{\text{cost}(S)}{\text{diam}(Q)}.$$

The Lemma follows by observing that $\text{cost}(S) \leq \text{diam}(Q)$ since Q can be covered by a single ball of radius $\text{diam}(Q)$. □

2.1 The Randomized Algorithm

We describe a recursive algorithm `BC-Compute` that takes as arguments a set $Q \subseteq P$ and an integer $0 \leq \kappa \leq n$ and returns with high probability an optimal κ-cover for Q. We begin by noting that we may restrict our attention to balls $B(x, r)$ whose radius r equals $d(x, q)$ for some $q \in P$. Henceforth in this section we only refer to this set of balls. For easing the description of the algorithm, it is convenient to add to this set of balls an element $\mathcal{I}$ whose cost is ∞. Any subset of this enlarged set of balls that includes $\mathcal{I}$ will also have a cost of ∞.

Algorithm 2. BC-Compute(Q, κ)

```
If |Q| = 0, return the empty set.
Otherwise, if κ = 0, return {I} (not possible to cover).
Otherwise, if diam(Q) ≤ 50, directly compute the optimal solution in polynomial
  time. In this case, the optimal solution has cost at most 50, so it consists of a set S
  of at most 50 balls of non-zero radius plus zero or more singleton balls. The number
  of such solutions is polynomial, and our algorithm checks them all.
for all 2 log_2 n iterations do
  Call Partition(Q) to obtain a partition of Q into two or more sets. Let
  Q_1, ..., Q_τ denote the nonempty sets in this collection.
  for all sets C of at most c log n balls, where c is the constant in Lemma 2 do
    Let Q'_i be the points in Q_i not covered by C. For each 1 ≤ i ≤ τ and 0 ≤
    κ_1 ≤ κ, recursively call BC-Compute(Q'_i, κ_1) and store the set returned in the
    local variable best(Q'_i, κ_1).
    For 0 ≤ i ≤ τ − 1, let R_i = ∪_{j=i+1}^{τ} Q'_j. Note that R_{τ−1} = Q'_τ and R_i =
    Q'_{i+1} ∪ R_{i+1} for 0 ≤ i ≤ τ − 2.
    for all i ← τ − 2 down to 0 and 0 ≤ κ_1 ≤ κ, do
      set local variable best(R_i, κ_1) to be the lowest cost solution among
      {best(Q'_{i+1}, κ') ∪ best(R_{i+1}, κ_1 − κ') | 0 ≤ κ' ≤ κ_1}.
      Let S denote the lowest cost solution best(R_0, κ − |C|) ∪ C over all choices
      of C tried above with |C| ≤ κ.
Return the lowest cost solution S obtained over the Θ(log n) trials.
```

Running time. To solve an instance (Q, κ) of the problem with $\mathrm{diam}(Q) \geq 50$, the algorithm makes $n^{O(\log n)}$ recursive calls to instances with diameter at most $\mathrm{diam}(Q)/2$. The additional book keeping takes $n^{O(\log n)}$ time. It follows that the running time of the algorithm invoked on the original instance (P, k) is $n^{O(\log n \cdot \log \Delta)}$.

Correctness. We will show that BC-Compute(P, k) computes an optimal k-cover for P with high probability. We begin by noting that the base case instances (Q, κ) are solved correctly with a probability of 1. We will show by induction on $|Q|$ that any instance (Q, κ) with $|Q| \geq 2$ is optimally solved with a probability of at least $1 - \frac{|Q|-1}{n^2}$.

If the (Q, κ) instance happens to fit in one of the base cases, we are done. Otherwise, consider an optimal κ-cover OPT for Q. It is enough to show that BC-Compute(Q, κ) returns a κ-cover of cost at most cost(OPT) with a probability of at least $1 - \frac{|Q|-1}{n^2}$.

By Lemma 2, the probability is at least $1 - \frac{1}{n^2}$ that one of the $2 \log_2 n$ calls to Partition(Q) returns a partition $(Q_1, \ldots, Q_\tau)$ of Q into $\tau \geq 2$ sets such that no more than $c \log n$ balls in OPT are cut by the partition. Assuming this good event happens, fix such a partition $(Q_1, \ldots, Q_\tau)$ of Q and consider the choice of C that exactly equals the balls in OPT that are cut by the partition. The balls in $\mathrm{OPT} \setminus C$ are not cut by the partition and can be partitioned into subsets $(\mathrm{OPT}_1, \ldots, \mathrm{OPT}_\tau)$ (some of these can be empty) such that for any ball $B \in \mathrm{OPT}_i$, we have $B \cap Q \subseteq Q_i$. It is easy to see that OPT_i must be an optimal $|\mathrm{OPT}_i|$-cover for Q'_i. By the induction hypothesis, BC-Compute(Q'_i, $|\mathrm{OPT}_i|$) returns an $|\mathrm{OPT}_i|$-cover for Q'_i with a probability of at least $1 - \frac{|Q'_i|-1}{n^2}$ if $|Q'_i| \geq 2$ and with a probability of 1 otherwise. The probability that BC-Compute(Q'_i, $|\mathrm{OPT}_i|$) returns an $|\mathrm{OPT}_i|$-cover for Q'_i for every i is at least

$$\prod_{i:|Q'_i|\geq 2} 1 - \frac{|Q'_i|-1}{n^2} \geq \prod_i 1 - \frac{|Q_i|-1}{n^2} \geq 1 - \frac{|Q|-2}{n^2}.$$

Assuming this second good event also happens, it follows from an easy backwards induction on i that $\mathrm{best}(R_i, \sum_{j>i}|\mathrm{OPT}_j|)$ is a $(\sum_{j>i}|\mathrm{OPT}|_j)$-cover for R_i with cost at most $\sum_{j>i}\mathrm{cost}(\mathrm{OPT}_j)$. Thus $\mathrm{best}(R_0, \kappa - |C|)$ is an $(\kappa - |C|)$-cover for $R_0 = \sum_{i=1}^{\tau} Q'_i$ with cost at most $\sum_{i=1}^{\tau}\mathrm{cost}(\mathrm{OPT}_i)$. Thus $\mathrm{best}(R_0, \kappa - |C|) \cup C$ is a κ-cover of Q with cost at most $\mathrm{cost}(\mathrm{OPT})$. The probability of this happening is at least the product of the probabilities of the two good events we assumed, which is at least $(1 - \frac{|Q|-1}{n^2})$. This completes the inductive step, because BC-Compute(Q, κ) returns the lowest cost κ-cover among the $2\log_2 n$ κ-covers that it sees.

Theorem 1. *There is a randomized algorithm that, given a set P of n points in a metric space and an integer k, runs in $n^{O(\log n \cdot \log \Delta)}$ time and returns, with probability at least 1/2, an optimal k-cover for P. Here Δ is an upper bound on the ratio between the maximum and minimum interpoint distances within P.*

3 NP-Hardness of Min-Cost k-Cover

A natural question is whether there is a quasipolynomial time algorithm in n for the case where the input metric has unbounded aspect ratio. This is unlikely to be the case because, as we show in this section, the general problem is NP-hard even in case of a planar metric. We give a reduction from a version of the planar 3-SAT problem - the *pn-planar* 3-SAT problem. This problem was shown to be NP-complete in [14]. Planar 3-SAT is defined as follows: Let $\Phi = (X, C)$ be an instance of 3SAT, with variable set $X = \{x_0, \ldots, x_{n-1}\}$ and clauses $C = \{c_1, \ldots, c_m\}$ such that each clause consists of exactly 3 literals. Define a *formula graph* $G_\Phi = (V, E)$ with vertex set $V = X \bigcup C$ and edges $E = E_1 \bigcup E_2$ where $E_1 = \{(x_i, x_{i+1}) | 0 \leq i \leq n-1\}$[1] and $E_2 = \{(x_i, c_j) | c_j \text{ contains } x_i \text{ or } \overline{x_i}\}$. A 3SAT formula Φ is called *planar* if the corresponding formula graph G_Φ is planar. The edge set E_1 defines a cycle on the vertices X, and thus divides the plane into exactly 2 faces. Each node $c_j \in C$ lies in exactly one of those two faces. In the *pn-planar* 3SAT problem, we have the additional restriction that there exists a planar drawing of G_Φ such that if c_j and $c_{j'}$ contain opposite occurrences of the same variable x_i, then they lie in opposite faces. In other words, all clauses with the literals x_i lie in one of the two faces and all clauses with $\overline{x_i}$ lie in the other face. We have to determine whether there exists an assignment of truth values to the variables in X that satisfies all the clauses in C.

[1] Here we assume that the arithmetic wraps around i.e. $(n-1)+1 = 0$.

We describe a simple transformation, easily seen to be effected by a polynomial time algorithm, from such a *pn-planar* 3SAT instance to an instance of finding an optimal k-cover in a metric induced by a weighted planar graph $G = (V, E)$. The transformation has the property that there is a k-cover in the metric of cost at most $2^k - 1$ if and only if the original *pn-planar* 3SAT instance is satisfiable.

We set $k = n$. The vertex set V of the graph is a union of $k + 2$ sets: (a) a set $X = \{x_0, \overline{x_0}, \ldots, x_{k-1}, \overline{x_{k-1}}\}$ that can be identified with the set of variables of the *pn-planar* 3SAT instance with each variable occurring twice - once as a positive literal and once as a negative literal, (b) a set $C = \{c_1, \ldots, c_m\}$ that can be identified with the set of clauses of the *pn-planar* 3SAT instance, and (c) sets $W^0, \ldots, W^{k-1}$, where each W^l consists of $k+1$ vertices. To obtain the edge set E, we add an edge between each vertex x_l and $\overline{x_l}$ in X with weight 2^l for $0 \leq l \leq k-1$. For each vertex $x_l \in X$ we add an edge between x_l and every vertex in W^l of weight 2^l for $0 \leq l \leq k-1$. Analogously, we add an edge between each vertex $\overline{x_l}$ and every vertex in W^l again of weight 2^l. In addition we add edges between every vertex $c_i \in C$ and every variable vertex x_l or its negation $\overline{x_l}$ whichever appears in it of weight 2^l. Note that this graph G is planar – this follows from the pn-planarity of the 3SAT instance. See Figure 1 for an illustration.

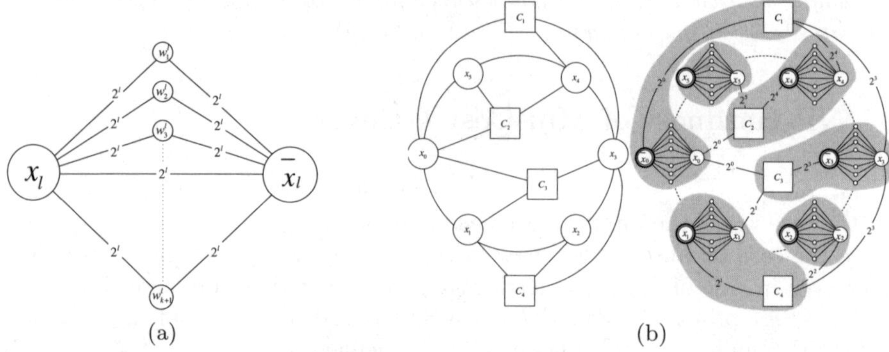

Fig. 1. (a) The gadget for variable x_l in Φ. (b) A planar embedding for Φ and construction of the corresponding instance of k-clustering problem. All "clause-literal" edges have weight 2^l for the variable x_l. The optimal cover is highlighted with grey "blobs". $\Phi = (\neg x_0 \vee x_3 \vee x_4) \wedge (x_0 \vee \neg x_4 \vee \neg x_5) \wedge (x_0 \vee \neg x_1 \vee \neg x_3) \wedge (x_1 \vee \neg x_2 \vee x_3)$. Satisfying assignment $X = (0, 1, 1, 0, 0, 1)$. Weight of the covering is exactly $2^6 - 1$.

Claim. Any k-cover of V whose cost is at most $2^k - 1$ includes, for each $0 \leq l \leq k - 1$, a ball centered at either x_l or $\overline{x_l}$ with radius at least 2^l.

Proof. Consider any k-cover of V and let t be the largest index such that there is no ball in the k-cover centered at either x_t or $\overline{x_t}$ and having radius at least 2^t. So for each $t + 1 \leq l \leq k - 1$, there is a ball B_l in the k-cover centered at either x_l or $\overline{x_l}$ and having radius at least 2^l. Since W^t has $k + 1$ points in it, there is point $a \in W^t$ that is not the center of any ball in the k-cover. Let B be some ball in the k-cover that covers a. If $B = B_l$ for some $t+1 \leq l \leq k-1$, then B_l has radius at least $2^l + 2 \cdot 2^t$. In this case the k-cover has cost at least $2^{k-1} + 2^{k-2} \cdots 2^{t+1} + 2 \cdot 2^t = 2^k$. If $B \neq B_l$ for any $t+1 \leq l \leq k-1$, then the radius of B is at least $2 \cdot 2^t$, since the distance of a from any point other than x_t and $\overline{x_t}$ is at least $2 \cdot 2^t$. Thus in this case too the k-cover has cost at least $2^{k-1} + 2^{k-2} \cdots 2^{t+1} + 2 \cdot 2^t = 2^k$. □

Now suppose the original *pn-planar* 3SAT instance is a yes instance. So there is an assignment of truth values to $x_0, \cdots, x_{k-1}$ such that all clauses in C are satisfied. Consider the set of k balls $B_0, \ldots, B_{k-1}$, where B_l is centered at x_l or $\overline{x_l}$ (whichever is satisfied by the assignment) and has radius 2^l. It is easily checked that these balls form a k-cover of V of cost $2^0 + 2^1 + \cdots 2^{k-1} = 2^k - 1$.

Now suppose the original *pn-planar* 3SAT instance is a no instance. We claim that any k-cover of V has cost strictly greater than $2^k - 1$ in this case. Suppose this is not the case and consider a k-cover of cost at most $2^k - 1$. As a consequence of the claim, such a k-cover must consist of balls $B_0, \ldots, B_{k-1}$ where B_l is centered at either x_l or $\overline{x_l}$ and has radius precisely 2^l. Since these balls must cover each vertex in C, it follows that the assignment of truth values to variables in X which comprises of x_l being true if the ball B_l is centered at x_l and false if it is centered at $\overline{x_l}$ satisfies all clauses in C. This contradicts the supposition that the original *pn-planar* 3SAT instance is a no instance.

Theorem 2. *The (decision version of the) problem of computing an optimal k-cover for an n-point planar metric (P, d) is NP-hard.*

4 The Doubling Metric Case

We now consider the k-cover problem when the input metric (P, d) has doubling dimension bounded by some constant $\rho \geq 0$. The doubling dimension of the metric (P, d) is said to be bounded by ρ if any ball $B(x, r)$ in (P, d) can be covered by 2^ρ balls of radius $r/2$ [12]. In this section, we show that for a large enough constant ρ, the k-cover problem for metrics of doubling dimension at most ρ is NP-hard.

The proof is by a reduction from a restricted version of 3SAT where each variable appears in at most 5 clauses [8]. Let Φ be such a 3-CNF formula with variables $x_0, \ldots, x_{n-1}$ and clauses $c_1, \ldots, c_m$. We describe a simple transformation, easily seen to be effected by a polynomial time algorithm, from such a 3SAT instance Φ to an instance of finding an optimal k-cover in a metric induced by a weighted graph $G = (V, E)$. The metric will have doubling dimension bounded by some constant. The transformation has the property that there is a k-cover in the metric of cost at most $2^k - 1$ if and only if the original 3SAT instance is satisfiable.

The transformation is similar to the one in the previous section with some modifications to ensure the doubling dimension property.

We set $k = n$. The vertex set V of the graph is a union of $k+2$ sets: (a) a set $X = \{x_0, \overline{x_0}, \ldots, x_{k-1}, \overline{x_{k-1}}\}$ that can be identified with the set of literals in

Φ, (b) a set $C = \{c_1, \ldots, c_m\}$ that can be identified with the set of clauses of Φ, and (c) sets $W^0, \ldots, W^{k-1}$, where each W^l consists of $n_l = 8(l+1)^2+1$ vertices $w_1^l, \ldots, w_{n_l}^l$. To obtain the edge set E, we add an edge between x_l and $\overline{x_l}$ with weight 2^l for $0 \leq l \leq k-1$. We add an edge between x_l and every vertex in W^l of weight 2^l for $0 \leq l \leq k-1$. Analogously, we add an edge between $\overline{x_l}$ and every vertex in W^l again of weight 2^l. In addition we add edges between every vertex $c_i \in C$ and every literal that appears in the clause c_i. If the literal is either x_l or $\overline{x_l}$, the weight of the corresponding edge is 2^l. Finally for each $0 \leq l \leq n-1$ and each $1 \leq i \leq n_l - 1$, we add an edge of weight $2^l/(l+1)^2$ between w_i^l and w_{i+1}^l. See Figure 2 for an illustration of the transformation.

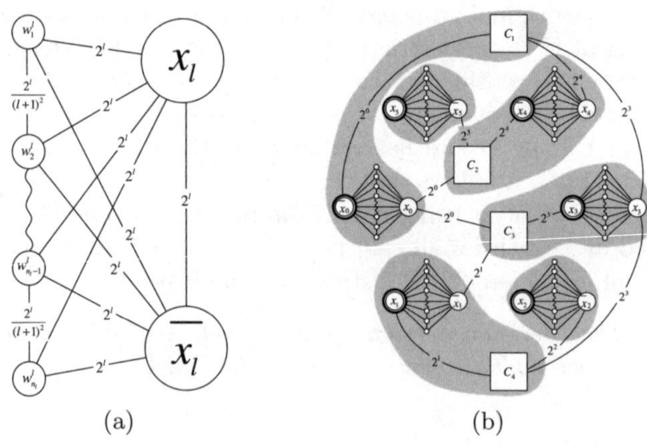

Fig. 2. (a) The gadget for the variable x_l in Φ. Each edge between w_i^l and w_{i+1}^l has weight exactly $2^l/(l+1)^2$ and the number of w_i^l's is $8(l+1)^2+1$. (b) A representation of an instance of k-clustering on a doubling metric constructed from an instance of Φ. All "clause-literal" edges have weight 2^l for variable x_l. The optimal cover is highlighted with grey "blobs". $\Phi = (\neg x_0 \vee x_3 \vee x_4) \wedge (x_0 \vee \neg x_4 \vee \neg x_5) \wedge (x_0 \vee \neg x_1 \vee \neg x_3) \wedge (x_1 \vee \neg x_2 \vee x_3)$. Satisfying assignment $X = (0, 1, 1, 0, 0, 1)$. Weight of the covering is exactly $2^6 - 1$.

Lemma 3. *There is a constant $\rho \geq 0$ so that the doubling dimension of the metric induced by the graph $G = (V, E)$ is bounded by ρ.*

Proof. Let $B(x, r)$ be some ball in the metric. If $r < 1$, then either (a) the ball consists of a singleton vertex, or (b) $B(x, r) \subseteq W^l$ for some l and the subgraph of G induced by $B(x, r)$ is a path. In either case, it is easily verified that $O(1)$ balls centered within $B(x, r)$ and having radius $r/2$ cover $B(x, r)$.

We therefore consider the case $r \geq 1$. Let t be the largest integer that is at most $n-1$ such that $2^t \leq r$. For each $s \in \{t-3, t-2, t-1, t\}$, we place balls of radius $r/2$ centered at (i) $\{x_s, \overline{x_s}\} \cap B(x, r)$, (ii) clause vertices incident to x_s or $\overline{x_s}$ that are in $B(x, r)$, and (iii) $O(1)$ points of $B(x, r) \cap W^s$ so that these balls cover $B(x, r) \cap W^s$ (this is possible because $B(x, r) \cap W^s$ induces a path of length at most 2^{s+3}.) In addition, if $x \in W^l$ for some l, we place $O(1)$ balls of radius $r/2$ at points of $B(x,r) \cap W^l$ so that these balls cover $B(x,r) \cap W^l$. Finally, we place a ball of radius $r/2$ at x. Clearly, we have placed $O(1)$ balls and we will show that these cover $B(x,r)$. Let C denote the set of centers at which we have placed balls.

Let $y \in B(x,r)$ be a point that is not in C or in W^s for $s \in \{t-3, t-2, t-1, t\}$ or in W^l (if $x \in W^l$). Fix a shortest path from x to y and let x' be the last vertex on this path that is in C. We first observe that none of the internal vertices on the path from x to y is in W^q for any q. Furthermore, if $x \in W^l$ for some l, then by assumption $y \notin W^l$. Thus all edges of the subpath from x' to y have weight 2^q for some $0 \leq q \leq n-1$. No such edge can have weight 2^{t+1} or greater because $2^{t+1} > r$ if $t \leq n-2$. No such edge can have weight 2^s for $s \in \{t-3, t-2, t-1, t\}$ because otherwise the endpoint of the edge closer to y would be in C. Thus every edge on the subpath from x' to y has weight at most 2^{t-4}. It is easy to see that the subpath contains at most 3 edges of weight 2^q for any $q \leq t-4$. Thus the weight of the subpath from x' to y is at most

$$3(2^{t-4} + 2^{t-5} + \cdots + 2^0) < 3 \cdot 2^{t-3} < 2^{t-1} < r/2.$$

So y is in the ball of radius $r/2$ centered at x'. □

Claim. Any k-cover of V whose cost is at most $2^k - 1$ includes, for each $0 \leq l \leq k-1$, a ball centered at either x_l or $\overline{x_l}$ with radius at least 2^l.

Proof. Consider any k-cover of V and let t be the largest index such that there is no ball in the k-cover centered at either x_t or $\overline{x_t}$ and having radius at least 2^t. So for each $t+1 \leq l \leq k-1$, there is a ball B_l in the k-cover centered at either x_l or $\overline{x_l}$ and having radius at least 2^l.

If some point in W^t is covered by some B_l for $t+1 \leq l \leq k-1$, then B_l has radius at least $2^l + 2 \cdot 2^t$. In this case the k-cover has cost at least $2^{k-1} + 2^{k-2} \ldots 2^{t+1} + 2 \cdot 2^t = 2^k$. If some point in W^t is covered by a ball B different from the B_l's and not centered at any of the points in W^t, then the radius of B is at least $2 \cdot 2^t$. (Note that by assumpion B can't be centered at x_t or $\overline{x_t}$.) Thus in this case too the k-cover has cost at least $2^{k-1} + 2^{k-2} \ldots 2^{t+1} + 2 \cdot 2^t = 2^k$.

The only remaining case is when each point in W^t is covered by some ball centered at a point in W^t. Since there can be at most $t+1$ balls in the k-cover centered within W^t, the sum of the radii of these balls is at least

$$\frac{1}{2}\left((n_t - 1)\frac{2^t}{(t+1)^2} - (t+1)\frac{2^t}{(t+1)^2}\right) > 2 \cdot 2^t.$$

The k-cover has cost at least $2^{k-1} + 2^{k-2} \ldots 2^{t+1} + 2 \cdot 2^t = 2^k$. □

We now argue that the transformation has the property that there is a k-cover in the metric of cost at most $2^k - 1$ if and only if the original 3SAT instance Φ is satisfiable.

Suppose that Φ is satisfiable. Then we can choose for each $0 \leq l \leq k-1$ exactly one of x_l or $\overline{x_l}$ such that within each clause of Φ there is a chosen literal. Consider the set of k balls $B_0, \ldots, B_{k-1}$ where B_l has radius 2^l and is centered at x_l or $\overline{x_l}$, whichever was chosen. These balls form a k-cover of V with cost $2^k - 1$.

For the reverse direction, consider a k-cover of the target metric space of cost at most $2^k - 1$. It follows from Claim 4 that the k-cover must consist of balls $B_0, \ldots, B_{k-1}$, where B_l is centered at either x_l or $\overline{x_l}$ and has radius precisely 2^l. Let us choose the literals corresponding to the centers of these balls. For each l, we clearly choose exactly one of x_l of $\overline{x_l}$. Consider any clause vertex c. It must be covered by at least one of the balls B_l. Given the radii of the balls, the only balls that can cover c are the ones centered at literals contained in the clause. It follows that our set of chosen literals contains, for each clause in Φ, at least one of the literals contained in the clause. Thus Φ is satisfiable.

Theorem 3. *For a large enough constant $\rho \geq 0$, the (decision version of the) k-cover problem for metrics of doubling dimension at most ρ is NP-hard.*

Acknowledgements

We thank Chandra Chekuri for his suggestion to study the problem.

References

1. Alt, H., Arkin, E.M., Brönnimann, H., Erickson, J., Fekete, S.P., Knauer, C., Lenchner, J., Mitchell, J.S.B., Whittlesey, K.: Minimum-cost coverage of point sets by disks. In: Amenta, N., Cheong, O. (eds.) Symposium on Computational Geometry, pp. 449–458. ACM, New York (2006)
2. Bartal, Y.: Probabilistic approximations of metric spaces and its algorithmic applications. In: FOCS, pp. 184–193 (1996)
3. Bilò, V., Caragiannis, I., Kaklamanis, C., Kanellopoulos, P.: Geometric clustering to minimize the sum of cluster sizes. In: Brodal, G.S., Leonardi, S. (eds.) ESA 2005. LNCS, vol. 3669, pp. 460–471. Springer, Heidelberg (2005)
4. Charikar, M., Panigrahy, R.: Clustering to minimize the sum of cluster diameters. J. Comput. Syst. Sci. 68(2), 417–441 (2004)
5. de la Vega, W.F., Kenyon, C.: A randomized approximation scheme for metric max-cut. J. Comput. Syst. Sci. 63(4), 531–541 (2001)
6. Doddi, S., Marathe, M.V., Ravi, S.S., Taylor, D.S., Widmayer, P.: Approximation algorithms for clustering to minimize the sum of diameters. Nord. J. Comput. 7(3), 185–203 (2000)
7. Fakcharoenphol, J., Rao, S., Talwar, K.: A tight bound on approximating arbitrary metrics by tree metrics. In: STOC, pp. 448–455. ACM, New York (2003)
8. Garey, M.R., Johnson, D.S.: Computers and Intractability: A Guide to the Theory of NP-Completeness. W. H. Freeman, New York (1979)
9. Gibson, M., Kanade, G., Krohn, E., Pirwani, I.A., Varadarajan, K.: On clustering to minimize the sum of radii. In: SODA, pp. 819–825. SIAM, Philadelphia (2008)
10. Hochbaum, D.S., Shmoys, D.B.: A best possible approximation algorithm for the k-center problem. Math. Oper. Res. 10, 180–184 (1985)

11. Kariv, O., Hakimi, S.L.: An algorithmic approach to network location problems. part II: The p-medians. SIAM J. Appl. Math. 37, 539–560 (1982)
12. Krauthgamer, R., Lee, J.R.: Navigating nets: simple algorithms for proximity search. In: SODA, pp. 798–807. SIAM, Philadelphia (2004)
13. Lev-Tov, N., Peleg, D.: Polynomial time approximation schemes for base station coverage with minimum total radii. Computer Networks 47(4), 489–501 (2005)
14. Lichtenstein, D.: Planar formulae and their uses. SIAM Journal on Computing 11(2), 329–343 (1982)

On Covering Problems of Rado

Sergey Bereg[1], Adrian Dumitrescu[2,⋆], and Minghui Jiang[3,⋆⋆]

[1] Department of Computer Science, University of Texas at Dallas, Box 830688, Richardson, TX 75083, USA
besp@utdallas.edu

[2] Department of Computer Science, University of Wisconsin–Milwaukee, WI 53201-0784, USA
ad@cs.uwm.edu

[3] Department of Computer Science, Utah State University, Logan, UT 84322-4205, USA
mjiang@cc.usu.edu

Abstract. T. Rado conjectured in 1928 that if $\mathcal{S}$ is a finite set of axis-parallel squares in the plane, then there exists an independent subset $\mathcal{I} \subseteq \mathcal{S}$ of pairwise disjoint squares, such that $\mathcal{I}$ covers at least $1/4$ of the area covered by $\mathcal{S}$. He also showed that the greedy algorithm (repeatedly choose the largest square disjoint from those previously selected) finds an independent set of area at least $1/9$ of the area covered by $\mathcal{S}$. The analogous question for other shapes and many similar problems have been considered by R. Rado in his three papers (1949, 1951 and 1968) on this subject. After 45 years (in 1973), Ajtai came up with a surprising example disproving T. Rado's conjecture. We revisit Rado's problem and present improved upper and lower bounds for squares, disks, convex sets, centrally symmetric convex sets, and others, as well as algorithmic solutions to these variants of the problem.

1 Introduction

Rado's problem on selecting disjoint squares is a famous unsolved problem in geometry [4, Problem D6]: What is the smallest number c such that, for any finite set $\mathcal{S}$ of axis-parallel squares in the plane, there exists an independent subset $\mathcal{I} \subseteq \mathcal{S}$ of pairwise disjoint squares with total area at least c times the union area of the squares in $\mathcal{S}$? T. Rado [12] observed that a greedy algorithm, which repeatedly selects the largest square disjoint from those previously selected, can find an independent subset $\mathcal{I}$ of disjoint squares with total area at least $1/9$ of the area of the union of all squares in $\mathcal{S}$. This lower bound was improved by R. Rado [9] to $1/8.75$ in 1949, and improved further by Zalgaller [15] to $1/8.6$ in 1960. On the other hand, an upper bound of $1/4$ for the area ratio is obvious: take four unit squares sharing a common vertex, then only one of them may be selected.

T. Rado conjectured that, for any finite set of axis-parallel squares, at least $1/4$ of the union area can be covered by a subset of disjoint squares. For congruent squares, the conjecture was confirmed by Norlander [8], Sokolin [13], and Zalgaller [15]. For the general case, Ajtai [1] came up with an ingenious construction with several hundred

⋆ Partially supported by NSF CAREER grant CCF-0444188.

⋆⋆ Partially supported by NSF grant DBI-0743670 and USU grant A13501.

J. Gudmundsson (Ed.): SWAT 2008, LNCS 5124, pp. 294–305, 2008.

squares and disproved T. Rado's conjecture in 1973! The problem of selecting disjoint squares has also been considered by R. Rado in a more general setting for various classes of convex sets, in his three papers entitled "Some covering theorems" [9,10,11].

We introduce some definitions. Throughout the paper, the term "convex set" refers to a closed compact convex set with nonempty interior. Denote by $|C|$ the Lebesgue measure of a convex set C in $\mathbb{R}^d$, i.e., the area in the plane, or the volume in the 3-space. For a finite set $\mathcal{S}$ of convex sets in $\mathbb{R}^d$, define its *union volume* (or *union area* when $d = 2$) as $|\mathcal{S}| = |\bigcup_{C \in \mathcal{S}} C|$. For a convex set S in $\mathbb{R}^d$, define

$$F(S) = \inf_{\mathcal{S}} \sup_{\mathcal{I}} \frac{|\mathcal{I}|}{|\mathcal{S}|},$$

where $\mathcal{S}$ ranges over all finite sets of convex sets in $\mathbb{R}^d$ that are homothetic to S, and $\mathcal{I}$ ranges over all independent subsets of $\mathcal{S}$. Also define

$$f(S) = \inf_{\mathcal{S}_1} \sup_{\mathcal{I}} \frac{|\mathcal{I}|}{|\mathcal{S}_1|},$$

where $\mathcal{S}_1$ ranges over all finite sets of convex sets in $\mathbb{R}^d$ that are homothetic and *congruent* to S, and $\mathcal{I}$ ranges over all independent subsets of $\mathcal{S}_1$.

The results of Zalgaller [15] and Ajtai [1], respectively, give lower and upper bounds of $1/8.6 \leq F(S) \leq 1/4 - 1/1728$, for a square S. In Section 2, we present a very simple algorithm A1 that improves the now 48 years old lower bound by Zalgaller [15]. Our algorithm is based on a novel idea that can be easily generalized to obtain improved lower bounds for hypercubes in *any* dimension $d \geq 2$. Let λ_d be the solution to the equation

$$3^d - (\lambda_d^{1/d} - 2)^d/2 = \lambda_d. \qquad (1)$$

For $d = 2$, $\lambda_2 = (\sqrt{46} + 2)^2/9 = 8.5699\ldots$.

Theorem 1. *Let $\mathcal{S}$ be a set of n axis-parallel hypercubes in $\mathbb{R}^d$. There exists an $O(dn^2)$ time algorithm A1 that computes an independent set $\mathcal{I} \subseteq \mathcal{S}$ such that $|\mathcal{I}|/|\mathcal{S}| \geq 1/\lambda_d$. That is, $F(S) \geq \lambda_d$ for a hypercube S in $\mathbb{R}^d$.*

In Section 3, we show that the algorithmic idea of A1 can be generalized to obtain an improved lower bound of $F(S) \geq 1/\lambda_2 > 1/8.5699$, for *any* centrally symmetric convex set S in the plane. The previous best lower bound of $F(S) \geq (1 + 1/200704)/9 = 1/8.999955\ldots$ was obtained by R. Rado [9, Theorem 8] in 1949.

Theorem 2. *For any centrally symmetric convex set S in the plane, $F(S) \geq 1/\lambda_2 > 1/8.5699$. In particular, for a disk S, $F(S) \geq 1/\lambda_{\text{disk}}$, where $\lambda_{\text{disk}} = 8.3539\ldots$.*

In Section 4, we present another algorithm A2 that achieves an even better lower bound for squares. The algorithm Z implicit in Zalgaller's lower bound [15] computes an independent set by repeatedly adding at most four disjoint squares at a time; our algorithm A2 adds at most three squares at a time.

Theorem 3. *Let $\mathcal{S}$ be a set of n axis-parallel squares in the plane. Then there is an $O(n^2)$ time algorithm A2 that computes an independent set $\mathcal{I} \subseteq \mathcal{S}$ such that $|\mathcal{I}|/|\mathcal{S}| \geq 1/\lambda_{\text{square}}$, where $\lambda_{\text{square}} = 8.4797\ldots$. That is, $F(S) \geq 1/\lambda_{\text{square}}$ for a square S.*

In Section 5, we present an improved upper bound for squares. Our construction refines Ajtai's idea [1] and consists of an infinite number of squares tiling the plane.

Theorem 4. *For a square S, $F(S) \leq \frac{1}{4} - \frac{1}{384}$.*

We know much more about $f(S)$ than about $F(S)$. For example, $f(S) = 1/4$ for a square S [4]. R. Rado [9] showed that $f(S) = 1/6$ for a triangle S, $f(S) = 1/4$ for a centrally symmetric hexagon S, $f(S) \geq \frac{\pi}{8\sqrt{3}} > 1/4.4106$ for a disk S, $f(S) \geq 1/16$ for any convex set S in the plane, and $f(S) \geq 1/7$ for any centrally symmetric convex set S in the plane. We improve the lower bounds on $f(S)$ for the two latter cases.

Theorem 5. *For any convex set S in the plane, $f(S) \geq 1/6$. This inequality cannot be improved. For any centrally symmetric convex set S in the plane, $f(S) \geq \delta(S)/4$ where $\delta(S)$ is the packing density of S; in particular, $f(S) > 1/4.4810$.*

We summarize our results for convex sets in $\mathbb{R}^2$, Theorems 2, 3, 4, and 5, in the following table.

Convex Set S in $\mathbb{R}^2$	Old Bound	New Bound
square	$F(S) < 1/4 - 1/1728$ [1]	$F(S) \leq 1/4 - 1/384$
square	$F(S) \geq 1/8.6$ [15]	$F(S) > 1/8.4797$
disk	$F(S) > 1/8.4898$ [2]	$F(S) > 1/8.3539$
centrally symmetric	$F(S) > 1/8.999955$ [9]	$F(S) > 1/8.5699$
centrally symmetric	$f(S) \geq 1/7$ [9]	$f(S) > 1/4.4810$
arbitrary	$f(S) \geq 1/16$ [9]	$f(S) \geq 1/6$

The problem of finding a maximum area independent set in a set of (axis-parallel) squares is NP-hard since the problem of finding a maximum independent set in a set of *unit* squares is already NP-hard [6]. On the other hand, maximum weight independent set admits polynomial-time approximation schemes [3,5,7] in square intersection graphs. Our algorithms come with a different guarantee: while they bound the weight of an independent set in terms of the maximum weight of any independent set, we bound the total area of an independent set in terms of the union area. Some proofs are omitted in this extended abstract.

2 Algorithm A1 and Lower Bounds for Squares and Hypercubes

In this section we prove Theorem 1. We first consider the planar case. Let $\mathcal{S} = \{S_1, \ldots, S_n\}$ be a set of n axis-parallel squares. For each square S_i in $\mathcal{S}$, denote by T_i the smallest axis-parallel square that contains all squares in $\mathcal{S}$ that intersect S_i (T_i contains S_i but is not necessarily concentric with S_i). Denote by x_i the side length of S_i. Denote by y_i the side length of T_i. Put $z_i = y_i - x_i$.

Let $\lambda = \lambda_2$. Recall that $\lambda_2 = (\sqrt{46}+2)^2/9 = 8.5699\ldots > (5/2)^2$. To construct an independent set $\mathcal{I}$, our algorithm A1 initializes $\mathcal{I}$ to be empty, then repeats the following *selection round* until $\mathcal{S}$ is empty:

1. Find the largest square S_l in $\mathcal{S}$. Assume without loss of generality[1] that $x_l = 1$.
2. If $y_l \le \sqrt{\lambda}$, add S_l to $\mathcal{I}$, delete from $\mathcal{S}$ the squares that intersect S_l, then return. Otherwise, set $k \leftarrow l$ and continue with the next step.
3. Let S_i and S_j be two squares that intersect S_k and touch two opposite sides of T_k. (We will prove later that S_i and S_j exist, are disjoint, and are different from S_k.) If both z_i and z_j are at most z_k, add S_i and S_j to $\mathcal{I}$, delete from $\mathcal{S}$ the squares that intersect S_i or S_j, then return. Otherwise, set $k \leftarrow i$ or j such that z_k increases, then repeat this step.

Analysis. In each selection round, step 3 is repeated at most n times since z_k is strictly increasing. So the algorithm terminates in $O(n^2)$ steps. Later in this section we will describe an efficient implementation of the algorithm and analyze its running time in more detail. We next prove that the algorithm achieves a lower bound of $1/\lambda$. Consider a selection round. If $y_l \le \sqrt{\lambda}$ in step 2, then

$$|S_l|/|T_l| = x_l^2/y_l^2 \ge 1/\lambda. \tag{2}$$

Now suppose that $y_l > \sqrt{\lambda}$. Then the algorithm proceeds to step 3. The two squares S_i and S_j clearly exist: even if no other squares in $\mathcal{S}$ intersect S_k, in which case $T_k = S_k$, we can still take $S_i = S_j = S_k$.

In every iteration of step 3, our choice of S_i and S_j implies that $x_i + x_j + x_k \ge y_k = x_k + z_k$. Therefore,

$$x_i + x_j \ge z_k. \tag{3}$$

Since $x_i \le x_l$, $x_j \le x_l$, and $x_l = 1$, we have $z_k \le 2$. In the first iteration, $z_k = y_l - x_l > \sqrt{\lambda} - 1$. The value z_k is strictly increasing. It follows that, in every iteration of step 3,

$$\sqrt{\lambda} - 1 < z_k \le 2. \tag{4}$$

Either S_i or S_j becomes S_k for the next iteration. It then follows from (3) and (4) that, in every iteration,

$$\sqrt{\lambda} - 2 < x_k \le 1. \tag{5}$$

S_i and S_j are disjoint (hence are both different from S_k) in every iteration because

$$y_k = x_k + z_k > (\sqrt{\lambda} - 2) + (\sqrt{\lambda} - 1) = 2\sqrt{\lambda} - 3 \ge 2(5/2) - 3 = 2 \ge x_i + x_j. \tag{6}$$

When the selection round ends, we have $z_i \le z_k$ and $z_j \le z_k$. Therefore,

$$|T_i| = y_i^2 = (x_i + z_i)^2 \le (x_i + z_k)^2, \qquad |T_j| = y_j^2 = (x_j + z_j)^2 \le (x_j + z_k)^2. \tag{7}$$

Since $S_k \subseteq T_i$ and $S_k \subseteq T_j$, we also have

$$|T_i \cap T_j| \ge |S_k| = x_k^2. \tag{8}$$

[1] This assumption mainly simplifies the analysis, and is not implemented in the algorithm.

Then,

$$\begin{aligned}
\frac{|T_i \cup T_j|}{|S_i| + |S_j|} &= \frac{|T_i| + |T_j| - |T_i \cap T_j|}{|S_i| + |S_j|} \\
&\leq \frac{(x_i + z_k)^2 + (x_j + z_k)^2}{x_i^2 + x_j^2} - \frac{x_k^2}{x_i^2 + x_j^2} \\
&= 1 + \frac{2z_k(x_i + x_j) + 2z_k^2}{x_i^2 + x_j^2} - \frac{x_k^2}{x_i^2 + x_j^2} \\
&\leq 1 + \frac{2z_k(x_i + x_j) + 2z_k^2}{(x_i + x_j)^2/2} - \frac{(\sqrt{\lambda} - 2)^2}{2} \\
&= 1 + \frac{4z_k}{x_i + x_j} + \frac{4z_k^2}{(x_i + x_j)^2} - \frac{(\sqrt{\lambda} - 2)^2}{2} \\
&\leq 9 - (\sqrt{\lambda} - 2)^2/2,
\end{aligned} \tag{9}$$

where the last inequality follows from (3). Recall that $\lambda = \lambda_2$ is the solution to the equation $9 - (\sqrt{\lambda} - 2)^2/2 = \lambda$. So we have

$$\frac{|S_i| + |S_j|}{|T_i \cup T_j|} \geq 1/\lambda. \tag{10}$$

From the two inequalities (2) and (10), it follows by induction that $|\mathcal{I}|/|\mathcal{S}| \geq 1/\lambda_2 > 1/8.5699$. Note that this bound already improves the previous best bound of $1/8.6$ by Zalgaller [15].

Generalization to higher dimensions. The algorithm can be easily generalized to any dimension $d \geq 2$ to achieve a bound of $1/\lambda_d$, where λ_d is the solution to (1). We omit the running time analysis. Note that $\lambda_d \geq (5/2)^d$. Set the threshold for y_l in the two cases to $\lambda_d^{1/d}$. Then the inequality $y_k > x_i + x_j$ in (6) still holds, and (9) becomes

$$\begin{aligned}
\frac{|T_i \cup T_j|}{|S_i| + |S_j|} &\leq \frac{(x_i + z_k)^d + (x_j + z_k)^d}{x_i^d + x_j^d} - \frac{x_k^d}{x_i^d + x_j^d} \\
&\leq \frac{2(z_k/2 + z_k)^d}{2(z_k/2)^d} - \frac{(\lambda^{1/d} - 2)^d}{2} \\
&= 3^d - (\lambda_d^{1/d} - 2)^d/2.
\end{aligned}$$

3 Lower Bounds for Centrally Symmetric Convex Sets in the Plane

We prove Theorem 2 in this section. Note that Theorem 1 implies a bound of $F(S) \geq 1/\lambda_2 > 1/8.5699$ for a square S. We extend this bound for any centrally symmetric convex set S in the plane. Let $\mathcal{S} = \{S_1, \ldots, S_n\}$ be a set of n homothetic copies of S. We make two changes to algorithm A1, and call the modified algorithm $\hat{\text{A}}1$.

The first change is in the definitions of T_i, x_i, and y_i. For each convex set S_i in $\mathcal{S}$, define T_i as the convex hull of the union of the convex sets in $\mathcal{S}$ that intersect S_i (T_i contains S_i). Define the *width of a convex set S along a line ℓ* as the distance between the pair of supporting lines of S perpendicular to ℓ. For each line ℓ through the center of S_i, denote by $w_i(\ell)$ the width of S_i along ℓ, and denote by $w_i'(\ell)$ the width of T_i along ℓ. Define $a_i = \max_\ell w_i'(\ell)/w_i(\ell)$, $x_i = \sqrt{|S_i|}$, and $y_i = a_i \cdot x_i$.

The second change is in step 3 of the selection round. We refer to Fig. 1. Let ℓ be a line through the center of S_k such that $w_k'(\ell)/w_k(\ell) = a_k$. Then, among the convex sets in $\mathcal{S}$ that intersect S_k, let S_i and S_j be any two convex sets that are tangent, respectively, to the two supporting lines of T_k perpendicular to ℓ.

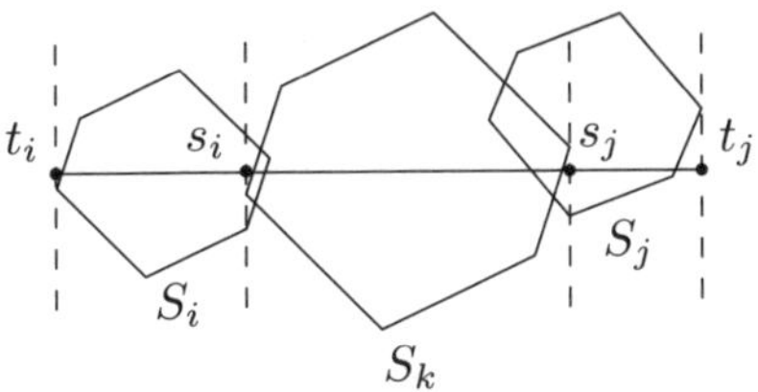

Fig. 1. Algorithm Â1 for centrally symmetric convex sets in the plane. The line ℓ is horizontal, and the four supporting lines are vertical in this example.

The analysis remains largely the same as that for squares. The following lemma is analogous to (3).

Lemma 1. *In each iteration of step 3 of the selection round, $x_i + x_j \geq z_k$.*

Proof. We refer back to Fig. 1. The two supporting lines of S_k perpendicular to the line ℓ intersect ℓ at the two points s_i and s_j. The two supporting lines of T_k perpendicular to the line ℓ intersect ℓ at the two points t_i and t_j. The supporting line through t_i is tangent to S_i. The supporting line through s_i, which is tangent to S_k, must also intersect S_i because otherwise S_i would be disjoint from S_k. Now S_i intersects the two supporting lines through t_i and s_i. On the other hand, S_k is tangent to the two supporting lines through s_i and s_j. It follows by similarity that $x_i/x_k \geq |t_i s_i|/|s_i s_j|$. A symmetric argument also shows that S_j intersects the two supporting lines through s_j and t_j, and satisfies $x_j/x_k \geq |s_j t_j|/|s_i s_j|$. Therefore,

$$x_i + x_j \geq \frac{|t_i s_i| + |s_j t_j|}{|s_i s_j|} x_k = \frac{w_k'(\ell) - w_k(\ell)}{w_k(\ell)} x_k = (a_k - 1)x_k = y_k - x_k = z_k. \qquad \square$$

The following lemma is analogous to the equality $|T_i| = y_i^2$ for squares, and maintains the inequalities in (2) and (7).

Lemma 2. *For each S_i in $\mathcal{S}$, $|T_i| \leq y_i^2$.*

Proof. Let T_i^* be the Steiner symmetrization of T_i with respect to the center of S_i. Then $|T_i| \leq |T_i^*|$ [14, Exercise 6-9]. Let S_i' be the concentric homothetic copy of S_i scaled by a_i. Then S_i' contains T_i^*. Therefore $|T_i| \leq |T_i^*| \leq |S_i'| = (a_i \cdot x_i)^2 = y_i^2$. $\square$

Following the same chain of reasoning from (2) to (10), we obtain a bound of $F(S) \geq 1/\lambda_2 > 1/8.5699$ for any centrally symmetric convex set S in the plane. For the special case that S is a disk, we can derive a better bound of $F(S) \geq 1/\lambda_{\text{disk}} > 1/8.3539$ by a tighter analysis.

4 Algorithm A2 and a New Lower Bound for Squares

We present a simple greedy algorithm for axis-parallel squares and prove Theorem 3. Let $\mathcal{S} = \{S_1, \ldots, S_n\}$ be a set of n axis-parallel squares. Denote by x_i the side length of S_i. For a square $S_i = [x, x+l] \times [y, y+l]$, we denote by S_i' the square $[x-1, x+l+1] \times [y-1, y+l+1]$, which contains all possible squares of side length at most 1 that intersect S_i. Note that S_i' is concentric with S_i.

Let s be a real number to be chosen later, $3/4 < s \leq 1$. To construct an independent set $\mathcal{I}$, our algorithm A2 initializes $\mathcal{I}$ to be empty, then repeats the following *selection round* until $\mathcal{S}$ is empty:

1. Let S_0 be the largest square in $\mathcal{S}$. Assume without loss of generality that S_0 is a unit square. Let $\mathcal{S}_0 \subseteq \mathcal{S} \setminus \{S_0\}$ be the set of squares of side length at least s that intersect S_0.
2. If $\mathcal{S}_0$ contains three disjoint squares S_1, S_2, and S_3, then add S_1, S_2, and S_3 to $\mathcal{I}$. Otherwise add S_0 to $\mathcal{I}$.
3. For each square S_i added to $\mathcal{I}$, remove from $\mathcal{S}$ the squares that intersect S_i.

In a selection round, let $\mathcal{J}$ be the set of selected squares, and let $\mathcal{T}$ be the set of squares in $\mathcal{S}$ that intersect the selected squares.

Lemma 3. *Suppose that the algorithm selects three squares S_1, S_2, and S_3, in a selection round. Then*

$$|\mathcal{T}|/|\mathcal{J}| \leq (8 + 3s^2 + 10s)/(3s^2).$$

Proof. It can be shown that the ratio of the area of the region $R = S_1' \cup S_2' \cup S_3'$ over the total area of the three squares S_1, S_2, and S_3 is maximized when each square intersects S_0 at a distinct corner as shown in Fig. 2(a) (possibly with a different correspondence between the squares and the corners), and when the three squares have equal side length $x_1 = x_2 = x_3 = s$. In this case the region is the union of 8 unit squares, 3 squares of side length s, and 10 rectangles of side lengths 1 and s. □

Lemma 4. *Suppose that the algorithm selects one square, S_0, in a selection round. Then*

$$|\mathcal{T}|/|\mathcal{J}| \leq 7 + 2s^2.$$

Proof. It can be shown that the maximum covered area in S_0' is the shaded area shown in Fig. 2(b), which contains 7 unit squares and 2 squares of side length s. □

Balancing the two bounds in Lemmas 3 and 4, we obtain a quartic equation $3s^4 + 9s^2 - 5s - 4 = 0$, which has only one positive root $s_0 = 0.8601\ldots$. Choose $s = s_0$, and we have $(8 + 3s^2 + 10s)/(3s^2) = 7 + 2s^2 = \lambda_{\text{square}} = 8.4797\ldots$.

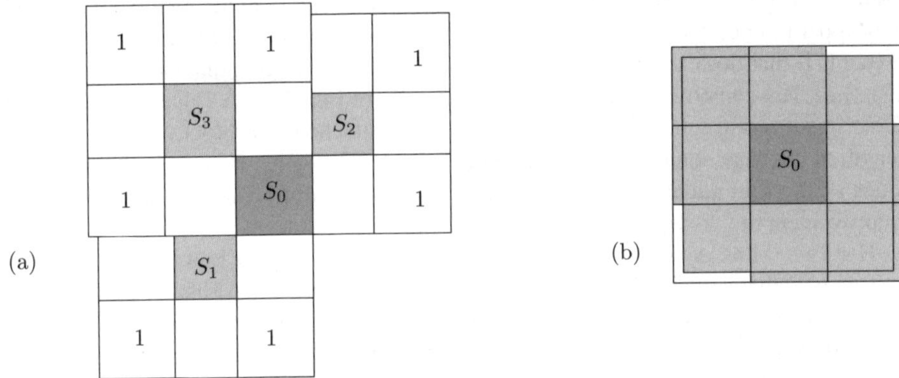

Fig. 2. (a) Maximum area of $R = S_1' \cup S_2' \cup S_3'$. (b) Maximum covered area in S_0'.

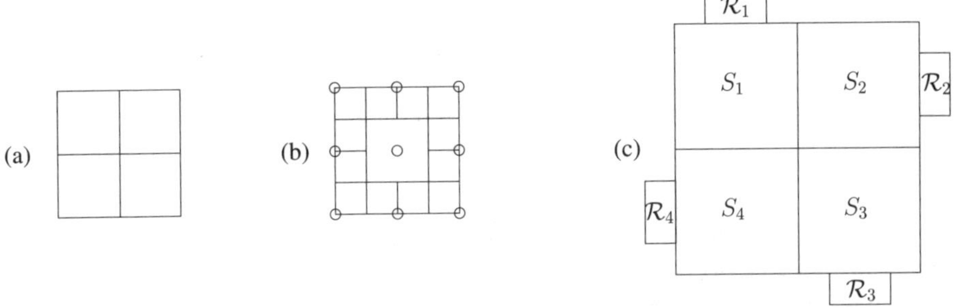

Fig. 3. (a) Starting point: a system of four congruent squares. (b) Ajtai's idea: an ambiguous system $\mathcal{Q}$ of 13 squares of sides 1 and 2. (c) Ajtai's construction shown schematically. $\mathcal{R}_i$, $i = 1$, 2, 3, 4, are rotated copies of a system of 66 squares.

5 A New Upper Bound for Squares: Proof of Theorem 4

We first describe briefly Ajtai's ingenious idea for the construction in [1]. The starting point is a system of 4 non-overlapping squares shown in Fig. 3(a). Now slightly enlarge each square with respect to its center by a small $\varepsilon > 0$. All constructions we discuss will be obtained in the same way, by starting from a system of non-overlapping (i.e., interior disjoint) squares and then applying the above transformation; the effect is that any pair of touching squares results in a pair of squares intersecting in their interior. Finally by letting ε tend to zero, one recovers the same upper bound for systems of intersecting squares. Alternatively, one can consider the squares as closed sets, to start with, and use non-overlapping squares in the construction.

The second step is to consider a system $\mathcal{Q}$ of 13 squares of side 1 and 2 such as that in Fig. 3(b), which can be also viewed as a system of four squares A_1, A_2, A_3, and A_4, the vertices of which are drawn as circles; these squares are only used in the analysis. $\mathcal{Q}$ has the nice property that any independent set can cover at most one quarter of (the area of) each A_i. Although $\mathcal{Q}$ by itself does not appear to be useful in reducing the conjectured $1/4$ upper bound, Ajtai found a more complicated system of 66 squares of sides 1 and 2 (see [1]) that does so, if used in combination with four larger squares arranged as in Fig. 3(a). His construction is shown schematically in Fig. 3(c). A calculation done for the original construction (where the length of the attached four blocks equals the side length of the large squares), yields an upper bound of $\frac{1}{4} - \frac{1}{1728}$. By using all eight outer sides of the four squares and eight blocks (an obvious optimization) yields a further improvement to $\frac{1}{4} - \frac{1}{1080}$.

Here we refine Ajtai's idea in several ways to obtain a better bound. First, we find a system $\mathcal{R}$ that serves the same purpose but uses fewer squares. Finally, we construct a tiling[2] of the plane, whose blocks (tiles) are made up from the previous pieces. To this end we first construct a variant $\mathcal{R}^*$ of $\mathcal{R}$ that admits a symmetry axis, which allows adjacent blocks in the tiling to share common parts in the original system $\mathcal{R}$. To implement the tiling idea, consider the *new* system $\mathcal{R}$ shown in Fig. 4(a), which borders two adjacent sides of a large square S. $\mathcal{R}$ consists of 38 unit squares and 16 squares of side 2. By replicating copies of $\mathcal{R}$, rotated by $0°$, $90°$, $180°$, and $270°$, we construct a tiling of the plane, see Fig. 5. We say that a square A_i is *not covered*, if 0% of its area is covered by $\mathcal{I}$.

Lemma 5. *Let $\mathcal{I}$ be an independent set of squares in the system $\mathcal{R} \cup \{S\}$ in Fig. 4(a). Assume that $S \in \mathcal{I}$, and consider the 10×4 rectangle Z_1 which borders S from above. Then $|\mathcal{I} \cap Z_1| \leq 9$.*

Proof. Observe that $\mathcal{R}$ has the property that any independent set can cover at most one quarter of (the area of) each A_i, conform with Fig. 4(b) and 4(d). By the assumption, the 10 unit squares in the bottom row of squares A_7 through A_{11} cannot be in $\mathcal{I}$. It is enough to show that at least one of the squares A_i ($i \in \{1, 2, 3, 4, 5\} \cup \{7, 8, 9, 10, 11\}$ is not covered. Observe that either $B_2 \in \mathcal{I}$ or $B_3 \in \mathcal{I}$ (otherwise A_9 is not covered and we are done). Since $\mathcal{R} \cap Z_1$ admits a vertical symmetry axis, we can assume that $B_2 \in \mathcal{I}$.

It follows that $e \in \mathcal{I}$ (otherwise A_{10} is not covered), and that $B_4 \in \mathcal{I}$ (else A_{11} is not covered). But then A_4 is not covered, since $f, B_3, C_3, C_4 \notin \mathcal{I}$. This completes the proof. □

Obviously, the property in the lemma holds also for Z_2 in place of Z_1. We now move to the final step—the tiling—which completes our construction. Take four squares S_1, S_2, S_3, and S_4, each of side 10, and arrange them as in Fig. 5(a). Place four rotated copies of $\mathcal{R}$ bordering the outer 8 sides of $S_1 \cup S_2 \cup S_3 \cup S_4$ as in Fig. 5(b), and obtain a block (cell) of the tiling. All the 92 2×2 squares shown in Fig. 4(d), for each of the four copies of $\mathcal{R}$ are assigned to the unique block containing the four large squares. It is important to notice that some of the 2×2 squares are shared among adjacent blocks in the tiling, however the 2×2 squares A_i used in the analysis are not shared, i.e., they are contained entirely in different blocks.

Let $\mathcal{T}$ be the infinite set of squares obtained by replicating the block in Fig. 5(b), as in Fig. 5(c). Let $\mathcal{I}$ be an independent set of squares in $\mathcal{T}$. Fix any given block σ in the

[2] Here we use this term in a broader sense, where the tile can have holes.

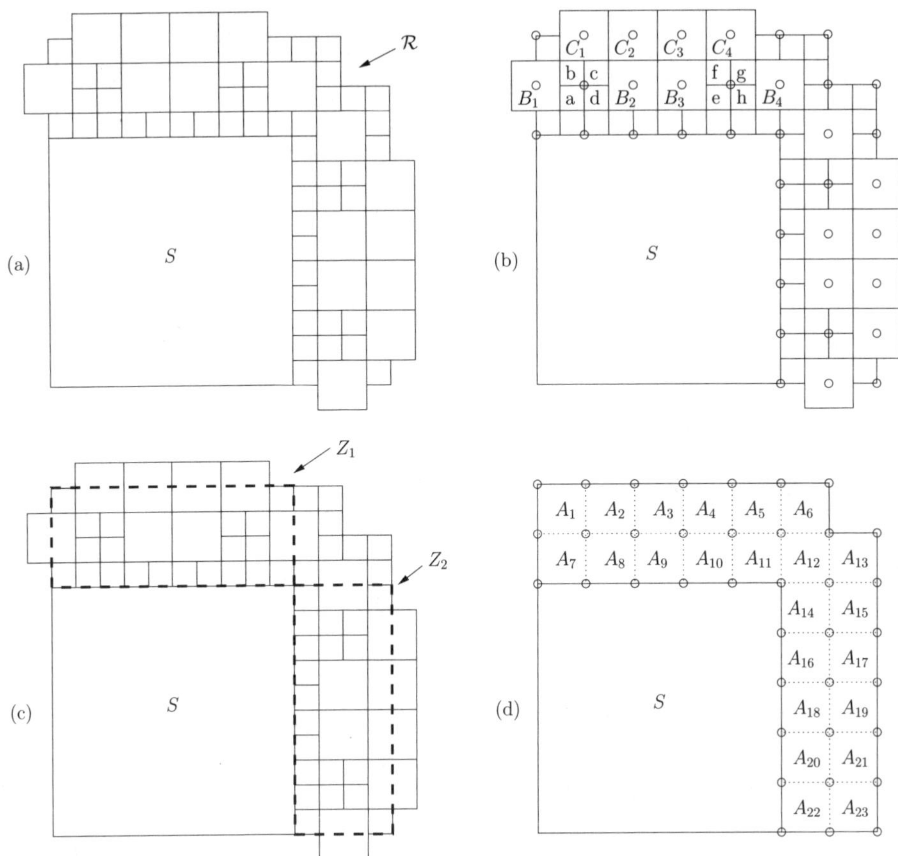

Fig. 4. (a) Preliminaries for the tiling: the new system $\mathcal{R}$ bordering two sides of a large square S. (b) The labeling of the squares used in the proof of the upper bound in Lemma 5. (c) Two rectangles Z_1 and Z_2 superimposed on $\mathcal{R}$. (d) A system of 23 squares of side 2, A_i, $i = 1, \ldots, 23$, superimposed on $\mathcal{R}$ (some of the squares in $\mathcal{R}$ are only partially covered by the squares A_i).

tiling. Observe that at most one of the S_i can be in $\mathcal{I}$, so at most one quarter of the area of $S_1 \cup S_2 \cup S_3 \cup S_4$ is covered by $\mathcal{I}$. Similarly $\mathcal{I}$ covers at most one quarter of the area in each of the 92 2×2 squares assigned to σ. Observe that if one of the four large squares, say S_2, is selected in an independent set $\mathcal{I}$, it forces the 10 unit squares in both, the bottom row of $\mathcal{R} \cap Z_1$ and the leftmost column of $\mathcal{R} \cap Z_2$ to be out of $\mathcal{I}$.

For the analysis, we can argue independently for each block. Fix any block σ in the tiling. The area covered by $\mathcal{T}$ in σ is

$$|\mathcal{T} \cap \sigma| = 4 \times 100 + 4 \times 92 = 768.$$

In the (easy) case that none of the S_i is in $\mathcal{I}$, the area covered by $\mathcal{I}$ in σ is

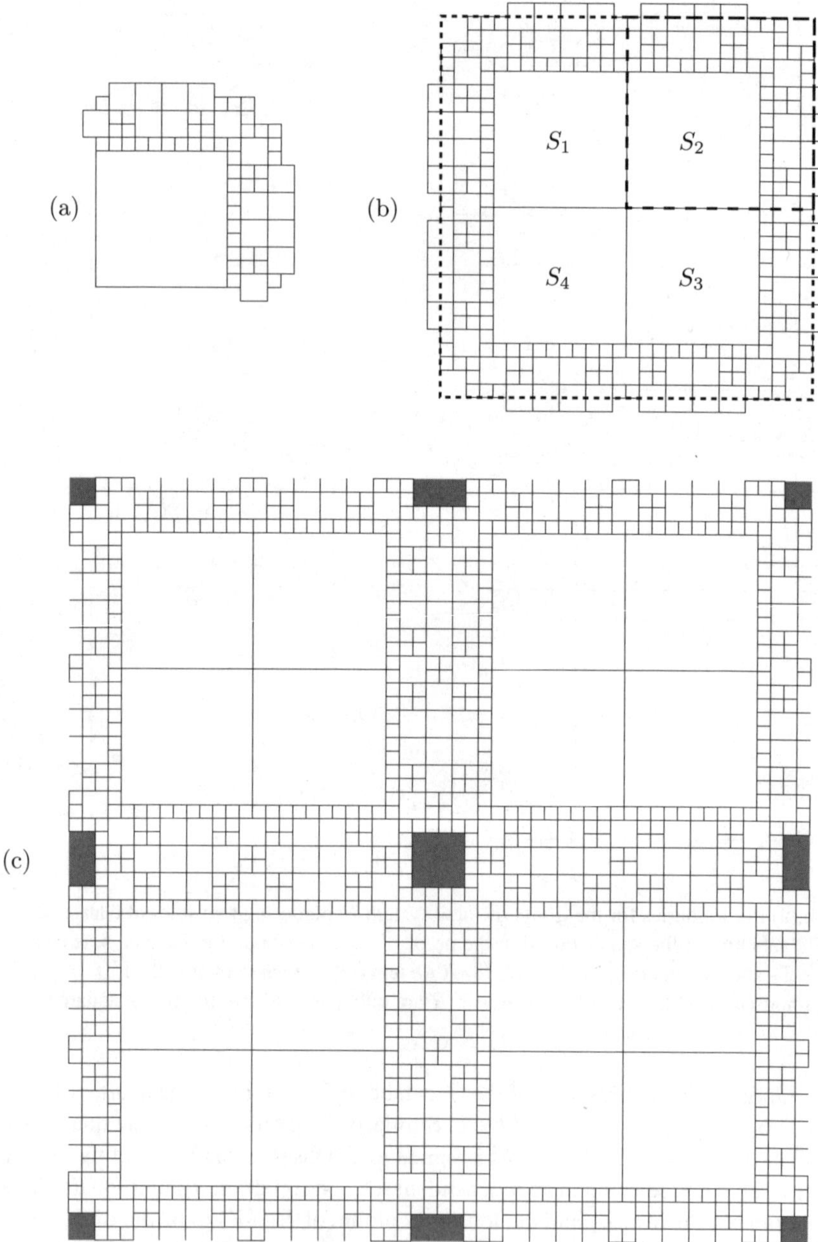

Fig. 5. (a) A large square of side 10 bordered by the system $\mathcal{R}$. (b) $S_1 \cup S_2 \cup S_3 \cup S_4$ bordered by 4 rotated copies of $\mathcal{R}$ (some squares are shared among adjacent copies). The block σ is the large dashed square containing $S_1 \cup S_2 \cup S_3 \cup S_4$. (c) Tiling of the plane with blocks composed of 4 large squares of side 10 bordered by 4 rotated copies of $\mathcal{R}$ (some squares are shared among adjacent blocks). The shaded rectangles in the figure represent holes in the tiling, and are not part of the square system.

$$|\mathcal{I} \cap \sigma| \leq 4 \times 23 = 92, \quad \text{thus } \frac{|\mathcal{I} \cap \sigma|}{|\mathcal{T} \cap \sigma|} \leq \frac{92}{768} = \frac{1}{4} - \frac{50}{384},$$

i.e., much smaller than required.

Assume now that one of the S_i, say S_2, belongs to $\mathcal{I}$. Observe that the 20 unit squares adjacent to top and right sides of S_2 do not belong to $\mathcal{I}$ (the same holds for the unit square in the corner, but this is irrelevant here). By Lemma 5,

$$|\mathcal{I} \cap \sigma| \leq 100 + 4 \times 23 - 2 = 190, \quad \text{thus } \frac{|\mathcal{I} \cap \sigma|}{|\mathcal{T} \cap \sigma|} \leq \frac{190}{768} = \frac{1}{4} - \frac{1}{384},$$

as desired. Of course, one can get arbitrarily close to this bound, by using a suitably large (square) section of the tiling instead—since the boundary effects are negligible.

References

1. Ajtai, M.: The solution of a problem of T. Rado. Bulletin de l'Académie Polonaise des Sciences, Série des Sciences Math. Astr. et Phys. 21, 61–63 (1973)
2. Bereg, S., Dumitrescu, A., Jiang, M.: Maximum area independent set in disk intersection graphs. International Journal of Computational Geometry & Applications (to appear)
3. Chan, T.: Polynomial-time approximation schemes for packing and piercing fat objects. Journal of Algorithms 46, 178–189 (2003)
4. Croft, H.T., Falconer, K.J., Guy, R.K.: Unsolved Problems in Geometry. Springer, New York (1991)
5. Erlebach, T., Jansen, K., Seidel, E.: Polynomial-time approximation schemes for geometric intersection graphs. SIAM Journal on Computing 34, 1302–1323 (2005)
6. Fowler, R.J., Paterson, M.S., Tanimoto, S.L.: Optimal packing and covering in the plane are NP-complete. Information Processing Letters 12, 133–137 (1981)
7. Hochbaum, D.S., Maass, W.: Approximation schemes for covering and packing problems in image processing and VLSI. Journal of ACM 32, 130–136 (1985)
8. Norlander, G.: A covering problem. Nordisk Mat. Tidskr. 6, 29–31 (1958)
9. Rado, R.: Some covering theorems (I). Proceedings of the London Mathematical Society 51, 241–264 (1949)
10. Rado, R.: Some covering theorems (II). Proceedings of the London Mathematical Society 53, 243–267 (1951)
11. Rado, R.: Some covering theorems (III). Journal of the London Mathematical Society 42, 127–130 (1968)
12. Rado, T.: Sur un problème relatif à un théorème de Vitali. Fund. Math. 11, 228–229 (1928)
13. Sokolin, A.: Concerning a problem of Rado. C.R. Acad. Sci. U.R.S.S (N.S.) 26, 871–872 (1940)
14. Yaglom, I.M., Boltyanskiǐ, V. G.: Convex Figures. Holt, Rinehart and Winston, New York (1961)
15. Zalgaller, V.A.: Remarks on a problem of Rado. Matem. Prosveskcheric 5, 141–148 (1960)

Packing Rectangles into 2 OPT Bins Using Rotations

Rolf Harren and Rob van Stee*

Max-Planck Institute for Computer Science (MPII),
Campus E1 4, D-66123 Saarbrücken, Germany
{rharren,vanstee}@mpi-inf.mpg.de

Abstract. We consider the problem of packing rectangles into bins that are unit squares, where the goal is to minimize the number of bins used. All rectangles can be rotated by 90 degrees and have to be packed non-overlapping and orthogonal, i.e., axis-parallel. We present an algorithm for this problem with an absolute worst-case ratio of 2, which is optimal provided $\mathcal{P} \neq \mathcal{NP}$.

Keywords: bin packing, rectangle packing, approximation algorithm, absolute worst-case ratio.

1 Introduction

In the rectangle packing problem, a list $I = \{r_1, \ldots, r_n\}$ of rectangles of width $w_i \leq 1$ and height $h_i \leq 1$ is given. An unlimited supply of unit sized bins is available to pack all items from I such that no two items overlap and all items are packed axis-parallel into the bins. The goal is to minimize the number of bins used. The problem is also known as two-dimensional orthogonal bin packing problem and has many applications, for instance in stock-cutting or scheduling on partitionable resources. In many applications, rotations are not allowed because of the pattern of the cloth or the grain of the wood. However, in other applications, it might be possible to rotate the items. In the current paper, we consider the problem with rotations, i.e., items might be rotated by 90 degrees.

Most of the previous work on rectangle packing has focused on the *asymptotic* approximation ratio, i.e., the long term behavior of the algorithm, and on packing *without rotations*. Caprara was the first to present an algorithm with an asymptotic approximation ratio less than 2 for rectangle packing without rotations. Indeed, he considered 2-stage packing, in which the items must first be packed into shelves that are then packed into bins, and showed that the asymptotic worst case ratio between rectangle packing and 2-stage packing is $T_\infty = 1.691\ldots$. Therefore the asymptotic $\mathcal{FPTAS}$ for 2-stage packing from Caprara, Lodi and Monaci [3] achieves an approximation guarantee arbitrary close to T_∞.

* Research supported by German Research Foundation (DFG). Work done while this author was at the University of Karlsruhe.

J. Gudmundsson (Ed.): SWAT 2008, LNCS 5124, pp. 306–318, 2008.

Recently, Bansal, Caprara & Sviridenko [1] presented a general framework to improve subset oblivious algorithms and obtained asymptotic approximation guarantees arbitrarily close to 1.525... for packing with or without rotations. These are the currently best-known approximation ratios for these problems. For packing squares into square bins, Bansal, Correa, Kenyon & Sviridenko [2] gave an asymptotic $\mathcal{PTAS}$. On the other hand, the same paper showed the $\mathcal{APX}$-hardness of rectangle packing without rotations, thus no asymptotic $\mathcal{PTAS}$ exists unless $\mathcal{P} = \mathcal{NP}$. Chlebík & Chlebíková [4] were the first to give explicit lower bounds of $1 + 1/3792$ and $1 + 1/2196$ on the asymptotic approximability of rectangle packing with and without rotations, respectively.

In the current paper we consider the absolute worst-case ratio. Attaining a good absolute worst-case ratio is more difficult than attaining a good asymptotic worst-case ratio, because in the second case an algorithm is allowed to waste a constant number of bins, which allows e.g. the classification of items followed by a packing where each class is packed separately. Zhang [15] presented an approximation algorithm with an absolute approximation ratio of 3 for the problem without rotations. For the special case of packing squares, van Stee [14] showed that an absolute 2-approximation is possible.

A related two-dimensional packing problem is the strip packing problem, where the items have to be packed into a strip of unit basis and unlimited height such that the height is minimized. Steinberg [13] and Schiermeyer [12] presented absolute 2-approximation algorithms for strip packing without rotations. Kenyon & Rémila [9] and Jansen & van Stee [7] gave asymptotic $\mathcal{FPTAS}$'s for the problem without rotations and with rotations, respectively. The additive constant of these algorithms was recently improved from $\mathcal{O}(1/\varepsilon^2)$ to 1 by Jansen & Solis-Oba [6]. Thus, most versions of the strip packing problem are now closed.

Our contribution. We present an approximation algorithm for rectangle packing with rotations with an absolute approximation ratio of 2. As Leung et al. [10] showed that it is strongly $\mathcal{NP}$-complete to decide wether a set of *squares* can be packed into a given square, this is best possible unless $\mathcal{P} = \mathcal{NP}$. The algorithm is based on a separation of large and small items according to their area. It is very efficient for inputs consisting of small items but uses a less efficient subroutine to deal with large items. Our main lemma on the packability of certain sets of small items is of independent interest.

We started our investigation on the problem with an algorithm of Jansen & Solis-Oba [6] that finds a packing of profit $(1-\delta)$OPT into a bin of size $(1, 1+\delta)$, where OPT denotes the optimum for packing into a unit bin. Using the area of the items as their profit gives an algorithm that packs almost everything into an δ-augmented bin. The algorithm can easily be generalized to a constant number of bins.

An immediate idea to transform such a packing into a packing into 2OPT bins is to remove all items that intersect a strip of height δ at the top or bottom of each bin. These items and the items that were not packed by the algorithm would have to be packed separately. In Figure 1 we present an instance where it is not immediately clear how the removed items can be packed separately.

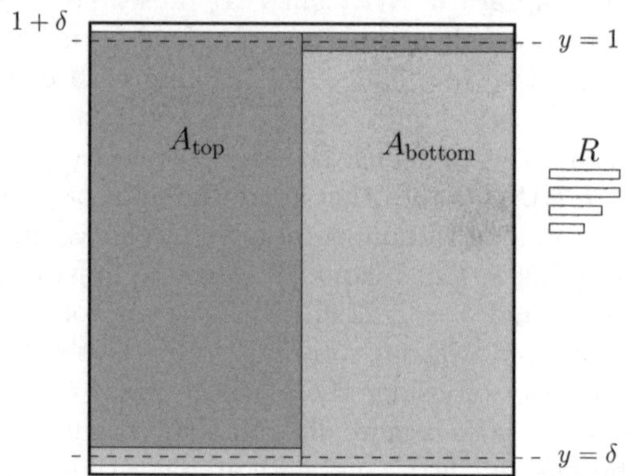

Fig. 1. Packing of Jansen & Solis-Oba's algorithm where it is not immediately clear how to derive a packing into 2 unit bins. The blocks in the packing might consist of several items and might contain small free spaces or items that are not in A_{top} or A_{bottom}. The items of A_{top} and the items of A_{bottom} have total area close to 1/2. Thus adding the additional items R and packing everything with Steinberg's algorithm is not possible. Furthermore, it is not obvious how to rearrange A_{top} or A_{bottom} such that there is suitable free space to pack R.

Organisation. The remainder of this article is organized as follows. In Section 2 we introduce notations and two algorithms for strip packing that we will use as subroutines for our rectangle packing algorithm: the algorithm of Steinberg and Next Fit Decreasing Height. We show that Steinberg's algorithm [13] yields an absolute 2-approximation for strip packing with rotations and an absolute 4-approximation for rectangle packing with rotations. Our main result is presented in Section 3. The algorithm is based on our main lemma that we prove in Section 4.

2 Steinberg's Algorithm and NFDH

We assume that all items are rotated such that $w_i \geq h_i$. Denote the total area of a given set T of items by $A(T) = \sum_{i \in T} w_i h_i$ and let $w_{\max} := \max_{r_i \in T} w_i$ and $h_{\max} := \max_{r_i \in T} h_i$. Steinberg [13] showed the following theorem.

Theorem 1 (Steinberg's algorithm [13]). *If the following inequalities hold,*

$$w_{\max} \leq a, \quad h_{\max} \leq b, \text{ and } \quad 2A(T) \leq ab - (2w_{\max} - a)_+(2h_{\max} - b)_+$$

where $x_+ = \max(x, 0)$, then it is possible to pack all items from T into $R = (a, b)$ in time $\mathcal{O}((n \log^2 n)/\log \log n)$.

In our algorithm, we will repeatedly use the following direct corollary of this theorem.

Corollary 1. *If $w_{\max} \leq a/2$ and $A(T) \leq ab/2$, then it is possible to pack all items from T into $R = (a, b)$ in time $\mathcal{O}((n \log^2 n)/\log\log n)$.*

The following theorem was already mentioned in [6].

Theorem 2. *Steinberg's algorithm gives an absolute 2-approximation for strip packing with rotations.*

Proof. Rotate all items $r_i \in I$ such that $w_i \geq h_i$ and let $b := \max(2h_{\max}, 2A(I))$. Use Steinberg's algorithm to pack I into the rectangle $(1, b)$. This is possible since $2A(I) \leq b$ and $(2h_{\max} - b)_+ = 0$. The claim on the approximation ratio follows from $\mathrm{OPT} \geq \max(h_{\max}, A(I)) = b/2$. □

It is well-known that a strip packing algorithm with an approximation ratio of δ directly yields a rectangle packing algorithm with an approximation ratio of 2δ. To see this, cut the strip packing of height h into slices of height 1 so as to get $\lceil h \rceil$ bins of the required size. The rectangles that are split between two bins can be packed into $\lfloor h \rfloor$ additional bins. The strip packing gives a lower bound for rectangle packing. Thus if $h \leq \delta \mathrm{OPT}_{\mathrm{strip}}$, then $\lceil h \rceil + \lfloor h \rfloor \leq 2\delta \mathrm{OPT}_{\mathrm{bin}}$. Accordingly, we get the following theorem.

Theorem 3. *Steinberg's algorithm yields an absolute 4-approximation algorithm for rectangle packing with rotations.*

Jansen & Zhang [8] showed a corollary of Steinberg's theorem which reads as follows if $w_i \geq h_i$ for all items.

Corollary 2 ([8]). *If the total area of a set T of items is at most $1/2$ and there is at most one item of height $h_i > 1/2$, then the items of T can be packed into a bin of unit size in time $\mathcal{O}((n \log^2 n)/\log\log n)$.*

The NEXT-FIT-DECREASING-HEIGHT algorithm (NFDH) was introduced for squares by Meir & Moser [11] and generalized to rectangles by Coffman, Garey, Johnson & Tarjan [5]. It is given as follows. Sort the items by non-increasing order of height. Pack the items one by one into shelfs. The height of a shelf is defined by its first item, further items are added left-aligned until an item does not fit. In this case this item opens a new shelf. The algorithm stops if it runs out of items or a new shelf does not fit into the designated area. The running time of the algorithm is $\mathcal{O}(n \log n)$. The following lemma is an easy generalization of the result from Meir & Moser.

Lemma 1. *If a given set T of items is packed into a rectangle $R = (a, b)$ by* NFDH, *then either a total area of at least $(a - w_{\max})(b - h_{\max})$ is packed or the algorithm runs out of items, i.e., all items are packed.*

3 Our Algorithm: Overview

As the asymptotic approximation ratio of the algorithm from Bansal et al. [1] is less than 2, there exists a constant k such that for any instance with optimal value larger than k, the asymptotic algorithm gives a solution of value at most 2OPT. We address the problem of approximating rectangle packing with rotations within an absolute factor of 2, provided that the optimal value of the given instance is less than k. Combined with the algorithm from [1] we get an overall algorithm with an absolute approximation ratio of 2.

We begin by applying the asymptotic algorithm from [1]. If OPT $> k$, then the algorithm outputs a solution of value $k' \leq$ 2OPT. Otherwise OPT $\leq k$ and we apply the following algorithm.

Let $\varepsilon := 1/68$. We separate the given input according to the area of the items, so we get a set of large items $L = \{r_i \in I \mid w_i h_i \geq \varepsilon\}$ and a set of small items $S = \{r_i \in I \mid w_i h_i < \varepsilon\}$. Since the number of large items in each bin is bounded by $1/\varepsilon$ and their total area is at most k, we can enumerate all possible packings of the large items. Take an arbitrary packing of the large items into a minimum number $\ell \leq k$ of bins.

If there are bins that contain items with a total area less than $1/2 - \varepsilon$, we greedily add small items such that the total area of items assigned to each of these bins is in $(1/2 - \varepsilon, 1/2]$. We use Corollary 2 to repack these bins including the newly assigned small items. There is at most one item of height $h_i > 1/2$ since otherwise the total area exceeds $1/2$, because $w_i \geq h_i$. If we run out of items in this step, we found an optimal solution. Assume that there are still small items left and each bin used so far contains items of a total area of at least $1/2 - \varepsilon$. The following crucial lemma shows that we can pack the remaining small items well enough to achieve an absolute approximation ratio of 2.

Lemma 2. *Let* $0 < \varepsilon \leq 1/68$. *Given a set* T *of items that all have area at most* ε *such that for all* $r \in T$ *the total area of* $T \setminus \{r\}$ *is less than* $1/2 + \varepsilon$. *We can find a packing of* T *into a unit bin in time* $\mathcal{O}((n \log^2 n)/\log\log n)$.

The lemma is proved in the next section. To apply Lemma 2 we consider the following partition of the remaining items.

Let $r_1, \ldots, r_m$ be the list of remaining small items, sorted by non-increasing order of size. Partition these small items into sets $S_1 = \{r_{t_1}, \ldots, r_{t_2-1}\}, S_2 = \{r_{t_2}, \ldots, r_{t_3-1}\}, \ldots, S_s = \{r_{t_s}, \ldots, r_{t_{s+1}-1}\}$ with $t_1 = 1$ and $t_{s+1} = m + 1$ such that

$$A(S_j \setminus \{r_{t_{j+1}-1}\}) < \frac{1}{2} + \varepsilon \qquad \text{and} \qquad A(S_j) \geq \frac{1}{2} + \varepsilon$$

for $j = 1, \ldots, s-1$. Obviously, each set S_i satisfies the precondition of Lemma 2 and can therefore be packed into a single bin. Only S_s might have a total area of less than $1/2 + \varepsilon$. The overall algorithm is given in Algorithm 1.

Note that if no packing of L into at most k bins exists, then OPT $\geq k$ and thus $k' \leq$ 2OPT by definition of k.

Algorithm 1. Approximate rectangle packing with rotations

apply the asymptotic algorithm from [1] to derive a packing P' into k' bins
if $k' \geq 2k$ **then**
  return P'
else
  let $\varepsilon := 1/68$
  partition I into $L = \{r_i \in I \mid w_i h_i \geq \varepsilon\}$ and $S = \{r_i \in I \mid w_i h_i < \varepsilon\}$
  if L cannot be packed in k or less bins **then**
    return P'
  else
    let P_ℓ be a packing of L into $\ell \leq k$ bins.
    while there exists a bin containing items of total area $< 1/2 - \varepsilon$ **do**
      assign small items to this bin until the total area exceeds $1/2 - \varepsilon$
      use Steinberg's algorithm (Corollary 2) to repack the bin
    order the remaining small items by non-increasing size
    greedily partition the remaining items into sets $S_1, \ldots, S_s$ such that

$$A(S_j \setminus \{r_{t_{j+1}-1}\}) < \frac{1}{2} + \varepsilon \quad \text{and} \quad A(S_j) \geq \frac{1}{2} + \varepsilon \quad \text{for } j = 1, \ldots, s-1$$

    use the method described in the proof of Lemma 2 to pack each set S_i into a bin
    let P be the resulting packing into $\ell + s$ bins
return the packing from P, P' that uses the least amount of bins

4 Packing Sets of Small Items

In this section we prove Lemma 2. We will use the following partition of a set T of items of area at most ε in the remainder of this section. Let

$$T_1 := \{r_i \in T \mid 2/3 < w_i\} \qquad T_2 := \{r_i \in T \mid 1/2 < w_i \leq 2/3\}$$
$$T_3 := \{r_i \in T \mid 1/3 < w_i \leq 1/2\} \qquad T_4 := \{r_i \in T \mid w_i \leq 1/3\}.$$

Since $w_i h_i \leq \varepsilon$ and $w_i \geq h_i$, the heights of the items in each set are bounded as follows.

$$h_i \leq 3/2 \cdot \varepsilon \quad \text{for } r_i \in T_1, \qquad h_i \leq 2 \cdot \varepsilon \quad \text{for } r_i \in T_2,$$
$$h_i \leq 3 \cdot \varepsilon \quad \text{for } r_i \in T_3 \text{ and} \quad h_i \leq \sqrt{\varepsilon} \quad \text{for } r_i \in T_4.$$

It turns out that packing the items in T_2 involves the most difficulties. We will therefore consider different cases for packing items in T_2, according to the total height of these items. For all cases we need to pack $T_1 \cup T_3 \cup T_4$ afterwards, using the following lemma.

Lemma 3. *Given a rectangle $R = (1, h)$ and a set T of items that all have area at most ε such that $T_2 = \emptyset$. We can find a packing of a selection $T' \subseteq T$ into R in time $\mathcal{O}(n \log n)$ such that $T' = T$ or*

$$A(T') \geq \frac{2}{3}(h - \sqrt{\varepsilon}) - \varepsilon.$$

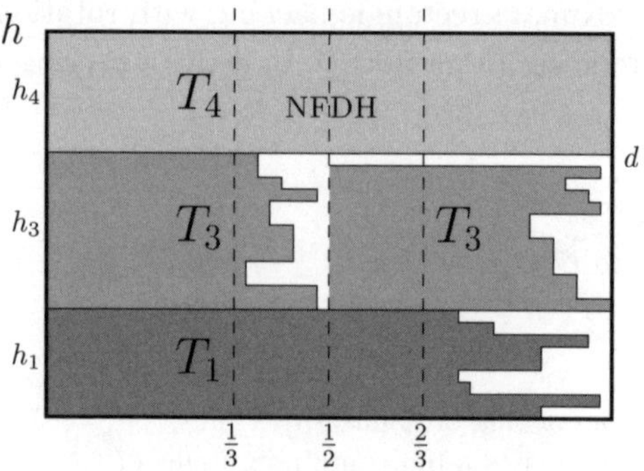

Fig. 2. Packing the sets T_1, T_3 and T_4 into a bin of width 1 and height h. The difference in height between the stacks of T_3 is denoted by d.

Proof. See Figure 2 for an illustration of the following packing. Stack the items of T_1 left-justified into the lower left corner of R. Stop if there is not sufficient space to accomodate the next item. In this case a total area of at least $A(T_1') \geq 2/3\,(h - 3/2 \cdot \varepsilon)$ is packed since $w_i > 2/3$ and $h_i \leq 3/2 \cdot \varepsilon$ for items in T_1.

Thus assume all items from T_1 are packed. Denote the height of the stack by h_1. Obviously, $A(T_1) \geq 2/3 \cdot h_1$.

Create two stacks of items from T_3 next to each other directly above the stack for T_1 by repeatedly assigning each item to the lower stack. Stop if an item does not fit into the rectangle. In this case both stacks have a height of at least $h - h_1 - 3\varepsilon$ as otherwise a further item could be packed. Therefore $A(T_1 \cup T_3') \geq 2/3(h - 3\varepsilon) \geq 2/3 \cdot (h - \sqrt{\varepsilon})$ since $3\varepsilon \leq \sqrt{\varepsilon}$ for $\varepsilon \leq 1/18$.

Otherwise denote the height of the higher stack by h_3 and the height difference by d. The total area of T_3 is at least $A(T_3) \geq 2/3(h_3 - d) + 1/3 \cdot d \geq 2/3 \cdot h_3 - 1/3 \cdot d \geq 2/3 \cdot h_3 - \varepsilon$ since $w_i \geq 1/3$ and $h_i \leq 3\varepsilon$ for $r_i \in T_3$.

Finally, let $h_4 := h - h_1 - h_3$ and add the items of T_4 by NFDH into the remaining rectangle of size $(1, h_4)$. Lemma 1 yields that either all items are packed, i.e., $T' = T$, or items $T_4' \subseteq T_4$ of total area at least $A(T_4') \geq 2/3(h_4 - \sqrt{\varepsilon})$ are packed. Thus the total area of the packed items T' is $A(T') \geq 2/3 \cdot h_1 + 2/3 \cdot h_3 - \varepsilon + 2/3(h_4 - \sqrt{\varepsilon}) \geq 2/3\,(h - \sqrt{\varepsilon}) - \varepsilon$.

The running time is dominated by the application of NFDH. □

If $T_4 = \emptyset$ then the last packing step is obsolete and the analysis above yields the following corollary.

Corollary 3. *Given a rectangle $R = (1, h)$ and a set T of items that all have area at most ε such that $T_2 \cup T_4 = \emptyset$. We can find a packing of a selection $T' \subseteq T$ into R in time $\mathcal{O}(n)$ such that $T' = T$ or*

$$A(T') \geq \frac{2}{3}h - 2\varepsilon.$$

The above packings are very efficient if there are no items of width within 1/2 and 2/3 as they essentially yield a width guarantee of 2/3 for the whole height, except for some wasted height that is suitably bounded. In order to pack items of T_2, we have to consider both possible orientations to achieve a total area of more than 1/2 in a packing. The following main lemma shows how sets of items including items of width within 1/2 and 2/3 are being processed.

Lemma 2. *Let $0 < \varepsilon \leq 1/68$. Given a set T of items that all have area at most ε such that for all $r \in T$ the total area of $T \setminus \{r\}$ is less than $1/2 + \varepsilon$. We can find a packing of T into a unit bin in time $\mathcal{O}((n \log^2 n)/\log\log n)$.*

Proof. Let h_2 be the total height of items in T_2. We present three methods for packing T depending on h_2. For each method we give a lower bound on the total area of items that are packed. Afterwards we show that there cannot be any items that remain unpacked. Throughout the proof, we assume that we do not run out of items while packing the items in T. This will eventually lead to a contradiction in all three cases.

Case 1: $h_2 \leq 1/3$
Stack the items of T_2 left-justified into the lower left corner of the bin. Use Lemma 3 to pack $T_1 \cup T_3 \cup T_4$ into the rectangle $(1, 1 - h_2)$ above the stack–see Figure 3. We get an overall packed area of

$$\begin{aligned} A \geq \frac{h_2}{2} + \frac{2}{3}\left(1 - h_2 - \sqrt{\varepsilon}\right) - \varepsilon &= \frac{2}{3} - \frac{h_2}{6} - \varepsilon - \frac{2}{3}\sqrt{\varepsilon} \\ &\geq \frac{11}{18} - \varepsilon - \frac{2}{3}\sqrt{\varepsilon} \qquad\qquad (\text{since } h_2 \leq \frac{1}{3}). \end{aligned}$$

Case 2: $h_2 \in (1/3, 2/3]$

Stack the items of T_2 left-justified into the lower left corner of the bin. Let $B = (1/3, h_2)$ be the free space to the right of the stack. We are going to pack items from $X = \{r_i \in T_3 \cup T_4 \mid w_i \leq h_2\}$ into B. Take an item from X and add it to X' as long as X is nonempty and $A(X') \leq h_2/6 - \varepsilon$. Rotate the items in X' and use Steinberg's algorithm (Corollary 1) to pack them into B. This is possible since the area of B is $h_2/3$, $A(X') \leq h_2/6$, and $h_i \leq h_2$ and $w_i \leq \sqrt{\varepsilon} \leq 1/6$ for $r_i \in X'$ (w_i and h_i are the rotated lengths of r_i). Use Lemma 3 to pack $(T_1 \cup T_3 \cup T_4) \setminus X'$ into the rectangle $(1, 1 - h_2)$ above the stack–see Figure 3. We distinguish two cases. If $A(X') \geq h_2/6 - \varepsilon$, then

$$A \geq \overbrace{\frac{h_2}{2}}^{T_2} + \overbrace{\frac{h_2}{6} - \varepsilon}^{X'} + \overbrace{\frac{2}{3}\left(1 - h_2 - \sqrt{\varepsilon}\right) - \varepsilon}^{(T_1 \cup T_3 \cup T_4)\setminus X'} = \frac{2}{3} - 2\varepsilon - \frac{2}{3}\sqrt{\varepsilon}.$$

Otherwise $A(X') < h_2/6 - \varepsilon$ and since no further item was added to X' we have $X' = X$. As $h_2 > 1/3$ we have $T_4 \subseteq X$ and we can apply Corollary 3 to get a total area of

$$\begin{aligned} A \geq \frac{h_2}{2} + \frac{2}{3}(1 - h_2) - 2\varepsilon &= \frac{2}{3} - \frac{h_2}{6} - 2\varepsilon \\ &\geq \frac{5}{9} - 2\varepsilon \qquad\qquad (\text{since } h_2 \leq \frac{2}{3}). \end{aligned}$$

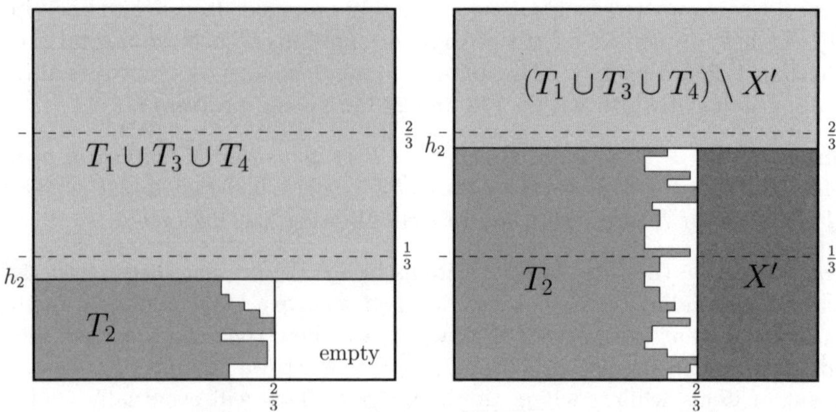

Fig. 3. Packing in Case 1 ($h_2 \leq 1/3$) and Case 2 ($1/3 < h_2 \leq 2/3$)

Case 3: $h_2 \in (2/3, 1 + 4\varepsilon]$
See Figure 4 for an illustration of the following packing and the notations. Order the items of T_2 by non-increasing order of width. Stack the items left-justified into the lower left corner of the bin while the current height h is less or equal to the width of the last item that was packed. In other words, the top right corner of the last item of this stack is above the line from $(1/2, 1/2)$ to $(2/3, 2/3)$, whereas the top right corners of all other items in the stack are below this line. Denote the height of the stack by h and the set of items that is packed into this stack by X_1. Let $r' = (w', h')$ be the last item on the stack. Clearly, $w_i \leq h$ for all items $r_i \in T_2 \setminus X_1$.

Consider the free space $B = (1/3, h)$ to the right of the stack. Rotate the items in $T_2 \setminus X_1$ and stack them horizontally, bottom-aligned into B. Stop if an item does not fit. We denote the items that are packed into B by X_2. Rotate the remaining items $T_2 \setminus (X_1 \cup X_2)$ back into their original orientation and stack them on top of the first stack X_1. Let this set of items be X_3 and the total height of the stack $X_1 \cup X_3$ be $\widehat{h}$. Use Lemma 3 to pack $T_1 \cup T_3 \cup T_4$ into the rectangle $(1, 1 - \widehat{h})$ above the stack $X_1 \cup X_3$.

Since $w_i \geq (h - h')$ for $r_i \in X_1 \setminus \{r'\}$ we have $A(X_1) \geq (h - h')^2 + h'/2$. Again we distinguish two cases for the analysis. If $X_3 = \emptyset$ (or equivalently $\widehat{h} = h$), then $A(X_2) \geq (h_2 - h)/2$ and therefore

$$\begin{aligned} A &\geq \overbrace{(h - h')^2 + \frac{h'}{2}}^{X_1} + \overbrace{\frac{h_2 - h}{2}}^{X_2} + \overbrace{\frac{2}{3}\left(1 - h - \sqrt{\varepsilon}\right) - \varepsilon}^{T_1 \cup T_3 \cup T_4} \\ &> (h - h')^2 + \frac{h'}{2} + \frac{1}{3} - \frac{h}{2} + \frac{2}{3}\left(1 - h - \sqrt{\varepsilon}\right) - \varepsilon =: A_1 \quad (\text{since } h_2 > \frac{2}{3}). \end{aligned}$$

To find a lower bound for the total packed area we consider the partial derivative of A_1 to h', which is $\frac{\partial A_1}{\partial h'} = 2h' - 2h + 1/2$. Since $2h' - 2h + 1/2 < 0$ for

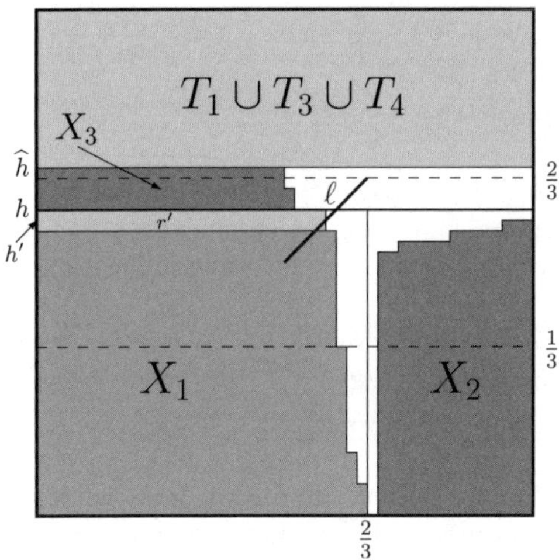

Fig. 4. Packing in Case 3 ($2/3 < h_2 \leq 1 + 4\varepsilon$). Item r' of height h' is dipicted larger than $\varepsilon \leq 1/68$ for the sake of visibility. The diagonal line ℓ shows the threshold at which the stack X_1 is discontinued.

$h' \leq 2\varepsilon$ and $h \geq 1/2$, the total packed area is minimized for the maximal value $h' = 2\varepsilon$ for any h in the domain. After inserting this value for h' we get $A_1 = (h - 2\varepsilon)^2 + \varepsilon + 1/3 - h/2 + 2/3(1 - h - \sqrt{\varepsilon}) - \varepsilon$ and $\frac{\partial A_1}{\partial h} = 2h - 7/6 - 4\varepsilon$. Thus the minimum is acquired for $h = 7/12 + 2\varepsilon$. We get

$$\begin{aligned} A_1 &\geq \left(\frac{7}{12}\right)^2 + \varepsilon + \frac{1}{3} - \frac{7}{24} - \varepsilon + \frac{2}{3}\left(\frac{5}{12} - 2\varepsilon - \sqrt{\varepsilon}\right) - \varepsilon \\ &= \frac{95}{144} - \frac{7}{3}\varepsilon - \frac{2}{3}\sqrt{\varepsilon}. \end{aligned}$$

Otherwise $X_3 \neq \emptyset$ (or equivalently $\widehat{h} > h$) and thus $A(X_2) \geq 1/2(1/3 - 2\varepsilon)$ as the stack X_2 leaves at most a width of 2ε of B unpacked. Furthermore, $\widehat{h} \leq 2/3 + 6\varepsilon$ since $h_2 \leq 1 + 4\varepsilon$ and a width of at least $1/3 - 2\varepsilon$ is packed into B. Since $A(X_3) \geq (\widehat{h} - h)/2$ and $\widehat{h} \leq 2/3 + 6\varepsilon$ we get

$$\begin{aligned} A &\geq \overbrace{(h - h')^2 + \frac{h'}{2}}^{X_1} + \overbrace{\frac{1}{2}\left(\frac{1}{3} - 2\varepsilon\right)}^{X_2} + \overbrace{\frac{\widehat{h} - h}{2}}^{X_3} + \overbrace{\frac{2}{3}\left(1 - \widehat{h} - \sqrt{\varepsilon}\right) - \varepsilon}^{T_1 \cup T_3 \cup T_4} \\ &\geq (h - h')^2 + \frac{h'}{2} + \frac{1}{2}\left(\frac{1}{3} - 2\varepsilon\right) - \frac{1}{9} - \varepsilon - \frac{h}{2} + \frac{2}{3}\left(1 - \sqrt{\varepsilon}\right) - \varepsilon := A_2. \end{aligned}$$

With an analysis similar to before we see that A_2 is minimal for $h = 1/2$ and $h' = 2\varepsilon$. We get

$$A_2 \geq \left(\frac{1}{2} - 2\varepsilon\right)^2 + \varepsilon + \frac{1}{6} - \varepsilon - \frac{1}{9} - \varepsilon - \frac{1}{4} + \frac{2}{3}\left(1 - \sqrt{\varepsilon}\right) - \varepsilon$$
$$\geq \frac{13}{18} + 4\varepsilon^2 - 4\varepsilon - \frac{2}{3}\sqrt{\varepsilon}.$$

If $h_2 > 1 + 4\varepsilon$ then $A(T_2) \geq 1/2 \cdot h_2 > 1/2 + 2\varepsilon$, which is a contradiction to the assumption of the lemma.Therefore the three cases cover all possibilities.

It is easy to verify that for $0 < \varepsilon \leq 1/68$ the following inequalities hold.

$$11/18 - \varepsilon - 2/3\sqrt{\varepsilon} \geq 1/2 + \varepsilon \qquad 2/3 - 2\varepsilon - 2/3\sqrt{\varepsilon} \geq 1/2 + \varepsilon$$
$$5/9 - 2\varepsilon \geq 1/2 + \varepsilon \qquad 95/144 - 7/3\varepsilon - 2/3\sqrt{\varepsilon} \geq 1/2 + \varepsilon$$
$$13/18 + 4\varepsilon^2 - 4\varepsilon - 2/3\sqrt{\varepsilon} \geq 1/2 + \varepsilon$$

Now let us assume that we do not run out of items while packing a set T with the appropriate method above. Then the packed area is at least $1/2 + \varepsilon$ as the inequalities above show. The contradiction follows from the precondition that removing an arbitrary item form T yields a remaining total area of less than $1/2 + \varepsilon$. Thus all items are packed. □

5 The Approximation Ratio

Theorem 4. *There is an approximation algorithm for rectangle packing with rotations with an absolute worst case ratio of 2.*

Proof. Recall that we denote the number of bins used for an optimal packing of the large items by ℓ. Obviously $\ell \leq \mathrm{OPT}$. Let s be the number of bins used for packing only small items. If $s \leq \ell$, then the total number of bins is $\ell + s \leq 2\ell \leq 2\mathrm{OPT}$. If $s > \ell$, then at least one bin is used for small items and thus all bins for large items contain items with a total area of at least $1/2 - \varepsilon$. According to the partition of the remaining small items, all but the last bin for the small items contain items with a total area of at least $1/2 + \varepsilon$. Let A be the total area of all items and let $f > 0$ be the area of the items contained in the last bin. Then

$$\mathrm{OPT} \geq A \geq \ell \cdot \left(\frac{1}{2} - \varepsilon\right) + (s-1) \cdot \left(\frac{1}{2} + \varepsilon\right) + f > (s + \ell - 1) \cdot \frac{1}{2}.$$

Thus $s + \ell < 2\mathrm{OPT} + 1$ and we get $s + \ell \leq 2\mathrm{OPT}$ which proves the theorem. □

6 Conclusion and Future Work

The algorithm we presented depends on the asymptotic approximation algorithm from [1], in particular, the constant k that follows from this algorithm. It would be interesting to design an approximation algorithm for rectangle packing with rotations with asymptotic approximation ratio strictly less than 2 and small additive term. This could also improve the efficiency of our algorithm.

We conjecture that every set of items of height at most 1/2 and total area at most 5/9 can be packed into a unit bin using rotations. This would again improve the efficiency of our algorithm and might be useful for other packing problems as well. Other interesting open questions for further investigation include the following.

1. Does an approximation algorithm for rectangle packing *without rotations* with an absolute worst case ratio of 2 exist? As we pointed out in the introduction, the best-known approximation ratio for this problem is 3 [15].
2. Does an approximation algorithm for strip packing with or without rotations with an absolute worst case ratio less than 2 exist? An answer to this question for strip packing without rotations would narrow the gap between the lower bound of 3/2 (as strip packing without rotations is a generalization of one-dimensional bin packing) and the upper bound of 2 from Steinberg's algorithm [13].

References

1. Bansal, N., Caprara, A., Sviridenko, M.: Improved approximation algorithms for multidimensional bin packing problems. In: FOCS: Proc. 47th IEEE Symposium on Foundations of Computer Science, pp. 697–708 (2006)
2. Bansal, N., Correa, J.R., Kenyon, C., Sviridenko, M.: Bin packing in multiple dimensions - inapproximability results and approximation schemes. Mathematics of Operations Research 31(1), 31–49 (2006)
3. Caprara, A., Lodi, A., Monaci, M.: Fast approximation schemes for two-stage, two-dimensional bin packing. Mathematics of Operations Research 30(1), 150–172 (2005)
4. Chlebík, M., Chlebíková, J.: Inapproximability results for orthogonal rectangle packing problems with rotations. In: CIAC: Proc. 6th Conference on Algorithms and Complexity, pp. 199–210 (2006)
5. Coffman Jr., E.G., Garey, M.R., Johnson, D.S., Tarjan, R.E.: Performance bounds for level-oriented two-dimensional packing algorithms. SIAM Journal on Computing 9(4), 808–826 (1980)
6. Jansen, K., Solis-Oba, R.: New approximability results for 2-dimensional packing problems. In: MFCS: Proc. 32nd International Symposium on Mathematical Foundations of Computer Science, pp. 103–114 (2007)
7. Jansen, K., van Stee, R.: On strip packing with rotations. In: STOC: Proc. 37th ACM Symposium on Theory of Computing, pp. 755–761 (2005)
8. Jansen, K., Zhang, G.: Maximizing the total profit of rectangles packed into a rectangle. Algorithmica 47(3), 323–342 (2007)
9. Kenyon, C., Rémila, E.: A near optimal solution to a two-dimensional cutting stock problem. Mathematics of Operations Research 25(4), 645–656 (2000)
10. Leung, J.Y.-T., Tam, T.W., Wong, C.S., Young, G.H., Chin, F.Y.: Packing squares into a square. Journal of Parallel and Distributed Computing 10(3), 271–275 (1990)
11. Meir, A., Moser, L.: On packing of squares and cubes. Journal of Combinatorial Theory 5, 126–134 (1968)

12. Schiermeyer, I.: Reverse-fit: A 2-optimal algorithm for packing rectangles. In: van Leeuwen, J. (ed.) ESA 1994. LNCS, vol. 855, pp. 290–299. Springer, Heidelberg (1994)
13. Steinberg, A.: A strip-packing algorithm with absolute performance bound 2. SIAM Journal on Computing 26(2), 401–409 (1997)
14. van Stee, R.: An approximation algorithm for square packing. Operations Research Letters 32(6), 535–539 (2004)
15. Zhang, G.: A 3-approximation algorithm for two-dimensional bin packing. Operations Research Letters 33(2), 121–126 (2005)

A Preemptive Algorithm for Maximizing Disjoint Paths on Trees

Yossi Azar[1,2,⋆], Uriel Feige[3], and Daniel Glasner[1]

[1] Tel Aviv University, Tel Aviv, 69978, Israel
[2] Microsoft Research, Redmond WA, 98052-6399, USA
[3] Weizmann Institute, Rehovot 76100, Israel
azar@tau.ac.il, uriel.feige@weizmann.ac.il, dglasner@gmail.com

Abstract. We consider the online version of the maximum vertex disjoint path problem when the underlying network is a tree. In this problem, a sequence of requests arrives in an online fashion, where every request is a path in the tree. The online algorithm may accept a request only if it does not share a vertex with a previously accepted request. The goal is to maximize the number of accepted requests. It is known that no online algorithm can have a competitive ratio better than $\Omega(\log n)$ for this problem, even if the algorithm is randomized and the tree is simply a line. Obviously, it is desirable to beat the logarithmic lower bound. Adler and Azar [SODA 1999] showed that if preemption is allowed (namely, previously accepted requests may be discarded, but once a request is discarded it can no longer be accepted), then there is a randomized online algorithm that achieves constant competitive ratio on the line. In the current work we present a randomized online algorithm with preemption that has constant competitive ratio on any tree. Our results carry over to the related problem of maximizing the number of accepted paths subject to a capacity constraint on vertices (in the disjoint path problem this capacity is 1). Moreover, if the available capacity is at least 4, randomization is not needed and our online algorithm becomes deterministic.

1 Introduction

We consider the online version of the maximum vertex disjoint paths problem, and of paths selection subject to congestion (a.k.a. capacity) constraints. Given a communication network which is a connected graph $G = (V, E)$ (where $|V| = n$), the on-line algorithm processes a sequence of call requests. Each request specifies a pair of vertices $(v, w) \in V \times V$. When a request arrives the algorithm can accept it by allocating a path connecting v and w in G, or reject it. The goal is to maximize the number of accepted calls in such a way that the allocated paths conform with the congestion constraints.

The performance of an on-line algorithm is measured by its competitive ratio. A deterministic or randomized on-line algorithm is ρ-competitive if for any input

⋆ Research supported in part by the Israel Science Foundation.

J. Gudmundsson (Ed.): SWAT 2008, LNCS 5124, pp. 319–330, 2008.

sequence its (expected) benefit is not less than $1/\rho$ times the benefit of an optimal off-line solution.

A preemptive algorithm is an algorithm which is allowed to preempt accepted calls. Such an algorithm may decide at any point to discard any number of calls which it has already accepted. These calls may not be recalled at a later time and do not count towards the algorithm's benefit.

Versions of this problem and its generalization, the *call control* problem, in which call requests also have varying *bandwidth* and *benefit* specifications, have been extensively studied. See for example [4,6,8], for surveys of the problem see [10] and [15].

Our results: We present a randomized preemptive algorithm for the on-line maximum vertex disjoint paths problem on trees, and show that it has constant competitive ratio. Our result is best possible in the sense that if one disallows either randomization or preemption, then every online algorithm cannot be better than $\Omega(\log n)$ competitive, even on line networks [12,7,17]. We also extend our result to maximizing the number of paths subject to a congestion bound of b for all $b > 1$. When $b \geq 4$, our algorithm can be made deterministic. For any b, preemption is still provably required if one is to achieve a constant competitive ratio.

Previously, $\Theta(\log D)$ competitive algorithms were known for trees where D is the diameter of the tree (see [8] and [16]). Those algorithms are non-preemptive. A constant competitive algorithm was known only for the line network [1], and as noted above, it is unavoidable that the online algorithm achieving this is preemptive and randomized. That algorithm can be made deterministic when a congestion bound of $b \geq 2$ is given.

Related work: There are numerous versions of the disjoint path problem, depending on whether graphs are directed or undirected (we consider undirected graphs), capacity constraints are on edges or vertices (we assume that they are on vertices), requests arrive as paths or as source-destination pairs and the algorithm may choose the path (for trees this does not matter, and for general graphs we assume that requests are source-destination pairs), algorithms are online or off-line (we consider the online case), algorithms are randomized or deterministic (we allow randomization), whether preemption is allowed in online settings (we allow preemption), and whether the underlying graph can be arbitrary or has some special structure (we consider trees). For lack of space, we shall mention only those results that we find most informative to our current setting.

Off-line setting: For small capacity bound b on edges, there is no polynomial time constant approximation [3] (unless NP has quasi-polynomial time algorithms) for a general network. In contrast, if the capacity bound is more than logarithmic then randomized rounding of a linear programming relaxation gives a $(1+\epsilon)$ approximation for maximizing the number of paths [18], and this holds regardless of whether capacity constraints are placed on edges or vertices.

On trees the maximum edge disjoint paths is solvable in polynomial time, and becomes NP-hard when edges have a capacity bound $b \geq 2$ [13]. We observe here that the maximum paths problem in trees with vertex capacity bound b is solvable using dynamic programming in time $n^{b+O(1)}$, and becomes NP-hard only when b grows as a function of n.

Online setting: When the capacity bound is $b \geq \log n$, the deterministic non-preemptive algorithm of [6] is $O(\log n)$ competitive on a general network. This is the best possible even among randomized algorithms, in the sense that there is a lower bound of $\Omega(\log n)$ for non-preemptive algorithms for any allowed congestion even for a line network. For the disjoint paths problem (i.e. congestion is $b = 1$) on a general network, an $\Omega(n^{\epsilon})$ lower bound was shown in [9] even for randomized preemptive algorithms. This lower bound is not known to extend to the case where $b \geq 2$. When the requests are paths rather than source-destination pairs and the capacity constraint is $b-1$, there is an $\Omega(n^{1/b}/b)$ lower bound on the competitive ratio of deterministic preemptive and randomized non-preemptive algorithms, and an $\Omega(n^{1/(2b)}/b)$ lower bound for randomized preemptive algorithms [2]. (Lower bounds for the case when requests are paths involve requests that need not resemble shortest paths.)

For some specific networks such as trees, meshes and classes of planar graphs (see [7,8,14]) there are known non-preemptive algorithms with $O(\log n)$ competitive ratios for the disjoint paths problem.

It still remains open whether a sub-logarithmic (randomized or deterministic) preemptive algorithm exists for general networks when we allow high congestion. Our result shows that this is possible for trees even when the congestion is low.

Overview of the paper: This paper is an abbreviated version, the full version can be found on the world wide web. In section 2 we introduce some definitions and notation. In section 3, which is the main section, we present a deterministic preemptive algorithm with constant competitive ratio. However, this algorithm assumes that the vertices have a capacity of 4 rather than 1. In section 4 we use randomization in order to remove the assumption on capacity, and thus derive a randomized preemptive algorithm for the disjoint paths problem. The extension of our results to the capacity b case is discussed in 5.

Our techniques: The approach followed in Section 3 is to decompose the tree problem to a sum of independent subproblems on line networks, and then on each subproblem to use the algorithm from [1] that has a constant competitive ratio on the line. Namely, in an online fashion our algorithm attempts to partition the requests into subsequences. With each subsequence it associates one path (hence, a line network) in the tree, and the subsequences have the property that all requests for the same subsequence intersect the path that is associated with the subsequence, and do not intersect any request from any other subsequence. Achieving a partition with the above property is in general impossible, so our online algorithm will need to drop some of the requests, so as to be able to partition the remaining sequence of requests into subsequences. Our analysis will show that the approximation ratio does not suffer much because of these dropped requests. An additional source of difficulty is that the line algorithm from [1] cannot be applied as is to a subsequence. The reason for this is that the partition to subsequences is dynamic and is not known in advance, and hence the path associated with a subsequence is also not fixed in advance. We overcome this problem by partitioning each subsequence into two groups. In one group, corresponding to the part of the path that is already fixed, we apply the line algorithm of [1]. In the other group, corresponding to the part of the path that may still grow dynamically, we apply a new on-line algorithm which follows the behavior of the off-line algorithm for the *activity selection* problem.

2 Preliminaries

We consider a network T which is a tree. By choosing an arbitrary vertex r we root the tree. A call request is characterized by two distinct nodes, since the underlying network is a tree a call request defines a single path. We denote the input sequence of call requests by σ and will refer to the call requests as paths.

Without loss of generality we can assume that all call requests are from a leaf to a leaf. This can be achieved by adding a new node for each internal node of T and connecting it to its corresponding node.

The *depth* of a node is the length of the path connecting it to the root. We define the *least depth node* of a path P, denoted by $ldn(P)$, as the node with the least depth in P. A *monotonic path*, is a path whose sequence of node depths is monotonic. Any non-monotonic path P is comprised of two monotonic paths which intersect at $ldn(P)$. Having fixed some arbitrary orientation, we call them $left(P)$ and $right(P)$. We define the *maximal depth node* of a monotonic path P, denoted by $mdn(P)$, as the node with the maximal depth in P. The notation $[v, w)$ will be used for the path connecting nodes v and w, excluding w.

The *congestion* created by a set of calls $C \subseteq \sigma$ on a node v is the number of calls in C which intersect v. The congestion on a subgraph $H \subseteq T$ is the maximal congestion on the nodes of H. The congestion created by an on-line algorithm $\mathcal{A}$ is the maximal congestion created by $\mathcal{A}(\sigma)$ on T for all input sequences σ, where $\mathcal{A}(\sigma) \subseteq \sigma$ are the calls accepted by $\mathcal{A}$. We say that an algorithm or a subset of calls is *b-congested* if the maximal congestion created by it on T is bounded by b.

The performance of a b-congested randomized on-line algorithm $\mathcal{A}$, is measured in terms of its competitive ratio, defined as follows. Let $\mathcal{OPT}_\sigma \subseteq \sigma$ be a maximal size b-congested subset. We say that randomized $\mathcal{A}$ is *ρ-competitive* if for all request sequences σ we have $E(|\mathcal{A}(\sigma)|) \geq \frac{1}{\rho}|\mathcal{OPT}_\sigma|$. During the analysis we will compare the performance of deterministic b-congested on-line algorithms on an input sequence σ to a maximal size 1-congested subset (that is, a subset of disjoint calls). We denote such a selection by $\mathcal{OPT}_\sigma^{(1)} \subseteq \sigma$ and say that $\mathcal{A}$ is *ρ-competitive against a 1-congested optimal selection* if for all request sequences σ we have $|\mathcal{A}(\sigma)| \geq \frac{1}{\rho}|\mathcal{OPT}_\sigma^{(1)}|$.

Some of the objects we will discuss evolve as a function of the input requests. We will use the notation O^* for such an object O, to denote its final state.

3 A 4-Congested Deterministic Algorithm

In this section we present a deterministic on-line algorithm whose maximal congestion does not exceed 4. We will also show that it is 6 competitive against a 1-congested optimal solution on the same request sequence.

Overview: The algorithm dynamically partitions the incoming calls into subsequences σ^i for $i = 1, \ldots, k$. The number of subsequences k, is not known in advance and increases over time. This partitioning is described in subsection 3.1. An algorithm for processing the calls in a single subsequence is given in subsection 3.2. The algorithm for combining the selections made on each subsequence into a global selection is discussed in subsection 3.3.

3.1 Partitioning σ into Subsequences

Definition 1. *Let $S \subseteq T$ be the subtree connecting the least depth nodes of the calls in σ and r, where r is the root of T. A* stem structure *for σ is a partition of S into node disjoint monotonic paths such that the maximal depth node in each path is a leaf of S. Each path (with one exception) is half open, i.e., it contains its maximal depth node but does not contain its least depth node. One path that contains the root r is closed, i.e., it contains both its maximal depth node and least depth node.*

Given a stem structure for σ we denote the closed path that contains r, by $stem^1$. We number the half open paths $2, \ldots, k$ and refer to the i'th such path as $stem^i$. The node incident in $stem^i$'s open edge which does not belong to $stem^i$ is called the *root* node of $stem^i$.

The stem structure has a tree hierarchy. Specifically, $stem^1$ is the root stem and for all other stems, a stem's parent is the stem that contains its root node.

Using a stem structure we can partition the calls in σ into subsequences. The calls whose least depth node lies in $stem^i$ are the calls in σ^i. Note that $stem^i$ is a monotonic path that connects the least depth nodes of the calls in σ^i, thus providing a line network structure.

We will use the procedure *StemStructure* described in figure 1 to create and maintain a stem structure for σ and partition the calls accordingly in an online fashion.

We can show (proof in the full version) that the procedure *StemStructure* maintains a stem structure for σ. In fact, the procedure generates a sequence of stem structures as a function of σ. We note that when further calls arrive the stem structure never "shrinks". In particular, for all i and for each arriving call, $stem^i$ before the arrival of the call is contained in $stem^i$ as modified by the call. This implies that $depth(mdn(stem^i))$ is a non decreasing sequence. Furthermore, existing stems are never removed, only new stems may be added. A stem's root node is fixed and does not change once the stem has been created. The stem's parent and ancestor stems are fixed at the moment of its creation but descendent stems may be created later on.

Procedure: *StemStructure*
Initialize: $i \leftarrow 1$, $\sigma^1 \leftarrow \emptyset$, $stem^1 \leftarrow r$
for each incoming call $P \in \sigma$
Starting at $ldn(P)$ traverse the path to r until reaching a node v belonging to $stem^j$ for some j
if $stem^j \cup [ldn(P), v)$ is a monotonic path
then $\sigma^j \leftarrow \sigma^j \cup P$, $stem^j \leftarrow stem^j \cup [ldn(P), v)$
else
$\sigma^{i+1} \leftarrow P$, $stem^{i+1} \leftarrow [ldn(P), v)$, $i \leftarrow i + 1$

Fig. 1. Algorithm for partitioning the calls and maintaining the stems

3.2 An Algorithm for Subsequence σ^i

In this subsection we consider the processing of the calls in a single subsequence σ^i competing against a 1-congested optimal selection on these calls only.

In an off-line setting, by considering the intersection of the calls in σ^{*i} (the final state of σ^i) with the appropriate stem, $stem^{*i}$ (the final state of $stem^i$) we can reduce the problem to a line network.

Lemma 1. *Let $C \subseteq \sigma^{*i}$, if all calls in C have a common (non-empty) intersection and $ldn(P)$ is of maximal depth in $\{ldn(Q)|Q \in C\}$ then $ldn(P) \in \bigcap_{Q \in C} Q$.*

Proof. Whenever two paths P and Q intersect $ldn(P \cap Q) = \max(ldn(P), ldn(Q))$. Since $ldn(P)$ is maximal in $\{ldn(Q)|Q \in C\}$, all calls in C intersect $ldn(P)$. □

Corollary 1. *A bound on the maximal congestion created by σ^i on $stem^i$ is also a bound on the congestion created by the calls in σ^i anywhere on T.*

We assume that we are given an algorithm *Line* for maximizing vertex disjoint paths on a line network. Specifically, a 2-congested algorithm for maximizing edge disjoint paths on a line was shown in [1]. It is 2 competitive against a 1-congested optimal selection.

The natural approach would be to reduce the tree problem to several line problems and apply the line algorithm on each one separately using the above corollary. A difficulty which arises in the on-line setting is that $stem^{*i}$ is not known in advance. Specifically, even after a call has been assigned to a subsequence its intersection with $stem^{*i}$ is not always known. This uncertainty rules out a straightforward reduction to an on-line algorithm for a line network.

For example, the known algorithm for the line has the property that it preempts a containing call in favor of the contained call. However, in reducing the tree to lines the containment relationship may become uncertain when the calls intersect $mdn(stem^i)$. We illustrate this difficulty in figure 2. Consider the calls P and Q, if $mdn(right(Q) \cap stem^{*i}) \leq mdn(right(P))$ then $Q \cap stem^{*i} \subseteq P \cap stem^{*i}$ and P should be preempted. Otherwise it should not be preempted.

To overcome this problem we make a further distinction between the calls. After a new call P has been assigned to a subsequence σ^i and the stem structure

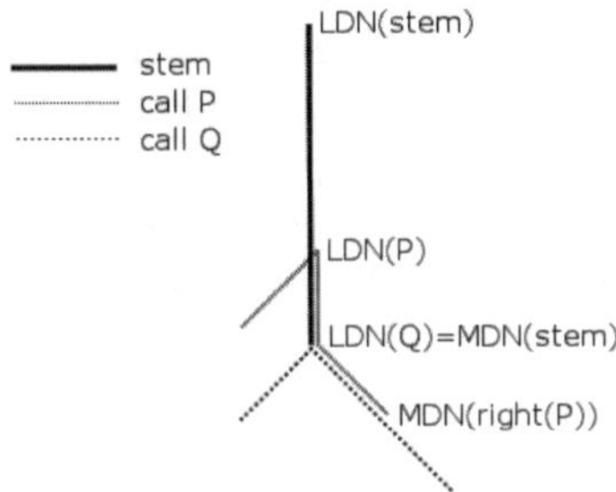

Fig. 2. An example illustrating the difficulty of determining containment relations of the intersections of the calls with the stem in an online setting

has been updated, we classify it as *determined* or *undetermined* depending on its relation to the stem structure. If $P \cap mdn(stem^i) = \emptyset$ we classify it as a determined call, otherwise it is an undetermined call. We denote by $\mathcal{D}$ the set of determined calls and by $\mathcal{U}$ the set of undetermined calls. Note that if P is classified as determined (i.e. $P \cap mdn(stem^i) = \emptyset$) then $P \cap (stem^{*i} \setminus stem^i) = \emptyset$. Hence, the intersection of each determined call is determined upon arrival. In contrast, the intersection of each undetermined call with its stem may change as further calls arrive.

Processing the Determined Calls: The procedure *Determined* described in figure 3 processes the determined calls by reducing the problem to a line network. It is applied to a call in $P \in \sigma^i \cap \mathcal{D}$ after the stem structure has been updated and P has been assigned to subsequence i.

```
Procedure: Determined
  for each incoming call P ∈ σ^i ∩ D
    Process P ∩ stem^i with Line
    Accept P if P ∩ stem^i was accepted by Line
    Preempt calls which were preempted by Line
```

Fig. 3. Algorithm for processing determined calls

Recall that *Line* is a 2-congested algorithm for maximizing edge disjoint paths on a line [1]. It is 2 competitive against a 1-congested optimal selection. To use this algorithm we reduce vertex disjointness to edge disjointness on a line by splitting each vertex into two vertices connected by an edge.

Lemma 2. *For all σ and i, the maximal congestion created by Determined($\sigma^i \cap \mathcal{D}$) on T is 2. Furthermore, Determined is 2 competitive on $\sigma^i \cap \mathcal{D}$ against a 1-congested optimal selection.*

The proof which can be found in the full version follows from the properties of *Line* and corollary 1.

Processing the undetermined calls: The undetermined calls will be processed by an on-line algorithm *UnDetermined* which follows the behavior of the optimal off-line algorithm for interval scheduling also called the activity-selection problem (see [11] chapter 17). In the off-line setting, optimal maximization of disjoint calls on a line can be achieved as follows. Sort the calls in ascending order by the depth of their maximal depth nodes. Accept the first call, discard all calls which intersect it and repeat for the remaining calls.

In the on-line setting the stem provides the line structure. The ordering of the calls is limited to lower bounds given by the current $mdn(stem^i)$. When an undetermined call P arrives, we can only say that $mdn(P \cap stem^{*i}) \geq mdn(stem^i)$. The on-line algorithm resolves this uncertainty by relaxing the congestion limitation for calls whose intersection with the final stem in still undetermined. We show that keeping three options is enough to ensure the correct selection.

Procedure *UnDetermined* described in figure 4 is applied to a call in $P \in \sigma^i \cap \mathcal{U}$ after the stem structure and specifically $stem^i$ have been updated and P has been assigned to subsequence i.

Procedure: *UnDetermined*
Initialize: $unfixed \leftarrow \emptyset$, $fixed \leftarrow \emptyset$
for each incoming call $P \in \sigma^i \cap \mathcal{U}$
(following the update of the stem and assignment of P to σ^i)
(1) **if** $P \cap fixed \neq \emptyset$ **then** reject P
(2) **elseif** $\exists Q \in unfixed, Q \cap mdn(stem^i) = \emptyset$
 (note: this happens only if P extended $stem^i$)
 Let $F \in unfixed$ such that $depth(mdn(F \cap stem^i))$ is minimal,
 $fixed \leftarrow fixed \cup F$
 Preempt all calls in $unfixed$
 $unfixed \leftarrow P$
 (P intersects $stem^i$ only at $mdn(stem^i)$ and hence does not intersect F)
(3) **elseif** $P \cup unfixed$ creates a congestion of 4 on $mdn(stem^i)$
 Let $Q_1 \in unfixed \cup P$ such that $ldn(Q_1) = mdn(stem^i)$
 and let $Q_2, Q_3 \in unfixed \cup P$ such that
 $depth(mdn(Q_2 \cap left(Q_1))), depth(mdn(Q_3 \cap right(Q_1)))$ are minimal
 (breaking ties arbitrarily)
 $unfixed \leftarrow \{Q_1, Q_2, Q_3\}$ (possibly $Q_2 = Q_3$)
 Discard the remaining call (or calls)
(4) **else** $unfixed \leftarrow unfixed \cup P$

Fig. 4. Algorithm for processing undetermined calls

Lemma 3. *For all σ and i, the maximal congestion created by UnDetermined $(\sigma^i \cap \mathcal{U})$ on T is 3. Furthermore, UnDetermined is 1-competitive on $\sigma^i \cap \mathcal{U}$ against a 1-congested optimal selection*

A detailed proof of this claim can be found in the full version. There we show that *UnDetermined* is able, without exceeding the congestion bound of 3, to imitate the behavior of the optimal algorithm for interval scheduling. We show that no matter how the stem structure will evolve, one of the three "optional calls" which $UnDetermined$ keeps in $unfixed$ is in fact the disjoint call with a minimal maximal depth node which the off-line algorithm would have chosen.

Processing the Calls in σ^i: The procedure *SubSeq* (figure 5) is applied to a call P after it has been assigned to subsequence i and $stem^i$ has been updated.

Procedure: *SubSeq*
for each incoming call $P \in \sigma^i$
if $(P \cap mdn(stem^i) = \emptyset)$, **then** Process P with *Determined*
else Process P with $UnDetermined$
Accept P if it was accepted by the algorithm it was assigned to
Preempt calls which were preempted by that algorithm

Fig. 5. Algorithm for processing calls in σ^i

Lemma 4. *For all σ and i, the maximal congestion created by $SubSeq(\sigma^i)$ on T is 4. Furthermore,*

$$|OPT^{(1)}_{\sigma^i}| \leq 2|SubSeq(\sigma^i) \cap \mathcal{D})| + |SubSeq(\sigma^i) \cap \mathcal{U}|$$

For the proof see the full version.

3.3 Combining the Calls from Subsequences

So far we have shown an algorithm for processing calls in each subsequence. Accepting the union of the selections made on each subsequence will result in a globally competitive algorithm. However, since calls in distinct subsequences may intersect, locally bounding the congestion created by σ^i on $stem^i$ does not ensure a global bound.

To attain the global bound we introduce the procedure *Global*. This procedure uses procedure *StemStructure* to partition σ and maintain the stem structure. It then simulates *SubSeq* on each subsequence σ^i. *Global* follows the decisions made by each instance of *SubSeq* but preempts any calls which intersect more than one stem. Note that a call may intersect two stems at the moment of its arrival, or it may come to intersect two stems after it has been accepted, when a new stem is created. In both cases these calls are discarded by *Global*, however the corresponding *SubSeq* algorithm is unaware of these changes and continues to behave as if the calls are there.

An incoming request P is handled by Procedure *Global* described in figure 6.

Lemma 5. *Let k be the number of subsequences, then $k \leq |Global(\sigma) \cap \mathcal{U}|$.*

A detailed proof can be found in the full version. There we show that there is at least one call in each subsequence. Namely, for all i there is an undetermined call $P \in unfixed$ such that $ldn(P) = mdn(stem^i)$ which is not preempted by *Global*.

Procedure: *Global*
for each incoming call $P \in \sigma$
Use procedure *StemStructure* to add P to a subsequence σ^i and update $stem^i$
Simulate *SubSeq* on σ^i and accept / discard calls
which were accepted / discarded by *SubSeq*
Preempt any calls which intersect two stems (do not update simulations)

Fig. 6. Global algorithm

Lemma 6. *The procedure Global discards at most* $3(k-1)$ *calls which were accepted by the instances of SubSeq, of which at most* $2(k-1)$ *are determined calls and* $k-1$ *are undetermined calls.*

Proof. For all $i = 2, \ldots, k$, $stem^i$ has a root node v_i. The set $\{v_i\}_{i=2}^k$ includes at most $k-1$ distinct vertices. We recall that a stem's root node is fixed at the moment of its creation and does not change.

Any call accepted by *SubSeq* running on σ^i, which is preempted by *Global* must intersect some root node $v_j \in stem^i$ such that $stem^i$ is the parent of $stem^j$. The selection made by *SubSeq* on σ^i may include at most two determined calls and one undetermined call that intersect some root node $v_j \in stem^i$.

Thus the total amount of calls which are discarded is bounded by $3(k-1)$ of which at most $2(k-1)$ are determined and $k-1$ are undetermined. □

Theorem 1. *For all* σ*, the maximal congestion created by Global*(σ) *on* T *is 4. Furthermore, Global is 6 competitive on* σ *against a 1-congested optimal selection (recall that Global is 4-congested).*

Proof. For all σ,

$$\begin{aligned}
|OPT_\sigma^{(1)}| = |OPT^{(1)}_{\cup_{i=1}^k \sigma^i}| &\le \textstyle\sum_{i=1}^k |OPT^{(1)}_{\sigma^i}| \\
&\le \textstyle\sum_{i=1}^k 2|SubSeq(\sigma^i) \cap \mathcal{D}| + |SubSeq(\sigma^i) \cap \mathcal{U}| \\
&\le 2|Global(\sigma) \cap \mathcal{D}| + 4(k-1) + |Global(\sigma) \cap \mathcal{U}| + (k-1) \\
&\le 2|Global(\sigma) \cap \mathcal{D}| + 6|Global(\sigma) \cap \mathcal{U}| \le 6|Global(\sigma)|.
\end{aligned}$$

For the bound on the congestion see the full version. The second inequality follows from lemma 4. The third inequality is due to lemma 6. The fourth follows from lemma 5 and the last is because $\sigma = \mathcal{D} \cup \mathcal{U}$. □

4 A Constant Competitive Randomized Algorithm for Disjoint Paths

In this section we present a 1-congested, randomized 24-competitive algorithm. We show that the calls accepted by *Global* can be assigned in an online manner into a small number of 1-congested sets. The randomized algorithm randomly chooses one of these sets and simulates *Global*. It accepts only the calls which are assigned to the chosen set and discards the rest.

Definition 2. *Let $\mathcal{A}$ be an on-line algorithm running on input σ. An on-line d-coloring of $A(\sigma)$ is an on-line assignment $\chi : A(\sigma) \mapsto \{1,\ldots,d\}$. When a call P is accepted the on-line coloring assigns it to some color class $\chi(P) \in \{1,\ldots,d\}$. The coloring is* valid *if intersecting calls have different colors.*

We note that whenever a determined call is accepted it intersects at most two other determined calls. The same is true for undetermined calls. Hence, allocating 3 colors for determined and 3 for undetermined calls we can define a valid on-line 6-coloring for the calls maintained by *Global*. This is shown formally in the full version.

Using the on-line coloring we construct a randomized 1-congested, algorithm *Rand* as described in figure 7.

```
Procedure: Rand
  Make a random selection i ∈ {1,...,6} with probability
  Pr[i] = 1/12 for i ∈ {1,2,3} and Pr[i] = 1/4 for i ∈ {4,5,6}
  for each incoming call P ∈ σ
    Simulate Global on P
    if P is accepted by Global and χ(P) = i, then Accept P
    else Reject P
    Preempt calls which were preempted by Global
```

Fig. 7. Disjoint path algorithm

Theorem 2. *Rand is a* 24 *competitive algorithm for the disjoint paths problem.*

The proof can be found in the full version.

5 A Constant Competitive Randomized Algorithm for Congestion b

In the full version of this paper we extend the setting and allow a bounded maximal congestion of $b > 1$. We use the general method of [5] for benefit problems. We also take advantage of its adaptation to handle preemption as presented in [1]. The following theorem is proved there

Theorem 3. *For all $b > 1$, There exists a b-congested, randomized, preemptive 25-competitive algorithm for the maximal vertex b-congested paths problem on trees. For $b \geq 4$ there is a deterministic constant competitive algorithm.*

References

1. Adler, R., Azar, Y.: Beating the logarithmic lower bound: randomized preemptive disjoint paths and call control algorithms. In: Proc. of the 10th ACM-SIAM Symposium on Discrete Algorithms, pp. 1–10 (1999)
2. Alon, N., Arad, U., Azar, Y.: Independent sets in hypergraphs with applications to routing via fixed paths. In: Hochbaum, D.S., Jansen, K., Rolim, J.D.P., Sinclair, A. (eds.) RANDOM 1999 and APPROX 1999. LNCS, vol. 1671, pp. 16–27. Springer, Heidelberg (1999)

3. Andrews, M., Chuzhoy, J., Khanna, S., Zhang, L.: Hardness of the undirected edge-disjoint paths problem with congestion. In: Proceedings 46th Annual IEEE Symposium on Foundations of Computer Science, pp. 226–244 (2005)
4. Aspnes, J., Azar, Y., Fiat, A., Plotkin, S., Waarts, O.: On-line routing of virtual circuits with applications to load balancing and machine scheduling. Journal of the ACM 44(3), 486–504 (1997); also in Proc. 25th ACM STOC, pp. 623-631 (1993)
5. Awerbuch, B., Azar, Y., Fiat, A., Leonardi, S., Rosen, A.: On-line competitive algorithms for call admission in optical networks. In: Proc. 4th Annual European Symposium on Algorithms, pp. 431–444 (1996)
6. Awerbuch, B., Azar, Y., Plotkin, S.: Throughput-competitive online routing. In: 34th IEEE Symposium on Foundations of Computer Science, pp. 32–40 (1993)
7. Awerbuch, B., Bartal, Y., Fiat, A., Rosén, A.: Competitive non-preemptive call control. In: Proc. of 5th ACM-SIAM Symposium on Discrete Algorithms, pp. 312–320 (1994)
8. Awerbuch, B., Gawlick, R., Leighton, T., Rabani, Y.: On-line admission control and circuit routing for high performance computation and communication. In: Proc. 35th IEEE Symp. on Found. of Comp. Science, pp. 412–423 (1994)
9. Bartal, Y., Fiat, A., Leonardi, S.: Lower bounds for on-line graph problems with application to on-line circuit and optical routing. In: Proc. 28th ACM Symp. on Theory of Computing, pp. 531–540 (1996)
10. Borodin, A., El-Yaniv, R.: Online Computation and Competitive Analysis. Cambridge University Press, Cambridge (1998)
11. Cormen, T.T., Leiserson, C.E., Rivest, R.L.: Introduction to Algorithms. MIT Press, Cambridge (1990)
12. Garay, J., Gopal, I., Kutten, S., Mansour, Y., Yung, M.: Efficient on-line call control algorithms. Journal of Algorithms 23, 180–194 (1993); In: Proc. 2'nd Annual Israel Conference on Theory of Computing and Systems (1993)
13. Garg, N., Vazirani, V.V., Yannakakis, M.: Primal-Dual Approximation Algorithms for Integral Flow and Multicut in Trees. ALGORITHMICA 18, 3–20 (1997)
14. Kleinberg, J., Tardos, E.: Disjoint paths in densely embedded graphs. In: Proc. 36th IEEE Symp. on Found. of Comp. Science, pp. 52–61 (1995)
15. Leonardi, S.: On-line network routing. In: Fiat, A., Woeginger, G. (eds.) Online Algorithms - The State of the Art, ch. 11, pp. 242–267. Springer, Heidelberg (1998)
16. Leonardi, S., Marchetti-Spaccamela, A., Presciutti, A., Rosén, A.: On-line randomized call control revisited. In: Proc. 9th ACM-SIAM Symp. on Discrete Algorithms, pp. 323–332 (1998)
17. Lipton, R.J., Tomkins, A.: Online interval scheduling. In: Proc. of the 5th ACM-SIAM Symposium on Discrete Algorithms, pp. 302–311 (1994)
18. Raghavan, P., Thompson, C.D.: Randomized rounding: a technique for provably good algorithms and algorithmic proofs. Combinatorica 7(4), 365–374 (1987)

Minimum Distortion Embeddings into a Path of Bipartite Permutation and Threshold Graphs*

Pinar Heggernes[1], Daniel Meister[1], and Andrzej Proskurowski[2]

[1] Department of Informatics, University of Bergen, Norway
Pinar.Heggernes@ii.uib.no, Daniel.Meister@ii.uib.no
[2] Department of Information and Computer Science, University of Oregon, USA
andrzej@cs.uoregon.edu

Abstract. The problem of computing minimum distortion embeddings of a given graph into a line (path) was introduced in 2004 and has quickly attracted significant attention with subsequent results appearing in recent STOC and SODA conferences. So far all such results concern approximation algorithms or exponential-time exact algorithms. We give the first polynomial-time algorithms for computing minimum distortion embeddings of graphs into a path when the input graphs belong to specific graph classes. In particular, we solve this problem in polynomial time for bipartite permutation graphs and threshold graphs.

1 Introduction

A metric space is defined by a set of points and a distance function between pairs of points. Given two metric spaces (U, d) and (U', d'), an *embedding* of the first into the second is a mapping $f : U \rightarrow U'$. The embedding has *distortion* c if for all $x, y \in U$, $d(x, y) \leq d'(f(x), f(y)) \leq c \cdot d(x, y)$. Low distortion embeddings between metric spaces are well-studied and have a long history. Embeddings of finite metric spaces into low dimensional geometric spaces have applications in various areas of computer science, like computer vision [18] and computational chemistry (see [9,10] for an introduction and a list of applications).

Minimum distortion embeddings are difficult to compute. It is NP-hard even to approximate by a ratio better than 3 a minimum distortion embedding between two given finite 3-dimensional metric spaces [15].

Every finite metric space can be represented by a matrix whose entries are the distances between pairs of points, and hence corresponds to a graph. Kenyon et al. [11] initiated the study of computing a minimum distortion embedding of a given graph onto[1] another given graph, and they gave a parameterized algorithm for computing a minimum distortion embedding between an arbitrary unweighted graph and a bounded-degree tree. Subsequently, Badoiu et al. [3] gave a constant-factor approximation algorithm for computing minimum distortion embeddings of arbitrary unweighted graphs into trees.

* This work is supported by the Research Council of Norway.

[1] They study a more restricted version of the problem where both graphs have the same number of vertices.

J. Gudmundsson (Ed.): SWAT 2008, LNCS 5124, pp. 331–342, 2008.

Since then, computing a minimum distortion embedding of a given graph on n vertices into a path was identified as a fundamental problem. This is exactly the problem that we study in this paper. Badoiu et al. [2] gave an exponential-time exact algorithm and a polynomial-time $\mathcal{O}(n^{1/2})$-approximation algorithm for arbitrary unweighted input graphs, along with a polynomial-time $\mathcal{O}(n^{1/3})$-approximation algorithm for unweighted trees. They also showed that the problem is hard to approximate within a constant factor. In another paper Badoiu et al. [1] showed that the problem is hard to approximate by a factor polynomial in n, even for weighted trees. They also gave a better polynomial-time approximation algorithm for general weighted graphs, along with a polynomial-time algorithm that approximates the minimum distortion embedding of a weighted tree into a path by a factor that is polynomial in the distortion.

We initiate the study of designing polynomial-time algorithms for exact computation of minimum distortion embeddings into a path for input graphs of specific graph classes. In particular we give polynomial-time algorithms for the solution of this problem on bipartite permutation graphs and on threshold graphs. These are widely studied graph classes with important theoretical applications, and they can both be recognized in linear time [5,6,13]. Minimum distortion into a path is very closely related to the widely known and extensively studied graph parameter bandwidth. The only difference between the two parameters is that a minimum distortion embedding has to be non-contractive, meaning that the distance in the embedding between two vertices of the input graph has to be at least their original distance, whereas there is no such restriction for bandwidth. Bandwidth is known to be one of the hardest graph problems; it is NP-hard even for very simple graphs like caterpillars of hair-length at most 3 [14], and it is hard to approximate by a constant factor even for trees [4]. Polynomial-time algorithms for the exact computation of bandwidth are known for very few graph classes, including bipartite permutation graphs [7] and threshold graphs (they are interval graphs) [12,17]. However, simple examples exist to show that these bandwidth algorithms cannot be used to generate minimum distortion embeddings into a path for these graph classes. In fact, there exist very simple bipartite permutation graphs, like $K_{3,4}$, for which no optimal bandwidth layout corresponds to a minimum distortion embedding into a path. It should be noted that the bandwidth and the minimum distortion into a path of a graph can be very different. For example, it is common knowledge that a cycle of length n has bandwidth 2, whereas its minimum distortion into a path is $\Omega(n)$.

The running times of the algorithms that we present are $\mathcal{O}(n^2)$ for bipartite permutation graphs and $\mathcal{O}(n)$ for threshold graphs, where n is the number of vertices of the input graph. As opposed to all non-trivial bandwidth algorithms, both of our algorithms compute the distortion into a path of the given graph *directly*, and not by deciding whether it is at most a given integer. In addition, we give a complete characterization of the distortion into a path of bipartite permutation graphs through their induced subgraphs. No such result is known for bandwidth, even for very small graph classes.

In this extended abstract most proofs and some intermediate technical results needed for those proofs are omitted. In addition to the results appearing in the main part, we report below on our following findings. A full version of this extended abstract with all details and formal proofs exists [8].

- The distortion into a path of a connected proper interval graph is equal to its bandwidth.
- The distortion into a path of a cycle on n vertices is $n-1$.
- The distortion into a path of a complete bipartite graph $K_{n,m}$ with $1 \leq n \leq m$ is $n+m-2$ if $n+m$ is odd, and $n+m-1$ if $n+m$ is even.
- The distortion into a path of a split graph where each of the n vertices of the clique is adjacent to each of the m vertices of the independent set is $n+m-2$, for $n, m \geq 2$.

2 Preliminaries

A graph is denoted by $G = (V, E)$, where V is the vertex set and E is the edge set of G. The set of neighbors of a vertex v is denoted by $N_G(v)$. The *degree* of a vertex v is $d_G(v) = |N_G(v)|$. (We will omit the subscripts when the graph is clear from the context.) The subgraph of G induced by the vertices in $S \subseteq V$ is denoted by $G[S]$. For any $v \in V$, $G{-}v$ denotes $G[V \setminus \{v\}]$. We study unweighted and connected input graphs. A u,v-path is a path between (and including) u and v. The *distance* $\mathrm{d}_G(u,v)$ between two vertices u and v in G is the number of edges in a shortest u,v-path in G. We say that a subgraph H of G is *distance-preserving* if $\mathrm{d}_H(u,v) \leq \mathrm{d}_G(u,v)$ for all $u, v \in V$. It follows directly that distances in H and G are then equal, since every path in H is a path in G. In particular, distance-preserving subgraphs are induced subgraphs. For any mapping f from V to (a subset of) $\mathbb{Z}$, the *distance* $\mathrm{d}_f(u,v)$ between u and v in f is $|f(u) - f(v)|$. We write $u \prec_f v$ when $f(u) < f(v)$. For a vertex v in G, every vertex u with $u \prec_f v$ is *to the left* of v, and every vertex w with $v \prec_f w$ is *to the right* of v in f. We will also informally write *leftmost* and *rightmost* vertices accordingly.

An *embedding into a path (line)* of a graph $G = (V, E)$ is a mapping $\mathcal{E} : V \to \mathbb{Z}$. In the rest of this paper we use simply *embedding* to mean an embedding into a path. An embedding $\mathcal{E}$ is *non-contractive* if $\mathrm{d}_{\mathcal{E}}(u,v) \geq \mathrm{d}_G(u,v)$ for every pair of vertices $u, v \in V$. The *distortion* $\mathrm{D}(G, \mathcal{E})$ of a non-contractive embedding is defined to be the smallest integer c such that $\mathrm{d}_{\mathcal{E}}(u,v) \leq c \cdot \mathrm{d}_G(u,v)$ for every pair of vertices $u, v \in V$. Since we consider only unweighted graphs, it is easy to see that $\mathrm{D}(G, \mathcal{E})$ is the smallest c such that $\mathrm{d}_{\mathcal{E}}(u,v) \leq c$ for every edge uv of G (see also [11]). A *minimum distortion embedding* is a non-contractive embedding for G of smallest possible distortion. In this paper, the *distortion* of G, denoted by $\mathrm{D}(G)$, is the distortion of a minimum distortion embedding for G. Hence, our purpose is to compute $\mathrm{D}(G)$ when G is a bipartite permutation graph or a threshold graph.

Each integer (*position*) in an embedding will be called a *slot*. Exactly n slots of a non-contractive embedding are occupied by the vertices of G, and the rest are called *empty slots*. For a given vertex v, we refer to the rightmost vertex to the left of v of a certain property by *the close vertex to the left of* v and specifying the property (*close vertex to the right* is defined symmetrically). The *vertex ordering underlying* $\mathcal{E}$, $\text{ord}(\mathcal{E})$, is an ordered list of the n vertices occupying the non-empty slots of $\mathcal{E}$ in increasing order of positions. In general, a *vertex ordering* for $G = (V, E)$ is a mapping $\sigma : V \to \{1, 2, \ldots, |V|\}$, thus a restricted embedding. Since every ordering can be considered as a permutation of V, we will also give an ordering as an ordered list of vertices $\sigma = \langle x_1, x_2, \ldots, x_n \rangle$. The following two results are of importance throughout the paper.

Lemma 1. *Let $G = (V, E)$ be a connected graph, and let $\mathcal{E}$ be an embedding for G with $\text{ord}(\mathcal{E}) = \langle x_1, \ldots, x_n \rangle$. If $\text{d}_{\mathcal{E}}(x_i, x_{i+1}) \geq \text{d}_G(x_i, x_{i+1})$ for every $1 \leq i < n$ then $\mathcal{E}$ is non-contractive.*

Note in particular that there is never a need for empty slots between consecutive vertices of $\mathcal{E}$ that are adjacent in G. We say that an embedding *does not contain unnecessary empty slots* if consecutive vertices in the embedding are at distance exactly their distance in the graph.

Lemma 2. *Let $G = (V, E)$ be a graph and let H be a subgraph of G. If H is a distance-preserving subgraph of G, then $\text{D}(G) \geq \text{D}(H)$.*

3 Distortion of Threshold Graphs

A graph is a *threshold graph* if and only if its vertex set can be partitioned into a clique X and an independent set I such that the vertices of I, and equivalently the vertices of X, can can be ordered by neighborhood inclusion [13]. Hence, for any partition $V = X \cup I$ of a threshold graph $G = (V, E) = (X, I, E)$, the I-vertices can be ordered as $a_1, a_2, \ldots, a_m$ such that $N(a_1) \subseteq N(a_2) \subseteq \cdots \subseteq N(a_m)$, and the X-vertices can be ordered as $b_1, b_2, \ldots, b_n$ such that $N(b_1) \supseteq N(b_2) \supseteq \cdots \supseteq N(b_n)$. Consequently, the given orderings correspond to a non-decreasing degree order for the I-vertices and a non-increasing degree order for the X-vertices. For simplicity, we will say *decreasing* instead of non-increasing, and *increasing* instead of non-decreasing. Every connected threshold graph has a universal vertex. Hence, every pair of vertices in a connected threshold graph is at distance at most 2. In $G = (X, I, E)$, if there is no X-vertex without a neighbor in I, there is an I-vertex a that is adjacent to all X-vertices. Then, $(X \cup \{a\}, I \setminus \{a\})$ is also a threshold partition for G. In the following, we assume that an X-vertex of smallest degree has no neighbors outside X.

In this section, we give an efficient algorithm for computing the distortion of threshold graphs.

Lemma 3. *Let $G = (X, I, E)$ be a connected threshold graph. There is a minimum distortion embedding for G without empty slots between X-vertices such that the X-vertices are ordered decreasingly by degree.*

With Lemma 3 we are ready to give an algorithm that computes the distortion of threshold graphs. Let $\mathcal{E}$ be a non-contractive embedding of the input graph G where the X-vertices are ordered by decreasing degree and let u be the leftmost universal vertex in $\mathcal{E}$. We define $R(\mathcal{E})$ to be the distance between u and the rightmost vertex in $\mathcal{E}$, and $L(\mathcal{E})$ to be the maximum taken over all distances between a vertex to the left of u and its rightmost neighbor in $\mathcal{E}$. Then, $\mathrm{D}(G, \mathcal{E}) = \max\{L(\mathcal{E}), R(\mathcal{E})\}$.

Algorithm. `thrg-distortion`
Input threshold graph $G = (X, I, E)$ with $I = \{y_1, \ldots, y_{|I|}\}$ and $d(y_1) \leq \cdots \leq d(y_{|I|})$

begin
  let $\mathcal{E}_0$ be the start embedding;
  let u be the leftmost vertex in $\mathcal{E}_0$;
  set $i = 0$;
  while $R(\mathcal{E}_i) \geq L(\mathcal{E}_i) + 2$ **and** $i < |I|$ **do**
    set $i = i + 1$;
    let $\mathcal{E}_i$ be obtained from $\mathcal{E}_{i-1}$ by moving y_i to the left of u
  end while;
  let $\mathcal{E}$ be obtained from $\mathcal{E}_i$ by moving the close I-vertex to the right of u to the right end;
  return $\min\{\mathrm{D}(G, \mathcal{E}), \mathrm{D}(G, \mathcal{E}_i)\}$ and the corresponding embedding
end.

For completing the algorithm we have to explain three operations. The start embedding is obtained by the following procedure. We only explain the underlying vertex ordering; the embedding then is obtained by adding the necessary empty slots. The X-vertices are ordered decreasingly by degree. The I-vertices are treated separately and in reverse given order, i.e., as $y_{|I|}, \ldots, y_1$, and are placed rightmost between two neighbors as long as possible, and when an I-vertex cannot be placed between two neighbors it is placed at the right end, particularly to the right of the rightmost X-vertex. Note that this embedding has no empty slots between X-vertices.

The second operation is the definition of embedding $\mathcal{E}_i$ inside the **while** loop. If y_i is between X-vertices then y_i is removed, all vertices between u and position $\mathcal{E}_{i-1}(y_i)$ are moved one position to the right, all vertices to the left are moved one position to the left and y_i is placed in the slot previously occupied by u. Note that $\mathcal{E}_i$ is a proper non-contractive embedding for G without unnecessary empty slots and without empty slots between X-vertices. If y_i is to the right of the rightmost X-vertex in $\mathcal{E}_{i-1}$ then y_i is removed, all vertices between u and position $\mathcal{E}_{i-1}(y_i)$ are moved two positions to the right and y_i is placed in the slot at position $\mathcal{E}_{i-1}(u) + 1$.

The third operation defines $\mathcal{E}$ after the **while** loop as follows. If the close I-vertex v to the right of u in $\mathcal{E}_i$ is to the left of the rightmost X-vertex, then remove v, move all vertices to the left of position $\mathcal{E}_i(v)$ one position to the right and place v at the right end at distance 2 to the close vertex to the left. If v is to the right of the rightmost X-vertex, the embedding remains unchanged.

Theorem 1. *There is an $\mathcal{O}(n)$-time algorithm that computes the distortion of a connected threshold graph on n vertices and outputs a corresponding embedding.*

Proof. We briefly sketch the proof. Let $G = (X, I, E)$ be a connected threshold graph and apply `thrg-distortion` to G. Let r be the number of iterations of the **while** loop. We show that $\mathcal{E}$ or $\mathcal{E}_r$ has smallest distortion among all non-contractive embeddings for G with the X-vertices ordered decreasingly by degree. Correctness then follows directly from Lemma 3. For each $i \in \{1, \ldots, r\}$, observe that $R(\mathcal{E}_{i-1}) - 2 \leq R(\mathcal{E}_i) \leq R(\mathcal{E}_{i-1}) - 1$ and $L(\mathcal{E}_{i-1}) + 1 \leq L(\mathcal{E}_i) \leq R(\mathcal{E}_{i-1})$. Hence, $\mathrm{D}(G, \mathcal{E}_i) \leq \mathrm{D}(G, \mathcal{E}_{i-1})$ and $L(\mathcal{E}_i) - R(\mathcal{E}_i) \leq 2$. Note in particular that $R(\mathcal{E}_i) \leq R(\mathcal{F})$ for any non-contractive embedding $\mathcal{F}$ for G with u the leftmost X-vertex and at least $|I| - i$ I-vertices to the right of u. Thus the algorithm stops with an embedding where L and R parameters are balanced and the R parameter is smallest possible. The full proof of correctness then analyzes the few possible differences between these parameters of the output embedding and shows that any change results in the same or larger distortion.

For an implementation of Algorithm `thrg-distortion` in $\mathcal{O}(n)$ time, observe that threshold graphs can be represented in $\mathcal{O}(n)$ space where each I-vertex keeps pointers to its leftmost and rightmost neighbor in the decreasingly sorted degree order of the X-vertices. Hence deciding adjacencies can be done in constant time. We can also in constant time find the position of y_i since it is either at the right end or it is the leftmost of the I-vertices between X-vertices. Hence with an $\mathcal{O}(n)$-time preprocessing we can gather enough information about positions to be able to implement each of the described operations in constant time.

4 Distortion of Bipartite Permutation Graphs

In this section, we show two main results about distortion of bipartite permutation graphs. We give a fast algorithm for computing the distortion of bipartite permutation graphs and we give a complete characterization of bipartite permutation graphs of bounded distortion by forbidden induced subgraphs.

A *bipartite graph* is a graph whose vertex set can be partitioned into two independent sets. We denote such a graph by $G = (A, B, E)$ where $A \cup B$ is the vertex set of G, and A and B are independent sets, also called *color classes*. The partition into color classes is unique for connected bipartite graphs. For each vertex v in a bipartite graph, we let $cc(v)$ denote the color class of vertex v and $\overline{cc}(v)$ denote the other color class.

Bipartite permutation graphs are permutation graphs that are bipartite. For the definition and properties of permutation graphs, we refer to [5]. Let $G = (A, B, E)$ be a bipartite graph. A *strong ordering* for G is a pair of orderings (σ_A, σ_B) on respectively A and B such that for every pair of edges ab and $a'b'$ in G with $a, a' \in A$ and $b, b' \in B$, $a \prec_{\sigma_A} a'$ and $b' \prec_{\sigma_B} b$ implies that $ab' \in E$ and $a'b \in E$. A bipartite graph is a *bipartite permutation graph* if and only if it has a strong ordering [16]. Spinrad et al. give a linear-time recognition algorithm for bipartite permutation graphs, that even produces a strong ordering [16]. It follows from the definition of a strong ordering that if $G = (A, B, E)$ is a connected bipartite permutation graph then any strong ordering (σ_A, σ_B) satisfies the following. For every vertex $a \in A$, the neighbors of a appear consecutively in σ_B.

We begin by defining and analyzing a special kind of bipartite permutation graphs that we will need for proving lower distortion bounds. A *clawpath* is a tree such that the set of its vertices that are not leaves induce a path, and each vertex of the path is adjacent to exactly one leaf. Hence, every vertex of the path has degree 3, except the end vertices of the path that have degree 2. The number of edges on this path is called the *length* of the clawpath. (Clawpaths are thus caterpillars where every vertex that is not a leaf has exactly one neighbor that is a leaf.) We define a *clawpath-like graph* to be a graph obtained from a clawpath by replacing each vertex by a (non-empty) independent set of new vertices. When replacing a vertex v with a set of new vertices $v_1, \ldots, v_\ell$ with $\ell \geq 1$, we give each v_i the same neighborhood as v had. (Thus, we can view this process as iteratively adding to the graph new false twins of chosen vertices.) The *underlying clawpath* of a clawpath-like graph is the clawpath from which the graph was obtained according to the definition. The *length* of a clawpath-like graph is the length of its underlying clawpath. Clawpath-like graphs are both bipartite and permutation. Hence, they are a subset of bipartite permutation graphs. Furthermore, they are connected and contain at least one edge. It can be shown, through a tedious but not difficult case analysis, that for a bipartite permutation graph, every induced subgraph that is a clawpath-like graph is distance-preserving. A second result gives a lower bound on the distortion of clawpath-like graphs.

Lemma 4. *Let $G = (V, E)$ be a clawpath-like graph of length r. Let $k \geq 1$ be an odd integer. If $|V| \geq \frac{1}{2}(rk + r + 2k + 6)$ then $\mathrm{D}(G) \geq k + 2$.*

By Lemmas 4 and 2 we can conclude that a connected bipartite permutation graph that contains a clawpath-like graph of length r on at least $\frac{1}{2}(rk+r+2k+6)$ vertices as induced subgraph has distortion at least $k + 2$. The main structural theorem of this section extends this result to an equivalence.

Theorem 2. *Let $k \geq 1$ be an odd integer. Then, a connected bipartite permutation graph G has distortion at most k if and only if G does not contain a clawpath-like graph of length r on $\frac{1}{2}(rk+r+2k+6)$ vertices as induced subgraph.*

The remaining part of the proof of Theorem 2 will be given through a series of structural and algorithmic results. The proof will be finalized through an algorithm that, on input a connected bipartite permutation graph, an embedding of restricted structure, and an odd integer k, decides whether the graph has distortion at most k or finds a clawpath-like graph. The algorithm will be fast, and it will even produce certificates for its decision. It will work on embeddings of restricted structure. Let $G = (A, B, E)$ be a bipartite permutation graph with strong ordering (σ_A, σ_B). We say that an embedding $\mathcal{E}$ for G is *normalized* (with respect to (σ_A, σ_B)) if it satisfies the following three conditions, where c denotes the leftmost A-vertex with respect to σ_A:

(C1) for every pair a, a' of vertices from A, $a \prec_{\sigma_A} a'$ implies $a \prec_{\mathcal{E}} a'$; and for every pair b, b' of vertices from B, $b \prec_{\sigma_B} b'$ implies $b \prec_{\mathcal{E}} b'$

(C2) for every triple u, v, w of vertices of G where $u \prec_{\mathcal{E}} v \prec_{\mathcal{E}} w$ and $uw \in E$: $uv \in E$ or $vw \in E$

(C3) for every A-vertex x, $\mathrm{d}_{\mathcal{E}}(c, x)$ is even; and for every B-vertex x, $\mathrm{d}_{\mathcal{E}}(c, x)$ is odd.

Thus, in a normalized embedding the slots (containing vertices or empty) are partitioned into "A-slots" and "B-slots": only A-slots can contain A-vertices, and only B-slots can contain B-vertices. We will show that every connected bipartite permutation graph has a minimum distortion embedding that is normalized with respect to a given strong ordering. From here on, embeddings are always assumed to be normalized.

Our algorithm is based on one single type of operations in embeddings: moving vertices. Vertex moving will appear in two different forms, depending on which vertices are moved in which direction. The two operations are denoted as `RightMove` and `DeleteTwo`. The latter operation, `DeleteTwo`, receives an embedding $\mathcal{E}$ and a position p as input and "deletes" the slots at position p and $p+1$ in $\mathcal{E}$, by moving all vertices to the right of position p two positions to the left. Note that the result is a proper embedding if the slots at position p and $p+1$ are empty. When we apply `DeleteTwo`, these two positions are empty.

The definition of operation `RightMove` is given as a small program. For the definition, we introduce the following notation. For an embedding $\mathcal{E}$, a vertex u and a position p, $\mathcal{E} - u$ denotes the embedding obtained from $\mathcal{E}$ by removing u (which leaves an empty slot), and $\mathcal{E} + (u \to p)$ is the embedding obtained from $\mathcal{E}$ by placing vertex u in the slot at position p (to obtain a proper embedding, we assume that u is not placed in $\mathcal{E}$ and that the slot at position p in $\mathcal{E}$ is empty). Operation `RightMove` mainly executes a right-shift for vertices of a single color class. It receives an embedding $\mathcal{E}$ and a vertex u as input and is defined as

Procedure `RightMove`
begin
  let $p = \mathcal{E}(u) + 2$; **set** $\mathcal{E} = \mathcal{E} - u$;
  while position p in $\mathcal{E}$ is occupied **do**
    let x be the vertex at position p in $\mathcal{E}$; **set** $\mathcal{E} = (\mathcal{E} - x) + (u \to p)$;
    set $u = x$; **set** $p = p + 2$
  end while;
  return $\mathcal{E} + (u \to p)$
end.

It can be verified that when we apply `RightMove` and `DeleteTwo` to normalized non-contractive embeddings, they produce normalized embeddings. For `DeleteTwo`, the non-contractiveness condition might be violated; however, whenever we apply this operation a violation does not occur. Hence, `RightMove` and `DeleteTwo` always produce normalized non-contractive embeddings.

To give a first outline, the mentioned algorithm iteratively takes a minimum distortion embedding for a connected induced subgraph, adds a new vertex to this embedding and determines on that basis the distortion of the extended graph. The new vertex is not an arbitrary vertex, but one with special properties. This incremental process defines a vertex ordering for the given graph. Let $G = (A, B, E)$ be a connected bipartite permutation graph on at least two vertices with strong ordering (σ_A, σ_B). We say that a vertex ordering $\sigma = \langle x_1, \ldots, x_n \rangle$ for G is *competitive* if it has the following properties:

- σ satisfies condition (C1)
- x_1 is the leftmost A-vertex in σ_A, and x_2 is the leftmost B-vertex in σ_B
- for every $i \in \{3, \ldots, n\}$, $N_G(x_i) \cap \{x_1, \ldots, x_{i-1}\} \subseteq N_G(w)$ where w is the $cc(x_i)$-vertex preceding x_i in σ_A or σ_B.

Observe that competitive vertex orderings exist for all connected bipartite graphs and given strong orderings: if the rightmost A-vertex has a neighbor that is not a neighbor of the preceding A-vertex then the last B-vertex has degree 1, and since G is connected the last A-vertex is adjacent to the last two B-vertices. It follows that all neighbors of the last B-vertex are neighbors of the previous B-vertex. Iteration proves existence of a competitive ordering. It is important to note that $G[\{x_1, \ldots, x_i\}]$ is connected for $2 \leq i \leq n$ and $\langle x_1, \ldots, x_n \rangle$ a competitive ordering.

We give the first step of our algorithm. We take an induced subgraph and a minimum distortion embedding and extend both by adding a new vertex, which is chosen according to a competitive ordering. For a graph $G = (V, E)$, an embedding $\mathcal{E}$ and an integer $k \geq 0$ we say that a vertex x is *$(G, \mathcal{E}, k)$-bad* if x has a neighbor y in G where $y \prec_{\mathcal{E}} x$ such that $d_{\mathcal{E}}(x, y) > k$. In particular, if x is a $(G, \mathcal{E}, k)$-bad vertex then its leftmost neighbor in $\mathcal{E}$ is at distance more than k in $\mathcal{E}$. If the context is clear we will write *$(\mathcal{E}, k)$-bad* or simply *k-bad*.

Lemma 5. *Let $G = (A, B, E)$ be a connected bipartite permutation graph on at least three vertices with competitive ordering σ. Let x be the rightmost vertex in σ. Let c be the $cc(x)$-vertex preceding x in σ, and let d be the leftmost neighbor of x in σ. Let $\mathcal{E}$ be a normalized minimum distortion embedding for $G{-}x$, and let $k =_{\text{def}} \mathrm{D}(G{-}x, \mathcal{E})$.*

1. *Let $c \prec_{\mathcal{E}} d$ and $\mathcal{F} =_{\text{def}} \mathcal{E} + (x \to \mathcal{E}(d) + 1)$.*
Then, $\mathcal{F}$ is a normalized minimum distortion embedding for G of distortion k.
2. *Let $d \prec_{\mathcal{E}} c$ and $\mathcal{F} =_{\text{def}} \mathcal{E} + (x \to \mathcal{E}(c) + 2)$.*
Then, $\mathcal{F}$ is a normalized non-contractive embedding for G of distortion k or $k + 2$, and if there is an $(\mathcal{F}, k)$-bad vertex then it is x.

For computing the distortion of G, it suffices to solve the question that is raised by the second case of Lemma 5, namely to decide whether the distortion of the graph in this case is k or $k + 2$. The main subroutine of our algorithm will do exactly this but requires an input embedding of a special form. Let $G = (A, B, E)$ be a connected bipartite permutation graph, $k \geq 1$ an integer and $\mathcal{E}$ a normalized non-contractive embedding for G. Denote the leftmost and rightmost $(G, \mathcal{E}, k)$-bad vertex in $\mathcal{E}$ as respectively b_l and b_r, and let a_r be the leftmost neighbor of b_r. We say that $\mathcal{E}$ has a *nice beginning* if all $(G, \mathcal{E}, k)$-bad vertices are $cc(b_r)$-vertices,

$d_{\mathcal{E}}(b_l, b_r) \leq k-1$, there is no empty $cc(b_r)$-slot between a_r and b_r and there is an empty $\overline{cc}(b_r)$-slot between a_r and b_l in $\mathcal{E}$. Note that b_l is to the right of a_r by the distance conditions. It can be shown that a nice beginning can be achieved by few modifications or it is easy to decide the distortion question already by looking at a small part of the given embedding. Furthermore, let d and x be vertices of G from different color classes where $d \prec_{\mathcal{E}} x$. We call a pair (v, w) of vertices for v a $cc(x)$-vertex and w a $\overline{cc}(x)$-vertex a *blocking pair* if $v \prec_{\mathcal{E}} w$, $d_{\mathcal{E}}(v, w) = 3$ and $vw \notin E$. We call vertex w for $d \prec_{\mathcal{E}} w \prec_{\mathcal{E}} x$ a *breakpoint vertex* between d and x if (v, w) is a blocking pair for some vertex v, there is no empty $cc(x)$-slot between d and v and no empty $\overline{cc}(x)$-slot between w and x in $\mathcal{E}$. The algorithm is then the following.

Algorithm. `bpg-distortion`
Input An embedding $\mathcal{E}$ and an integer k
begin
  while there is an $(\mathcal{E}, k)$-bad vertex **do**
    let x be the rightmost $(\mathcal{E}, k)$-bad vertex in $\mathcal{E}$;
    let d be the leftmost neighbor of x in $\mathcal{E}$;
    if there is no empty $\overline{cc}(x)$-slot between d and x in $\mathcal{E}$ **then** **reject**;
    let $\mathcal{F}$ = `RightMove`($\mathcal{E}, d$);
    if slot at position $\mathcal{F}(d) - 1$ is not occupied in $\mathcal{F}$ **then** **accept**;
    if there is no breakpoint vertex between d and x in $\mathcal{F}$ **and**
      there is an empty $cc(x)$-slot between d and x in $\mathcal{F}$ **then** **accept**;
    set $\mathcal{E} = \mathcal{F}$
  end while;
  accept
end.

The next lemma shows the main property of `bpg-distortion`, that it can be used for deciding the question of Lemma 5.

Lemma 6. *Let $G = (A, B, E)$ be a connected bipartite permutation graph on at least three vertices. Let $k \geq 1$ be an odd integer. Let $\mathcal{E}$ be a normalized non-contractive embedding for G of distortion $k + 2$ with a nice beginning. Then, $\mathrm{D}(G) \leq k$ or G contains a clawpath-like graph of length r on $\frac{1}{2}(rk + r + 2k + 6)$ vertices as induced subgraph.*

We can now complete the proof of the main structural theorem, Theorem 2. We prove the missing direction by induction. If G contains at most two vertices, then $\mathrm{D}(G) \leq 1$. So, let G have $n \geq 3$ vertices. Assume that the claim holds for all graphs on at most $n - 1$ vertices. Let σ be a competitive ordering for G, and let x be the last vertex in σ. If $\mathrm{D}(G-x) \geq k + 2$, then $G-x$ contains a clawpath-like graph of length r on $\frac{1}{2}(rk+r+2k+6)$ vertices as induced subgraph, and thus G. Now, let $\mathrm{D}(G-x) \leq k$, and let $\mathcal{F}$ be the embedding obtained as in Lemma 5 on input $\mathcal{E}$, σ and x. Assume that $\mathrm{D}(\mathcal{F}) = k + 2$. Lemma 6 completes the proof of Theorem 2. In addition, this gives a simple algorithm for computing the distortion of a bipartite permutation graph.

Theorem 3. *There is an $\mathcal{O}(n^2)$-time algorithm that computes the distortion of a connected bipartite permutation graph on n vertices. The algorithm certifies the computed distortion by a normalized non-contractive embedding as an upper bound and an induced clawpath-like subgraph as a lower bound.*

Proof. Let $G = (A, B, E)$ be a bipartite permutation graph with competitive ordering $\sigma = \langle x_1, \ldots, x_n \rangle$. Let $G_i =_{\text{def}} G[\{x_1, \ldots, x_i\}]$ for $1 \leq i \leq n$. Iteratively, normalized minimum distortion embeddings for $G_1, \ldots, G_n$ are computed applying the algorithm of Lemma 5 and the computation of an embedding with nice beginning as preprocessing and `bpg-distortion` as the main procedure. If the distortion of G_{i+1} is larger than the distortion of G_i then the algorithms even output an induced clawpath-like graph as certificate. The computed minimum distortion embedding for G_i serves as input for computing the distortion of G_{i+1}. For the running time, it mainly suffices to observe that `bpg-distortion` does not move a vertex twice. Storing the information about the number of vertices to the right of a position, it can be checked in constant time whether there are empty slots between two vertices. Consecutive vertices in the embeddings are at distance at most 3, so that at most $3n$ slots are used. Hence, `bpg-distortion` has an $\mathcal{O}(n)$-time implementation. The two preprocessing algorithms require only $\mathcal{O}(n)$ time, and a competitive ordering is obtained in linear time. Since the main algorithm has $\mathcal{O}(n)$ iterations, the $\mathcal{O}(n^2)$ running time follows.

5 Final Remarks

We gave an $\mathcal{O}(n^2)$-time implementation of the algorithm for computing the distortion of connected bipartite permutation graphs. The input embedding to `bpg-distortion` in our implementation can be very arbitrary. However, the actual input embedding is of specific form. Is it possible to give a linear-time implementation of our distortion algorithm using information about the embedding gained during computation for the smaller graphs? We leave this question as a task for future work.

Acknowledgments

The authors thank Fedor V. Fomin for suggesting the study of minimum distortion embeddings of specific graph classes, and Dieter Kratsch for preliminary discussions on the topic.

References

1. Badoiu, M., Chuzhoy, J., Indyk, P., Sidiropoulos, A.: Low-distortion embeddings of general metrics into the line. In: Proceedings of STOC 2005, pp. 225–233. ACM Press, New York (2005)
2. Badoiu, M., Dhamdhere, K., Gupta, A., Rabinovich, Y., Räcke, H., Ravi, R., Sidiropoulos, A.: Approximation algorithms for low-distortion embeddings into low-dimensional spaces. In: Proceedings of SODA 2005, pp. 119–128. ACM and SIAM (2005)

3. Badoiu, M., Indyk, P., Sidiropoulos, A.: A constant-factor approximation algorithm for embedding unweighted graphs into trees. In: AI Lab Technical Memo AIM-2004-015, MIT Press, Cambridge (2004)
4. Blache, G., Karpinski, M., Wirtgen, J.: On approximation intractability of the bandwidth problem. Technical report TR98-014, University of Bonn (1997)
5. Brandstädt, A., Le, V.B., Spinrad, J.P.: Graph Classes: A Survey. SIAM Monog. Disc. Math. Appl. (1999)
6. Golumbic, M.C.: Algorithmic Graph Theory and Perfect Graphs, 2nd edn. Ann. Disc. Math. vol. 57. Elsevier, Amsterdam (2004)
7. Heggernes, P., Kratsch, D., Meister, D.: Bandwidth of bipartite permutation graphs in polynomial time. In: Proceedings of LATIN 2008. LNCS, vol. 4957, pp. 216–227. Springer, Heidelberg (2008)
8. Heggernes, P., Meister, D., Proskurowski, A.: Computing minimum distortion embeddings into a path for bipartite permutation graphs and threshold graphs. In: Reports in Informatics, University of Bergen, Norway, vol. 369 (2008)
9. Indyk, P.: Algorithmic applications of low-distortion geometric embeddings. In: Proceedings of FOCS 2001, pp. 10–35. IEEE, Los Alamitos (2005)
10. Indyk, P., Matousek, J.: Low-distortion embeddings of finite metric spaces. In: Handbook of Discrete and Computational Geometry, 2nd edn. pp. 177–196. CRC Press, Boca Raton (2004)
11. Kenyon, C., Rabani, Y., Sinclair, A.: Low distortion maps between point sets. In: Proceedings of STOC 2004, pp. 272–280. ACM Press, New York (2004)
12. Kleitman, D.J., Vohra, R.V.: Computing the bandwidth of interval graphs. SIAM J. Disc. Math. 3, 373–375 (1990)
13. Mahadev, N., Peled, U.: Threshold graphs and related topics. In: Ann. Disc. Math. vol. 56, North-Holland, Amsterdam (1995)
14. Monien, B.: The Bandwidth-Minimization Problem for Caterpillars with Hair Length 3 is NP-Complete. SIAM J. Alg. Disc. Meth. 7, 505–512 (1986)
15. Papadimitriou, C., Safra, S.: The complexity of low-distortion embeddings between point sets. In: Proceedings of SODA 2005, pp. 112–118. ACM and SIAM (2005)
16. Spinrad, J., Brandstädt, A., Stewart, L.: Bipartite permutation graphs. Disc. Appl. Math. 18, 279–292 (1987)
17. Sprague, A.P.: An $O(n \log n)$ algorithm for bandwidth of interval graphs. SIAM J. Disc. Math. 7, 213–220 (1994)
18. Tenenbaum, J.B., de Silva, V., Langford, J.C.: A global geometric framework for nonlinear dimensionality reduction. Science 290, 2319–2323 (2000)

On a Special Co-cycle Basis of Graphs

Telikepalli Kavitha

Indian Institute of Science, Bangalore, India
kavitha@csa.iisc.ernet.in

Abstract. In this paper we consider the problems of computing a minimum co-cycle basis and a minimum *weakly fundamental* co-cycle basis of a directed graph G. A co-cycle in G corresponds to a vertex partition $(S, V \setminus S)$ and a $\{-1, 0, 1\}$ edge incidence vector is associated with each co-cycle. The vector space over $\mathbb{Q}$ generated by these vectors is the co-cycle space of G. Alternately, the co-cycle space is the orthogonal complement of the cycle space of G. The minimum co-cycle basis problem asks for a set of co-cycles that span the co-cycle space of G and whose sum of weights is minimum. Weakly fundamental co-cycle bases are a special class of co-cycle bases, these form a natural superclass of strictly fundamental co-cycle bases and it is known that computing a minimum weight strictly fundamental co-cycle basis is NP-hard. We show that the co-cycle basis corresponding to the cuts of a Gomory-Hu tree of the underlying undirected graph of G is a minimum co-cycle basis of G and it is also weakly fundamental.

1 Introduction

Let $G = (V, E)$ be a directed graph with m edges, n vertices, and weight function $w : E \to \mathbb{R}^+$. A *co-cycle* in G is a set of edges of the form $E \cap [(S \times S^c) \cup (S^c \times S)]$, where S is a nontrivial subset of V and $S^c = V \setminus S$; that is, a co-cycle is a cut in the underlying undirected graph. The edge directions are captured by assigning a $\{-1, 0, 1\}$ edge incidence vector to a co-cycle. The vector C assigned to the co-cycle $E \cap [(S \times S^c) \cup (S^c \times S)]$ is defined as follows. For each $e \in E$, we have

$$C(e) = \begin{cases} 1 & \text{if } e \text{ is from } S \text{ to } S^c, \\ -1 & \text{if } e \text{ is from } S^c \text{ to } S, \\ 0 & \text{otherwise.} \end{cases}$$

Note that the incidence vector C is only determined up to a factor ± 1 as the definition of a co-cycle is symmetric in S and S^c. For ease of exposition, we often refer to the partition (S, S^c) as the above co-cycle C rather than the vector associated with this partition. The weight of a co-cycle C is the sum of the weights of those edges e such that $C(e) = \pm 1$. The vector space over $\mathbb{Q}$ spanned by the $\{-1, 0, 1\}$ vectors of its co-cycles is the *co-cycle space* of G.

Definition 1. *A co-cycle basis of G is a set of linearly independent co-cycles that span the co-cycle space.*

J. Gudmundsson (Ed.): SWAT 2008, LNCS 5124, pp. 343–354, 2008.

We can assume without loss of generality that the underlying undirected graph of G is connected; then the co-cycle space of G has dimension $n-1$. It is easy to see that the set of $n-1$ co-cycles $\{C_e, e \in T\}$, where T is a spanning tree of the underlying undirected graph and C_e is the co-cycle corresponding to the partition of V created by deleting edge e from T, forms a co-cycle basis of G.

Problem Definition. The first problem that we consider in this paper is to compute a co-cycle basis such that the sum of the weights of the co-cycles in this basis is minimum. The minimum co-cycle basis problem has already been studied in *undirected* graphs [3]. In undirected graphs, a $\{0,1\}$ edge incidence vector C is assigned to a co-cycle; $C(e) = 1$ for edges e crossing the partition corresponding to this co-cycle and $C(e) = 0$ for the remaining edges e. Thus a co-cycle is a cut in an undirected graph. De Pina [3] showed that the $n-1$ cuts of a Gomory-Hu tree yield a minimum cut basis of an undirected graph.

The main problem that we consider in this paper is that of computing a minimum weight co-cycle basis of G that is *weakly fundamental.*

Definition 2. *A co-cycle basis $\mathcal{C}$ is called weakly fundamental if the co-cycles in $\mathcal{C}$ can be labeled as $C_1, \ldots, C_{n-1}$ so that for every $1 \le i \le n-2$ we have: $C_i \setminus (C_{i+1} \cup \cdots \cup C_{n-1}) \neq \emptyset$, where C_i here is the set of edges e such that $C_i(e) = \pm 1$.*

That is, each C_i has an edge e_i such that $C_i(e_i) = \pm 1$ while $C_j(e_i) = 0$ for $j > i$. A related class of co-cycle bases is the class of strictly fundamental co-cycle bases. A co-cycle basis $\mathcal{C}$ of G is *strictly fundamental* if the co-cycles in $\mathcal{C}$ can be labeled as $C_1, \ldots, C_{n-1}$ so that we have an edge e_i corresponding to each C_i such that $C_i(e_i) = \pm 1$ while $C_j(e_i) = 0$ for $j \neq i$. Equivalently, a strictly fundamental co-cycle basis $\mathcal{C}$ is a set of $n-1$ co-cycles corresponding to a spanning tree T, where each co-cycle corresponds to the partition of V created by deleting an edge from T. It is known that the problem of computing a minimum weight strictly fundamental co-cycle basis is NP-hard (see [2]).

Weakly fundamental co-cycle bases form a natural superclass of strictly fundamental co-cycle bases. Here we consider the problem of computing a minimum weight weakly fundamental co-cycle basis of a given directed/undirected graph. Other interesting classes of co-cycle bases are *integral* co-cycle bases and *totally unimodular* co-cycle bases (refer to Section 4 for the definitions).

Background. The co-cycle space of G is also the orthogonal complement of the *cycle space* of G. A *cycle* B in G is actually a cycle in the underlying undirected graph, i.e., edges in B are traversable in both directions. Associated with each cycle is a $\{-1,0,1\}$ edge incidence vector and the vector space over $\mathbb{Q}$ generated by these vectors is the cycle space $\mathcal{C}$ of G. The cycle space $\mathcal{C}$ has dimension $m-n+1$ when the underlying undirected graph of G is connected. We have $\langle B, C\rangle = 0$ for every cycle B and co-cycle C, where $\langle x, y\rangle$ denotes the dot product of the vectors x and y in $\mathbb{Q}^m$. The orthogonality between a cycle B and a co-cycle C is due to the fact that any cycle vector B has to traverse an even cardinality subset of the edges that cross the vertex partition (S, S^c) corresponding to C. For half of the edges e in $B \cap C$, we have $B(e) = C(e)$ and for the remaining half we have $B(e) = -C(e)$. Thus $\langle B, C\rangle = 0$. We have $\mathbb{Q}^m = \mathcal{C} \oplus \mathcal{C}^\perp$ where $\mathcal{C}^\perp$ is the co-cycle space of G.

Minimum Cycle Bases. The problem of computing a set of linearly independent cycles that span the cycle space and whose sum of weights is minimum is the minimum cycle basis problem. The minimum cycle basis problem has been well-studied and there are many polynomial time algorithms for this problem in undirected and directed graphs[10,3,6,1,13,12,11,17,9]. While there are polynomial time algorithms to compute a minimum cycle basis in a given graph, it has recently been shown that the problem of computing a minimum weight *weakly fundamental cycle basis* (this definition is analogous to Definition 2) is APX-hard [19]. Weakly fundamental cycle bases were first investigated in 1935 by Whitney [21] and recent interest in weakly fundamental cycle bases is due to the practical relevance of low weight weakly fundamental cycle bases in applications like the periodic event scheduling problem [16,15].

Liebchen and Rizzi [18] studied various classes of cycle bases for general graphs; this refined classification of cycle bases was of strong relevance for practical applications and they identified several new variants of the minimum cycle basis problem. More precisely, they showed that for general graphs, computing a minimum cycle basis for each class of cycle bases is different from computing a minimum cycle basis among *any* of the other classes. In this paper, we explore this question for minimum co-cycle basis problems.

Our results. We show that the minimum cycle basis and the minimum co-cycle basis problems exhibit marked dissimilarities. We show that the co-cycle basis corresponding to the cuts of a Gomory-Hu tree of the underlying undirected graph of G is a "special co-cycle basis". We first extend de Pina's argument to show that this co-cycle basis is a minimum co-cycle basis of the directed graph G. Our main result, which is based on an interesting structural property of a Gomory-Hu tree, is that this co-cycle basis is also weakly fundamental. This implies that this co-cycle basis is also integral and using known facts, it can be easily shown that this co-cycle basis is also totally unimodular. This is a surprising result when contrasted with the analogous minimum cycle basis problems since there is no such special cycle basis.

Regarding bounds on the weights of cycle bases and co-cycle bases, it has recently been shown in [4] that any weighted 2-connected undirected graph G admits a cycle basis (in fact, a weakly fundamental cycle basis) of weight $O(W \cdot \log n \log\log n)$ where W is the sum of edge weights in G and n is the number of vertices in G. Our results on co-cycle bases show that any weighted graph G admits a weakly fundamental co-cycle basis, that is also totally unimodular, of weight at most $2W$, since it can be shown that the sum of edge weights of a Gomory-Hu tree of G is at most $2W$, where W is the sum of edge weights in G.

Organization of the paper. Section 2 contains preliminaries and Section 3 contains our main results. Section 4 discusses several classes of co-cycle bases.

2 Preliminaries

For a directed graph, we obtain the *underlying undirected graph* by removing the directions from the edges of the given directed graph. In undirected graphs, cycles and co-cycles are 0-1 vectors and the cycle space and co-cycle space are vector spaces over $\mathbb{Z}_2$. There are examples of co-cycle bases in directed graphs which do not project on to co-cycle bases in the underlying undirected graph. Note that every co-cycle basis of the underlying undirected graph is always a co-cycle basis of the directed graph, when the 0-1 vectors of the co-cycles in the undirected graph are interpreted appropriately as $\{-1, 0, 1\}$ vectors in the directed graph.

2.1 Gomory-Hu Tree

A classical result in graph connectivity, due to Gomory and Hu [7], states that the edge connectivity between all pairs of vertices in an undirected graph can be computed using $n - 1$ (rather than the naïve $\binom{n}{2}$) max-flow computations. The Gomory-Hu algorithm computes a weighted tree $\mathcal{T}$, known as the Gomory-Hu tree, on V, with the following property:

The edge connectivity of any two vertices s and t in the graph exactly equals the weight on the lightest edge in the unique s-t path in $\mathcal{T}$. Further, the partition of the vertices produced by removing this edge from $\mathcal{T}$ is a minimum s-t cut in the graph.

The following theorem is used in the Gomory-Hu algorithm.

Theorem 1. *If $(X, V \setminus X)$ is a minimum s-t cut in G and $u, v \in X$, then there exists a minimum u-v cut $(X^*, V \setminus X^*)$ such that $X^* \subset X$.*

The Gomory-Hu tree construction algorithm [7] initializes the tree $\mathcal{T}$ to a single node that contains the entire vertex set. At any step of the algorithm, pick a node X of $\mathcal{T}$ containing more than one vertex and choose any two vertices s and t in X. Contract the entire subtree subtended at each neighbor of X into a single node and perform a max flow computation from s to t in the new graph. Theorem 1 ensures that the minimum s-t cut thus obtained (we call it C) is also a minimum s-t cut in the original graph. Now, in $\mathcal{T}$, the node X is split into X_1 and X_2 according to C and the two nodes thus formed are joined by an edge of weight equal to the size of C. Further, all the neighboring subtrees of X become neighboring subtrees of X_1 or X_2 depending upon which side of C they lie on. The algorithm terminates when all the nodes of $\mathcal{T}$ become singleton sets. Thus $\mathcal{T}$ is a weighted tree whose nodes are the vertices of V. It can be shown that $\mathcal{T}$ captures all-pairs minimum cuts.

Note that the Gomory-Hu tree $\mathcal{T}$ need not be a spanning tree of G. That is, the edges of $\mathcal{T}$ need not be edges of G. Fig. 1 illustrates such a graph and a Gomory-Hu tree of this graph. The a-b minimum cut is 3 and the s-t minimum cut for $s \in \{x, y, z\}$ and $t \in \{a, b\}$ is 2. Thus a and b, which are non-adjacent in G, have to be adjacent (with an edge of weight 3) in any Gomory-Hu tree of G.

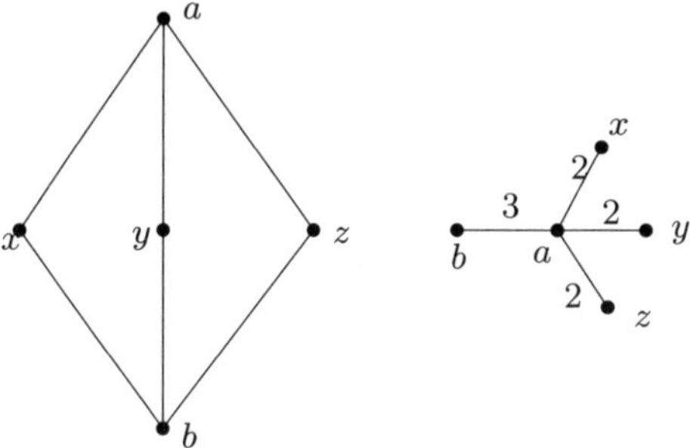

Fig. 1. An example of a graph with unit edge weights and a Gomory-Hu tree of that graph. Note that (a, b) is an edge of the Gomory-Hu tree although (a, b) is not an edge in the given graph.

3 Minimum Co-cycle Bases and Algorithms

In this section we first present a proof from [3] by de Pina that for any undirected graph G, the $n-1$ cuts defined by a Gomory-Hu tree form a minimum cut basis. We then extend this proof to show that for any directed graph, the $n-1$ cuts of a Gomory-Hu tree of the underlying undirected graph form a minimum co-cycle basis. We then show that this co-cycle basis is also weakly fundamental.

We will assume without loss of generality that the underlying undirected graph of the given graph is connected (note that the co-cycle space of any graph is the direct sum of the co-cycle spaces of its connected components). Then the co-cycle space has dimension $n-1$.

Lemma 1. *For any undirected graph G, the $n-1$ cuts defined by a Gomory-Hu tree form a minimum cut basis.*

Proof (from [3]). Let $\mathcal{T}$ be a Gomory-Hu tree of G. Let $\mathcal{B} = \{B_{e_1}, \ldots, B_{e_{n-1}}\}$ be the $n-1$ cuts defined by the $n-1$ edges $e_1, \ldots, e_n$ of $\mathcal{T}$. We will first show that these cuts are linearly independent over $\mathbb{Z}_2$.

Consider an edge $e_i = (u, v)$ of $\mathcal{T}$. First, note that B_{e_i} is the only cut in $\mathcal{B}$ that separates u and v. Let P_i be any u-v path in G; encode P_i as a 0-1 edge incidence vector. For any cut B and path P, $\langle B, P\rangle$ is the number of edges that belong to both B and P and $\langle B, P\rangle$ is an odd number if and only if B separates the endpoints of P. Since B_{e_i} is the only cut in $\mathcal{B}$ that separates u and v, we have $\langle B_{e_i}, P_i\rangle = 1 \pmod 2$ while $\langle B_{e_j}, P_i\rangle = 0 \pmod 2$ for all $j \neq i$. So the $n-1$ cuts in $\mathcal{B}$ are linearly independent over $\mathbb{Z}_2$. Thus $\mathcal{B}$ forms a cut basis.

Suppose $\mathcal{B}$ is not a minimum cut basis. Let $\mathcal{B}'$ be a minimum cut basis of G that has the maximum intersection with $\mathcal{B}$ among all minimum cut bases of G. That is, $|\mathcal{B}' \cap \mathcal{B}|$ is maximum for $\mathcal{B}'$ among all minimum cut bases of G. Let B_{e_i} be a cut of $\mathcal{B}$ that is not present in $\mathcal{B}'$. Since $\mathcal{B}'$ is a cut basis, there are cuts $A_1, \ldots, A_t$ in $\mathcal{B}'$ such that

$$B_{e_i} = A_1 + A_2 + \cdots + A_t \text{ (mod2)}.$$

Since $\langle B_{e_i}, P_i\rangle = 1 \pmod 2$, there must exist a $k \in \{1, \ldots, t\}$ such that $\langle A_k, P_i\rangle = 1 \pmod 2$, that is, A_k separates u and v. This implies that weight(A_k) $\geq$ weight(B_{e_i}) since B_{e_i} is a minimum u-v cut (recall the Gomory-Hu tree property that for any pair of vertices s, t the partition of V produced by removing the lightest edge on the s-t path from $\mathcal{T}$ is a minimum s-t cut in the graph).

Note that A_k cannot be a cut in $\mathcal{B}$ since all the cuts B_{e_j} in $\mathcal{B}$ other than B_{e_i} satisfy $\langle B_{e_j}, P_i\rangle = 0 \,(\mathrm{mod} 2)$. It is easy to see that $\mathcal{B}'' = \mathcal{B}' \setminus \{A_k\} \cup \{B_{e_i}\}$ is a minimum cut basis such that $|\mathcal{B}'' \cap \mathcal{B}| > |\mathcal{B}' \cap \mathcal{B}|$, contradicting the definition of $\mathcal{B}'$. □

We now generalize the above result to directed graphs.

Lemma 2. *Let $G = (V, E)$ be the given directed graph with n vertices and m edges. The set of $n - 1$ co-cycles corresponding to the $n - 1$ cuts (interpreted as $\{-1, 0, 1\}^m$ vectors) of a Gomory-Hu tree of the underlying undirected graph of G forms a minimum co-cycle basis of G.*

Proof. Let $C_1, \ldots, C_{n-1}$ be the co-cycles corresponding to the $n - 1$ cuts of the Gomory-Hu tree of the underlying undirected graph of G. It follows from the proof of Lemma 1 that the $\{-1, 0, 1\}^m$ incidence vectors of these co-cycles are linearly independent over $\mathbb{Q}$ (since their residues modulo 2 are linearly independent over $\mathbb{Z}_2$). Thus $\mathcal{C} = \{C_1, \ldots, C_{n-1}\}$ forms a co-cycle basis of G.

We now need to show that $\mathcal{C}$ is a minimum co-cycle basis of G. Suppose not. Let $\mathcal{B}$ be a minimum co-cycle basis such that $|\mathcal{B} \cap \mathcal{C}|$ is maximum. Let C_i be a co-cycle such that $C_i \in \mathcal{C} \setminus \mathcal{B}$. Since $\mathcal{B} = \{B_1, \ldots, B_{n-1}\}$ is a co-cycle basis, there exist rational numbers $\alpha_1, \ldots, \alpha_{n-1}$ such that

$$C_i = \alpha_1 B_1 + \alpha_2 B_2 + \cdots + \alpha_{n-1} B_{n-1}. \tag{1}$$

The co-cycle C_i corresponds to the vertex partition $(S, V \setminus S)$ determined by an edge (u, v) of the Gomory-Hu tree of the underlying undirected graph. So, in the underlying undirected graph, the edges of C_i form a minimum u-v cut. Let P_i be a u-v path in the underlying undirected graph and interpret P_i as a $\{-1, 0, 1\}^m$ incidence vector in the given directed graph G. We make the following claim which is easy to show.

Claim. Let C and P be elements of $\{-1, 0, 1\}^m$ such that C is a co-cycle and P is a path in the underlying undirected graph. Then the dot product $\langle C, P\rangle$ can take only one of three values: 0, 1, or -1. If the co-cycle C corresponds to the vertex partition (Y, Y^c) (recall that $Y^c = V \setminus Y$) and if a and b are the endpoints of P, then

$$\langle C, P\rangle = \begin{cases} 1 & \text{if } a \in Y \text{ and } b \in Y^c, \\ -1 & \text{if } a \in Y^c \text{ and } b \in Y, \\ 0 & \text{otherwise.} \end{cases}$$

Since the edges of C_i form a u-v cut in the underlying undirected graph and u, v are the endpoints of P_i, we have $\langle C_i, P_i\rangle = 1$ by the above claim. Now, using Equation (1), it follows that there is some $B_k \in \mathcal{B}$ with $\alpha_k \neq 0$ such that $\langle B_k, P_i\rangle \neq 0$. However this implies by our above claim that $\langle B_k, P_i\rangle = \pm 1$. In other words, the edges of B_k form a u-v cut in the underlying undirected graph. But we already know that C_i is a minimum u-v cut in the underlying undirected graph. Thus the weight of C_i is at most the weight of B_k. Since $\alpha_k \neq 0$, B_k can be written as a linear combination of C_i and the other co-cycles in $\mathcal{B}$ (refer to Equation (1)). Hence $\mathcal{D} = \mathcal{B} \setminus \{B_k\} \cup \{C_i\}$ is a minimum co-cycle basis.

Also, each of the co-cycles $C \in \{C_1, \ldots, C_{n-1}\} \setminus \{C_i\}$ satisfies: $\langle C, P_i \rangle = 0$ by the above claim since none of the partitions corresponding to these co-cycles separates u and v (the endpoints of P_i). Hence B_k is not any of the co-cycles in $\{C_1, \ldots, C_{n-1}\}$. Thus $|\mathcal{D} \cap \mathcal{C}| > |\mathcal{B} \cap \mathcal{C}|$ and $\mathcal{D}$ is a minimum co-cycle basis - contradicting the definition of $\mathcal{B}$. □

Since the Gomory-Hu co-cycle basis, which is a minimum co-cycle basis of the underlying undirected graph (by Lemma 1), is also a minimum co-cycle basis of the given directed graph (by Lemma 2), we can now conclude the following theorem.

Theorem 2. *For any directed graph G, a minimum cut basis of the underlying undirected graph of G, with cuts interpreted as $\{-1, 0, 1\}^m$ vectors, forms a minimum co-cycle basis of G.*

Lemma 2 also gives us the following efficient algorithm to compute a minimum co-cycle basis of a given directed graph G: compute a Gomory-Hu tree $\mathcal{T}$ of the underlying undirected graph of G. Each edge e_i in $\mathcal{T}$ defines a partition of the vertex set $(S_i, V \setminus S_i)$. Let $C_i \in \{-1, 0, 1\}^m$ be the co-cycle corresponding to the partition $(S_i, V \setminus S_i)$. Return the co-cycles $C_1, \ldots, C_{n-1}$.

Since a Gomory-Hu tree can be computed using $n - 1$ max-flows computations [7,8] in the underlying undirected graph, it follows that a minimum co-cycle basis in a directed graph on n vertices and m edges can be computed in time $\tilde{O}(mn^2)$ where $|E| = m, |V| = n$. When the edge weights are integers in the range $[1, \ldots, U]$, then a minimum co-cycle basis can be computed in time $O(mn \cdot \min(\sqrt{m}, n^{2/3}) \log(n^2/m) \log U)$, using the Goldberg-Rao max-flow algorithm [5].

3.1 Computing a Minimum Weakly Fundamental Co-cycle Basis

We will next show our result on computing a minimum weakly fundamental co-cycle basis. Let $G = (V, E)$ be an *undirected* connected graph on $n \geq 2$ vertices. We will assume here that all edge weights in G are positive[1]. Let $\mathcal{T}$ be a Gomory-Hu tree of G. We will show an ordering $X_1, X_2 \ldots, X_{n-1}$ of the $n - 1$ cuts of $\mathcal{T}$ and edges $\{e_1, \ldots, e_{n-1}\} \subseteq E$ such that $e_i \in X_i$ and $e_i \notin X_{i+1} \cup \cdots \cup X_{n-1}$. This shows that $X_1, \ldots, X_{n-1}$ is weakly fundamental (see Definition 2).

[1] Edges with weight 0 can as well be deleted from G since they do not contribute to the weight of any cut, but deleting such edges may leave the graph disconnected. However this problem can be easily circumvented by working on each connected component with positive edge weights and connecting the Gomory-Hu trees of these components with the 0 weight edges.

Refer to Fig. 1, where some edges in the Gomory-Hu tree were not edges of the given graph. However, there were some edges of the Gomory-Hu tree (for instance, $(a, x), (a, y), (a, z)$) which were edges of the given graph. The following lemma shows that there is always at least one *real* edge in a Gomory-Hu tree.

Lemma 3. *Let $G = (V, E)$ be an undirected connected graph on $n \geq 2$ vertices with positive edge weights and let $\mathcal{T}$ be a Gomory-Hu tree of G. Then there exists an edge (x, y) in $\mathcal{T}$ where x is a leaf in $\mathcal{T}$, that is a* real *edge, that is $(x, y) \in E$.*

Proof. Root the Gomory-Hu tree $\mathcal{T}$ at any arbitrary vertex r. Let y be a vertex such that all its children, call them $x_1, \ldots, x_k$, are leaves in $\mathcal{T}$. We claim that at least one of the edges (x_i, y), where $1 \leq i \leq k$, is in E. Suppose not. We will consider two cases here.

CASE (1). The vertex y is not the root r. Let z be the parent of y. The Gomory-Hu tree $\mathcal{T}$ tells us that the vertex partition $(S, V \setminus S)$ caused by deleting (y, z) from $\mathcal{T}$ is a minimum y-z cut in G. The vertex y and its children $x_1, \ldots, x_k$ are on one side of this bipartition, call this S, and the vertex z and the remaining vertices constitute $V \setminus S$ (see Fig. 2).

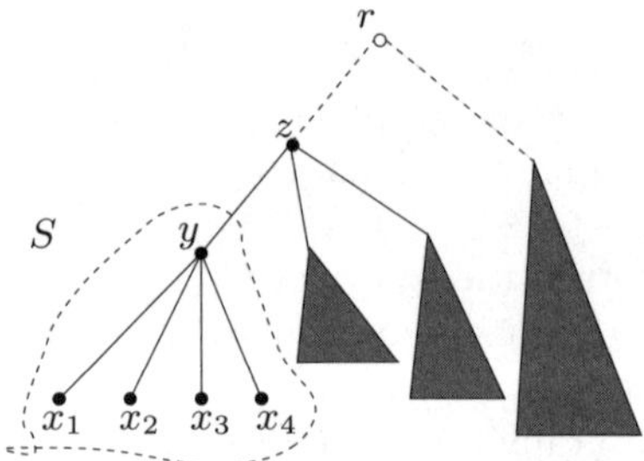

Fig. 2. The vertices $x_1, \ldots, x_k$ are leaves in $\mathcal{T}$

Since we assumed that none of $x_1, \ldots, x_k$ is adjacent to y, some of the vertices in $\{x_1, \ldots, x_k\}$ have to be adjacent to vertices of $V \setminus S$ since the given graph G is connected. The weight of the cut $(S, V \setminus S)$ is $\alpha + \beta$, where α is the sum of weights of edges between y and $V \setminus S$ and β is the sum of weights of edges between $\{x_1, \ldots, x_k\}$ and $V \setminus S$. Now consider the partition $(\{y\}, V \setminus \{y\})$ obtained by moving the vertices $x_1, \ldots, x_k$ from the side of y to the side of $V \setminus S$. This cut has weight only α, that is, the sum of weights of edges between y and $V \setminus S$, since we assumed that there are no edges between $\{x_1, \ldots, x_k\}$ and y. Since all edge weights are positive, this is a y-z cut whose weight is strictly smaller than $(S, V \setminus S)$, contradicting that $(S, V \setminus S)$ is a minimum y-z cut.

CASE (2). The vertex y is the root r. Since $x_1, \ldots, x_k$ are leaves in $\mathcal{T}$, this implies that the Gomory-Hu tree $\mathcal{T}$ is a tree of depth 1, the vertex y is the root and y's children $x_1, \ldots, x_k$ are all the remaining vertices in $V \setminus \{y\}$. If none of the vertices of $V \setminus \{y\}$ is adjacent to y in G, then y is an isolated vertex in G, contradicting that G is connected. Hence there is at least one x_i that is adjacent to y in G. □

We now need to have a method to use the above lemma repeatedly and our next lemma gives us such a method. Note that *contracting* an edge (u, v) implies identifying the vertices u and v into a single vertex $\{u, v\}$ and this vertex $\{u, v\}$ retains all edges incident on u, all edges incident on v, and self-loops are discarded.

Lemma 4. *Let $\mathcal{T}_i$ be a Gomory-Hu tree of the undirected connected graph $G_i = (V_i, E_i)$. Let (u_i, v_i) be an edge that is present in* both *$\mathcal{T}_i$ and G_i, where u_i is a leaf vertex in $\mathcal{T}_i$ and v_i is u_i's neighbor in $\mathcal{T}_i$ (Lemma 3 tells us there is such a u_i). The tree $\mathcal{T}_{i+1}$ obtained by contracting the edge (u_i, v_i) in $\mathcal{T}_i$, is a Gomory-Hu tree of the graph G_{i+1} which is obtained by contracting the edge (u_i, v_i) in G_i.*

Proof. The tree $\mathcal{T}_i$ is a Gomory-Hu tree of G_i and u_i is a leaf in $\mathcal{T}_i$ and v_i is the (only) neighbor of u_i in $\mathcal{T}_i$. Thus except for the pair u_i-v_i, the edge (u_i, v_i) cannot be an intermediate edge in the path in $\mathcal{T}_i$ between any other pair of vertices. This means that there is a minimum x-y cut for every pair of vertices $x, y \in V_i \setminus \{u_i\}$ that has both u_i and v_i on the same side of the cut. Thus contracting the edge (u_i, v_i) in $\mathcal{T}_i$ preserves the minimum cuts between all the pairs of vertices in $V_i \setminus \{u_i\}$. Hence the edges of $\mathcal{T}_{i+1}$ capture minimum x-y cuts for all pairs of vertices in $V_i \setminus \{u_i\}$.
The vertex set of $\mathcal{T}_{i+1}$ is the same as the vertex set of the graph G_{i+1}. We claim that each of the edges (x, y) in $\mathcal{T}_{i+1}$ defines a minimum x-y cut in G_{i+1}. This is because the value of a minimum x-y cut in G_{i+1}, for any two vertices in G_{i+1}, is at least the value of the minimum x-y cut in G_i and we saw that the edges of $\mathcal{T}_{i+1}$ capture minimum x-y cuts in G_i for all pairs of vertices x, y in $V_i \setminus \{u_i\}$. Thus the edges of $\mathcal{T}_{i+1}$ capture minimum x-y cuts for all pairs of vertices x, y in G_{i+1}. Thus $\mathcal{T}_{i+1}$ is a Gomory-Hu tree of the graph G_{i+1}. □

Let $G = (V, E)$ be the input graph. Lemmas 3 and 4 indicate an obvious strategy to construct a weakly fundamental co-cycle basis of G. However we have to be careful of the following: at any stage when we choose an edge e_i in the Gomory-Hu tree $\mathcal{T}_i$ of graph G_i, which is obtained by contracting several edges in G, it is not enough for us to claim that $e_i \in \mathcal{T}_i$ is a real edge in G_i. We need e_i to be a real edge in G. However this is easily done.

Each vertex of G_i is a subset of vertices of G. The vertex set V of G is partitioned into disjoint subsets which form the vertices of G_i and (x_i, y_i) is an edge in G_i only if there is an edge (u_i, v_i) in E between some vertex $u_i \in x_i$ and some vertex $v_i \in y_i$. Thus if (x_i, y_i) is a real edge in $\mathcal{T}_i$, then there is an original edge (u_i, v_i) in G between some $u_i \in x_i$ and $v_i \in y_i$. The edge e_i is always such an original edge (u_i, v_i) in G. The partition in $\mathcal{T}_i$ created by deleting the edge (x_i, y_i) from $\mathcal{T}_i$ is the partition corresponding to co-cycle C_i. Note that this co-cycle C_i corresponds to some edge (a, b) in the original Gomory-Hu tree, where the vertex a got identified with several other vertices to form the set x_i and similarly, b got identified with several other vertices to form the set y_i. Initially, (a, b) need not be a real edge, however after these identifications, due to some $u_i \in x_i$ and $v_i \in y_i$ being adjacent in G, we get (x_i, y_i) as a real edge in $\mathcal{T}_i$.

An efficient algorithm to compute a minimum weight weakly fundamental co-cycle basis in a given directed/undirected graph now follows easily. If the input graph is directed, then work with the underlying undirected graph, call this $G = (V, E)$, and we will finally interpret the cuts returned by the algorithm as $\{-1, 0, 1\}^m$ vectors.

The algorithm constructs a Gomory-Hu tree $\mathcal{T}$ of G and initializes $\mathcal{T}_1$ to $\mathcal{T}$. Then the following steps are run for $i = 1$ to $n-1$ and the co-cycles $X_1, \ldots, X_{n-1}$ are returned.

1. Find an edge $e_i = (u_i, v_i)$ that is present in G which is between a leaf $x_i \in \mathcal{T}_i$ and its neighbor y_i in $\mathcal{T}_i$.
2. Let X_i be the cut in G defined by the partition of V caused by removing the edge (x_i, y_i) from $\mathcal{T}_i$.
3. Contract the edge (x_i, y_i) in $\mathcal{T}_i$ and call the resulting tree $\mathcal{T}_{i+1}$.

Lemma 5, which is easy to show, proves the correctness of our algorithm.

Lemma 5. *Let $X_1, \ldots, X_{n-1}$ be the cuts returned by the above algorithm. For each $1 \le i \le n-2$ we have:*

$$X_i \setminus (X_{i+1} \cup \cdots \cup X_{n-1}) \neq \emptyset.$$

The bottleneck in the above algorithm is the time taken to build a Gomory-Hu tree. We conclude the following theorem.

Theorem 3. *The co-cycle basis corresponding to the $n-1$ cuts of the Gomory-Hu tree $\mathcal{T}$ of an undirected connected graph $G = (V, E)$ on n vertices is weakly fundamental. The ordered set $\{X_1, \ldots, X_{n-1}\}$ of weakly fundamental co-cycles corresponding to the $n-1$ edges of $\mathcal{T}$ can be computed in time in time $O(mn \cdot \min(\sqrt{m}, n^{2/3}) \log(n^2/m) \log U)$, where $|E| = m$ and the edge weights are integers in the range $[1, \ldots, U]$.*

4 Classes of Co-cycle Bases

Liebchen and Rizzi [18] studied various classes of cycle bases for general graphs. We consider the analogous classes of co-cycle bases now and discuss the minimum co-cycle basis problems in these classes. Recall that we defined strictly fundamental co-cycle bases in Section 1 and it is known that the problem of computing a minimum weight strictly fundamental co-cycle basis is NP-hard [2]. So here we will consider the minimum co-cycle basis problem in the five classes: directed co-cycle bases, undirected co-cycle bases, integral co-cycle bases, weakly fundamental co-cycle bases, and totally unimodular co-cycle bases.

Let $G = (V, E)$ be a directed graph on n vertices whose underlying undirected graph is connected. We have already discussed directed co-cycle bases, undirected co-cycle bases, and weakly fundamental co-cycle bases in G. It is easy to see that any weakly fundamental co-cycle basis is also an undirected co-cycle basis and any undirected co-cycle basis is also a directed co-cycle basis. We have seen so far (Lemma 1, Theorem 2, and Theorem 3) that the co-cycle basis corresponding to a Gomory-Hu tree of the underlying undirected graph answers the minimum co-cycle basis question in these three classes. Let us define the other two classes below.

• *integral co-cycle bases*: a set of co-cycles $\mathcal{C} = \{C_1, \ldots, C_{n-1}\}$ of G is an integral co-cycle basis if any co-cycle X of G can be written as an *integral* linear combination of the co-cycles in $\mathcal{C}$.

It is easy to show that any weakly fundamental co-cycle basis is integral. Thus the Gomory-Hu co-cycle basis is also a minimum weight integral co-cycle basis.

• *totally unimodular co-cycle bases*: a directed co-cycle basis $\mathcal{C} = \{C_1, \ldots, C_{n-1}\}$ of G forms a *totally unimodular* co-cycle basis if the $m \times (n-1)$ edge co-cycle incidence matrix M of $\mathcal{C}$ is totally unimodular.

Let $C_1, \ldots, C_{n-1}$ be the co-cycles corresponding to the $n-1$ edges of a Gomory-Hu tree of the underlying undirected graph. Let $(S_i, V \setminus S_i)$ be the partition corresponding to the co-cycle C_i, for each i. It is easy to see that the family of sets S_i for $i = 1, \ldots, n-1$ is *cross-free* (a family $\mathcal{F}$ of subsets of V is called *cross-free* if $S, T \in \mathcal{F}$, then $S \subseteq T$, or $T \subseteq S$, or $S \cap T = \emptyset$, or $S \cup T = V$). Given a directed graph $G = (V, E)$ and a cross-free family of subsets $\mathcal{F}$ of the vertex set V, the $\{-1, 0, 1\}$ edge incidence matrix M of these subsets is totally unimodular (see Schrijver [20], Section 19.3, Example 5). Hence the $m \times (n-1)$ edge co-cycle incidence matrix M of $C_1, \ldots, C_{n-1}$ is totally unimodular. Thus the Gomory-Hu co-cycle basis is a totally unimodular co-cycle basis, in particular a minimum weight totally unimodular co-cycle basis. We can now conclude the following theorem.

Theorem 4. *Let G be a directed graph. The co-cycle basis $\mathcal{C}$ corresponding to a Gomory-Hu tree of the underlying undirected graph of G is a minimum directed co-cycle basis, a minimum undirected co-cycle basis, a minimum integral co-cycle basis, a minimum weakly fundamental co-cycle basis, and a minimum totally unimodular co-cycle basis of G.*

Remark. The above theorem immediately implies results for certain minimum cycle basis problems in planar graphs. It is known that $\mathcal{X} = \{X_1, \ldots, X_d\}$ is a cycle basis in a planar graph $G = (V, E)$ if and only if $\mathcal{X}' = \{X'_1, \ldots, X'_d\}$ is a co-cycle basis in the planar *dual* graph $G' = (F, E')$ where for each i, $1 \le i \le d$, the vectors $X_i \in \{-1, 0, 1\}^{|E|}$ and $X'_i \in \{-1, 0, 1\}^{|E'|}$ are the same. It is easy to show that $\mathcal{X}$ is an integral/weakly fundamental/totally unimodular cycle basis iff $\mathcal{X}'$ is an integral/weakly fundamental/totally unimodular co-cycle basis. Thus Theorem 4 implies that the minimum integral/weakly fundamental/totally unimodular cycle basis problems in planar graphs can be solved in polynomial time. However, the polynomial time solvability of the minimum integral/weakly fundamental cycle basis problems in planar graphs was already shown in [18] using the results of Leydold and Stadler in [14].

Acknowledgments. I thank Alex Golynski for asking me about minimum co-cycle bases and the referees for their helpful comments.

References

1. Berger, F., Gritzmann, P., de Vries, S.: Minimum Cycle Bases for Network Graphs. Algorithmica 40(1), 51–62 (2004)
2. Bunke, F., Hamacher, H.W., Maffioli, F., Schwahn, A.M.: Minimum cut bases in undirected networks Report in Wirtschaftsmathematik 56, Fachbereich Mathematik, TU Kaiserslautern (2007)
3. de Pina, J.C.: Applications of Shortest Path Methods. PhD thesis, University of Amsterdam, Netherlands (1995)
4. Elkin, M., Liebchen, C., Rizzi, R.: New length bounds for cycle bases. Information Processing Letters, 104(5): 186-193 (2007)
5. Goldberg, A.V., Rao, S.: Beyond the flow decomposition barrier. Journal of the ACM, 45(5): 783-797 (1998)
6. Golynski, A., Horton, J.D.: A polynomial time algorithm to find the minimum cycle basis of a regular matroid. In: Penttonen, M., Schmidt, E.M. (eds.) SWAT 2002. LNCS, vol. 2368, pp. 200–209. Springer, Heidelberg (2002)
7. Gomory, R., Hu, T.C.: Multi-terminal network flows. SIAM Journal on Computing 9, 551–570 (1961)
8. Gusfield, D.: Very simple methods for all pairs network flow analysis. SIAM Journal on Computing 19(1), 143–155 (1990)
9. Hariharan, R., Kavitha, T., Mehlhorn, K.: A Faster Deterministic Algorithm for Minimum Cycle Bases in Directed Graphs. In: Bugliesi, M., Preneel, B., Sassone, V., Wegener, I. (eds.) ICALP 2006. LNCS, vol. 4051, pp. 273–284. Springer, Heidelberg (2006)
10. Horton, J.D.: A polynomial-time algorithm to find a shortest cycle basis of a graph. SIAM Journal on Computing 16, 359–366 (1987)
11. Kavitha, T.: An $\tilde{O}(m^2n)$ Randomized Algorithm to compute a Minimum Cycle Basis of a Directed Graph. In: Caires, L., Italiano, G.F., Monteiro, L., Palamidessi, C., Yung, M. (eds.) ICALP 2005. LNCS, vol. 3580, pp. 273–284. Springer, Heidelberg (2005)
12. Kavitha, T., Mehlhorn, K.: Algorithms to compute Minimum Cycle Bases in Directed Graphs. Theory of Computing Systems 40, 485–505 (2007)
13. Kavitha, T., Mehlhorn, K., Michail, D., Paluch, K.: A faster algorithm for Minimum Cycle Bases of graphs. In: Díaz, J., Karhumäki, J., Lepistö, A., Sannella, D. (eds.) ICALP 2004. LNCS, vol. 3142, pp. 846–857. Springer, Heidelberg (2004)
14. Leydold, J., Stadler, P.F.: Minimum cycle bases of outerplanar graphs. Electronic Journal of Combinatorics 5 (1998)
15. Liebchen, C.: Finding Short Integral Cycle Bases for Cyclic Timetabling. In: Di Battista, G., Zwick, U. (eds.) ESA 2003. LNCS, vol. 2832, pp. 715–726. Springer, Heidelberg (2003)
16. Liebchen, C., Peeters, L.: On Cyclic Timetabling and Cycles in Graphs. Technical Report 761/2002, TU Berlin (2002)
17. Liebchen, C., Rizzi, R.: A Greedy Approach to compute a Minimum Cycle Basis of a Directed Graph. Information Processing Letters 94(3), 107–112 (2005)
18. Liebchen, C., Rizzi, R.: Classes of Cycle Bases. Discrete Applied Mathematics 155(3), 337–355 (2007)
19. Rizzi, R.: Minimum Weakly Fundamental Cycle Bases are hard to find. Algorithmica (published online October 2007)
20. Schrijver, A.: Theory of Linear and Integer Programming, 2nd edn. Wiley, Chichester (1998)
21. Whitney, H.: On the abstract properties of linear dependence. American Journal of Mathematics 57, 509–533 (1935)

A Simple Linear Time Algorithm for the Isomorphism Problem on Proper Circular-Arc Graphs

Min Chih Lin[1,⋆], Francisco J. Soulignac[1,⋆⋆], and Jayme L. Szwarcfiter[2,⋆⋆⋆]

[1] Universidad de Buenos Aires, Facultad de Ciencias Exactas y Naturales, Departamento de Computación, Buenos Aires, Argentina
[2] Universidade Federal do Rio de Janeiro, Instituto de Matemática, NCE and COPPE, Caixa Postal 2324, 20001-970 Rio de Janeiro, RJ, Brasil
{oscarlin,fsoulign}@dc.uba.ar, jayme@nce.ufrj.br

Abstract. A circular-arc model $\mathcal{M} = (C, \mathcal{A})$ is a circle C together with a collection $\mathcal{A}$ of arcs of C. If no arc is contained in any other then $\mathcal{M}$ is a proper circular-arc model, and if some point of C is not covered by any arc then $\mathcal{M}$ is an interval model. A (proper) (interval) circular-arc graph is the intersection graph of a (proper) (interval) circular-arc model. Circular-arc graphs and their subclasses have been the object of a great deal of attention in the literature. Linear time recognition algorithms have been described both for the general class and for some of its subclasses. For the isomorphism problem, there exists a polynomial time algorithm for the general case, and a linear time algorithm for interval graphs. In this work we develop a linear time algorithm for the isomorphism problem in proper circular-arc graphs, based on uniquely encoding a proper circular-arc model. Our method relies on results about uniqueness of certain PCA models, developed by Deng, Hell and Huang in [6]. The algorithm is easy to code and uses only basic tools available in almost every programming language.

Keywords: isomorphism problems, proper circular-arc graphs, proper circular-arc canonization.

1 Introduction

Interval graphs, circular-arc graphs and its subclasses are interesting families of graphs that have been receiving much attention recently. Proper interval

⋆ Partially supported by UBACyT Grants X184 and X212 and by PICT ANPCyT Grant 1562.

⋆⋆ Partially supported by UBACyT Grant X184, PICT ANPCyT Grant 1562 and by a grant of the YPF Foundation.

⋆⋆⋆ Partially supported by the Conselho Nacional de Desenvolvimento Científico e Tecnológico, CNPq, and Fundação de Amparo à Pesquisa do Estado do Rio de Janeiro, FAPERJ, Brasil.

J. Gudmundsson (Ed.): SWAT 2008, LNCS 5124, pp. 355–366, 2008.

and proper circular-arc graphs are two of the most studied subclasses of interval and circular-arc graphs [4,8,21]. Booth and Lueker found the first linear time algorithm for recognizing interval graphs using a data structure called PQ-trees [3]. Since then a lot of effort has been put into simplifying this algorithm and avoiding the use of PQ-trees [5,9,10,14]. For circular-arc graphs there is also a great amount of work focused into finding a simple linear time recognition algorithm [12]. The first linear time algorithm is due to McConnell [18] and is not simple to implement.

Algorithms for proper circular-arc recognition in linear time are also known, and they were always much easier to implement than those for the general case [6,13]. Deng, Hell and Huang exploit their proper interval graph recognition algorithm to develop a linear time algorithm for proper circular-arc graphs, based on local tournaments [6]. They also described results about the uniqueness of connected proper circular-arc graphs and proper circular-arc models, that we use to build our isomorphism testing algorithm.

The isomorphism problem is a hard to solve NP problem, although it is not known whether it is NP-hard. Nevertheless, this problem is known to be polynomial or even linear for several classes of graphs [7]. For interval graphs, labeled PQ-trees can be used to test for isomorphism in linear time [17], while for circular-arc graph the best known algorithm runs in $O(mn)$ time [11].

In this work we present a simple algorithm for the isomorphism problem restricted to proper circular-arc graphs. This algorithm runs in $O(n)$ time, when a proper circular-arc model is given. The objective is to uniquely encode a "canonical" proper circular-arc model of the graph. This canonical model is obtained by rotating, reflecting and sorting each (co-)component of the input model.

Let $G = (V(G), E(G))$ be a graph, $n = |V(G)|$ and $m = |E(G)|$. Denote by $\overline{G}$ the *complement* of G. Graph G is *co-connected* when $\overline{G}$ is connected. A *(co-)component* is a maximal (co-)connected subgraph of G. For $v \in V(G)$, denote by $N(v)$ the set of vertices adjacent to v, and write $N[v] = N(v) \cup \{v\}$ and $\overline{N}(v) = V(G) \setminus N[v]$. A vertex v of G is *universal* if $N[v] = V(G)$.

A *circular-arc* (CA) model $\mathcal{M}$ is a pair $(C, \mathcal{A})$, where C is a circle and $\mathcal{A}$ is a collection of arcs of C. When traversing the circle C, we will always choose the clockwise direction, unless explicitly stated. If s, t are points of C, write (s, t) to mean the arc of C defined by traversing the circle from s to t. Call s, t the *extremes* of (s, t), while s is the *start point* and t the *end point* of the arc. The *extremes* of $\mathcal{A}$ are those of all arcs $A \in \mathcal{A}$. The *reverse model* of $\mathcal{M}$ is denoted by $\mathcal{M}^{-1}$, i.e. $\mathcal{M}^{-1}$ is the reflection of $\mathcal{M}$ with respect to some chord of the circle. Unless otherwise stated, we always assume that $\mathcal{A} = \{A_1, \ldots, A_n\}$ where $A_i = (s_i, t_i)$. Moreover, in a traversal of C the order in which the start points appear is $s_1, \ldots, s_n$. The set $s_i, \ldots, s_j$ $(t_i, \ldots, t_j)$, $1 \leq i < j \leq n$, is called an *s-range (t-range)* with $\emptyset$ being the empty s-range (t-range). Similarly, a set of contiguous extremes is an *st-range*, and $A_i, \ldots, A_j$ is a *range*. Without loss of generality, all arcs of C are considered as open arcs, no two extremes of distinct arcs of $\mathcal{A}$ coincide and no single arc entirely covers C.

When no arc of $\mathcal{A}$ contains any other, $(C, \mathcal{A})$ is a *proper* circular-arc (PCA) model. A (proper) interval model is a (proper) CA model where $\bigcup_{A \in \mathcal{A}} A \neq C$. A CA (PCA) graph is the intersection graph of a CA (PCA) model. A (proper) interval graph is the intersection graph of a (proper) interval model. We may use the same terminology used for vertices when talking about arcs (intervals). For example, we say an arc in a CA model is *universal* when its corresponding vertex in the intersection graph is universal. Similarly, a *connected* model is one whose intersection graph is connected.

Let Σ be an alphabet. A *string S (over Σ)* is a sequence $S(1), \ldots, S(|S|)$ where $|S|$ is the *length* of S. The set $\{1, \ldots, |S|\}$ is the set of *positions* of S. For positions $i < j$ we denote by $S[i; j]$ the substring $S(i), \ldots, S(j)$. If $<$ is a total order over Σ, then $<_{lex}$ denotes the *lexicographical order* of strings, i.e. $S <_{lex} T$ if and only if there exists $k < |T|$ such that $S(i) = T(i)$ for all $1 \leq i \leq k$ and either $k = |S|$ or $S(k+1) < T(k+1)$. The *rotation* $S \ll i$ is the string $S[i; |S|]$ followed by string $S[1; i-1]$. Position i is *canonical* if $S \ll i \leq_{lex} S \ll j$ for every $1 \leq j \leq |S|$. Observe that since $<$ is a total order, then $S \ll i = S \ll j$ for every pair i, j of canonical positions.

2 PCA Representations

Let $\mathcal{M} = (C, \mathcal{A})$ be a PCA model of a graph G and fix some arc $A_i = (s_i, t_i) \in \mathcal{A}$. The *arc representation* $R^i(\mathcal{M})$ of $\mathcal{M}$ is a string obtained by transversing C from s_i and writing the character '$\mathrm{a_{j+1}}$' ('$\mathrm{b_{j+1}}$') when s_{i+j} (t_{i+j}) is reached. Thus, the j-th start (end) point that appears after s_i (t_i) is designated with the character '$\mathrm{a_{j+1}}$' ('$\mathrm{b_{j+1}}$'). Observe that we consider the order $s_1, \ldots, s_n$ to be fixed for $\mathcal{M}$, but in a computer program we do not have access to this order. What we have is an arc representation that allows us to gain access to that order.

It is clear that there are at most n different arc representations of $\mathcal{M}$, one for each arc A_i. Algorithms on (proper) CA graphs usually perform a linear time preprocessing on its input, given as an arc representation. For instance, by simply transversing the representation it is possible to build a data structure where t_i can be found in $O(1)$ time, given s_i.

Arc representations have a lot of redundant information for encoding PCA models. Fix some arc $A_i \in \mathcal{M}$. The *extreme sequence (of $\mathcal{M}$) from s_i* is the string $E^i(\mathcal{M})$ that is obtained by replacing '$\mathrm{a_j}$' ('$\mathrm{b_j}$') with 'a' ('b') in $R^i(\mathcal{M})$ for every $1 \leq j \leq n$. In other words, $E^i(\mathcal{M})$ is the string obtained by transversing C from s_i and writing the character 'a' ('b') when a start point (an end-point) is reached. The *mark point (of $\mathcal{M}$) from s_i* is the position $t^i(\mathcal{M})$ where '$\mathrm{b_1}$' appears in $R^i(\mathcal{M})$. Define the function r, such that $r(E^i(\mathcal{M}), t^i(\mathcal{M}))$ is the string obtained from $E^i(\mathcal{M})$ by replacing the j-th character 'a' with '$\mathrm{a_j}$' and the j-th character 'b' that appears from position $t^i(\mathcal{M})$ with character '$\mathrm{b_j}$'.

Remark 1. Function r is a bijection between $r(E^i(\mathcal{M}), t^i(\mathcal{M}))$ and $R^i(\mathcal{M})$. Moreover, r and r^{-1} can both be computed in $O(n)$ time.

Remark 2. For $1 \leq i, j \leq n$, $R^i(\mathcal{M}) = R^j(\mathcal{M})$ if and only if $E^i(\mathcal{M}) = E^j(\mathcal{M})$ and $t^i(\mathcal{M}) = t^j(\mathcal{M})$.

From now on we may not write the superscripts if we want to refer to any representation of $\mathcal{M}$. Also, when $\mathcal{M}$ is understood, we do not write it explicitly as a parameter. Let $\mathcal{M}$ be a PCA model. *Extreme representation* (E,t) uses only $O(n)$ bits while R uses $O(n \log n)$ bits. However, some operations, as taking the end point of some arc when the start point is given, are not (a-priori) fast enough when using (E,t). This is why in this work and others (see e.g. [1,16]) arc representations are taken as the input of the algorithms. When we say that these algorithms run in $O(n)$ time when an arc representation is given what we mean is that they run in $O(n)$ time where n is the length of codification R. But if we instead use (E,t) as input, we have to build R first, so the algorithms take $O(n \log n)$ time where n is the length of (E,t). With this in mind, when we say that $\mathcal{M}$ is given as input, we mean that an arc representation (or some linear-time preprocessing of it) is given as input.

Let $\mathcal{M},\mathcal{N}$ be two PCA models with n arcs. We write $\mathcal{M} =_M \mathcal{N}$ ($\mathcal{M}$ is *equal* to $\mathcal{N}$) if $R^i(\mathcal{M}) = R^j(\mathcal{N})$ for some $1 \leq i,j \leq n$. This is what one would intuitively assume as equality of models, i.e., do they have the extremes in the same order? Clearly, if two PCA models are equal then their intersection graphs are isomorphic, but the converse is not always true. Testing if two PCA models are equal can be trivially done in $O(n^2)$ time by fixing some $1 \leq i \leq n$ and then testing whether $R^i(\mathcal{M}) = R^j(\mathcal{N})$ for every $1 \leq j \leq n$. However this can also be verified in $O(n)$ time.

3 Basic Algorithms

In this section we describe linear-time algorithms that we use several times in the paper. Below is the list of problems we need to solve throughout the paper:

- Is a PCA graph co-bipartite? If so, give a unique co-bipartition of its co-components.
- Compute the components of a PCA graph.
- Is the intersection graph of a PCA model an interval graph? If so, output a proper interval model [15].
- Find all canonical positions of a circular string [2,20].

In the next subsections, we show how to solve each of the above problems and discuss the complexities of the corresponding algorithms. The proposed solutions are easy to implement. Furthermore, we believe they are of interest on their own.

3.1 Co-bipartitions of the Co-components of a PCA Graph

The first problem we need to solve is to determine whether a PCA graph is co-bipartite and, if so, output the co-bipartitions of its co-components. The following algorithm determines the co-component of a graph G, containing a given vertex $v \in V(G)$.

Algorithm 1. *Co-component containing v in a graph G*

1. *Unmark all vertices and define* $V_1^0 := \{v\}$, $V_2^0 := \emptyset$ *and* $k = 0$.
2. *While there exists an unmarked vertex* $w \in V_1^k \cup V_2^k$, *perform the following operation. Let* $i, j \in \{1, 2\}$, *such that* $v \in V_i^k$ *and* $j \neq i$. *Mark* v *and compute* $V_i^{k+1} := V_i^k$, $V_j^{k+1} := V_j^k \cup \overline{N}(w)$ *and* $k := k + 1$.
3. *Output* $V_1 := V_1^k, V_2 := V_2^k$

Lemma 1. *$V_1 \cup V_2$ is the co-component containing v in G. Moreover, V_1, V_2 is a co-bipartition if and only if $V_1 \cap V_2 = \emptyset$.*

For the general case this algorithm can be implemented in $O(n^2)$ time. Next, we consider that G is a PCA graph given by a PCA model $\mathcal{M}$. Define a subset of $V(G)$ to be a *range* whenever their corresponding arcs in $\mathcal{M}$ form a range. The following lemma is relevant to our purposes.

Lemma 2. *At every step k, V_1^k and V_2^k are ranges.*

Proof. Clearly V_1^0 and V_2^0 are ranges. Now consider the k-th iteration and let i, j and $w \in V_i^k$ be as in Step 2. Since G is a PCA graph, $\overline{N}(w)$ is a range. If $\overline{N}(w) \cap V_j^k = \emptyset$, then $w = v$ and $k = 0$, thus $V_j^1 = \overline{N}(w)$ corresponds to a range. If $\overline{N}(w) \cap V_j^k \neq \emptyset$ then it follows that $k > 0$ and V_j^k is a range by the inductive hypothesis. Hence $\overline{N}(w) \cup V_j^k$ is also a range, because the union of two intersecting ranges is a range. □

Now we consider the complexity of the algorithm when a PCA ordering of the vertices is given. Let i, j and k be as in Step 2, $V_i = V_i^k$ and $V_j = V_j^k$. By Lemma 2, V_i is a range. The invariant we use is that V_i is partitioned into three ranges L_i, C_i and R_i, where L_i, C_i, and R_i appear in this order. The set V_j has an analogous partition L_j, C_j and R_j. The set of marked vertices of V_i is C_i, while $L_i \cup R_i$ is the set of unmarked vertices. To maintain the invariant, vertex w can be selected from V_i if and only if $C_i \cup \{w\}$ is also a range, thus there are at most four unmarked vertices that could be selected at each step. Suppose w is chosen from L_i. For the next iteration we have to modify each of the ranges to reflect the inclusion of $\overline{N}(w)$.

For the implementation, each range in the algorithm can be represented by a pair of integers, the *low index* and the *high index*, corresponding to the first and the last vertices of the range. Before applying the algorithm, range $\overline{N}(v)$ can be found for every vertex v in $O(n)$ time. Hence, $V_j \cup \overline{N}(w)$ in Step 2 can be computed in constant time. For this, the low index of L_j is updated to the smallest of the low indices of L_j and $\overline{N}(w)$. Similarly, for the high index of R_j we would choose the greatest of the two high indices. Finally, the mark of w is done by decreasing by one the high index of L_i and increasing by one the low index of C_i when $w \in L_i$. The case when $w \in R_i$ is analogous. With such an implementation, each iteration of Step 2 takes $O(1)$ time, thus each component C can be found in $O(|C|)$ time. That is in overall $O(n)$ time the algorithm finds the co-bipartitions of all co-components.

3.2 Components of a PCA Graph

The second problem we solve is how to find the components of a PCA graph, when the input is a PCA model $\mathcal{M}$. A *leftmost* arc is an arc whose start point is not contained in any arc. When the graph is not connected, every model has at least two leftmost arcs. It is easy to find every leftmost arc in $O(n)$ time by traversing twice the circle. In the first traversal mark A_i when s_i is crossed, and then unmark A_i when t_i is crossed. In the second traversal, the start points having no marks when crossed correspond to leftmost arcs. Conversely, the start points that are crossed when there are marks are not from leftmost arcs. Now, if A_i and A_j are leftmost arcs with no leftmost arc between them then $A_i, \ldots, A_{j-1}$ is the range corresponding to the component of A_i. To sum up, the components of a PCA graph can be found in $O(n)$ time.

3.3 PCA Representation of Interval Graphs [15]

The third problem is to determine whether the intersection graph G of a PCA model $\mathcal{M} = (C, \mathcal{A})$ is an interval graph. If affirmative, then we need to construct a proper interval model. We can check if $\mathcal{M}$ is an interval model in $O(n)$ time by checking if it contains any leftmost arc. In this case the output is $\mathcal{M}$. But if G is an interval graph and $\mathcal{M}$ is not an interval model, we can transform it into an interval model in $O(n)$ time, employing the algorithm described in [15]. There, it is shown that $\mathcal{M}$ must have three arcs covering the circle. Moreover, one of these arcs, say (s, t), must be universal. Then $\mathcal{M}' = (C, (\mathcal{A} \setminus \{(s, t)\}) \cup \{(t, s)\})$ is a proper interval model of G.

3.4 Minimum Circular String [2,20]

Finally we need to find the minimum of a circular string. The minimum circular string problem is to find every canonical position of S. For this it is enough to find one canonical position i and a period w such that $i + kw$ is canonical for every $k \geq 0$. This problem can be solved in $O(n)$ comparisons over the alphabet Σ [2,20].

4 Canonical Representation of PCA Models

In this section we describe how to canonize a representation of a PCA model, so that equality of models can be tested by equality of representations. What we want is a function C from models to arc representations, so that $C(\mathcal{M}) = C(\mathcal{N})$ if and only if $\mathcal{M} =_M \mathcal{N}$. The idea is to take the "minimum" arc representation as the canonical representation. Fix a PCA model $\mathcal{M}$ and let 'a' < 'b'. Define $\prec$ as the order over the arcs, where $A_i \prec A_j$ if and only if $E^i <_{lex} E^j$. Define also, $<_R$ as the total order over arc representations where $R^i(\mathcal{M}) <_R R^j(\mathcal{N})$ if and only if either $E^i(\mathcal{M}) <_{lex} E^j(\mathcal{N})$ or $E^i(\mathcal{M}) = E^j(\mathcal{N})$ and $t^i(\mathcal{M}) < t^j(\mathcal{N})$. Arc A_i is *canonical* if A_i is minimum with respect to $\prec$ and R^i is a *canonical representation* when R^i is minimum with respect to $<_R$. Since R^i can be uniquely determined from (E^i, t^i) then $<_R$ is a total order and therefore the canonical representation of $\mathcal{M}$ is unique. That is, if R^i and R^j are canonical representations then $R^i = R^j$.

Proposition 1. *Let $\mathcal{M}$ be a PCA model and $1 \leq i, i+j \leq n$. Then $E^{i+j} = E^i \ll j$ and $t^j = b_j - a_j + 1$, where a_j (b_j) is the position of the j-th 'a' ('b') in R^i.*

Theorem 1. *Let $\mathcal{M}$ be a PCA model. Then A_i is a canonical arc if and only if R^i is a canonical representation of $\mathcal{M}$.*

Proof. If A_i is not a canonical arc, then there exists $A_j \prec A_i$. Consequently $E^j <_{lex} E^i$ which implies that $R^j <_R R^i$, so R^i is not a canonical representation.

Now suppose that A_i is a canonical arc, and let R^{i+j} be a canonical representation of $\mathcal{M}$. Then both E^i and E^{i+j} are minimum sequences with respect to $<_{lex}$, so $E^i = E^{i+j}$. By Remark 2, it is enough to see that $t^i = t^{i+j}$. Let a_k be the position of '$\mathrm{a_k}$' in R^i and b_k be the position of '$\mathrm{b_k}$' in R^i for every $1 \leq k \leq n$. By Proposition 1, $E^{i+j} = E^i \ll a_j$ and $t^{i+j} = b^j - a_j + 1$.

Since $E^i = E^{i+j} = E^i \ll a_j$ and there is the same quantity of symbols 'a' and 'b' in E^i, then in $E^i[1+k; a_j+k-1]$ there is also the same quantity of 'a' and 'b' for every $1 \leq k \leq 2n - a_j + 1$. Moreover, this quantity must be $j-1$ because in $E^i[1; a_j - 1]$ there are $j-1$ characters 'a'. Then, in $E[b_1; a_j + b_1 - 2]$ there are $j-1$ characters 'b' and $E^i(a_j + b_1 - 1)$ is also a 'b'. Consequently $b_j = a_j + b_1 - 1$, because b_j is the position of the j-th character 'b' after b_1. Hence $t^{i+j} = b^j - a_j + 1 = b_1 = t^i$ as required. □

From now on, we denote by $C(\mathcal{M})$ the unique canonical representation of $\mathcal{M}$. The algorithm we present below finds $C(\mathcal{M})$ for a PCA model $\mathcal{M}$ using some arc representation R^i as input.

Algorithm 2. *Canonical representation of model $\mathcal{M}$*

1. *Compute E^i and t^i.*
2. *Find some canonical position a_c of E^i that corresponds to some character '$\mathrm{a_c}$' in R^i, and let b_c be the position of character '$\mathrm{b_c}$' in R^i.*
3. *Output $r(E^i \ll a_c, a_c - b_c + 1)$.*

The algorithm finds $C(\mathcal{M})$ by Remark 1, Proposition 1 and Theorem 1. With respect to its time complexity, Step 1 can be done in $O(n)$ time by Remark 1, Step 2 takes $O(n)$ time as shown in Subsection 3.4, and Step 3 takes $O(n)$ time by Remark 1. Thus the time complexity is $O(n)$.

In the next section we show how to find a unique canonical model $\mathcal{M}(G)$ of a PCA graph G, so that $C(\mathcal{M}(G))$ is the unique canonical representation of a PCA graph.

5 Canonical Models of PCA Graphs

We divide the canonization of PCA graphs in three non-disjoint cases. These are the connected PCA graphs which are co-connected or non co-bipartite, the proper interval graphs, and the co-bipartite PCA graphs.

In this section we need to sort models according to their canonical representations. Define $<_M$ as the total order between models where $\mathcal{M}_1 <_M \mathcal{M}_2$ if and only if $C(\mathcal{M}_1) <_R C(\mathcal{M}_2)$. Note that $\mathcal{M}_1 =_M \mathcal{M}_2$ if and only if $\mathcal{M}_1 \not<_M \mathcal{M}_2$ and $\mathcal{M}_2 \not<_M \mathcal{M}_1$ because $C(\mathcal{M}_1)$ is unique. Although $<_M$ corresponds to a natural way to compare models, it does not behave so well for the sorting. Nevertheless, all the information in $C(\mathcal{M})$ can be encoded nicely into a somehow compressed string by combining $E(\mathcal{M})$ and $t(\mathcal{M})$. Let $(E(\mathcal{M}), t(\mathcal{M}))$ be the extreme representation $r^{-1}(C(\mathcal{M}))$ for a model $\mathcal{M}$. Define $S(\mathcal{M})$ as the string that is obtained from $E(\mathcal{M})$ by replacing the character 'b' at position $t(\mathcal{M})$ with a character 'm'. Extend $<$ so that 'a' $<$ 'm' $<$ 'b'.

Proposition 2. *$\mathcal{M}_1 <_M \mathcal{M}_2$ if and only if $S(\mathcal{M}_1) <_{lex} S(\mathcal{M}_2)$.*

Now, if $\{\mathcal{M}_1, \ldots, \mathcal{M}_k\}$ is a multiset of models, we can lexicographically sort it in $O(\sum_{i=1}^{k} |\mathcal{M}_i|)$ time using the well known most significant digit (MSD) radix sort algorithm.

5.1 Connected PCA Graphs Which Are Co-connected or Non Co-bipartite

We start first with the connected PCA graphs which are co-connected or non co-bipartite. The motivation for considering this case is the following theorem.

Theorem 2 ([6]). *Connected PCA graphs which are co-connected or non co-bipartite have at most two non-equal models, one being the reverse of the other.*

Let $\mathcal{M}$ be a PCA model of a connected PCA graph which is co-connected or non co-bipartite G. Define $\mathcal{M}(G)$ as the minimum of $\mathcal{M}$ and $\mathcal{M}^{-1}$ with respect to $<_M$; $\mathcal{M}(G)$ can be computed in $O(n)$ time.

Corollary 1. *Let G_1, G_2 be two connected PCA graphs which are co-connected or non co-bipartite. Then G_1 is isomorphic to G_2 if and only if $\mathcal{M}(G_1) =_M \mathcal{M}(G_2)$*

5.2 Proper Interval Graphs

The second class is that of proper interval graphs. As for PCA graphs, we employ a basic theorem.

Theorem 3 ([6,19]). *Connected proper interval graphs have at most two non-equal proper interval models, one being the reverse of the other.*

Extend $\mathcal{M}(G)$ to connected proper interval graphs, i.e. if $\mathcal{M}$ is a proper interval model of a connected graph G, define $\mathcal{M}(G)$ as the minimum between $\mathcal{M}$ and $\mathcal{M}^{-1}$.

Corollary 2. *Let G_1, G_2 be two connected proper interval graphs. Then G_1 is isomorphic to G_2 if and only if $\mathcal{M}(G_1) =_M \mathcal{M}(G_2)$.*

Now, let G be a proper interval graph and $G_1, \ldots, G_k$ be its components, where $\mathcal{M}(G_i) \leq_M \mathcal{M}(G_{i+1})$. Extend $\mathcal{M}(G)$ to the model where the circle is partitioned into k consecutive segments $S_1, \ldots, S_k$ and $\mathcal{M}(G_i)$ is contained in the i-th segment that appears in a traversal of the circle (see Figure 5.2). Clearly, $\mathcal{M}(G)$ is a proper interval model of G and is uniquely defined.

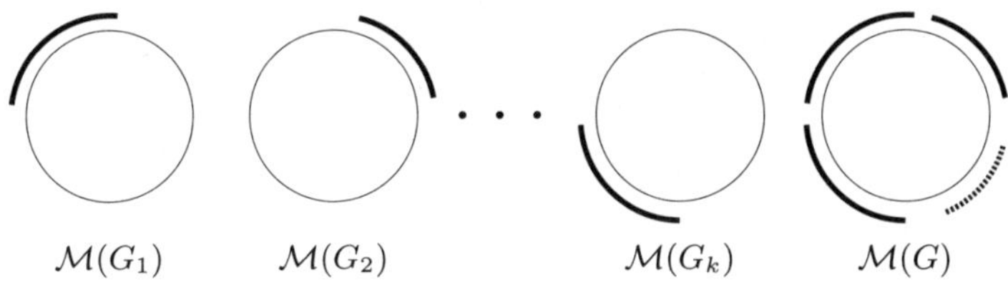

Fig. 1. The figures, except the last one, show the k segments whose corresponding models lie in $\mathcal{M}(G)$, whereas the last figure depicts $\mathcal{M}(G)$ itself

Theorem 4. *Let G_1, G_2 be two proper interval graphs. Then G_1 is isomorphic to G_2 if and only if $\mathcal{M}(G_1) =_M \mathcal{M}(G_2)$.*

We describe below the algorithm to find $\mathcal{M}(G)$ for proper interval graphs when the input is (any arc representation of) $\mathcal{M}$.

Algorithm 3. *Canonical model of a proper interval graph G.*

1. *Let $\mathcal{M}$ be some proper interval model of G*
2. *Find the components $\mathcal{M}_1, \ldots, \mathcal{M}_k$ of $\mathcal{M}$.*
3. *Define $\mathcal{M}(i)$ as the minimum of $\mathcal{M}_i$ and $\mathcal{M}_i^{-1}$ for $1 \leq i \leq k$.*
4. *Sort the multiset $\{\mathcal{M}(1), \ldots, \mathcal{M}(k)\}$ so that $\mathcal{M}(i) \leq_M \mathcal{M}(i+1)$ for every $1 \leq i < k$.*
5. *Output the model with k segments where $\mathcal{M}(i)$ is contained in the i-th segment.*

We now consider the complexity of the algorithm when $\mathcal{M}$ is given as in Step 1. Step 2 can be solved in $O(n)$ time as in Subsection 3.2, where we obtain a range representing each component. Step 3 is done by reversing $\mathcal{M}_i$ and then comparing $\mathcal{M}_i$ and $\mathcal{M}_i^{-1}$. Both the reversal and the comparison can be computed in $O(|\mathcal{M}_i|)$ for $1 \leq i \leq k$. Step 4 can done as explained at the beginning of this section in $O(n)$ time. For Step 5 traverse the range corresponding to $\mathcal{M}_i$ that was obtained in Step 1 and insert it into the new circle. The algorithm then runs in $O(n)$ time.

5.3 Canonization of Co-bipartite PCA Graphs

Finally we consider co-bipartite PCA graphs. The algorithm for this class of graphs is quite similar in its concept to the one for proper interval graphs. For co-connected PCA graphs $\mathcal{M}(G)$ has already been defined in Subsection 5.1.

Consider a co-connected co-bipartite PCA graph G and denote by $\mathcal{A}_1, \mathcal{A}_2$ the co-bipartition of $\mathcal{M}(G)$. By Lemma 2, both $\mathcal{A}_1$ and $\mathcal{A}_2$ are ranges. Assume w.l.o.g. that A_1 is the first arc of range $\mathcal{A}_1$ and A_i is the first arc of range $\mathcal{A}_2$. Moreover, assume that s_1 is represented by 'a$_1$' in $C(\mathcal{M}(G))$. In the segment (s_{i-1}, t_1) there is no start point of arcs in $\mathcal{A}_2$, because otherwise every arc of $\mathcal{A}_1$ would contain these start points. Consequently, (s_{i-1}, t_1) is a segment contained by all the arcs of $\mathcal{A}_1$ that is not crossed by any arcs of $\mathcal{A}_2$. Moreover, (s_{i-1}, t_1) is the unique maximal segment in these conditions. The same argument can be applied interchanging $\mathcal{A}_1$ with $\mathcal{A}_2$. But in the case where $\mathcal{A}_2$ is empty and $\mathcal{A}_1$ has only one universal arc A_1, then (t_1, s_1) is the required segment. This means that $X = \{t_i, \ldots, t_n\} \cup \{s_1, \ldots, s_{i-1}\}$ and $Y = \{t_1, \ldots, t_{i-1}\} \cup \{s_i, \ldots, s_n\}$ are two st-ranges that define $\mathcal{M}(G)$ (see Figure 2(a)). We call these two st-ranges as *co-bipartition ranges*, where X is the *low co-bipartition range* and Y is the *high co-bipartition range*. Observe that low and high are well defined for $\mathcal{M}(G)$, because X contains 'a$_1$' in $C(\mathcal{M}(G))$.

Now we show a unique way to accommodate the co-components when the PCA graph is not co-connected (see also [6]). This is rather similar to what we did in the previous section. Let G be a non interval co-bipartite PCA graph and $G_1, \ldots, G_k$ be its co-components where $\mathcal{M}(G_i) \leq_M \mathcal{M}(G_{i+1})$ for $1 \leq i \leq k$. Let X_i, Y_i be the respective low and high co-bipartitions ranges of $\mathcal{M}(G_i)$ for $1 \leq i < k$. Define $\mathcal{M}(G)$ as the model where the circle is partitioned into $2k$ consecutive segments. The i-th segment in a traversal of the circle contains X_i and the $i+k$ segment contains Y_i for $1 \leq i \leq k$ (see Figure 2). It is not hard to see that $\mathcal{M}(G)$ is a PCA model of G, because the model induced by the arcs in segments i and $i+k$ is precisely $\mathcal{M}(G_i)$ and every arc with one extreme in segment i intersects every arc with one extreme in segment j for $1 \leq i < j \leq k$.

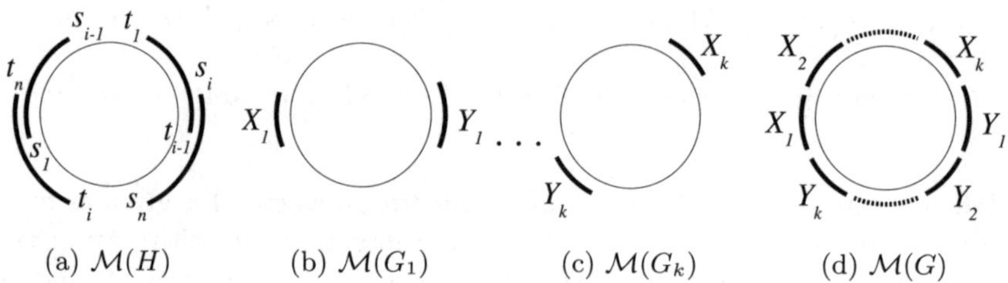

Fig. 2. Figure (a) shows the high co-bipartition range $\{t_i, \ldots, t_n\} \cup \{s_1, \ldots, s_{i-1}\}$ and the low co-bipartition range $\{t_1, \ldots, t_{i-1}\} \cup \{s_i, \ldots, s_n\}$ of a co-connected co-bipartite graph. Figures (b) and (c) show the co-bipartition ranges of the co-components of G in their corresponding segments. In (d) the whole picture of $\mathcal{M}(G)$ is shown.

Theorem 5. *Let G_1, G_2 be two non interval co-bipartite PCA graphs. Then G_1 is isomorphic to G_2 if and only if $\mathcal{M}(G_1) =_M \mathcal{M}(G_2)$.*

The algorithm to find a canonical representation of a co-bipartite PCA graph is very similar to the one for a proper interval graph. The two main changes are that components are replaced by co-components in Steps 2-5, and that the circle is partitioned into $2k$ segments, where segments from 1 to k contain the low co-bipartition ranges and segments from $k+1$ to $2k$ contain the high co-bipartition ranges (Step 6). Since both the co-components and the co-bipartition ranges can be found in $O(n)$ time, as in Section 3.1, the whole algorithm for this case takes $O(n)$ time.

6 Putting It All Together

Function $\mathcal{M}$ as defined in the previous section maps every PCA graph to a PCA model. However, it should be mentioned that if G is both proper interval and co-bipartite, then $\mathcal{M}(G)$ is computed as in Subsection 5.2. The complete algorithm is depicted below.

Algorithm 4. *Canonical representation of a PCA graph G*

1. *If G is an interval graph, then compute $\mathcal{M}(G)$ as in Subsection 5.2.*
2. *Else if G is a co-bipartite model then compute $\mathcal{M}(G)$ as in Subsection 5.3.*
3. *Otherwise, compute $\mathcal{M}(G)$ as in Subsection 5.1.*

Finally we discuss the complexity of the entire algorithm. The input of the algorithm is a PCA model as in Step 1. This model is encoded as an arc representation, which is obtained as the output of the recognition algorithm for PCA graphs [6]. We can check if G is an interval graph as in Subsection 3.3 in $O(n)$ time. If so, we obtain a proper interval model that we can use in Step 1 to find $\mathcal{M}(G)$ in $O(n)$ time. The rest of the algorithm takes $O(n)$ time as explained in the previous section. When the input is G instead of $\mathcal{M}$, the algorithm takes $O(n+m)$ time by first computing a PCA model [6].

Theorem 6. *Let G and H be two PCA graphs. Then the following are equivalent:*

1. *G and H are isomorphic,*
2. *$\mathcal{M}(G) =_M \mathcal{M}(H)$,*
3. *$C(\mathcal{M}(G)) = C(\mathcal{M}(G))$.*

Proof. It is a direct consequence of Corollary 1 and Theorems 4 and 5, and the fact that $\mathcal{M}$ is a well defined function. □

Corollary 3. *The isomorphism problem for PCA graphs can be solved in $O(n)$ time when a PCA models model is given as input, or in $O(n+m)$ time when the input is a graph given by its sets of vertices and edges.*

References

1. Bhattacharya, B., Hell, P., Huang, J.: A linear algorithm for maximum weight cliques in proper circular arc graphs. SIAM J. Discrete Math. 9(2), 274–289 (1996)
2. Booth, K.S.: Lexicographically least circular substrings. Inform. Process. Lett. 10(4-5), 240–242 (1980)
3. Booth, K.S., Lueker, G.S.: Testing for the consecutive ones property, interval graphs, and graph planarity using *PQ*-tree algorithms. J. Comput. System Sci. 13(3), 335–379 (1976)

4. Brandstädt, A., Le, V.B., Spinrad, J.P.: Graph classes: a survey. SIAM, Philadelphia (1999)
5. Corneil, D.G., Olariu, S., Stewart, L.: The ultimate interval graph recognition algorithm (extended abstract). In: 9th Annual ACM-SIAM Symposium on Discrete Algorithms, pp. 175–180. ACM, New York (1998)
6. Deng, X., Hell, P., Huang, J.: Linear-time representation algorithms for proper circular-arc graphs and proper interval graphs. SIAM J. Comput. 25(2), 390–403 (1996)
7. Garey, M.R., Johnson, D.S.: Computers and intractability. W. H. Freeman and Co., San Francisco (1979)
8. Golumbic, M.C.: Algorithmic Graph Theory and Perfect Graphs, 2nd edn. North–Holland Publishing Co, Amsterdam (2004)
9. Habib, M., McConnell, R.M., Paul, C., Viennot, L.: Lex-BFS and partition refinement, with applications to transitive orientation, interval graph recognition and consecutive ones testing. Theoret. Comput. Sci. 234(1-2), 59–84 (2000)
10. Hsu, W.: A simple test for interval graphs. In: Mayr, E.W. (ed.) WG 1992. LNCS, vol. 657, pp. 11–16. Springer, Heidelberg (1993)
11. Hsu, W.: $O(m.n)$ algorithms for the recognition and isomorphism problems on circular-arc graphs. SIAM J. Comput. 24(3), 411–439 (1995)
12. Kaplan, H., Nussbaum, Y.: A simpler linear-time recognition of circular-arc graphs. In: Arge, L., Freivalds, R. (eds.) SWAT 2006. LNCS, vol. 4059, pp. 41–52. Springer, Heidelberg (2006)
13. Kaplan, H., Nussbaum, Y.: Certifying algorithms for recognizing proper circular-arc graphs and unit circular-arc graphs. In: Fomin, F.V. (ed.) WG 2006. LNCS, vol. 4271, pp. 289–300. Springer, Heidelberg (2006)
14. Korte, N., Möhring, R.H.: An incremental linear-time algorithm for recognizing interval graphs. SIAM J. Comput. 18(1), 68–81 (1989)
15. Lin, M.C., Soulignac, F.J., Szwarcfiter, J.L.: Proper Helly circular-arc graphs. In: Brandstädt, A., Kratsch, D., Müller, H. (eds.) WG 2007. LNCS, pp. 248–257. Springer, Heidelberg (2007)
16. Lin, M.C., Szwarcfiter, J.L.: Unit Circular-Arc Graph Representations and Feasible Circulations. SIAM J. Discrete Math. 22(1), 409–423 (2008)
17. Lueker, G.S., Booth, K.S.: A linear time algorithm for deciding interval graph isomorphism. J. Assoc. Comput. Mach. 26(2), 183–195 (1979)
18. McConnell, R.M.: Linear-time recognition of circular-arc graphs. Algorithmica 37(2), 93–147 (2003)
19. Roberts, F.S.: Indifference graphs. In: Proof Techniques in Graph Theory (2nd Ann Arbor Graph Theory Conf.), pp. 139–146. Academic Press, New York (1969)
20. Shiloach, Y.: Fast canonization of circular strings. J. Algorithms 2(2), 107–121 (1981)
21. Spinrad, J.P.: Efficient graph representations. American Mathematical Society, Providence (2003)

Spanners of Additively Weighted Point Sets*

Prosenjit Bose, Paz Carmi, and Mathieu Couture

School of Computer Science, Carleton University, Herzberg Building
1125 Colonel By Drive, Ottawa, Ontario, Canada

Abstract. We study the problem of computing geometric spanners for (additively) weighted point sets. A weighted point set is a set of pairs (p, r) where p is a point in the plane and r is a real number. The distance between two points (p_i, r_i) and (p_j, r_j) is defined as $|p_i p_j| - r_i - r_j$. We show that in the case where all r_i are positive numbers and $|p_i p_j| \geq r_i + r_j$ for all i, j (in which case the points can be seen as non-intersecting disks in the plane), a variant of the Yao graph is a $(1 + \epsilon)$-spanner that has a linear number of edges. We also show that the Additively Weighted Delaunay graph (the face-dual of the Additively Weighted Voronoi diagram) has constant spanning ratio. The straight line embedding of the Additively Weighted Delaunay graph may not be a plane graph. Given the Additively Weighted Delaunay graph, we show how to compute a plane embedding with a constant spanning ratio in $O(n \log n)$ time.[1]

1 Introduction

Let G be a complete weighted graph where edges have positive weight. Given two vertices u, v of G, we denote by $\delta_G(u, v)$ the length of a shortest path in G between u and v. A spanning subgraph H of G is a *t-spanner* of G if $\delta_H(u, v) \leq t\delta_G(u, v)$ for all pair of vertices u and v. The smallest t having this property is called the *spanning ratio* of the graph H with respect to G. Thus, a graph with spanning ratio t approximates the $\binom{n}{2}$ distances between the vertices of G within a factor of t. Let P be a set of n points in the plane. A *geometric graph* with vertex set P is an undirected graph whose edges are line segments that are weighted by their length. The problem of constructing t-spanners of geometric graphs with $O(n)$ edges for any given point set has been studied extensively; see the book by Narasimhan and Smid [2] for an overview.

In this paper, we address the problem of computing geometric spanners with additive constraints on the points. More precisely, we define a weighted point set as a set of pairs (p, r) where p is a point in the plane and r is a real number. The distance between two points (p_i, r_i) and (p_j, r_j) is defined as $|p_i p_j| - r_i - r_j$. The problem we address is to compute a spanner of a complete graph on a weighted point set. To the best of our knowledge, the problem of constructing a geometric spanner in this context has not been previously addressed. We show how the Yao

* Research partially supported by NSERC, MRI, CFI, and MITACS.

[1] Due to space constraints, some proofs have been omitted. All missing proofs can be found in the technical report version of this paper [1].

J. Gudmundsson (Ed.): SWAT 2008, LNCS 5124, pp. 367–377, 2008.

graph can be adapted to compute a $(1 + \epsilon)$-spanner in the case where all r_i are positive real numbers and $|p_i p_j| \geq r_i + r_j$ for all i, j (in which case the points can be seen as non-intersecting disks in the plane). In the same case, we also show how the Additively Weighted Delaunay graph (the face-dual of the Additively Weighted Voronoi diagram) provides a plane spanner that has the same spanning ratio as the Delaunay graph of a set of points. Since $|p_i p_j| < r_i + r_j$ implies that the distance is negative, we believe that the restriction $|p_i p_j| \geq r_i + r_j$ is reasonable because the t-spanner problem does not make sense when there are negative distances.

2 Related Work

Well known examples of geometric t-spanners include the Yao graph [3], θ-graphs [4], the Delaunay graph [5], and the Well-Separated Pair Decomposition [6]. Let $\theta < \pi/4$ be an angle such that $2\pi/\theta = k$, where k is an integer. The Yao graph with angle θ is defined as follows. For every point p, partition the plane into k cones $C_{p,1}, \ldots, C_{p,k}$ of angle θ and apex p. Then, there is an oriented edge from p to q if and only if q is the closest point to p in some cone $C_{p,i}$. For Yao graphs [3], the spanning ratio is at most $1/(\cos\theta - \sin\theta)$ provided that $\theta < \pi/4$. For θ-graphs, the spanning ratio is at most $1/(1 - 2\sin\frac{\theta}{2})$ provided that $\theta < \pi/3$ [4].

Given a set of points in the plane, there is an edge between p and q in the Delaunay graph if and only if there is an empty circle with p and q on its boundary [5]. The spanning ratio of the Delaunay triangulation is at most 2.42 [5]. The *Voronoi diagram* [7] of a finite set of points P is a partition of the plane into $|P|$ regions such that each region contains exactly those points having the same nearest neighbor in P. The points in P are also called *sites*. It is well known that the Voronoi diagram of a set of points is the face dual of the Delaunay graph of that set of points [7], i.e. two points have adjacent Voronoi regions if and only if they share an edge in the Delaunay graph.

3 Definitions and Notation

Definition 1. *A set $P = \{(p_1, r_1), \ldots, (p_n, r_n)\}$ of ordered pairs, where each p_i is a point in the plane and each r_i is a real number, is called a* weighted point set. *The notation $p_i \in P$ means that there exists an ordered pair (p_i, r_i) such that $(p_i, r_i) \in P$. The* additive distance *from a point $p \notin P$ in the plane to a point $p_i \in P$, noted $d(p, p_i)$, is defined as $|pp_i| - r_i$, where $|pp_i|$ is the Euclidean distance from p to p_i. The* additive distance *between two points $p_i, p_j \in P$, noted $d(p_i, p_j)$, is defined as $|p_i p_j| - r_i - r_j$, where $|p_i p_j|$ is the Euclidean distance from p_i to p_j.*

The problem we address in this paper is the following:

Problem 1. Let P be a weighted point set and let $K(P)$ be the complete weighted graph with vertex set P and edges weighted by the additive distance between

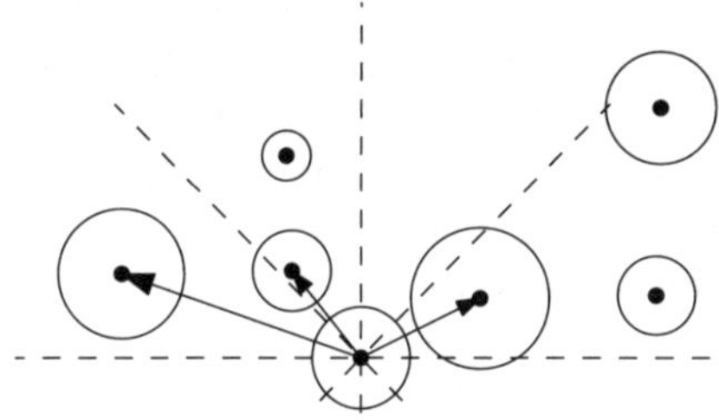

Fig. 1. A straightforward generalization of the Yao graph

their endpoints. Compute a t-spanner with $O(n)$ edges of $K(P)$ for a fixed constant $t > 1$.

Notice that in the case where all r_i are positive numbers, the pairs (p_i, r_i) can be viewed as disks D_i in the plane. If, for all i, j we also have $d(p_i, p_j) \geq 0$, then the disks are disjoint. In that case, the distance $d(D_i, D_j) = d(p_i, p_j) = |p_i p_j| - r_i - r_j$ is also equal to $\min\{|q_i q_j| : q_i \in D_i \text{ and } q_j \in D_j\}$, where the notation $q_i \in D_i$ means $|p_i q_i| \leq r_i$. To compute a spanner of an additively weighted point set is then equivalent to computing a spanner of a set of disks in the plane. **From now to the end of this paper, it is assumed that all r_i are positive numbers and $d(p_i, p_j) \geq 0$ for all i, j.** If $\mathcal{D}$ is a set of disks in the plane, then a *spanner* of $\mathcal{D}$ is a spanner of the complete graph whose vertex set is $\mathcal{D}$ and whose edges (D_i, D_j) are given weights equal to $d(D_i, D_j)$.

Notice also that the additive distance may not be a metric since the triangle inequality does not necessarily hold. Although this may seem counter-intuitive, this makes sense in some networks, since a direct communication is not always easier than routing through a common neighbor. For example, in wireless networks, the amount of energy that is needed to transmit a message is a power of the Euclidean distance between the sender and the receiver. Therefore, using several small hops can be more energy efficient than a direct communication over one long-distance link.

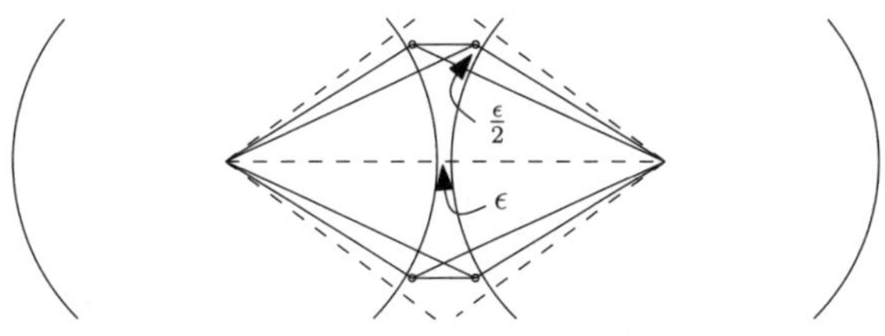

Fig. 2. The straightforward generalization of the Yao graph does not have constant spanning ratio

Figure 1 shows how the Yao graph can be generalized using the additive distance: every node keeps an outgoing edge with the closest disk that intersects each cone. However, this graph is not a spanner. Figure 2 shows how to construct an example with four disks that has an arbitrarily large spanning ratio. Nonetheless, in Section 4, we see that a minor adjustment to the Yao graph can be made in order to compute a $(1+\epsilon)$-spanner of a set of disjoint disks that has $O(n)$ edges.

The Delaunay graph in the additively weighted setting is computable in time $O(n \log n)$ [8]. To the best of our knowledge, its spanning properties have not been previously studied. In Section 6, we show that it is a spanner and that its spanning ratio is the same as that of the standard Delaunay graph.

4 The Additively Weighted Yao Graph

As we saw in the previous section, a straightforward generalization of the Yao graph fails to provide a graph with bounded spanning ratio. In this section, we show how a few subtle modifications to the construction, provide an approach to build a $(1+\epsilon)$-spanner. We define the modified Yao construction below.

Definition 2. *Let $\mathcal{D}$ be a finite set of disjoint disks and $\theta \leq 0.228$ be an angle such that $2\pi/\theta = k$, where k is an integer. The* Yao$(\theta, \mathcal{D})$ *graph is defined as follows. For every disk $D = (p, r)$, partition the plane into k cones $C_{p,1}, \ldots, C_{p,k}$ of angle θ and apex p. A disk* blocks *a cone $C_{p,i}$ provided that the disk intersects both rays of $C_{p,i}$. Let $F \in \mathcal{D}$ be a disk different from D with center in $C_{p,j}$. Add an edge from D to F in* Yao$(\theta, \mathcal{D})$ *if and only if one of the two following conditions is met:*

1. *among all blocking disks that have their center in $C_{p,j}$, F is the one that is the closest to D;*
2. *among all disks that have their center in $C_{p,j}$ and are at a distance of at least r from D, F is the one that is the closest to D.*

Notice that there are two main changes. Within each cone, we now add potentially two edges as opposed to only one edge in the case of unweighted points. Next, in the second condition to add an edge, we do not add an edge to the closest disk within a cone but to the closest disk whose distance is at least r from the disk centered at the apex with radius r. We now prove that these two modifications imply that the resulting graph is a $(1+\epsilon)$-spanner.

Lemma 1. *Let p_1, p_2, p_3 such that the angle $\angle p_3 p_1 p_2 = \alpha \leq \theta < \pi/4$ and $|p_1 p_3| \leq |p_1 p_2|$. Then $|p_2 p_3| \leq |p_1 p_2| - (\cos\theta - \sin\theta)|p_1 p_3|$.*

Theorem 1. *Let $\mathcal{D}$ be a finite set of disjoint disks and $\theta \leq 0.228$. Then $Y(\theta, \mathcal{D})$ is a t-spanner of $\mathcal{D}$, where $t = 1/(\cos 2\theta - \sin 2\theta - 2\sin(\theta/2))$.*

Proof: We proceed by induction on the rank of the weighted distances between the pairs of disks D_1 and D_2.

Base case: The disks D_1 and D_2 form a closest pair. In that case, the edge (D_1, D_2) is in $\text{Yao}(\theta, \mathcal{D})$. To see this, let $r_1 \leq r_2$. If D_2 is blocking the cone centered at p_1 that contains it, then it is in $\text{Yao}(\theta, \mathcal{D})$ by Case 1 of Definition 2. Otherwise, then it is at distance at least r_1 from D_1 and therefore it is in $\text{Yao}(\theta, \mathcal{D})$ by Case 2 of Definition 2.

Induction case: Let $D_1 = (p_1, r_1)$ and $D_2 = (p_2, r_2)$. Without loss of generality, $r_1 \leq r_2$. If the edge (D_1, D_2) is in $\text{Yao}(\theta, \mathcal{D})$, then there is nothing to prove. Otherwise, there are two cases to consider depending on whether or not the shortest path from D_1 to D_2 in the complete graph on $\mathcal{D}$ is the edge (D_1, D_2). If the shortest path is not the edge (D_1, D_2), then all edges on the shortest path must have length less than $d(D_1, D_2)$. By applying the induction hypothesis on each of those edges, we conclude that the distance from D_1 to D_2 in $\text{Yao}(\theta, \mathcal{D})$ is at most t times the length of the shortest path D_1 to D_2 in the complete graph on $\mathcal{D}$, as required.

We now consider the case when the edge (D_1, D_2) 1) is not in $\text{Yao}(\theta, \mathcal{D})$ and 2) is the shortest path from D_1 to D_2 in the complete graph. Observe that the conjunction of those two facts imply that the disk D_2 does not block the cone whose apex is p_1 and contains p_2: If D_2 was blocking the cone, then since (D_1, D_2) is not an edge in $\text{Yao}(\theta, \mathcal{D})$, there must be a disk D_3 that is also blocking the cone and is closer to D_1 than D_2. However, this implies that the shortest path from D_1 to D_2 in the complete graph is not the edge (D_1, D_2) (see Figure 3).

The conjunction of the following three facts:

1. $r_1 \leq r_2$;
2. $\theta \leq 0.228 < \sin^{-1}(1/3)$ and
3. D_2 does not block the cone,

imply that $d(D_1, D_2) > r_1$. Since (D_1, D_2) is not an edge, there is another disk whose distance is at least r that is closer to D_1. Let $D_3 = (p_3, r_3)$ be the closest disk to D_1 such that p_3 is in the same θ-cone with apex at p_1 as p_2 and $d(D_1, D_3) \geq r_1$. By definition, the edge (D_1, D_3) is in $\text{Yao}(\theta, \mathcal{D})$. Observe that $d(D_2, D_3) < d(D_1, D_2)$. To see this, let $a := d(D_1, D_2) - r_1$. We have that

$$d(D_2, D_3) \leq a + 4r_1 \sin(\theta/2) \leq a + 4r_1 \sin(0.114) < a + r_1 = d(D_1, D_2).$$

Let p_1' be the point of D_1 that is the closest to D_3, p_1'' be the point of D_1 that is

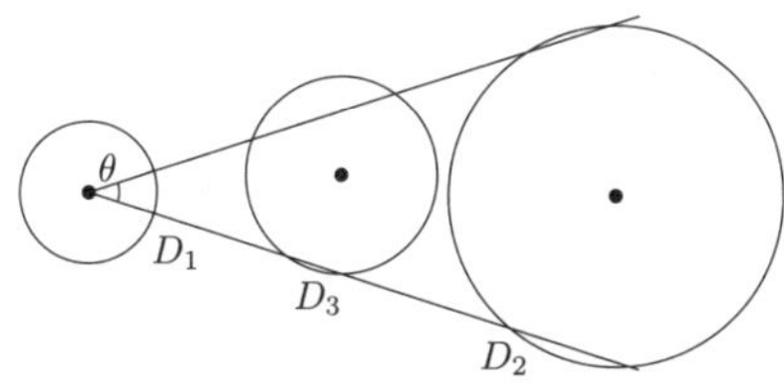

Fig. 3. If D_2 blocks the cone but the edge (D_1, D_2) is not in $\text{Yao}(\theta, \mathcal{D})$, then there exists D_3 such that $d(D_1, D_3) + d(D_3, D_2) < d(D_1, D_2)$

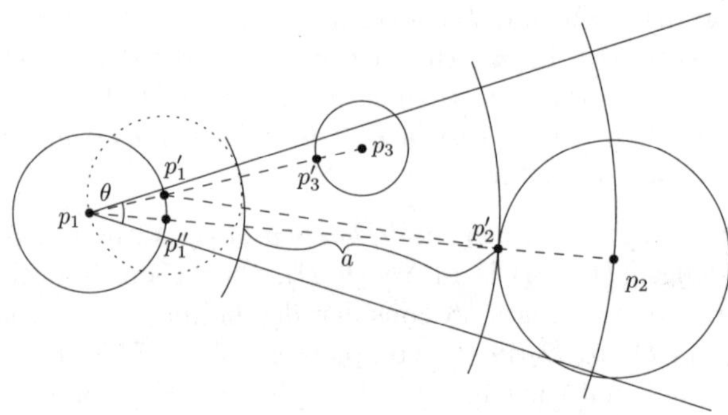

Fig. 4. Illustration of the proof of Theorem 1

the closest to D_2, p_2' be the point of D_2 that is the closest to D_1, and p_3' be the point of D_3 that is the closest to D_1 (see Figure 4). Notice that $|p_1'p_3'| \leq |p_1'p_2'|$ and that since $d(D_1, D_2) \geq d(D_1, D_3) \geq r_1$, then the angle $\angle p_2'p_1'p_3'$ is at most $2\theta < \pi/4$. Therefore, we can apply Lemma 1 to conclude that

$$|p_2'p_3'| \leq |p_1'p_2'| - (\cos 2\theta - \sin 2\theta)|p_1'p_3'|,$$

which implies that

$$d(D_2, D_3) \leq d(D_1, D_2) + |p_1'p_1''| - (\cos 2\theta - \sin 2\theta)d(D_1, D_3).$$

Also, since $|p_1'p_1''| \leq 2\sin(\theta/2)r_1 \leq 2\sin(\theta/2)d(D_1, D_3)$, we have

$$d(D_2, D_3) \leq d(D_1, D_2) - (\cos 2\theta - \sin 2\theta - 2\sin(\theta/2))d(D_1, D_3).$$

Finally, since $d(D_2, D_3) < d(D_1, D_2)$, the induction hypothesis tells us that $\mathrm{Yao}(\theta, \mathcal{D})$ contains a path from D_2 to D_3 whose length is at most $td(D_2, D_3)$. This means that the distance from D_1 to D_2 in $\mathrm{Yao}(\theta, \mathcal{D})$ is at most

$$d(D_1, D_3) + td(D_2, D_3) \leq d(D_1, D_3) + t(d(D_1, D_2) - \frac{1}{t}d(D_1, D_3)) = td(D_1, D_2).$$

Using Maple, we verified that the value 0.228 is an upper bound on the values of θ such that $t > 0$. □

Corollary 1. *For any $\epsilon > 0$ and any set $\mathcal{D}$ of n disjoint disks, it is possible to compute a $(1 + \epsilon)$-spanner of $\mathcal{D}$ that has $O(n)$ edges.*

Proof: The bound on the number of edges comes from the fact that each cone contains at most two edges, and the stretch factor of $1 + \epsilon$ comes from the fact that $\lim_{\theta \to 0} 1/(\cos 2\theta - \sin 2\theta - 2\sin(\theta/2)) = 1$. □

5 Quotient Graphs and Quotient Spanners

The main idea in the remainder of this paper is the following: we show how to compute a set of points from each D_i such that the (standard) Delaunay graph of those points is *equivalent* to the Additively Weighted Delaunay graph. By choosing the appropriate equivalence relation as well as the appropriate point set, we can then show that the spanning ratio of the Additively Weighted Delaunay graph is bounded by the spanning ratio of the standard Delaunay graph. The reduction of one graph to another is done by means of a quotient:

Definition 3. *Let P_1 and P_2 be non-empty sets of points in the plane. The* distance *between P_1 and P_2, denoted by $|P_1P_2|$, is defined as the minimum $|p_1p_2|$ over all pairs of points such that $p_1 \in P_1$ and $p_2 \in P_2$.*

Definition 4. *Let $G = (V, E)$ be a geometric graph and $\mathcal{V}$ be a partition of V. The* quotient graph *of G by $\mathcal{V}$, denoted $G/\mathcal{V}$, is the graph having $\mathcal{V}$ as vertices and there is an edge (U, W) (where U and W are in $\mathcal{V}$) if and only if there exists an edge $(u, w) \in E$ with $u \in U$ and $w \in W$. The weight of the edge (U, W) is equal to $|UW|$.*

If P is a (non-weighted) point set and $\mathcal{P}$ is a partition of P, then the notation $P/\mathcal{P}$ designates the quotient of the complete Euclidean graph on P by $\mathcal{P}$. If $\mathcal{S}$ is a set of pairwise disjoint sets of points in the plane such that $P \subseteq \bigcup \mathcal{S}$, then the notation $P/\mathcal{S}$ designates the quotient of the complete Euclidean graph on P by the partition of P induced by $\mathcal{S}$.

Lemma 2. *Let $G = (V, E)$ be a complete geometric graph, $\mathcal{V}$ be a partition of V and S be a t-spanner of G. Then $S/\mathcal{V}$ is a t-spanner of $G/\mathcal{V}$.*

6 The Additively Weighted Delaunay Graph

Lee and Drysdale [9] studied a variant of the Voronoi diagram called the Additively Weighted Voronoi diagram, which is defined as follows: Let P be a weighted point set. The *Additively Weighted Voronoi diagram* of P is a partition of the plane into $|P|$ regions such that each region contains exactly the points in the plane having the same closest neighbor in P according to the additive distance. In other words, the Voronoi cell of a pair (p_i, r_i) contains the points p such that $d(p, p_i)$ is minimum over all other pairs in P. The *Additively Weighted Delaunay graph* (AW-Delaunay graph) is defined as the face-dual of the Additively Weighted Voronoi diagram.

Alternatively, if all r_i are positive and for all i, j, we have $|p_ip_j| \geq r_i + r_j$, then the pairs (p_i, r_i) can be seen as disks D_i of radius r_i centered at p_i and $d(p, D_i)$ is the minimum $|pq|$ over all $q \in D_i$. For a set $\mathcal{D}$ of disks in the plane, we denote the AW-Delaunay graph computed from $\mathcal{D}$ as $\text{Del}(\mathcal{D})$. When no two disks intersect, the AW-Delaunay graph is a natural generalization of the Delaunay graph of a set of points. We say that two intersecting disks A and B *properly* intersect if $|A \cap B| > 1$ (i.e. they are not tangent).

Proposition 1. *Let $\mathcal{D}$ be a set of disjoint disks in the plane, and $A, B \in \mathcal{D}$. The edge (A, B) is in* Del($\mathcal{D}$) *if and only if there is a disk C that is tangent to both A and B and does not properly intersect any other disk in $\mathcal{D}$.*

Proof: Suppose (A, B) is in Del($\mathcal{D}$), and let c be a point on the boundary of the Voronoi cells of A and B and r be the distance from c to A. Since c is equidistant from A and B, it is also at distance r from B. This means that the disk C centered at c is tangent to both A and B. This disk cannot properly intersect any other disk of $\mathcal{D}$, since this would contradict the fact that c is in the Voronoi cells of A and B. Similarly, if there is a disk that is tangent to both A and B but does not properly intersect any other disk of $\mathcal{D}$, then A and B are Voronoi neighbors. □

Note that the Additively Weighted Delaunay graph is not necessarily isomorphic to the Delaunay graph of the centers of the disks. When all radii are equal, however, the two graphs coincide. We now show that if $\mathcal{D}$ is a set of disks in the plane, then Del($\mathcal{D}$) is a spanner of $\mathcal{D}$. The intuition behind the proof is the following: we show the existence of a finite set of points P such that $K(P)/\mathcal{D}$ (where $K(P)$ is the complete graph with vertex set P) is isomorphic (i.e. there is a one-to-one relation between the nodes that preserves the lengths of the edges) to the complete graph on $\mathcal{D}$ and Del$(P)/\mathcal{D}$ is a subgraph of Del($\mathcal{D}$). Then, we use Lemma 2 to prove that Del$(P)/\mathcal{D}$ is a spanner of $\mathcal{D}$, which implies that Del($\mathcal{D}$) is a spanner of $\mathcal{D}$.

Definition 5. *Let A, B be disjoint disks and S a set of points such that $A \cap S = \emptyset$ and $B \cap S = \emptyset$. A set of points R* represents *S with respect to A and B if for every disk F that is tangent to both A and B, we have $F \cap S \neq \emptyset \Rightarrow F \cap R \neq \emptyset$. If $\mathcal{D}$ is a set of disjoint disks, then a set of points $\mathcal{R}$* represents *$\mathcal{D}$ if for all $A, B, C \in \mathcal{D}$, there is a subset of $\mathcal{R}$ that represents C with respect to A and B.*

Lemma 3. *Let $\mathcal{D}$ be a set of n disjoint disks. There exists a set of at most $2\binom{n}{3}$ points that represents $\mathcal{D}$.*

Lemma 4. *Let A and B be two disjoint disks and C be a disk intersecting both of them. Then there exists a disk G inside C that is tangent to both A and B.*

Proof: We show how to construct G. Let a, b, c and r_A, r_B, r_C respectively be the centers and radii of A, B and C. Without loss of generality, assume $|ac| - r_C \leq |bc| - r_B$. Let F be the disk centered at c and having radius $r_F = |bc| - r_B$. The disk F is tangent to B. If F is also tangent to A, then let $G = F$ and we are done. Otherwise, F is properly intersecting A. In that case, let p be the tangency point of F and B, l be the line through b and c, and G be the disk through p having its center on l and tangent to A. The result follows from the fact that G is tangent to B and inside C. □

Definition 6. *Let A and B be two disks in the plane. The* distance points *of A and B are the two ends of the shortest line segment between A and B.If $\mathcal{D}$ is a set of disjoint disks, then the set of* distance points *of $\mathcal{D}$ is the set containing the distance points of every pair of disks in $\mathcal{D}$.*

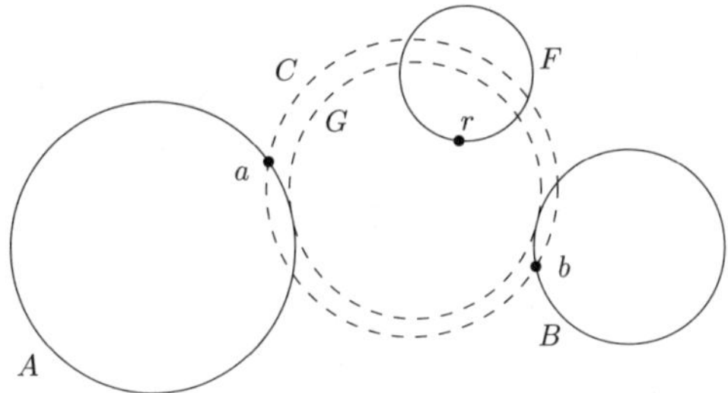

Fig. 5. Illustration of the proof of Theorem 2

Theorem 2. *Let $\mathcal{D}$ be a set of n disjoint disks. Then* Del($\mathcal{D}$) *is a t-spanner of $\mathcal{D}$, where t is the spanning ratio of the Delaunay triangulation of a set of points.*

*P*roof: By Lemma 3, let R be a set of size at most $2\binom{n}{3}$ that represents $\mathcal{D}$, let S be the set of distance points of $\mathcal{D}$, and let $P = R \cup S$. Since $\mathrm{Del}(P)$ is a t-spanner of P, by Lemma 2, we have $\mathrm{Del}(P)/\mathcal{D}$ is a t-spanner of $K(P)/\mathcal{D}$, where $K(P)$ is the complete graph with vertex set P. Since P contains the distance points of $\mathcal{D}$, $K(P)/\mathcal{D}$ is isomorphic to the complete graph defined on $\mathcal{D}$. We show that each edge (A, B) of $\mathrm{Del}(P)/\mathcal{D}$ is in $\mathrm{Del}(\mathcal{D})$. Let (A, B) be an edge of $\mathrm{Del}(P)/\mathcal{D}$. This means that in P, there are two points a and b with $a \in A, b \in B$ such that there is an empty circle C through a and b. By Lemma 4, C contains a disk G that is tangent to both A and B. The disk G is a witness of the presence of the edge (A, B) in $\mathrm{Del}(\mathcal{D})$. If that was not the case, this would mean that there exists a disk $F \in \mathcal{D}$ such that $G \cap F \neq \emptyset$. By definition of R, this implies that $G \cap R \neq \emptyset$ and thus $C \cap P \neq \emptyset$, which contradicts the fact that C is an empty circle. Therefore, the edge (A, B) is in $\mathrm{Del}(\mathcal{D})$. Since $\mathrm{Del}(P)/\mathcal{D}$ is a t-spanner of $\mathcal{D}$ and a subgraph of $\mathrm{Del}(\mathcal{D})$, $\mathrm{Del}(\mathcal{D})$ is a t-spanner of $\mathcal{D}$. □

7 Computing a Plane Embedding

Note that the embedding of the AW-Delaunay graph that consists of straight line segments between the centers of the disks is not necessarily a plane graph. However, the Voronoi diagram of a set of disks $\mathcal{D}$, denoted $\mathrm{Vor}(\mathcal{D})$, is planar [10]. Since $\mathrm{Del}(\mathcal{D})$ is the face-dual of $\mathrm{Vor}(\mathcal{D})$, it is also planar. An important characteristic of the Delaunay graph of a set of points regarded as a spanner is that it is a plane graph. Therefore, a natural question is whether $\mathrm{Del}(\mathcal{D})$ has a plane embedding that is also a spanner.

The proof of Theorem 2 suggests the existence of an algorithm allowing to compute such an embedding: compute the Delaunay triangulation of the set P that contains the distance points and the representative of $\mathcal{D}$. The graph $\mathrm{Del}(P)$ can be regarded as a multigraph whose vertex set is $\mathcal{D}$. Then, for each pair of disks that share one or more edges, just keep the shortest of those edges. This simple algorithm allows us to compute a plane embedding of $\mathrm{Del}(\mathcal{D})$ that is also a spanner of $\mathcal{D}$. However, its running time is $O(n^3 \log n)$.

On the other hand, it is also possible to compute in time $O(n \log n)$ a plane spanner of $\mathcal{D}$ whose spanning ratio is t^2, where t is the spanning ratio of the Delaunay graph of a set of points. Here is how to do this: First, compute Del($\mathcal{D}$). Then, let P be the set of distance points of all pairs of disks that share an edge in Del($\mathcal{D}$). Compute Del(P). Since P has size $O(n)$, this can be done in time $O(n \log n)$. Also, Del(P) is a plane graph. As in the above paragraph, the graph Del(P) can be regarded as a multigraph whose vertex set is $\mathcal{D}$. Again, for each pair of disks that share one or more edges, just keep the shortest of those edges. All that remains to explain is why the resulting graph is a t^2-spanner of $\mathcal{D}$. Let $D_1, D_2 \in \mathcal{D}$. The straight line embedding of Del($\mathcal{D}$) contains a t-spanning path between D_1 and D_2. The endpoints of the edges of that path are the distance points between the disks. In Del(P), each of those edges is approximated within a factor of t, leading to a spanning ratio of t^2. Therefore, we showed the following:

Theorem 3. *Let $\mathcal{D}$ be a set of n disjoint disks and t be the spanning ratio of the Delaunay triangulation of a set of points. In time $O(n^3 \log n)$, it is possible to compute a plane t-spanner of $\mathcal{D}$, and in time $O(n \log n)$, it is possible to compute a plane t^2-spanner of $\mathcal{D}$.*

8 Conclusion

In this paper, we showed how, given a weighted point set where weights are positive and $|p_i p_j| \geq r_i + r_j$ for all $i \neq j$, it is possible to compute a $(1+\epsilon)$-spanner of that point set that has a linear number of edges. We also showed that the Additively Weighted Delaunay graph is a t-spanner of an additively weighted point set in the same case. The constant t is the same as for the Delaunay triangulation of a point set (the best current value is 2.42 [5]). We could not see how the Well-Separated Pair Decomposition (WSPD) can be adapted to solve that problem. The first difficulty resides in the fact that it is not even clear that, given a weighted point set, a WSPD of that point set always exists. Other obvious open questions are whether our results still hold when some weights are negative or $|p_i p_j| < r_i + r_j$ for some $i \neq j$. Also, we did not verify whether our variant of the Yao graph can be computed in time $O(n \log n)$. Finally, whether or not it is possible to compute a plane embedding of Del($\mathcal{D}$) that has the same spanning ratio than the Delaunay graph of a set of points in time $O(n \log n)$ remains a open question.

References

1. Bose, P., Carmi, P., Couture, M.: Spanners of additively weighted point sets. CoRR abs/0801.4013 (2008)
2. Narasimhan, G., Smid, M.: Geometric Spanner Networks. Cambridge University Press, New York (2007)
3. Yao, A.C.C.: On constructing minimum spanning trees in k-dimensional spaces and related problems. SIAM J. Comput. 11(4), 721–736 (1982)

4. Ruppert, J., Seidel, R.: Approximating the d-dimensional complete euclidean graph. In: CCCG 1991: Proceedings of the 3rd Canadian Conference on Computational Geometry, pp. 207–210 (1991)
5. Keil, J.M., Gutwin, C.A.: Classes of graphs which approximate the complete Euclidean graph. Discrete Comput. Geom. 7(1), 13–28 (1992)
6. Callahan, P.B., Kosaraju, S.R.: A decomposition of multidimensional point sets with applications to k-nearest-neighbors and n-body potential fields. Journal of the ACM 42, 67–90 (1995)
7. de Berg, M., van Kreveld, M., Overmars, M., Schwarzkopf, O.: Computational Geometry: Algorithms and Applications. Springer, Heidelberg (1997)
8. Fortune, S.: A sweepline algorithm for voronoi diagrams. Algorithmica 2, 153–174 (1987)
9. Lee, D.T., Drysdale, R.L.: Generalization of voronoi diagrams in the plane. SIAM Journal on Computing 10(1), 73–87 (1981)
10. Okabe, A., Boots, B., Sugihara, K.: Spatial tessellations: concepts and applications of Voronoi diagrams, 2nd edn. John Wiley & Sons, Inc. New York (2000)

The Kinetic Facility Location Problem*

Bastian Degener[1,2], Joachim Gehweiler[2], and Christiane Lammersen[2]

[1] International Graduate School Dynamic Intelligent Systems,
[2] Heinz Nixdorf Institute, Computer Science Department
Paderborn University, 33095 Paderborn, Germany
{bastian.degener,joge,christiane.lammersen}@upb.de

Abstract. We present a deterministic kinetic data structure for the facility location problem that maintains a subset of the moving points as facilities such that, at any point of time, the accumulated cost for the whole point set is at most a constant factor larger than the optimal cost. Each point can change its status between client and facility and moves continuously along a known trajectory in a d-dimensional Euclidean space, where d is a constant.

Our kinetic data structure requires $\mathcal{O}(n(\log^d(n) + \log(nR)))$ space, where $R := \frac{\max_{p_i \in \mathcal{P}} f_i \cdot \max_{p_i \in \mathcal{P}} d_i}{\min_{p_i \in \mathcal{P}} f_i \cdot \min_{p_i \in \mathcal{P}} d_i}$, $\mathcal{P} = \{p_1, p_2, \ldots, p_n\}$ is the set of given points, and f_i, d_i are the maintenance cost and the demand of a point p_i, respectively. In case that each trajectory can be described by a bounded degree polynomial, we process $\mathcal{O}(n^2 \log^2(nR))$ events, each requiring $\mathcal{O}(\log^{d+1}(n) \cdot \log(nR))$ time and $\mathcal{O}(\log(nR))$ status changes. To the best of our knowledge, this is the first kinetic data structure for the facility location problem.

Keywords: facility location, kinetic data structure, approximation.

1 Introduction

The facility location problem is a fundamental combinatorial problem in computer science. In its classical interpretation, the goal is to find an optimal placement of industrial facilities, such that the combined cost for the maintenance of the facilities and the transportation cost for the customers are minimized.

We consider a scenario of continuously moving objects. Each object can either be a facility or a client. Applications for this scenario are for example in sensor networks and mobile ad-hoc networks. In these networks, nodes move continuously and interact with each other. Often they are organized in a hierarchical way, where the upper layer offers the lower layer a certain service, such as a routing infrastructure. Each node can act as a server, but, at any time, cost arises for each node that is set up or maintained as a server. This additional overhead for a server is caused by a higher energy consumption due to message

* Partially supported by the EU within FP7-ICT-2007-1 under contract no. 215270 (FRONTS), DFG-project "Smart Teams" within the SPP 1183 "Organic Computing", and DFG grant Me 872/8-3.

J. Gudmundsson (Ed.): SWAT 2008, LNCS 5124, pp. 378–389, 2008.

passing, storing of routing tables etc. Since each node should be able to access a service as fast as possible, there is also a cost for each client, namely the delay time which depends on the distance to the nearest server. Now imagine that, to decrease the total cost for the system, nodes are allowed to change their status from server to client or vice versa. This is the kinetic facility location problem.

Kinetic Data Structures. The kinetic data structure (KDS) framework is well-suited to maintain a combinatorial structure of continuously moving objects and common in the field of computational geometry [2,4,12]. In this framework, we are given a set of objects and a *flight plan*, i.e., each object moves continuously along a known trajectory. It is possible to change the flight plan by performing a *flight plan update*, which means that one object changes its trajectory. The main idea is now that the continuous motion of the objects is utilized in a way that updates take place only at discrete points of time and can be processed fast. As a result, a lot of computational effort can be saved maintaining the KDS compared to handling just a series of instances of the corresponding static problem. To guarantee that the required properties of the combinatorial structure are satisfied at any point of time, a KDS ensures that certain *certificates* are always hold up. Whenever a certificate fails, we call this an *event*, and an update is required.

There are four important properties to measure the quality of a KDS. The worst-case amount of time to process an update is called *responsiveness*. The *compactness* is given by the ratio between the maximum size of the event queue and the complexity of the moving objects. The *locality* addresses the maximum number of events in the queue, in which one object can be involved in. The fourth property, the *efficiency* of a KDS, is the ratio between the number of total events processed by the KDS and the minimum number of events that would have been sufficient to maintain a solution for the given kinetic problem. We call a KDS responsive, compact, local, and efficient, respectively, if the associated value is at most poly-logarithmic in the complexity of the moving objects.

Our Contribution. We present a KDS for the facility location problem that gets as input a set of n points in $\mathbb{R}^d$, where d is a constant, and for each given point a trajectory. At any point of time, each point is either a facility or a client. The cost that arises for a facility persist during the entire time it is open. Analogously, a client has to pay some cost for its connection to a facility permanently. Our KDS maintains a subset of the moving points as facilities such that, at any time, the sum of the maintenance cost for the facilities and the connection cost for the clients is at most a constant factor larger than the current optimal cost.

The challenge is to construct a KDS whose underlying combinatorial structure is stable. To be able to ensure this, we keep up the invariant that, on the one hand, for each client there exists a facility in a certain local neighborhood and, on the other hand, no facility has another facility in a certain local neighborhood. The problem is now that restoring the invariant at one point (by changing the status of the point from open to close or vice versa) can lead to a violation of the invariant at many other points. Our main technical contribution is a technique that allows us to restore the invariant in poly-logarithmic time.

Related Work. The facility location problem has been extensively studied in combinatorial optimization and operations research [16,17,19]. In general, the problem is known to be $\mathcal{NP}$-complete. For the Euclidean case, there exists a randomized PTAS [18]. However, the problem has been investigated in different settings, for instance, in distributed [11] and dynamic settings [15], but so far no algorithm is known for the kinetic setting. Unfortunately, it does not seem that the only known $(1+\varepsilon)$-approximation given in [18] can be translated to this setting, since the authors use the Arora-scheme including dynamic programming techniques, which does not well comply with kinetization.

The KDS framework was introduced and applied on the convex hull problem by Basch et al. [4]. Later KDSs for measuring various descriptors of the extent of point sets have been designed [1,2]. Several further problems have been considered in the KDS framework, but only some results are known for problems related to clustering, which the facility location problem belongs to. For instance, Gao et al. [10] provided a KDS to maintain an expected constant factor approximation for the minimal number of centers to cover all points for a given radius. Bespamyatnikh et al. [6] studied k-center problems for $k = 1$ in the KDS framework, where the centers are not necessarily located at the moving points. Another algorithm for the kinetic k-center problem can be found in [9]. Har-Peled [13] considered the k-center problem in a mobile setting different from the KDS framework. Hershberger [14] proposed a kinetic algorithm for maintaining a covering of the moving points in $\mathbb{R}^d$ by unit boxes such that the number of boxes is always within a factor of 3^d of the optimal static covering at any instance. Recently, Czumaj et al. [7] presented a KDS for the Euclidean MaxCut problem. For other work on KDSs, we refer to the survey by Guibas [12].

2 Preliminaries

We define the kinetic facility location problem as follows. Let $\mathcal{P} = \{p_1, p_2, \ldots, p_n\}$ be a set of n independently moving points in $\mathbb{R}^d$, where d is a constant. Let $p_i(t)$ denote the position of p_i at time t and let $\mathcal{P}(t) = \{p_1(t), p_2(t), \ldots, p_n(t)\}$. At any point of time t, the set $\mathcal{P}(t)$ is divided into the current set of *facilities* $\mathcal{F}(t)$ and the current set of *clients* $\mathcal{G}(t) = \mathcal{P}(t) \backslash \mathcal{F}(t)$. For each point $p_i(t) \in \mathcal{P}(t)$, there exists a non-negative *maintenance cost* f_i, that has to be paid at time t if $p_i(t)$ is a facility, and a non-negative *demand* d_i. Note that both the maintenance cost and the demand of a point do not change over time. The problem is now to maintain, at each point of time t, a subset $\mathcal{F}(t) \subseteq \mathcal{P}(t)$, such that

$$cost(\mathcal{F}(t)) := \sum_{p_i(t) \in \mathcal{F}(t)} f_i + \sum_{p_j(t) \in \mathcal{G}(t)} d_j \cdot D(p_j(t), \mathcal{F}(t))$$

is minimized. Here, $D(p_j(t), \mathcal{F}(t))$ is the minimum Euclidean distance from $p_j(t)$ to a facility in $\mathcal{F}(t)$. We let $\mathcal{F}_{\mathrm{OPT}}(t)$ denote an optimal set of facilities at time t.

In the following, we introduce some basics required for our approach. The main idea is to use a set of nested cubes around each point and to update the KDS each time a point enters or leaves a cube of another point.

Cubes. For a point $p_i(t) \in \mathcal{P}(t)$ and a non-negative value r, we define $\mathcal{C}(p_i(t), r)$ to be the axis-parallel cube whose center is the point $p_i(t)$ and whose side length is $2r$. Given such a cube $\mathcal{C}(p_i(t), r)$, we let $weight(\mathcal{C}(p_i(t), r))$ denote the sum of the demands of all the points in $\mathcal{P}(t)$ that are located in the cube $\mathcal{C}(p_i(t), r)$, i.e., we define

$$weight(\mathcal{C}(p_i(t), r)) := \sum_{p_j(t) \in \mathcal{P}(t) \cap \mathcal{C}(p_i(t), r)} d_j \,.$$

Radius Associated with a Point. For each point $p_i(t) \in \mathcal{P}(t)$, we calculate a special radius $r_i^*(t)$ which is an approximation for the radius $r_i(t)$ of the ball with center $p_i(t)$ that is used in [19] and satisfies

$$\sum_{p_j(t) \in \mathcal{P}(t) | D(p_i(t), p_j(t)) \leq r_i(t)} d_j \cdot (r_i(t) - D(p_i(t), p_j(t))) = f_i \,.$$

Due to this definition, $r_i(t)$ ranges from $\frac{\min_{p_j \in \mathcal{P}} f_j}{n \cdot \max_{p_j \in \mathcal{P}} d_j}$ to $\frac{\max_{p_j \in \mathcal{P}} f_j}{\min_{p_j \in \mathcal{P}} d_j}$. To obtain a constant factor approximation for $r_i(t)$, we define $r_i^*(t)$ to be the value 2^{k^*}, such that $k^* = k_0 + \lceil \log(4\sqrt{d}) \rceil$ and k_0 is the minimum integer k, $\log(\frac{\min_{p_j \in \mathcal{P}} f_j}{n \cdot \max_{p_j \in \mathcal{P}} d_j}) \leq k \leq \log(\frac{\max_{p_j \in \mathcal{P}} f_j}{\min_{p_j \in \mathcal{P}} d_j})$, for which $weight(\mathcal{C}(p_i(t), 2^{k_0})) \geq f_i \cdot 2^{-k_0}$ holds. The choice of k^* is explained in Section 4. Hence, due to our definition, we have to consider only $\mathcal{O}(\log(nR))$ possible values for the radii, where $R := \frac{\max_{p_i \in \mathcal{P}} f_i \cdot \max_{p_i \in \mathcal{P}} d_i}{\min_{p_i \in \mathcal{P}} f_i \cdot \min_{p_i \in \mathcal{P}} d_i}$.

Walls around a Point. We consider a set of $\mathcal{O}(\log(nR))$ nested cubes for each point $p_i(t) \in \mathcal{P}(t)$. In particular, there is the cube $\mathcal{C}(p_i(t), 2^k)$ with radius 2^k for each $k \in \{\lceil \log(\frac{\min_{p_j \in \mathcal{P}} f_j}{n \cdot \max_{p_j \in \mathcal{P}} d_j}) \rceil + \lceil \log(4\sqrt{d}) \rceil, \lceil \log(\frac{\min_{p_j \in \mathcal{P}} f_j}{n \cdot \max_{p_j \in \mathcal{P}} d_j}) \rceil + 1 + \lceil \log(4\sqrt{d}) \rceil, \ldots, \lfloor \log(\frac{\max_{p_j \in \mathcal{P}} f_j}{\min_{p_j \in \mathcal{P}} d_j}) \rfloor + \lceil \log(4\sqrt{d}) \rceil\}$. The side faces of the cube defined by $\mathcal{C}(p_i(t), 2^k)$ form a wall around $p_i(t)$, which we call $W_{i,k}(t)$. Hence, there exists a set of $\mathcal{O}(\log(nR))$ walls for $p_i(t)$. We use this set of walls to determine the points of time when an update of p_i in our KDS is required. In general, an event occurs each time when any point crosses any wall of another point.

Range Trees. We maintain two $(d+1)$-dimensional dynamic range trees denoted by T_1 and T_2. At any time, T_1 is used to manage the current set of facilities (which we call open points), and T_2 stores the current set of clients (which we call closed points). Apart from the fact that the two data structures contain different point sets, they are constructed in the same way. In the first d levels, the points are handled according to their coordinates and in the $(d+1)$-st level according to their special radii. Additionally, with each node v in every binary search tree of the $(d+1)$-st level, we store the sum of the demands of all the points contained in the subtree of v. Beside the two range trees, we maintain a binary search tree T that contains, for each point in $\mathcal{P}$, a pair consisting of the point's index and its current status. T is sorted according to the indices.

The dynamic data structure described in [5] supports all required properties of T_1 and T_2 efficiently. For a set of n points in $\mathbb{R}^{d+1}$, it has size $\mathcal{O}(n \log^d(n))$, can be built in $\mathcal{O}(n \log^{d+1}(n))$ time, and can be maintained in $\mathcal{O}(\log^{d+1}(n))$ worst case time per insertion and deletion. Given any orthogonal range in $\mathbb{R}^{d+1}$, we can output the points inside this range in $\mathcal{O}(\log^{d+1}(n) + N)$ time, where N is the output size. Due to the additional information, we can also compute the sum of the demands of all the points in a certain range in $\mathcal{O}(\log^{d+1}(n))$ time. Finally, we can output the status of a point in $\mathcal{O}(\log(n))$ time by querying T.

The movement of the points is reflected by insertion and deletion operations on T_1 and T_2 upon an event. That means that the actual position of any point p_i is represented by its coordinates at the latest event it was involved in.

Initialization. To get an initial set of facilities, we apply the algorithm of Mettu and Plaxton [19] on the input points. Unfortunately, we cannot use this greedy approach to obtain a KDS with poly-logarithmic update time. The reason is that keeping up the solution provided by the Mettu-Plaxton algorithm is not stable, so that a slight perturbation of the input might result in $\Omega(n)$ status changes.

3 The Kinetic Data Structure

3.1 Event Queue

We perform an update each time a point $p_j(t)$ crosses a wall $W_{i,k}(t)$, where $\lceil \log(\frac{\min_{p_j \in \mathcal{P}} f_j}{n \cdot \max_{p_j \in \mathcal{P}} d_j}) \rceil + \lceil \log(4\sqrt{d}) \rceil \leq k \leq \lfloor \log(\frac{\max_{p_j \in \mathcal{P}} f_j}{\min_{p_j \in \mathcal{P}} d_j}) \rfloor + \lceil \log(4\sqrt{d}) \rceil$, of another point $p_i(t)$. In order to keep track of these events, we need another data structure beside the two range trees: For each dimension ℓ, $1 \leq \ell \leq d$, we store all n points and all $\mathcal{O}(n \cdot \log(nR))$ wall faces that are orthogonal to the ℓ-th coordinate axis in a list sorted by the ℓ-coordinate. For each consecutive pair in each of the d lists, we keep up one certificate to certify the sorted order of the lists. We define the failure time of the certificate for any pair of consecutive objects to be the first future time when these objects swap their places in their sorted list. The failure times of all certificates are maintained in one event queue.

For simplicity we assume that the points are in general position. Then at most two events occur at the same time, which are handled in an arbitrary order. The event queue has the following complexity (for details cf. [8]):

Lemma 1. *The event queue for the kinetic facility location problem has size $\mathcal{O}(n \log(nR))$, can be initialized in time $\mathcal{O}(n \log^2(nR))$, and updated in time $\mathcal{O}(\log(nR))$. Provided that each trajectory can be described by a bounded degree polynomial, the total number of events is $\mathcal{O}(n^2 \log^2(nR))$. A flight plan update involves $\mathcal{O}(\log(nR))$ certificates and requires $\mathcal{O}(\log^2(nR))$ time.*

3.2 Handling an Update

Now we describe how an event, occurring at any point of time t, is handled. As the first step, the event queue is updated. All further steps are performed to keep up one invariant consisting of the following conditions:

a) for each closed point $p_i(t) \in \mathcal{G}(t)$ there is an open point $p_j(t) \in \mathcal{F}(t)$ with $r^*_j(t) \le r^*_i(t)$ in $\mathcal{C}(p_i(t), 4 \cdot r^*_i(t))$ and
b) for each open point $p_i(t) \in \mathcal{F}(t)$ there is no other open point $p_j(t) \in \mathcal{F}(t)$ with $r^*_j(t) \le r^*_i(t)$ in $\mathcal{C}(p_i(t), 2 \cdot r^*_i(t))$.

We say that a point $p_i(t)$ violates the invariant in the following case: Either $p_i(t)$ is closed and condition a) does not hold for $p_i(t)$ or $p_i(t)$ is open and condition b) does not hold for $p_i(t)$. We will show that, if the invariant is satisfied, then $cost(\mathcal{F}(t))$ is at most a constant factor larger than $cost(\mathcal{F}_{\mathrm{OPT}}(t))$. Moreover, the asymmetric choice of condition a) and b) enables our KDS to be stable.

Now, let us assume that the invariant is satisfied by the time when an event e occurs. Then the only way that the invariant can be violated is that e indicates that one point crosses a wall of another point. If this is not the case, handling e is finished after updating the event queue. To detect wall crossings, we check if one of the involved objects is a point and if the other one is the face of a wall. Then we update both associated points, the point that might cross the wall and the point whose wall might be crossed, in the range trees and check if the first point really crosses a wall of the second point.

In case that any point $p_j(t)$ crosses a wall of any other point $p_i(t)$ at any time t, we first update the radius $r^*_i(t) = 2^{k^*}$, such that $k^* = k_0 + \lceil \log(4\sqrt{d}) \rceil$ and k_0 is the minimum integer k with $\log(\frac{\min_{p_j \in \mathcal{P}} f_j}{n \cdot \max_{p_j \in \mathcal{P}} d_j}) \le k \le \log(\frac{\max_{p_j \in \mathcal{P}} f_j}{\min_{p_j \in \mathcal{P}} d_j})$, for which $weight(\mathcal{C}(p_i(t), 2^{k_0})) \ge f_i \cdot 2^{-k_0}$ holds. Note that the new value of k_0 differs from its value before e by at most 1. Thus there are three possible values for k_0, where each value can be tested by running one range query on both T_1 and T_2. Afterwards, we test if $p_i(t)$ violates the invariant by using a range query on T_1. If this is the case, we change the status of $p_i(t)$. For simplicity of description, we assume that the range trees are always up to date. We will show in Section 4 that our KDS works as desired, although the position of a point in the range tree can slightly deviate from its real current position. As an effect of changing the radius or the status of one point, the invariant may be violated by many other points (e.g., their facility has been closed). In the following, we will show how to deal with this problem.

Algorithm RESTORE. Suppose that, due to an event at any point of time t, the radius or the status of a point $p_e(t)$ changed and its new radius is $r^*_e(t) = 2^{k^*}$. First, we restore the invariant at all points with radius 2^{k^*-1}, to ensure that no point with radius less than or equal to 2^{k^*-1} violates the invariant. Then we handle the points with radius 2^{k^*}, then the ones with radius 2^{k^*+1}, ..., up to the biggest possible radius. Now, we describe the procedure in general for any radius 2^k.

We define two cubes $S_1 := \mathcal{C}(p_e(t), 4 \cdot 2^{k+1})$ and $S_2 := \mathcal{C}(p_e(t), 6 \cdot 2^{k+1})$ and divide them into equally sized cubelets with radius 2^k. Figure 1 (a) illustrates this decomposition in the plane for k and the next iteration $k+1$. To guarantee that no open point with radius 2^k violates the invariant, we perform the following test for each cubelet in S_1: Let m be the center point of the considered cubelet. If

Algorithm 3.1. RESTORE($p_e(t), k^*$)

```
1: for k ← k* − 1 to ⌊log(max_{p_j∈P} f_j / min_{p_j∈P} d_j)⌋ + ⌈log(4√d)⌉ do
2:     define cubes S_1 := C(p_e(t), 4 · 2^{k+1}) and S_2 := C(p_e(t), 6 · 2^{k+1})
3:     for each cubelet C with center m_C and radius 2^k in S_1 do
4:         if ∃ facility with radius < 2^k in C(m_C, 3 · 2^k) then
5:             close all facilities with radius 2^k in C
6:     for each cubelet C with center m_C and radius 2^k in S_2 do
7:         if ∄ facility with radius ≤ 2^k in C(m_C, 3 · 2^k) then
8:             open one point with radius 2^k in C (if existing)
```

there is a facility with radius less than 2^k in $\mathcal{C}(m, 3 \cdot 2^k)$, then close all facilities with radius 2^k in $\mathcal{C}(m, 2^k)$. Note that there is at most one such facility. The considered area around a cubelet is illustrated in Figure 1 (b). Then, we perform a similar test for each cubelet in S_2 (cf. Algorithm 3.1, line 6 ff), to guarantee that the certificate of every closed point with radius 2^k holds.

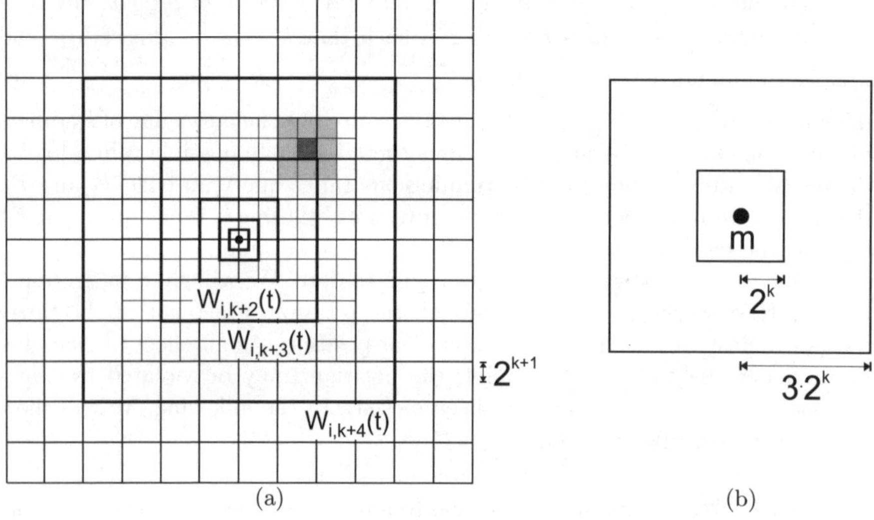

Fig. 1. (a) Decomposition into cubelets. (b) Tested area.

4 Quality and Complexity of the Kinetic Data Structure

At first, we prove that the invariant is satisfied each time our KDS has handled an update. Then, this fact is used to show that, at any point of time t, we have $cost(\mathcal{F}(t)) = \mathcal{O}(cost(\mathcal{F}_{\text{OPT}}(t)))$. Finally, we analyze the complexity of our KDS.

Maintenance of the Invariant. The difficulty in proving the correctness of maintaining the invariant is that both range trees contain out-dated information. It is guaranteed that the radius of each point stored in the range trees is equal to its current special radius, but this is not true for the position. For any point of time t and any $p_i(t) \in \mathcal{P}(t)$, let $p_i^T(t)$ be the position of p_i stored in the range trees at time t. The following proposition shows that, at any time, every point is stored in the correct range with respect to the walls of all other points.

Proposition 1. *Let p_i and p_j be any two points in $\mathcal{P}$, let k be an integer in $\{\lceil\log(\frac{\min_{p_j\in\mathcal{P}} f_j}{n\cdot\max_{p_j\in\mathcal{P}} d_j})\rceil + \lceil\log(4\sqrt{d})\rceil, \lceil\log(\frac{\min_{p_j\in\mathcal{P}} f_j}{n\cdot\max_{p_j\in\mathcal{P}} d_j})\rceil + 1 + \lceil\log(4\sqrt{d})\rceil, \ldots, \lfloor\log(\frac{\max_{p_j\in\mathcal{P}} f_j}{\min_{p_j\in\mathcal{P}} d_j})\rfloor + \lceil\log(4\sqrt{d})\rceil\}$, and let t be any point of time between two successive events which involve p_i and p_j. If and only if we have $p_j^T(t) \in \mathcal{C}(p_i^T(t), 2^k)$, then $p_j(t) \in \mathcal{C}(p_i(t), 2^k)$ is true as well.*

Proof. Let $t_1 < t$ be the latest point of time when p_i and p_j have been involved in one event. Furthermore, $p_j^T(t) \in \mathcal{C}(p_i^T(t), 2^k)$ implies that we have updated p_i and p_j at time t_1, such that $p_j(t_1) \in \mathcal{C}(p_i(t_1), 2^k)$ and $p_j^T(t_1) \in \mathcal{C}(p_i^T(t_1), 2^k)$. Now let us assume that we have $p_j^T(t) \in \mathcal{C}(p_i^T(t), 2^k)$ but $p_j(t) \notin \mathcal{C}(p_i(t), 2^k)$. Thus, there must be a point of time t_2 with $t_1 < t_2 < t$ when the point $p_j(t_2)$ crosses the wall $W_{i,k}(t_2)$. Then t_1 could not be the latest point of time when p_i and p_j have been involved in one event, a contradiction. Analogously, we can show that $p_j^T(t) \notin \mathcal{C}(p_i^T(t), 2^k)$ implies $p_j(t) \notin \mathcal{C}(p_i(t), 2^k)$. □

We can easily obtain the following result.

Proposition 2. *The invariant is satisfied as long as the KDS does not call algorithm* RESTORE.

The proof of Proposition 2 as well as further omitted proofs can be found in [8]. The following propositions show that the invariant is restored after each call of algorithm RESTORE.

Proposition 3. *Let $p_e(t)$ be a point whose radius or status changed due to an event e. Let $r_e^*(t) = 2^{k^*}$ be the updated radius of $p_e(t)$. If no point with radius less than or equal to 2^{k^*-2} violates the invariant before e, then this holds after e as well.*

Proof. Due to the definition of the special radii and the fact that only one point has crossed one wall of $p_e(t)$, the radius of p_e has been at least 2^{k^*-1} before e. Now, the proposition follows due to the fact that we only change the status of points with radius larger than or equal to 2^{k^*-1} while processing event e. □

Proposition 4. *Let $p_e(t)$ be a point whose radius or status changed due to an event e. Let $r_e^*(t) = 2^{k^*}$ be the updated radius of $p_e(t)$. If the invariant is satisfied before e and no open point with radius less than or equal to $2^{\ell-1}$ violates the invariant before running the outer **for**-loop of algorithm* RESTORE *for $k = \ell$, where $k^* - 1 \le \ell \le \lfloor\log(\frac{\max_{p_j\in\mathcal{P}} f_j}{\min_{p_j\in\mathcal{P}} d_j})\rfloor + \lceil\log(4\sqrt{d})\rceil$, then, after running this **for**-loop, no open point with radius 2^ℓ violates the invariant.*

Proof. The proof is by contradiction. Let us assume that after running the outer **for**-loop of algorithm RESTORE for $k = \ell$ there is an open point $p_i(t)$ with radius $r_i^*(t) = 2^\ell$ that has another open point $p_j(t)$ with radius $r_j^*(t) \le r_i^*(t)$ in $\mathcal{C}(p_i(t), 2 \cdot r_i^*(t))$. We have to consider the cases i) $p_i^T(t) \in S_2$ and ii) $p_i^T(t) \notin S_2$.

Case i). Subcase $r_j^*(t) < r_i^*(t)$: Due to the fact that $r_j^*(t) < 2^\ell$, we have opened p_j before running the outer **for**-loop for $k = \ell$. It follows that $p_i^T(t) \in \mathcal{C}(m, 2^\ell)$ and $p_j^T(t) \notin \mathcal{C}(m, 3 \cdot 2^\ell)$ for one center m of a considered cubelet, because otherwise we either would have closed $p_i(t)$ or would not have opened $p_i(t)$. As a consequence, $p_j^T(t) \notin \mathcal{C}(p_i^T(t), 2^{\ell+1}) = \mathcal{C}(p_i^T(t), 2 \cdot r_i^*(t))$. Now, due to Proposition 1, we have $p_j(t) \notin \mathcal{C}(p_i(t), 2 \cdot r_i^*(t))$, which is a contradiction.

Subcase $r_j^*(t) = r_i^*(t)$: We have to consider the case that neither p_i nor p_j is opened while running the outer **for**-loop for $k = \ell$ and the case that at least one of p_i and p_j is opened during this **for**-loop. In the first case, it follows that p_i and p_j must have been open before running the outer **for**-loop for $k = \ell$. As a consequence, both points have been open before e or one point is p_e. Then either the invariant was violated before e or changing the status of p_e violated the invariant, a contradiction. In the latter case, we have opened p_i or p_j or both while running the outer **for**-loop for $k = \ell$. W.l.o.g., let us assume that we have opened p_i before we have opened p_j. Then we must have that $p_j^T(t) \in \mathcal{C}(m, 2^\ell)$ and $p_i^T(t) \notin \mathcal{C}(m, 3 \cdot 2^\ell)$ for one center m of a considered cubelet. It follows that $p_j^T(t) \notin \mathcal{C}(p_i^T(t), 2^{\ell+1}) = \mathcal{C}(p_i^T(t), 2 \cdot r_i^*(t))$. Due to Proposition 1, we have $p_j(t) \notin \mathcal{C}(p_i(t), 2 \cdot r_i^*(t))$, which is a contradiction.

Case ii). Subcase $r_j^*(t) < r_i^*(t)$: Due to the fact that $r_j^*(t) < 2^\ell$, we have opened p_j before running the outer **for**-loop for $k = \ell$. Furthermore, it follows from $p_i^T(t) \notin S_2$ that we must have opened p_i before running the outer **for**-loop for $k = \ell$ as well. Hence, we have that both p_i and p_j have been open before running this **for**-loop. Thus, the invariant must have been violated at point $p_j(t)$ with $r_j^*(t) \le 2^{\ell-1}$ before running the outer **for**-loop for $k = \ell$, a contradiction.

Subcase $r_j^*(t) = r_i^*(t)$: We can use the same argumentation as for case i) in subcase $r_j^*(t) = r_i^*(t)$ with the modification that we know that p_i has been opened before running the outer **for**-loop for $k = \ell$. The reason is that $p_i^T(t) \notin S_2$, so that we do not change its status while running this **for**-loop. □

In a similar way, we can get the following result for the closed points:

Proposition 5. *Let $p_e(t)$ be a point whose radius or status changed due to an event e. Let $r_e^*(t) = 2^{k^*}$ be the updated radius of $p_e(t)$. If the invariant is satisfied before e and no closed point with radius less than or equal to $2^{\ell-1}$ violates the invariant before running the outer* ***for****-loop of algorithm* RESTORE *for $k = \ell$, where $k^* - 1 \le \ell \le \lfloor \log(\frac{\max_{p_j \in \mathcal{P}} f_j}{\min_{p_j \in \mathcal{P}} d_j}) \rfloor + \lceil \log(4\sqrt{d}) \rceil$, then, after running this* ***for****-loop, no closed point with radius 2^ℓ violates the invariant.*

Now, we can combine the obtained results to the following lemma:

Lemma 2. *The invariant is satisfied after the KDS has handled an event.*

Proof. Due to Proposition 2, the invariant is satisfied as long as we do not call algorithm RESTORE. Let $p_e(t)$ be the point whose radius or status changed due to an event e, and let $r_e^*(t) = 2^{k^*}$ be its updated radius. Because of the precondition given above and Proposition 3 the lemma is true for all points $p_i(t)$ with radius $r_i(t)^* = 2^\ell$ where $\ell \le k^* - 2$. Due to Propositions 4 and 5 it also follows for $\ell \ge k^* - 2$. Hence, it is true for all points. □

The Special Radii. For any point $p_i(t) \in \mathcal{P}(t)$, we define $\mathcal{B}(p_i(t), r)$ to be the ball with center $p_i(t)$ and radius r. Given such a ball $\mathcal{B}(p_i(t), r)$, we let $weight(\mathcal{B}(p_i(t), r))$ denote the sum of the demands of all the points in $\mathcal{P}(t)$ that lie inside the ball $\mathcal{B}(p_i(t), r)$. Now, we can prove that, at any point of time t, the special radius $r_i^*(t)$ is a constant factor approximation for the value $r_i(t)$.

For the uniform facility location problem, the authors in [3] showed how to approximate $r_i(t)$ by counting the number of points in a certain ball with center $p_i(t)$. We generalize their result to the non-uniform case (for details cf. [8]):

Lemma 3. *For any point of time t, let k_1 be the minimum integer k with $\lceil \log(\frac{\min_{p_j \in \mathcal{P}} f_j}{n \cdot \max_{p_j \in \mathcal{P}} d_j}) \rceil \le k \le \lfloor \log(\frac{\max_{p_j \in \mathcal{P}} f_j}{\min_{p_j \in \mathcal{P}} d_j}) \rfloor$, such that $weight(\mathcal{B}(p_i(t), 2^k)) \ge f_i \cdot 2^{-k}$. Then $\frac{1}{2} \cdot r_i(t) \le 2^{k_1} \le 2 \cdot r_i(t)$ holds.*

Our algorithm uses the approach of [3], but we approximate the sum of the demands of all the points in a distance 2^k by a cube with radius 2^k:

Lemma 4. *For any point of time t, let k_0 be the minimum integer k with $\lceil \log(\frac{\min_{p_j \in \mathcal{P}} f_j}{n \cdot \max_{p_j \in \mathcal{P}} d_j}) \rceil \le k \le \lfloor \log(\frac{\max_{p_j \in \mathcal{P}} f_j}{\min_{p_j \in \mathcal{P}} d_j}) \rfloor$, such that $weight(\mathcal{C}(p_i(t), 2^k)) \ge f_i \cdot 2^{-k}$. Then $\frac{1}{4\sqrt{d}} \cdot r_i(t) \le 2^{k_0} \le 2 \cdot r_i(t)$ holds.*

Proof. Let k_1 be defined as in Lemma 3. Then the radius of $\mathcal{C}(p_i(t), 2^{k_0})$ is at most 2^{k_1}, since each point in $\mathcal{P}(t)$, that is located in $\mathcal{B}(p_i(t), 2^{k_1})$, is also located in $\mathcal{C}(p_i(t), 2^{k_1})$, so that $weight(\mathcal{C}(p_i(t), 2^{k_1})) \ge f_i \cdot 2^{-k_1}$. Furthermore, the radius of $\mathcal{C}(p_i(t), 2^{k_0})$ is larger than $\frac{1}{\sqrt{d}} \cdot 2^{k_1 - 1}$ since $weight(\mathcal{B}(p_i(t), 2^{k_1 - 1})) < f_i \cdot 2^{-(k_1 - 1)}$ and $weight(\mathcal{C}(p_i(t), \frac{1}{\sqrt{d}} 2^{k_1 - 1})) \le weight(\mathcal{B}(p_i(t), 2^{k_1 - 1}))$, so that $weight(\mathcal{C}(p_i(t), \frac{1}{\sqrt{d}} \cdot 2^{k_1 - 1})) < f_i \cdot 2^{-(k_1 - 1)} < f_i \cdot 2^{-(k_1 - 1 - \log(\sqrt{d}))}$. Now, the lemma follows due to $2^{k_0} > \frac{1}{\sqrt{d}} \cdot 2^{k_1 - 1}$ and Lemma 3. □

Due to Lemma 4 and by setting $r_i^*(t) = 2^{k^*} = 2^{k_0 + \lceil \log(4\sqrt{d}) \rceil}$, we get $r_i(t) \le r_i^*(t) \le 2^{3 + \lceil \log(\sqrt{d}) \rceil} \cdot r_i(t)$.

Lemma 5. *The KDS for the facility location problem for an arbitrary but fixed dimension d maintains, at any point of time t, a subset $\mathcal{F}(t) \subseteq \mathcal{P}(t)$ such that $cost(\mathcal{F}(t)) \le (64d + 1) \cdot cost(\mathcal{F}_{\mathrm{OPT}}(t))$.*

Proof. For each point $p_i(t) \in \mathcal{P}(t)$, there is a facility $p_j(t) \in \mathcal{F}(t)$ with radius $r_j^*(t) \leq r_i^*(t)$ in $\mathcal{C}(p_i(t), 4 \cdot r_i^*(t))$. Furthermore, we know that $r_i(t) \leq r_i^*(t) \leq 2^{3+\lceil \log(\sqrt{d}) \rceil} \cdot r_i(t)$. It follows that we have $D(p_i(t), p_j(t)) \leq \sqrt{d} \cdot 4 \cdot r_i^*(t) \leq \sqrt{d} \cdot 4 \cdot 2^{3+\lceil \log(\sqrt{d}) \rceil} \cdot r_i(t) \leq 64d \cdot r_i(t)$. Now, the lemma follows from the analysis in [19]. Details can be found in [8]. □

Complexity. Due to Lemma 1, we have already proven that our KDS is compact and local. Now we show that the requirement for responsiveness is also fulfilled.

Lemma 6. *Each update operation requires $\mathcal{O}(\log^{d+1}(n) \cdot \log(nR))$ time and $\mathcal{O}(\log(nR))$ status changes.*

Proof. Due to Lemma 1, the time to update the event queue is $\mathcal{O}(\log(nR))$. Except for algorithm Restore, all further steps to handle an event require $\mathcal{O}(\log^{d+1}(n))$ time. Next we examine the time that algorithm Restore needs to restore the invariant at points with radius 2^k. The number of cubelets with radius 2^k in S_2 is 12^d. The query of open or closed points for one cubelet can be answered by T_1 and T_2 in time $\mathcal{O}(\log^{d+1}(n))$. Afterwards, there has to be at most one point inserted and deleted in T_1 and T_2. This requires $\mathcal{O}(\log^{d+1}(n))$ time. Summation over all radii leads to a total running time of $\mathcal{O}(\log^{d+1}(n) \cdot \log(nR))$.

There is at most one facility with radius 2^k in a cubelet with radius 2^k, otherwise the invariant is violated. Hence, algorithm Restore closes a constant number of facilities with radius 2^k. Obviously, the number of opened facilities with radius 2^k is also constant. Due to the fact that we handle $\mathcal{O}(\log(nR))$ radii, there are $\mathcal{O}(\log(nR))$ status changes per event. □

Due to Lemmas 1 and 6, the total processing time is $\mathcal{O}(n^2 \log^{d+1}(n) \cdot \log^3(nR))$. Thus, we get the following result:

Theorem 1. *Let $\mathcal{P}$ be a set of n independently moving points in $\mathbb{R}^d$, where d is a constant. Then there is a KDS for the facility location problem that maintains, at any point of time t, a set $\mathcal{F}(t) \subseteq \mathcal{P}(t)$, such that cost$(\mathcal{F}(t)) \leq (64d+1) \cdot$ cost$(\mathcal{F}_{\mathrm{OPT}}(t))$. The KDS has a space requirement of $\mathcal{O}(n(\log^d(n) + \log(nR)))$, where $R = \frac{\max_{p_i \in \mathcal{P}} f_i \cdot \max_{p_i \in \mathcal{P}} d_i}{\min_{p_i \in \mathcal{P}} f_i \cdot \min_{p_i \in \mathcal{P}} d_i}$. Each update operation requires $\mathcal{O}(\log(nR))$ status changes and $\mathcal{O}(\log^{d+1}(n) \cdot \log(nR))$ time. In case each trajectory can be described by a bounded degree polynomial, the number of updates is $\mathcal{O}(n^2 \log^2(nR))$, which results in a total processing time of $\mathcal{O}(n^2 \log^{d+1}(n) \cdot \log^3(nR))$. A flight plan update involves $\mathcal{O}(\log(nR))$ certificates and requires $\mathcal{O}(\log^2(nR))$ time.*

5 Conclusion

In this paper, we initiated the study on the kinetic facility location problem. In particular, we proposed a KDS that maintains a subset of the moving input points as facilities such that, at any point of time, the associated total cost is at most a constant factor larger than the current optimal cost. We showed that our KDS is compact, local, and responsive. We also consider our algorithm to be efficient, although we cannot prove this in the formal sense of KDSs, because it is hard to lower bound the number of mandatory events in a non-trivial way.

References

1. Agarwal, P., Guibas, L., Hershberger, J., Veach, E.: Maintaining the Extent of a Moving Point Set. Discrete & Computational Geometry 26(3), 353–374 (2001)
2. Agarwal, P., Har-Peled, S., Varadarajan, K.: Approximating Extent Measures of Points. Journal of the ACM 51(4), 606–635 (2004)
3. Bădoiu, M., Czumaj, A., Indyk, P., Sohler, C.: Facility Location in Sublinear Time. In: Caires, L., Italiano, G.F., Monteiro, L., Palamidessi, C., Yung, M. (eds.) ICALP 2005. LNCS, vol. 3580, pp. 866–877. Springer, Heidelberg (2005)
4. Basch, J., Guibas, L., Hershberger, J.: Data Structures for Mobile Data. Journal of Algorithms 31(1), 1–28 (1999)
5. Basch, J., Guibas, L., Zhang, L.: Proximity Problems on Moving Points. In: Proc. 13th Symposium on Computational Geometry, pp. 344–351 (1997)
6. Bespamyatnikh, S., Bhattacharya, B., Kirkpatrick, D., Segal, M.: Mobile Facility Location. In: Proc 4th International Workshop on Discrete Algorithms and Methods for Mobile Computing and Communications, pp. 46–53 (2000)
7. Czumaj, A., Frahling, G., Sohler, C.: Efficient Kinetic Data Structures for MaxCut. In: Proc. 19th Canadian Conference on Computational Geometry, pp. 157–160 (2007)
8. Degener, B., Gehweiler, J., Lammersen, C.: The Kinetic Facility Location Problem. Technical Report tr-ri-08-2880, University of Paderborn (2008)
9. Gao, J., Guibas, L., Nguyen, A.: Deformable Spanners and Applications. In: Proc. 20th Symposium on Computational Geometry, pp. 190–199 (2004)
10. Gao, J., Guibas, L., Hershberger, J., Zhang, L., Zhu, A.: Discrete Mobile Centers. Journal of Discrete and Computational Geometry 30(1), 45–63 (2003)
11. Gehweiler, J., Lammersen, C., Sohler, C.: A Distributed $O(1)$-Approximation Algorithm for the Uniform Facility Location Problem. In: Proc. 18th ACM Symposium on Parallelism in Algorithms and Architectures, pp. 237–243 (2006)
12. Guibas, L.: Kinetic Data Structures: A State of the Art Report. In: Proc. 3rd Workshop on Algorithmic Foundations of Robotics, pp. 191–209 (1998)
13. Har-Peled, S.: Clustering Motion. In: Proc. 42nd IEEE Symposium on Foundations of Computer Science, pp. 84–93 (2001)
14. Hershberger, J.: Smooth Kinetic Maintenance of Clusters. In: Proc. Symposium on Computational Geometry, pp. 48–57 (2003)
15. Indyk, P.: Algorithms for Dynamic Geometric Problems over Data Streams. In: Proc. 36th ACM Symposium on Theory of Computing, pp. 373–380 (2004)
16. Jain, K., Mahdian, M., Saberi, A.: A New Greedy Approach for Facility Location Problems. In: Proc. 34th ACM Symposium on Theory of Computing, pp. 731–740 (2002)
17. Jain, K., Vazirani, V.: Approximation Algorithms for Metric Facility Location and k-Median Problems Using the Primal-Dual Schema and Lagrangian Relaxation. Journal of the ACM 48(2), 274–296 (2001)
18. Kolliopoulos, S., Rao, S.: A Nearly Linear-Time Approximation Scheme for the Euclidean k-Median Problem. SIAM Journal on Computing 37(3), 757–782 (2007)
19. Mettu, R.R., Plaxton, C.G.: The Online Median Problem. SIAM J. Comput. 32(3), 816–832 (2003)

Computing the Greedy Spanner in Near-Quadratic Time

Prosenjit Bose, Paz Carmi, Mohammad Farshi, Anil Maheshwari, and Michiel Smid

School of Computer Science, Carleton University, Ottawa, ON, K1S 5B6, Canada
jit@scs.carleton.ca, {paz,mfarshi}@cg.scs.carleton.ca, {anil,michiel}@scs.carleton.ca

Abstract. It is well-known that the greedy algorithm produces high quality spanners and therefore is used in several applications. However, for points in d-dimensional Euclidean space, the greedy algorithm has cubic running time. In this paper we present an algorithm that computes the greedy spanner (spanner computed by the greedy algorithm) for a set of n points from a metric space with bounded doubling dimension in $\mathcal{O}(n^2 \log n)$ time using $\mathcal{O}(n^2)$ space. Since the lower bound for computing such spanners is $\Omega(n^2)$, the time complexity of our algorithm is optimal to within a logarithmic factor.

1 Introduction

A *network* on a point set V is a connected graph $G(V, E)$. When designing a network several criteria are taken into account. In particular, in many applications it is important to ensure a fast connection between every pair of points. For this it would be ideal to have a direct connection between every pair of points—the network would then be a complete graph—but in most applications this is unacceptable due to the very high costs associated with constructing such networks. This leads to the concept of spanners, as defined below.

Let $(V, \mathbf{d})$ be a metric space and $G(V, E)$ be a network on V such that the weight of each edge $e \in E$ is equal to the distance between its endpoints. We say that G is a *t-spanner* of V, for some constant $t > 1$, if for each pair of points $u, v \in V$, there exists a path in G between u and v of length at most $t \cdot \mathbf{d}(u, v)$. The *dilation* or *stretch factor* of G is the minimum t for which G is a t-spanner of V. Spanners were introduced by Peleg and Schäffer [12] in the context of distributed computing and by Chew [3] in the geometric context. Since then spanners have received a lot of attention, see [10, 11].

A classical algorithm for computing a geometric spanner for any set V of n points in $\mathbb{R}^d$, where $\mathbf{d}$ is the Euclidean metric, and for any fixed $t > 1$, is the *greedy algorithm*, proposed independently by Bern in 1989 and Althöfer *et al.* [1]. The main steps of this algorithm are the following: first sort all the pairs of points in V with respect to their distances in increasing order and initialize the greedy graph $G(V, E)$ so that its edge set is empty. Next, the pairs are processed in

J. Gudmundsson (Ed.): SWAT 2008, LNCS 5124, pp. 390–401, 2008.

sorted order. Processing a pair (u, v) entails a shortest path query in G between u and v. If there is no t-path between u and v (a path of length at most $t \cdot \mathbf{d}(u, v)$) in G then (u, v) is added to G, otherwise it is discarded. We will refer to the graph G generated by the greedy algorithm as the *greedy spanner*.

The greedy algorithm as stated above performs $\binom{n}{2}$ single-source shortest path queries. By employing Dijkstra's single-source shortest path algorithm, the time complexity is $\mathcal{O}(mn^2 + n^3 \log n)$ using $\mathcal{O}(n^2)$ space, where n is the number of points and m is the number of edges in the spanner. It has been shown that for any set V of n points in $\mathbb{R}^d$ and for any fixed $t > 1$ in the Euclidean metric, the greedy spanner has $\mathcal{O}(n)$ edges, bounded degree, and its total weight is $\mathcal{O}(wt(MST(V)))$, where $wt(MST(V))$ is the weight of the minimum spanning tree of V [4, 10]. Unfortunately, the naïve implementation of the greedy algorithm runs in cubic time.

Due to the high time complexity of computing the greedy spanner, researchers have proposed algorithms for computing other types of sparse t-spanners, see [10]. But it turns out that in practice the greedy algorithm produces t-spanners of high quality in comparison to other spanners [5, 6]. For example, they have been used for protein visualization as a low-weight data structure, which is used as a contact map, that allows approximate reconstruction of the full distance matrix [13].

For points in the plane under the Euclidean metric, Farshi and Gudmundsson [5, 6] introduced a speed-up strategy that generates the greedy spanner much faster in practice. They conjectured that their algorithm runs in $\mathcal{O}(n^2 \log n)$ time. However, as we will show in this paper, this conjecture is incorrect. They also showed that the greedy algorithm produces graphs whose size, weight, maximum degree and number of crossings are superior to the graphs produced from the other approaches which have near linear time complexity. For $t = 2$, $t = 1.1$ and $t = 1.05$ the number of edges in the greedy t-spanner is approximately $2n$, $4n$ and $6n$ respectively, which is surprisingly small. For comparison it is interesting to note that the Delaunay triangulation has approximately $3n$ edges and dilation bounded by 2.42 [9]. Also the maximum degree of the greedy 1.1-spanner, generated on a uniformly distributed set, of 8000 points is 14 and its weight is 11 times the weight of the minimum spanning tree of the point set. To have a rough comparison, the Θ-graph algorithm, which runs in $\mathcal{O}(n \log n)$ time, generated a graph containing 370K edges, the maximum degree was 144 and the weight was 327 times the weight of the minimum spanning tree.

For general metric spaces, there are cases where the complete graph is the only t-spanner of a point set. For example, assume V is a set of points from a metric space where the distance between any two distinct points is equal to 1. For any $1 < t < 2$, the complete graph is the only t-spanner of V. Therefore, in general metric spaces, we can not guarantee that the generated graph is sparse. The doubling dimension is defined as follows. Let λ be the smallest integer such that for each real number r, any ball of radius r can be covered by at most λ balls of radius $r/2$. The *doubling dimension* of V is defined to be $\log \lambda$.

The doubling dimension is a generalization of the Euclidean dimension, as the doubling dimension of d-dimension Euclidean space is $\Theta(d)$.

1.1 Main Results and Organization of the Paper

The main result of this paper is that for any metric space V of bounded doubling dimension, the greedy spanner of V has linear size and can be computed in $\mathcal{O}(n^2 \log n)$ time, where $n = |V|$. The organization of the remainder of this paper is as follows. In Sect. 2, we review the improved greedy algorithm of [5, 6] and give a counterexample to the conjecture that this algorithm only performs $\mathcal{O}(n)$ shortest path queries. In Sect. 3, we present an algorithm that generates the greedy spanner in near-quadratic time for some special cases. These results are generalized to metric spaces of bounded doubling dimension in Sect. 4.

2 The FG-Greedy Algorithm

As mentioned above the running time of a naïve implementation of the greedy algorithm is $\mathcal{O}(mn^2 + n^3 \log n)$. Farshi and Gudmundsson [5, 6] introduced an improved version of the algorithm and showed that it improves the running time for constructing the greedy spanner considerably in practice on point sets in the plane with the Euclidean metric. We call this algorithm as the FG-greedy algorithm. The FG-greedy algorithm is the same as the original greedy algorithm except that it uses a matrix to save the length of the shortest path between every two points and updates the matrix only when it is required. Thus the matrix is not always up to date. Instead of computing a shortest path for each pair, it first checks the matrix to see if there is a t-path; if the answer is no, then it performs a shortest path query and updates the matrix which enables us to answer the distance queries correctly. They conjectured that the FG-greedy algorithm performs only $\mathcal{O}(n)$ shortest path queries, which would imply a running time of $\mathcal{O}(n^2 \log n)$.

2.1 A Counterexample

We give an example which shows that the FG-greedy algorithm may perform $\Theta(n^2)$ shortest path queries. Consider a set $S = \{p_0, p_1, \dots, p_{n-1}\}$ of n points on the real line such that $p_i = 2^i$. The algorithm sorts all pairs of points based on their distance. We also assume that for each pair (p_i, p_j) the index of the first point in the pair is less than the index of the second point, i.e. $i < j$. It is easy to see that the algorithm performs a shortest path query for each pair of points.

Note that if we change the algorithm such that when a new edge is added to the greedy spanner, it performs a shortest path query for both endpoints of the new edge and updates the weight matrix then the new algorithm performs only $\mathcal{O}(n)$ shortest path queries on the above counterexample. However, in the full version of this paper, we give an example where we still need to perform $\Omega(n \log n)$ shortest path queries.

3 A Preliminary Algorithm

Let V be a set of n points in a metric space with distance function $\mathbf{d}$. As mentioned before, for generating the greedy t-spanner, we start with a graph $G(V, E)$ with no edges and we go through all the pairs of points in V (in increasing order based on their distances). For each pair we check if there exists a t-path between them in G, if not we add the edge between them in G.

In the new algorithm we basically do the same thing as the FG-greedy algorithm in the sense that we use the weight matrix to answer the shortest path queries. The differences are the following:

- We process the pairs whose distance is less than L, for a real number L, exactly in the way as in the original (or FG-) greedy algorithm.
- We divide the remaining pairs into buckets such that the ith bucket contains all the pairs whose distance is between $2^{i-1}L$ and $2^i L$.
- During the processing of each bucket, we keep the shortest path between the pairs in the bucket up to date. To do this we update the shortest path between all pairs in the bucket before processing the pairs in the bucket and during the process we update it when we add an edge to the graph. Note that we keep the shortest path length between pairs in a weight matrix.

Without loss of generality we assume that the diameter of the point set is one. Let $0 < L < 1$ be a real number to be specified later. We split the pairs of points into $k = \mathcal{O}(\log(\frac{1}{L}))$ buckets such that the ith bucket, i.e. E_i, contains all the pairs with distance in $[2^{(i-1)}L, 2^i L)$. Let E_0 contain all the pairs with distance less than L.

The algorithm starts with the pairs in E_0. It process all the pairs in the set E_0 in the same manner as the original (or FG-) greedy algorithm does. Therefore, if E_0 contains $\mathcal{O}(n^\beta)$ pairs then processing it takes $\mathcal{O}(n^{\beta+1} \log n)$ time. Let G denote the current greedy spanner after processing E_0.

Now we show how to process the remaining buckets. Assume that we have processed buckets $E_1, E_2, \ldots, E_{i-1}$ and we need to process bucket E_i. Before processing the edges in this bucket, we compute the single-source shortest path with source at each point p of V and update the weight matrix. Then we make "*local*" updates when we add an edge to the graph. By "local" update, we mean we update the weight matrix for all points nearby each of the endpoints of the new edge. Since the weight matrix is up to date for all pairs in the current bucket we can answer the t-path queries in constant time using the weight matrix. For details see Algorithm 3.1.

Theorem 1. *Algorithm 3.1 generates the greedy spanner of the input point set.*

Proof. To prove the correctness of the algorithm, we need to prove that the t-path queries (line 17 of Algorithm 3.1) are answered correctly.

Let (p, q) be an arbitrary pair in E_i with $\mathbf{d}(p, q) \in [L_i, 2L_i)$ which is about to be processed in the algorithm. If there is no t-path between p and q in G then the algorithm answers the t-path query correctly since the entry in the weight matrix corresponding to the pair is at least equal to the shortest path length between p and q in G. Assume that there is a t-path between p and q in G. We have two cases:

Case 1: The shortest path between p and q in G does not pass through any edges that were added during processing of pairs in E_i. In this case we are done because before processing the pairs in E_i, we updated all-pair shortest paths and adding new edges to the graph does not change the shortest path length between p and q.

Case 2: The shortest path π in G between p and q passes through some edges of E_i. Among all edges of $E_i \cap \pi$, let (u, v) be the one that was added last by the algorithm. We may assume without loss of generality that, starting at p, the path π goes to u, then traverses (u, v), and then continues to q. We define

$$S_{(u,v)} = \{x \in V : \mathbf{d}(x, u) < (t - 1/2)L_i \text{ or } \mathbf{d}(x, v) < (t - 1/2)L_i\}.$$

We claim (and show below) that p or q belongs to $S_{(u,v)}$. This will imply that, in the iteration in which (u, v) is added to the graph, the algorithm computes the exact shortest-path length between p and all vertices of V, or between q and all vertices of V. Therefore, at the moment when (p, q) is processed, the value of $weight(p, q)$ is equal to the shortest-path length in G between p and q.

It remains to prove the claim. Assume that neither p nor q is contained in $S_{(u,v)}$. Then $\mathbf{d}(p, u) \geq (t - \frac{1}{2})L_i$ and $\mathbf{d}(q, v) \geq (t - \frac{1}{2})L_i$. Thus, if we denote the shortest-path length between p and q by $\mathbf{d}_G(p, q)$, then

$$\mathbf{d}_G(p, q) \geq \mathbf{d}(p, u) + \mathbf{d}(u, v) + \mathbf{d}(v, q) \geq 2(t - 1/2)L_i + L_i = 2tL_i > t \cdot \mathbf{d}(p, q),$$

which contradicts the fact that there is a t-path in G between p and q. □

Algorithm 3.1. NEW-GREEDY(V, t, L)

Input: V, $t > 1$ and $L > 0$.
Output: t-spanner $G = (V, E')$.

foreach $(u, v) \in V^2$ **do** $weight(u, v) := \infty$;
$E :=$ sorted list of all the pairs in V^2 by increasing distance; /*ties are broken arbitrarily*/
$E_0 :=$ all the point pairs in E with distance in $[0, L)$;
$j := 1$;
while $E \setminus (\bigcup_{k=1}^{j-1} E_k) \neq \emptyset$ **do**
  $E_j :=$ all the point pairs in $E \setminus (\bigcup_{k=1}^{j-1} E_k)$ with distance in $[2^{j-1}L, 2^j L)$;
  $j := j + 1$;
$l := j - 1$;
$E' := \emptyset$;
$G := (V, E')$;
Process pairs in E_0 in the same way as the original (or FG-) greedy algorithm;
for $i := 1, \ldots, l$ **do**
  $L_i := 2^{i-1}L$;
  foreach $u \in V$ **do**
    Compute single-source shortest paths with source at u and update $weight$;
  foreach $(u, v) \in E_i$; /* in sorted order */
  do if $weight(u, v) > t \cdot \mathbf{d}(u, v)$ **then** $E' := E' \cup \{(u, v)\}$;
  foreach $p \in V$ **do**
    if $\mathbf{d}(p, u) < (t - \frac{1}{2})L_i$ *or* $\mathbf{d}(p, v) < (t - \frac{1}{2})L_i$ **then**
      Compute single-source shortest paths with source at p and update $weight$;
return $G(V, E')$;

Now we show that the algorithm runs in near quadratic time in some special cases. First we need to recall the well-separated pair decomposition developed by Callahan and Kosaraju [2] in d-dimensional Euclidean space.

Definition 1. *Let $s > 0$ be a real number, referred to as the* separation constant. *We say that two point sets A and B in $\mathbb{R}^d$ are* well-separated *with respect to s, if there are two disjoint balls D_A and D_B of the same radius, r, such that*

(i) D_A contains A and D_B contains B,
(ii) the distance between D_A and D_B is at least $s \cdot r$.

Lemma 1 ([2]). *Let A and B be two finite sets of points that are well-separated with respect to s, let x and p be points of A, and let y and q be points of B. Then (i) $|xy| \leq (1 + 4/s) \cdot |pq|$ and (ii) $|px| \leq (2/s) \cdot |pq|$.*

Definition 2. *Let V be a set of n points in $\mathbb{R}^d$ and let $s > 0$ be a real number. A* well-separated pair decomposition *(WSPD) for V with respect to s is a collection $\mathcal{W} := \{(A_1, B_1), ..., (A_m, B_m)\}$ of pairs of non-empty subsets of V such that*

1. *A_i and B_i are well-separated with respect to s, for all $i = 1, \ldots, m$.*
2. *for any two distinct points p and q of V, there is exactly one pair (A_i, B_i) in the collection, such that (i) $p \in A_i$ and $q \in B_i$ or (ii) $q \in A_i$ and $p \in B_i$.*

The number of pairs, m, is called the *size* of the WSPD. Callahan and Kosaraju[2] have shown that any set $V \subseteq \mathbb{R}^d$ admits a WSPD of size $m = \mathcal{O}(s^d n)$. Har-Peled and Mendel [8] generalized the results to metric spaces with doubling dimension λ. They showed that any set of n points from a metric space with doubling dimension λ admits a WSPD with respect to $s > 1$, of size $\mathcal{O}(s^{\mathcal{O}(\lambda)} n)$. In the rest of the paper, we assume that V is a set of n points from a metric space with doubling dimension λ.

Observation 1. *If $\{(A_i, B_i)\}_i$ is a WSPD of a point set V with respect to $s = \frac{4(t+1)}{t-1}$, then for each i, there is at most one greedy edge between A_i and B_i in the t-spanner generated by the greedy algorithm.*

Proof. Assume that we have a pair (A, B) in the WSPD such that there exist two edges (a_1, b_1) and (a_2, b_2) in the greedy t-spanner where $a_1, a_2 \in A$ and $b_1, b_2 \in B$. Without loss of generality assume the greedy algorithm process the pair (a_1, b_1) before the pair (a_2, b_2).

Because A and B are s-well-separated pair, by Lemma 1, we have $\mathbf{d}(a_1, a_2) \leq \frac{2}{s}\mathbf{d}(a_2, b_2) < \mathbf{d}(a_2, b_2)$. By the same argument $\mathbf{d}(b_1, b_2) < \mathbf{d}(a_1, b_1)$. Therefore, there exists a t-path between a_1 and a_2 and a t-path between b_1 and b_2 when the greedy algorithm processes the pair (a_2, b_2). Let G' be the graph just before processing (a_2, b_2) and let P be a path in G' between a_2 and b_2 generated by concatenating a t-path between a_2 and a_1, the edge (a_1, b_1) and a t-path between b_1 and b_2. If $|P|$ denotes the length of the path P, then

$$\begin{aligned}|P| &= \mathbf{d}_{G'}(a_2, a_1) + \mathbf{d}(a_1, b_1) + \mathbf{d}_{G'}(b_1, b_2)\\ &\le t\cdot \mathbf{d}(a_2, a_1) + \mathbf{d}(a_1, b_1) + t\cdot \mathbf{d}(b_1, b_2)\\ &\le t\cdot \frac{2}{s}\,\mathbf{d}(a_2, b_2) + (1+\frac{4}{s})\,\mathbf{d}(a_2, b_2) + t\cdot \frac{2}{s}\,\mathbf{d}(a_2, b_2) \qquad \text{(by Lemma 1)}\\ &= (\frac{4t}{s} + 1 + \frac{4}{s})\,\mathbf{d}(a_2, b_2)\\ &= t\cdot \mathbf{d}(a_2, b_2).\end{aligned}$$

This means that the greedy algorithm does not add (a_2, b_2) to the spanner since there already exist a t-path between them in G'. □

As a corollary, since there exists a linear size WSPD for any point set in a metric space with bounded doubling dimension, the size of a greedy t-spanner on such a point set is linear.

Lemma 2. *While processing the pairs in E_i, for each point p, line 20 in Algorithm 3.1 is executed $\mathcal{O}(\frac{1}{(t-1)^{\mathcal{O}(\lambda)}})$ times.*

Proof. For simplicity, we first prove the lemma in the 2-dimensional Euclidean case. Assume the distance between the pairs in E_i is in $[L, 2L)$. Algorithm 3.1 performs a single-source shortest path computation with source at p, after adding an edge (u, v) to the graph, if $\mathbf{d}(p, u) < (t - \frac{1}{2})L$ or $\mathbf{d}(p, v) < (t - \frac{1}{2})L$. So if we draw a ball C with center at p and radius $(2t + 1)L$, then all the edges which affect p lies inside the ball C. So the number of times we need to execute line 20 for p is at most the number of edges in the greedy spanner with length between L and $2L$ which lie inside C. Now we show that the number of such edges is at most $\mathcal{O}(\frac{1}{(t-1)^2})$.

To show this, assume B is a square with side length $2(2t + 1)L$ which includes C. We cover the square B with cells of side length $\ell = \frac{L}{\sqrt{2}(s+4)}$ where $s = \frac{4(t+1)}{t-1}$. The number of such cells inside B is $\left(2\sqrt{2}(2t + 1)(s + 4)\right)^2 = \mathcal{O}(\frac{1}{(t-1)^2})$. Let (u, v) be a greedy edge with distance in the interval $[L, 2L)$. First we show that the grid cells which contains u and v are s-well-separated. Assume C_1 and C_2 are balls with radius $\sqrt{2}\ell$ which contain the grid cell of u and the grid cell of v, respectively— see Fig. 1. Since $\mathbf{d}(u, v) \ge L$ and the radius of the circles are $\sqrt{2}\ell$, the distance between C_1 and C_2 is at least $L - 4\sqrt{2}\ell$. Therefore

$$d(C_1, C_2) \ge L - 4\sqrt{2}\ell = L - 4\sqrt{2}\frac{L}{\sqrt{2}(s+4)} = L(\frac{s}{s+4}) = s(\frac{L}{s+4}) = s \times \sqrt{2}\ell$$

which means the cells are s-well-separated.

Therefore, by Observation 1, we have at most one greedy edge between grid cells which are well-separated. This means that the number of the greedy edges with distance in $[L, 2L)$ inside the circle C is bounded by the number of cell pairs which is $\mathcal{O}(\frac{1}{(t-1)^4})$.

For the general case, the same argument works. In this case we use the property of doubling dimension which guarantees that each ball of radius $r > 0$ can be covered by 2^λ balls of radius $r/2$. This means that the number of balls with radius $\sqrt{2}\ell$ inside the ball C centered at p is $\mathcal{O}(\frac{1}{(t-1)^{\mathcal{O}(\lambda)}})$. □

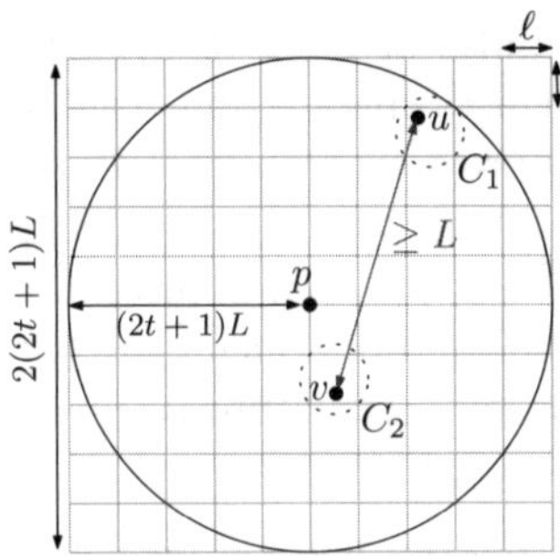

Fig. 1. Illustration for the proof of Lemma 2

Now we are ready to compute the time complexity of Algorithm 3.1. Clearly lines 1–11 of the algorithm take $\mathcal{O}(n^2 \log n)$ time. For line 12, if the size of E_0 is β then it takes $\mathcal{O}(\beta(m + n \log n))$ since for each pair it performs a shortest path query.

For each interval, computing all-pairs shortest path, lines 15–17, takes $\mathcal{O}(mn + n^2 \log n)$ time, and by Lemma 2, the update procedure takes at most $\mathcal{O}(\frac{1}{(t-1)^{\mathcal{O}(\lambda)}}(mn + n^2 \log n))$. Since the number of intervals is $\mathcal{O}(\log(1/L))$, processing all intervals takes $\mathcal{O}\left(\frac{\log(1/L)}{(t-1)^{\mathcal{O}(\lambda)}}(mn + n^2 \log n)\right)$ time. For a metric space with doubling dimension λ, the size of the generated graph is $\mathcal{O}(\frac{n}{(t-1)^{\mathcal{O}(\lambda)}})$, the total running time of Algorithm 3.1 is $\mathcal{O}\left(\frac{\beta n + \log(1/L) n^2 \log n}{(t-1)^{\mathcal{O}(\lambda)}}\right)$. Therefore for a point set V with the property that there exists a real number L such that $\frac{1}{L} = \mathcal{O}(n^c)$ (c is a constant) and $\beta = \mathcal{O}(n \log^2 n)$, the greedy spanner can be computed in $\mathcal{O}(n^2 \log^2 n)$ time.

Points Sets with Polynomial Aspect Ratio. If the input point set has aspect ratio less than n^c, for some constant c, then by scaling the point set such that the longest distance is 1 and setting $L = \frac{1}{n^c}$, we have no pair of points in the scaled point set with distance less than L. Therefore the running time of Algorithm 3.1 in this case is $\mathcal{O}(\frac{n^2 \log^2 n}{(t-1)^{\mathcal{O}(\lambda)}})$.

Uniformly Distributed Point Set. Assume we have a set of n points uniformly distributed inside the unit square and let $L = \frac{1}{\sqrt{n}}$. Since the points are uniformly distributed, for each point p, the expected number of points inside the ball with center at p and radius L is $L^2 n$. So the expected number of pairs with distance less than L is $L^2 n^2/2 = \mathcal{O}(n)$. Therefore, the expected running time of Algorithm 3.1 in this case is $\mathcal{O}(\frac{n^2 \log^2 n}{(t-1)^{\mathcal{O}(\lambda)}})$.

4 An Improved Algorithm

To generalize the results of the previous section to bounded doubling dimension, we have to overcome obstacles. First, we need to speedup processing the pairs in the first set (i.e. the set E_0). The second problem is that if we decrease the number L to bound the number of pairs in the first interval, it increases the number of buckets which causes higher time complexity. We overcome these difficulties by modifying the previous algorithm in the following way.

- we partition the $\binom{n}{2}$ pairs into a linear number of buckets,
- we maintain a data structure for each point during the algorithm. When we need to update a point, instead of doing a single-shortest path computation from scratch, we use the data structure to update just the necessary part and use it to update the weight matrix.

First we claim that for updating the shortest path lengths with source at a point p, performing a length-bounded Dijkstra's algorithm is sufficient. More precisely, if we are working on point pairs in a bucket with distances in $[L, 2L)$ and we need to update the shortest path lengths with source at p, it is sufficient to update the distance between p and all the points q such that $\mathbf{d}_G(p, q) < 2tL$, where G is the current graph. The reason is that if $\mathbf{d}_G(p, q) \geq 2tL$ then either the pair (p, q) is outside the current bucket or there is no t-path between p and q in G.

So, in the new algorithm, we maintain a data structure for each point, which is the same as the data structure used in the Dijkstra's single-source shortest paths algorithm. When we perform a shortest path query with bound U, we execute Dijkstra's algorithm but stop when the key of the element on the top of the priority queue (heap) is bigger than U. We also maintain a stack storing all changes that are made to the heap in this process, so that we can undo the procedure, if required.

Note that our graph is dynamic and we add edge(s) to the graph, in increasing order of length. We use the "undo facility" to fix the part of the execution of Dijkstra's algorithm that is affected by the insertion of an edge. For the details of the algorithm, see Algorithm 4.1.

To complete the correctness proof, we show that in the algorithm, the shortest path queries are answered correctly. To do this, we show that the length-bounded Dijkstra's algorithm on a subgraph of the greedy spanner works exactly same as Dijkstra's algorithm on the greedy spanner.

Lemma 3. *Let G_e be the subgraph of the greedy spanner G that contains all the edges added to the graph up to the processing of the pair $e = (p, q)$ in the greedy algorithm. Let u be an arbitrary vertex of G. As long as the key of the element on the top of the heap in Dijkstra's algorithm with source at u is less than $\mathbf{d}(p, q)$ the algorithm works the same on G and G_e.*

Proof. Let x be a vertex of the graph G and assume the shortest path between u and x passes through at least one edge in $E \setminus E_e$. Since the length of each edge in $E \setminus E_e$ is at least $\mathbf{d}(p, q)$, the key of the point x in the heap is at least $\mathbf{d}(p, q)$. This completes the proof. □

4.1 Running Time

Now, we show that Algorithm 4.1 runs in $\mathcal{O}(n^2 \log n)$ time. To this end, we show that for each point $p \in V$ the overall time spent is proportional to running Dijkstra's single-source shortest paths algorithm with source p.

Algorithm 4.1. QUAD. GREEDY(V, t)

Input: V and $t > 1$.
Output: t-spanner $G = (V, E')$.
1 **foreach** $(u, v) \in V^2$ **do** $weight(u, v) := \infty$;
2 $E :=$ sorted list of all the pairs in V^2 by increasing distance; `/*ties are broken arbitrarily*/`
3 $L_1 :=$ the distance between the closest pair in E;
4 $E_1 :=$ all the pairs in E with distance in $[L_1, 2L_1)$;
5 $j := 2$;
6 **while** $E \setminus (\bigcup_{k=1}^{j-1} E_k) \neq \emptyset$ **do**
7     $L_j :=$ the distance between the closest pair in $E \setminus (\bigcup_{k=1}^{j-1} E_k)$;
8     $E_j :=$ all the pairs in $E \setminus (\bigcup_{k=1}^{j-1} E_k)$ with distance in $[L_j, 2L_j)$;
9     $j := j + 1$;
10 $l := j - 1$;
11 $E' := \emptyset$; $G := (V, E')$;
12 **foreach** $u \in V$ **do**
13   Initialize PQ_u required for executing Dijkstra's algorithm with source at u;
14 **for** $i := 1, \ldots, l$ **do**
15   **foreach** $u \in V$;
16   **do**
17         $\tau_u := \emptyset$;
18   DIJKSTRA-BOUNDED$(G, u, 2tL_i, PQ_u, \tau_u)$;
19   **foreach** $(u, v) \in E_i$; `/* in sorted order */`
20   **do**
21   **if** $weight(u, v) > t \cdot \mathbf{d}(u, v)$ **then**
22             $E' := E' \cup \{(u, v)\}$;
23   **foreach** $p \in V$ **do** **if** $\mathbf{d}(p, u) < (t - \frac{1}{2})L_i$ *or* $\mathbf{d}(p, v) < (t - \frac{1}{2})L_i$ **then**
24   UPDATE$(G, p, u, v, 2tL_i, PQ_p, \tau_p)$;
25 **return** $G(V, E')$;

Algorithm 4.2. UPDATE$(G, s, u, v, L, PQ, \tau)$

if $weight(s, v) < weight(s, u) + \mathbf{d}(u, v)$ *and* $weight(s, u) < weight(s, v) + \mathbf{d}(u, v)$; `/* This means adding` (u, v) `does not change any shortest path with source at` s `*/`
then
    return ;
else
    if $weight(s, v) < weight(s, u) + \mathbf{d}(u, v)$ **then**
        Swap u and v;
DIJKSTRA-UNDO$(\tau, PQ, weight(s, u) + \mathbf{d}(u, v))$;
Decrease the key of v to $weight(s, u) + \mathbf{d}(u, v)$ in PQ;
DIJKSTRA-BOUNDED(G, s, L, PQ, τ)

Algorithm 4.1 basically performs Dijkstra's algorithm with source at each point of the graph. The only differences are

- it performs bounded Dijkstra's algorithm and
- it fixes the process after adding edge(s) to the graph by undoing some parts and redoing it.

The following lemma follows from [7]:

Lemma 4. *The value of l computed in line 10 of Algorithm 4.1 is $\mathcal{O}(n)$.*

Now we show that for any point p, the time we spend to update p in the whole process is proportional to the time for running Dijkstra's algorithm with source at p. Assume we are processing the point pairs in E_i and let G be the current graph. As one can see in Algorithm 4.1, when we process the pairs in E_i, for

Algorithm 4.3. DIJKSTRA-BOUNDED(G, s, L, PQ, τ)

Input: Graph G, a vertex s, a real number L and a priority queue PQ.
Output: Shortest path length between s and all other vertices in G which has length at most L.

while *The key of the element on the top of* PQ *is at most* L **do**
  $u :=$ pop the element with minimum key from PQ;
  $weight(s, u) := weight(u, s) :=$ the key of u;
  foreach *node* v *adjacent to* u *in* G **do**
    if $weight(s, u) + wt(u, v) < weight(s, v)$ **then**
      Decrease the key of v in PQ to $weight(s, u) + wt(u, v)$ and add all the changes in PQ to the stack τ ; /* To be used in the undo procedure. */

Algorithm 4.4. DIJKSTRA-UNDO(τ, PQ, L)

Input: a stack τ, a priority queue PQ and a real number L.

while *the key of the element on the top of* τ *is at most* L **do**
  Pop the top element from τ;
  Undo the changes on PQ based on the information in the element;

updating followed by adding an edge, we undo the execution of Dijkstra's algorithm until the key of the points on the top of the heap is less than the length of the new edge and redo the execution until the key of the point on the top of the heap is more than $2tL_i$. But the length of the new edge is at least L_i which means that the undo process never goes further when the key on the top of the heap is less than L_i. We say that a vertex q is in the undo area of E_i if $L_i \leq \mathbf{d}_G(p, q) \leq 2tL_i$. The claim is that each point q appears in the undo area for a constant number of sets. To prove this let q be in the undo area of E_i. This means

$$L_i \leq \mathbf{d}_G(p, q) \leq 2tL_i. \tag{1}$$

We show that q can not be in the undo area of E_j if $j > i + \log t$. Let G' be the graph when we process the pairs in E_j. Since we add edge(s) to the graph, we know that for each pair (p, q) of points $\mathbf{d}_{G'}(p, q) \leq \mathbf{d}_G(p, q)$. Since for each i, $L_{i+1} > 2L_i$ we have $L_{i+k} > 2^k L_i$, for each $k > 0$. Using Equation (1), we have

$$\mathbf{d}_{G'}(p, q) \leq \mathbf{d}_G(p, q) \leq 2tL_i < \frac{2t}{2^{\log t}} L_j \leq L_j,$$

which means q is not in the undo area of E_j.

On the other hand, by Lemma 2, we update the shortest paths with source p at most $\mathcal{O}(\frac{1}{(t-1)^{\mathcal{O}(\lambda)}})$ times in each set E_i. This means in total we need to recompute the shortest path between p and any point q at most $\mathcal{O}(\frac{\log t}{(t-1)^{\mathcal{O}(\lambda)}})$ times. Therefore, the whole process for a fixed point p takes $\mathcal{O}(\frac{\log t}{(t-1)^{\mathcal{O}(\lambda)}} n \log n)$ time. So we have the following theorem.

Theorem 2. *For each point set V of n points from a metric space with doubling dimension λ, we can compute the greedy t-spanner of V in $\mathcal{O}(\frac{\log t}{(t-1)^{\mathcal{O}(\lambda)}} n^2 \log n)$ time using $\mathcal{O}(n^2)$ space.*

5 Conclusion

In this paper we presented an algorithm which, given a set of n points from a metric space with bounded doubling dimension, computes the greedy spanner of the point set in $\mathcal{O}(n^2 \log n)$ time. In the greedy spanner, every point is connected to its nearest neighbor. Therefore, we can compute all nearest neighbors of the input point set using the greedy spanner in $\mathcal{O}(n)$ time. Har-Peled and Mendel [8] showed that the all nearest neighbor problem has an $\Omega(n^2)$ lower bound for metric spaces with bounded doubling dimension. This implies that computing the greedy t-spanner also has an $\Omega(n^2)$ lower bound. There is a logarithmic gap between the running time of the greedy algorithm presented in this paper and the lower bound. Another interesting problem is finding a $o(mn^2)$ time algorithm for constructing the greedy graph on a set of n points from a metric space, where m is the number of edges in the greedy graph. In the special case where the points are a subset of $\mathbb{R}^2$, is there a $o(n^2)$ algorithm for constructing the greedy graph?

References

1. Althöfer, I., Das, G., Dobkin, D.P., Joseph, D., Soares, J.: On sparse spanners of weighted graphs. Discrete and Computational Geometry 9(1), 81–100 (1993)
2. Callahan, P.B., Kosaraju, S.R.: A decomposition of multidimensional point sets with applications to k-nearest-neighbors and n-body potential fields. Journal of the ACM 42, 67–90 (1995)
3. Chew, L.P.: There is a planar graph almost as good as the complete graph. In: SCG 1986: Proceedings of the 2nd Annual ACM Symposium on Computational Geometry, pp. 169–177 (1986)
4. Das, G., Narasimhan, G.: A fast algorithm for constructing sparse Euclidean spanners. Int. J. of Computational Geometry & Applications 7, 297–315 (1997)
5. Farshi, M., Gudmundsson, J.: Experimental study of geometric t-spanners. In: Brodal, G.S., Leonardi, S. (eds.) ESA 2005. LNCS, vol. 3669, pp. 556–567. Springer, Heidelberg (2005)
6. Farshi, M., Gudmundsson, J.: Experimental study of geometric t-spanners: A running time comparison. In: Demetrescu, C. (ed.) WEA 2007. LNCS, vol. 4525, pp. 270–284. Springer, Heidelberg (2007)
7. Har-Peled, S.: A simple proof? (2006), `http://valis.cs.uiuc.edu/blog/?p=362`
8. Har-Peled, S., Mendel, M.: Fast construction of nets in low-dimensional metrics and their applications. SIAM Journal on Computing 35(5), 1148–1184 (2006)
9. Keil, J.M., Gutwin, C.A.: Classes of graphs which approximate the complete Euclidean graph. Discrete and Computational Geometry 7, 13–28 (1992)
10. Narasimhan, G., Smid, M.: Geometric spanner networks. Cambridge University Press, Cambridge (2007)
11. Peleg, D.: Distributed Computing: A Locality-Sensitive Approach. SIAM, Philadelphia (2000)
12. Peleg, D., Schäffer, A.: Graph spanners. Journal of Graph Theory 13, 99–116 (1989)
13. Russel, D., Guibas, L.J.: Exploring protein folding trajectories using geometric spanners. In: Pacific Symposium on Biocomputing, pp. 40–51 (2005)

Parameterized Computational Complexity of Dodgson and Young Elections

Nadja Betzler*, Jiong Guo**, and Rolf Niedermeier

Institut für Informatik, Friedrich-Schiller-Universität Jena, Ernst-Abbe-Platz 2, D-07743 Jena, Germany
{betzler,guo,niedermr}@minet.uni-jena.de

Abstract. We show that, other than for standard complexity theory with known NP-completeness results, the computational complexity of the Dodgson and Young election systems is completely different from a parameterized complexity point of view. That is, on the one hand, we present an efficient fixed-parameter algorithm for determining a Condorcet winner in Dodgson elections by a minimum number of switches in the votes. On the other hand, we prove that the corresponding problem for Young elections, where one has to delete votes instead of performing switches, is W[2]-complete. In addition, we study Dodgson elections that allow ties between the candidates and give fixed-parameter tractability as well as W[2]-hardness results depending on the cost model for switching ties.

1 Introduction

Computational social choice and, more specifically, the computational complexity of election systems has become an increasingly important field of interdisciplinary research [3,7]. Besides the obvious application in computational politics where the fact that people have varying preferences concerning whom to elect leads to preference aggregation demands via some election system, it also has become very important in multiagent systems: In groups of software agents often a common decision has to be found, again taking into account different preferences about which decision is to be made. Thus, election systems for instance play a central role in planning (artificial intelligence in general) or page ranking systems for Internet search engines.

We study the following classic scenario for election systems: There is a set of candidates and a set of vote(r)s and each voter chooses an order of preference (total order) among the candidates. The well-known Condorcet principle from 1785 then requires that a winner of an election is the candidate who is preferred to each other candidate in more than half of the votes. Unfortunately, a Condorcet winner does not always exist. Hence, several voting systems have been proposed which always choose the Condorcet winner if it exists, and, otherwise,

* Supported by the DFG, research project DARE, GU 1023/1.
** Supported by the DFG, Emmy Noether research group PIAF, NI 369/4.

J. Gudmundsson (Ed.): SWAT 2008, LNCS 5124, pp. 402–413, 2008.

pick a candidate that is in some sense closest to being a Condorcet winner. In other words, these election systems deal with certain "editing problems". In this work, we focus on two classic editing problems from social choice theory [15], that is, the one due to C. L. Dodgson[1] from 1876 and the one due to H. P. Young from 1977. In Dodgson elections, the editing operation is to switch neighboring candidates in the voters' preference lists and the goal is to minimize the overall number of switches needed in order to end up with a Condorcet winner. In Young elections, the editing operation is to remove a vote, again trying to minimize the number of removals in order to end up with a Condorcet winner.

In their seminal work, Bartholdi et al. [1] initiated the study of the computational complexity of election systems. They showed that to decide whether a distinguished candidate can be made a Condorcet winner by performing no more than a given number of editing operations is NP-complete for both Dodgson and Young elections. In a further breakthrough, for Dodgson elections Hemaspaandra et al. [11] and later for Young elections Rothe et al. [18] showed that the corresponding winner and ranking problems are even complete for Θ_2^p, the class of problems that can be solved via parallel access to NP. Thus, Faliszewski et al. [7] concluded that "since checking whether a given candidate has won should be in polynomial time in any system to be put into actual use, these results show that Dodgson and Young elections are unlikely to be useful in practice". This is the point of view of classical, "one-dimensional" computational complexity analysis.[2] By way of contrast, we propose the framework of parameterized computational complexity theory [6,9,16] for studying election systems.[3]

For Dodgson and Young elections, we consider the question whether the NP-hard problems become fixed-parameter tractable with respect to the parameter number of editing operations. We choose this standard parameterization[4] as a natural first step towards a systematic (future) study using further parameterizations. As we can show, other than in the classical context, the parameterized complexity of Dodgson and Young elections completely differs. For n votes and m candidates, for Dodgson elections we can determine in $O(2^k \cdot nk + nm)$ time whether a distinguished candidate can be made a Condorcet winner by performing at most k switches, that is, the problem is fixed-parameter tractable with respect to the parameter k. In contrast, for Young elections the corresponding problem with the parameter denoting either the number of deleted votes or the number of remaining votes becomes W[2]-complete. Our results imply that Dodgson elections can be put in actual use whenever the input instances are close to having a Condorcet winner. This answers an open question of Christian et al. [4][5] and refutes a parameterized hardness conjecture of McCabe-Dansted [13].

---

1 Also known as the writer Lewis Carroll.

2 For more classical complexity results w.r.t. election systems we refer to [5,10].

3 In companion work [2], we also do so for Kemeny elections.

4 In parameterized algorithmics [6,9,16] the solution size typically is the "standard parameter".

5 In the meantime, Fellows et al. independently showed that DODGSON SCORE is fixed-parameter tractable with a worse running time [8].

Moreover, this complements recent work on a simple greedy heuristic for finding Dodgson winners with a guaranteed frequency of success [12] and some work on the approximability of Dodgson and Young elections [14,17]. In particular, Procaccia et al. [17] gave (randomized) approximation algorithms for Dodgson elections and show that it is hard to approximate Young elections by any factor. Moreover, for Dodgson elections we can show that allowing ties (that is, votes may remain undecided between certain candidates), depending on the choice between two switching mechanisms, we either obtain fixed-parameter tractability or W[2]-completeness.

Due to lack of space, several details had to be deferred to the full version of this paper.

2 Preliminaries

Throughout this work, an *election* (V, C) consists of a set V of n votes and a set C of m candidates.[6] A vote is a *preference list* of the candidates, that is, for each voter the candidates are ordered by preference. For instance, in case of three candidates a, b, c, the ordering $a < b < c$ would mean that candidate c is the best liked one and candidate a is the least liked one for this voter. We also consider the case where *ties* are allowed in the votes, that is, instead that a vote consists of a totally ordered list of candidates it then may only be partially ordered. In an election (V, C), a candidate $c \in C$ is called *Condorcet winner* if c wins against every other candidate from C, that is, for each $d \in C \setminus \{c\}$, candidate c is better liked than d in at least $\lfloor n/2 \rfloor + 1$ votes, or, having ties, the number of votes in which c is better liked than d is higher than the number of votes in which d is better liked than c. Observe that a Condorcet winner does not always exist. As a consequence, several specific election systems have been introduced to remedy this situation. Here, we study two popular ones, that is, Dodgson and Young elections. To this end, define as a *switch* the swapping of the two positions of two neighboring candidates in a vote. Based on this, we now can introduce the basic computational problems of this work:

DODGSON SCORE:
Given: An election (V, C), a distinguished candidate $c \in C$, and an integer $k \geq 0$.
Question: Can c be made a Condorcet winner by at most k switches?

In other words, for DODGSON SCORE, we ask whether the *Dodgson score* of c is at most k. The *Young score* is defined by the number of remaining votes:

YOUNG SCORE:
Given: An election (V, C), a distinguished candidate $c \in C$, and an integer $l \geq 0$.
Question: Is there a subset $V' \subseteq V$ of size at least l such that (V', C) has the Condorcet winner c?

[6] Note that we identify votes and voters.

The *dual Young score* is defined by the number of removed votes:

Dual Young Score:
Given: An election (V, C), a distinguished candidate $c \in C$, and an integer $k \geq 0$.
Question: Is there a subset $V' \subseteq V$ of size at most k such that $(V \backslash V', C)$ has the Condorcet winner c?

Note that all three problems are NP-complete [1,18].

To present our results, an important concept is the one of the *deficit* of a candidate $d \in C \backslash \{c\}$ against the distinguished candidate c: Let N_d denote the number of votes from V in which d *defeats* c, that is, in which d is better positioned than c. Then, the *Dodgson deficit* of d is $\lfloor (N_d - (n - N_d))/2 \rfloor + 1$, that is, the minimum number of votes in which the relative order of c and d has to be reversed such that c defeats d in strictly more than half of the votes. Analogously, the *Young deficit* is defined as $N_d - (n - N_d)$. Moreover, we call a candidate with a positive Dodgson deficit *dirty.*

Finally, we briefly introduce the relevant notions of parameterized complexity theory [6,9,16]. Parameterized algorithmics aims at a multivariate (at least two-dimensional) complexity analysis of problems. This is done by studying relevant problem parameters and their influence on the computational complexity of problems. The hope lies in accepting the seemingly inevitable combinatorial explosion for NP-hard problems, but to confine it to the parameter. Hence, the decisive question is whether a given parameterized problem is *fixed-parameter tractable (FPT)* with respect to the parameter, say k. In other words, here we ask for the existence of a solving algorithm with running time $f(k) \cdot \text{poly}(n, m)$ for some computable function f. Unfortunately, not all parameterized problems are fixed-parameter tractable. Downey and Fellows [6] developed a theory of parameterized intractability by means of devising a completeness program with complexity classes. The first two levels of (presumable) parameterized intractability are captured by the complexity classes W[1] and W[2]. We will show several W[2]-completeness results. There is good reason to believe that the corresponding problems thus are not fixed-parameter tractable. To this end, a reduction concept is needed. A *parameterized reduction* reduces a problem instance (I, k) in $f(k) \cdot \text{poly}(|I|)$ time to an instance (I', k') such that (I, k) is a yes-instance iff (I', k') is a yes-instance and k' only depends on k but not on $|I|$.

3 Dodgson Score

In this section, we describe an efficient fixed-parameter algorithm based on dynamic programming for the Dodgson Score problem with respect to the parameter score, answering an open question of Christian et al. [4]. The algorithm can not only decide whether a given Dodgson Score instance is a "yes"-instance, but for a "yes"-instance also constructs a set of at most k switches which lead to a modified input instance where the distinguished candidate c becomes a Condorcet winner. The following two observations are used for the design of the algorithm:

First, it is easy to see (McCabe-Dansted [13, Lemma 2.19]) that there is always an optimal solution that considers only switches that move c in a vote to better positions (Observation 1). Making use of this, our algorithm only considers switches of such kind. Second, since a switch never increases any deficit, we only consider candidates with positive deficit (dirty candidates). With one switch, we can decrease the deficit of exactly one candidate by one. Therefore, with at most k switches allowed, in a yes-instance, the sum of the deficits of the dirty candidates is upper-bounded by k (Observation 2). This fact is crucial for the analysis of the algorithm when bounding the size of the dynamic programming table.

The basic idea of the algorithm is that a solution can be decomposed into subsolutions. In each subsolution the deficit of each dirty candidate is decreased by a certain amount, the *partial decrement*. More precisely, our dynamic programming considers a linear number of subsets of votes, beginning with the subset that contains only one vote and then extending it by adding the other votes one by one. For each of these vote subsets, we consider all possible combinations of partial decrements of deficits.

Definitions for the Algorithm. Let c be the distinguished candidate and let $C_d = (c_1, c_2, \ldots, c_p)$ denote the list of candidates with positive deficit in an arbitrary but fixed order. Let $D = (d_1, d_2, \ldots, d_p)$ be the corresponding *deficit list.*

The dynamic programming table is denoted by T, each row corresponding to a vote v_i for $i = 1, \ldots, n$ and each column corresponding to a *partial deficit list* $(d'_1, d'_2, \ldots, d'_p)$ with $0 \leq d'_j \leq d_j$ for $1 \leq j \leq p$. The entry $T(v_i, (d'_1, d'_2, \ldots, d'_p))$ stores the minimum number of switches within the votes $\{v_j \mid 1 \leq j \leq i\}$ that result in a new instance where the deficits of the p dirty candidates are *at most* $d'_1, d'_2, \ldots, d'_p$, respectively.[7] If a deficit list $(d'_1, d'_2, \ldots, d'_p)$ cannot be achieved by switching within the set of votes $\{v_j \mid 0 \leq j \leq i\}$, we set $T(v_i, (d'_1, d'_2, \ldots, d'_p)) := +\infty$.

Let switch(v_i, c_j) denote the minimum number of switches needed such that in vote v_i candidate c defeats candidate c_j. If c already defeats c_j in v_i, then switch$(v_i, c_j) := 0$. For a deficit list $D' = (d'_1, d'_2, \ldots, d'_p)$ and a subset of indices $S \subseteq \{1, \ldots, p\}$, we use $D' + S$ to denote a deficit list $(e_1, \ldots, e_p)$ where $e_i := d'_i + 1$ for $i \in S$ and $d'_i < d_i$, and $e_i := d'_i$, otherwise. Analogously, for the original deficit list $D = (d_1, \ldots, d_p)$, $D - S$ denotes the list $(f_1, \ldots, f_p)$ where $f_i := d_i - 1$ if $i \in S$ and $f_i := d_i$, otherwise. Let best(S, i) denote the candidate c_j with $j \in S$ that is best in vote v_i.

Algorithm. The dynamic programming algorithm for DODGSON SCORE is given in Figure 1. We assume that we already have the deficits of the candidates and the sum of the deficits of the dirty candidates is at most k as argued in Observation 2. In the initialization of the first row of the dynamic programming table (Figure 1, lines 4–6), the algorithm considers all possible combinations of deficit decrements that can be achieved by switches within the first vote, and

[7] Note that with this definition of table entries, we do not have to consider deficit lists $(d'_1, \ldots, d'_p)$ where $d'_i < 0$ for some i. In this way, the case that an optimal solution may decrease the deficit of a dirty candidate to a negative value is also covered.

Algorithm. *DodScore*
Input: Set of votes $V = \{v_1, \ldots, v_n\}$, set of candidates C, set of dirty candidates $C_d = \{c_1, \ldots, c_p\} \subseteq C$, distinguished candidate c, positive integer k with $p \leq k$, deficit list $D = (d_1, \ldots, d_p)$ of dirty candidates
Output: Yes, if c can become a Condorcet winner with at most k switches

Initialization:

01 for all $D' = (d'_1, \ldots, d'_p)$ with $0 \leq d'_j \leq d_j$ for $0 \leq j \leq p$

02   for $i = 1, \ldots, n$

03 $T(v_i, D') := +\infty$

04 for all $S \subseteq \{1, \ldots, p\}$

05   *if* c_j defeats c in v_1 *for all* $j \in S$ *then*

06 $T(v_1, D - S) := \text{switch}(v_1, \text{best}(S, 1))$

Recursion:

07 for $i = 2, \ldots, n$

08   for all $D' = (d'_1, \ldots, d'_p)$ with $0 \leq d'_j \leq d_j$ for $0 \leq j \leq p$

09   for all $S \subseteq \{1, \ldots, p\}$

10   *if* c_j defeats c in v_i *for all* $j \in S$ *then*

11 $T(v_i, D') := \min\{T(v_i, D'),\ T(v_{i-1}, D' + S) + \text{switch}(v_i, \text{best}(S, i))\}$

Output:

12 if $T(v_n, (0, 0, \ldots, 0)) \leq k$ *then*

13   *return* "Yes"

Fig. 1. Algorithm for DODGSON SCORE

stores the minimum number of switches needed for each of them. In the recursion (lines 7–11), the subset of votes is extended by a new vote v_i and for the new subset $\{v_1, \ldots, v_i\}$ a solution for all partial deficit lists is computed by combining a number of switches within the new vote v_i with information already stored in the table.

Lemma 1. *The algorithm DodScore (Figure 1) is correct.*

Lemma 2. *The algorithm DodScore (Figure 1) runs in $O(4^k \cdot nk + nm)$ time.*

Proof. (Sketch) It is easy to see that the deficit list D can be computed in $O(nm)$ time by iterating over all votes and counting the deficits for all candidates. Now, we consider the size of the dynamic programming table.

A deficit d'_i can have values ranging from 0 to d_i. Hence, the number of partial deficit lists, that is, the number of columns in the table, is $\prod_{i=1}^{p}(d_i + 1)$. Clearly, for a potential "yes"-instance, we have the constraints $p \leq k$ and $\sum_{i=1}^{p} d_i \leq k$ (see Observations 1 and 2). It is not hard to see that 2^k is a tight upper bound on $\prod_{i=1}^{p}(d_i + 1)$. Thus, the overall table size is $n \cdot 2^k$.

For computing an entry $T(v_i, D')$, the algorithm iterates over all 2^p subsets of $\{1, \ldots, p\}$. For each such subset S, it computes the "distance" in v_i between the best of the dirty candidates with indices in S and c, that is, the number of switches needed to make c better than this best dirty candidate. This distance can be computed in $O(k)$ time and, hence, the computation of $T(v_i, D')$ can be done in $O(2^k \cdot k)$ time. The initialization clearly needs $O(2^k \cdot n)$ time. Hence, table T can be computed in $O(2^k \cdot n \cdot 2^k \cdot k + 2^k \cdot n) = O(4^k \cdot nk)$ time. □

By making use of a "monotonicity property" of the table, we can improve the running time of *DodScore* as shown in the following theorem.

Theorem 1. Dodgson Score *can be solved in* $O(2^k \cdot nk + nm)$ *time.*

Proof. The improvement is achieved by replacing the innermost for-loop (lines 9–11) of the recursion which computes a table entry and needs $O(2^k \cdot k)$ time as shown in Lemma 2 by an instruction running in time linear in k.

Let $S_i(d)$ denote the set of the dirty candidates that are better than the distinguished candidate c but not better than the candidate d in vote v_i. This set is empty if d is worse than c in v_i. We replace lines 9–11 in Figure 1 by the following recurrence:

$$T(v_i, D') := \min_{1 \leq r \leq p} \{T(v_{i-1}, D' + S_i(c_r)) + \text{switch}(v_i, c_r)\}.$$

To prove the correctness of the recurrence, on the one hand, observe that, for every r with $1 \leq r \leq p$, there exists a subset $S \subseteq \{1, \ldots, p\}$ satisfying the if-condition in line 10 of *DodScore* such that $S = S_i(c_r)$ and $\text{best}(S, i) = c_r$. Thus,

$$\begin{aligned} &\min_{S \subseteq \{1,\ldots,p\}} \{T(v_{i-1}, D' + S) + \text{switch}(v_i, \text{best}(S, i))\} \\ \leq\ &\min_{1 \leq r \leq p} \{T(v_{i-1}, D' + S_i(c_r)) + \text{switch}(v_i, c_r)\}. \end{aligned}$$

On the other hand, for every $S \subseteq \{1, \ldots, p\}$ satisfying the if-condition in line 10, there exists an r with $1 \leq r \leq p$ such that $S \subseteq S_i(c_r)$. For instance, let r be the index of the candidate in S that is the best in v_i; we then have $\text{best}(S, i) = c_r$ and, thus, $\text{switch}(v_i, \text{best}(S, i)) = \text{switch}(v_i, c_r)$. Moreover, from the definition of table entries, the following monotonicity of the table is easy to verify:

$$T(v_i, (d_1, \ldots, d_i, \ldots, d_p)) \geq T(v_i, (d_1, \ldots, d_i + 1, \ldots, d_p))$$

Thus, from $S \subseteq S_i(c_r)$, we conclude that $T(v_i, D' + S) \geq T(v_i, D' + S_i(c_r))$. Since $S_i(c_r) \subseteq \{1, \ldots, p\}$ and, by the definition of $S_i(c_r)$, this subset satisfies the if-condition in line 10, we then have

$$\begin{aligned} &\min_{1 \leq r \leq p} \{T(v_{i-1}, D' + S_i(c_r)) + \text{switch}(v_i, c_r)\} \\ \leq\ &\min_{S \subseteq \{1,\ldots,p\}} \{T(v_{i-1}, D' + S) + \text{switch}(v_i, \text{best}(S, i))\}. \end{aligned}$$

The time for computing a table entry in the improved version is clearly $O(k)$: Before looking for the minimum, we can compute $S_i(c_r)$ for all $1 \leq r \leq p$ by iterating one time over v_i. Then, based on Lemma 2, the overall running time becomes $O(2^k \cdot nk + nm)$. □

Allowing Ties. As noted by Hemaspaandra et al. [11], there are two natural ways going from total to partial orders. In the first model, transforming $c < a = b$ into $a = b < c$ costs one switch and in the second model it costs two switches. We denote the problem based on the first model by DODGSON TIE SCORE 1 (DTS1) and the problem based on the second one by DODGSON TIE SCORE 2 (DTS2). In the second model we allow c to improve upon an arbitrary subset of equally ranked candidates in order to prevent for "paying" for all equally ranked candidates in a vote even if some of the candidates have already negative deficits.[8] Note that, in the case with ties, the deficit of a candidate $d \neq c$ is defined as $\lfloor (N_d - N_{\overline{d}})/2 \rfloor + 1$, where N_d is the number of votes in which d defeats c and $N_{\overline{d}}$ is the number of votes in which c defeats d. Further, one can choose upon which candidate one likes to improve c with the first switch. Then, the order within a set of equally ranked candidates including c does matter, for example, starting with $c < a = b < d$, by one switch we can either achieve $a < c < b < d$ or $b < c < a < d$. To improve c upon d, we need two additional switches in both cases. Moreover, with ties, we have now two sorts of switches: The first sort swaps the positions of c and other candidates, as in the case without ties, and the second sort breaks and builds ties between c and other candidates. Since these two sorts of switches decrease the deficits of candidates by different values, we introduce the notion of "switch cost". We associate a switch of the first sort with cost 2, while a switch of the second sort has cost 1. For example, given $c < a < b$, by one switch between c and a, we can get $c = a < b$ or $a < c < b$. However, the switch in the first case has cost 1 but the second case needs a cost-2 switch. Finally, the parameter we consider in the following is the total cost of the switches needed to make c a Condorcet winner.

Whereas the both variants DTS1 and DTS2 remain NP-complete (which easily follows from the NP-completeness of the case without ties [1]), their parameterized complexity differs. The problem DTS2 is fixed-parameter tractable, but DTS1 becomes W[2]-complete.[9] Note that compared to Theorem 1 (without ties) we obtain a slightly worse running time for DTS2. More precisely, due to a slight modification of algorithm *DodScore* from Figure 1 we can state the following.

Theorem 2. DODGSON TIE SCORE 2 *can be solved in* $O(4^k \cdot nk + nm)$ *time.*

Theorem 3. DODGSON TIE SCORE 1 *is* $W[2]$*-complete with respect to the parameter* k*.*

4 Young Score

In this section, we show that YOUNG SCORE and DUAL YOUNG SCORE are W[2]-complete with respect to their corresponding solution size bounds l and k,

[8] If we only allow to improve upon whole sets of equally ranked candidates, we can use the improved version of the algorithm *DodScore* by treating the whole set as one possibility. In this way, we achieve a running time of $O(2^k \cdot nk + nm)$.

[9] Interestingly, Hemaspaandra et al. [11] have shown that the ranking and the winner versions remain Θ_2^p-complete in both cases considering ties—no differentiation in the classical complexity can be observed.

respectively. We start with the proof of the W[2]-hardness of Dual Young Score where two parameterized reductions are given, the first from the W[2]-hard Red Blue Dominating Set (RBDS) [6] to an intermediate problem, a variant of Red Blue Dominating Set, and then the second one from the intermediate problem to Dual Young Score. Red Blue Dominating Set (RBDS) is defined as follows: Given a bipartite graph $G = (R \cup B, E)$ with R and B being the two disjoint vertex sets, $E \subseteq R \times B$ and an integer $k \geq 0$, the question is whether there is a subset $D \subseteq R$ of size at most k such that every vertex in B has at least one neighbor in D. The $k/2$-Red Blue Dominating Set ($k/2$-RBDS) problem has a bipartite graph $G = (R \cup B, E)$ with R and B being the two disjoint vertex sets, $E \subseteq R \times B$, and an integer $k \geq 0$ as input, and asks whether there is a subset $D \subseteq R$ of size at most k such that every vertex in B has at least $\lfloor k/2 \rfloor + 1$ neighbors in D.

Lemma 3. *$k/2$-RBDS is W[2]-hard.*

Next, we give a parameterized reduction from $k/2$-RBDS to Dual Young Score.

Lemma 4. Dual Young Score *is W[2]-hard.*

Proof. Given a $k/2$-RBDS-instance $(G = (B \cup R, E), k)$ with $B = \{b_1, ..., b_m\}$ and $R = \{r_1, ..., r_n\}$, we first consider the case that k is odd. The corresponding Dual Young Score instance is constructed as follows. For each blue vertex we add a candidate to the candidate set C. After that, three additional candidates a, b, and c are added to C, where c is the distinguished candidate of the Dual Young Score instance. Thus, $C = \{c_i \mid b_i \in B\} \cup \{a, b, c\}$.

Let $N_C(r_i) := \{c_j \in C \mid \{r_i, b_j\} \in E\}$ and $\overline{N_C(r_i)} := C \setminus (\{a, b, c\} \cup N_C(r_i))$, that is, the elements in $N_C(r_i)$ correspond to the neighbors of r_i in G and $\overline{N_C(r_i)}$ corresponds to the rest of the vertices in B. We construct three disjoint subsets of votes, V_1, V_2, and V_3.

- The votes in V_1 correspond to the red vertices in R. For every red vertex r_i, we add a vote v_i to V_1 by placing the candidates in $N_C(r_i) \cup \{a, b\}$ better than c and the candidates in $\overline{N_C(r_i)}$ worse than c, that is,

$$V_1 := \{\overline{N_C(r_i)} < c < N_C(r_i) < a < b \mid i = 1, \ldots, n\}.$$

  Note that, here and in the following, if we write a set in a vote, then the order of the elements in the set is irrelevant and can be fixed arbitrarily.
- The set V_2 also contains n votes which guarantee that the total deficit of each candidate is zero in $V_1 \cup V_2$. However, there is an exception with b which has deficit $2k - 2$ in $V_1 \cup V_2$.

$$V_2 := \{a < b < N_C(r_i) < c < \overline{N_C(r_i)} \mid i = 1, \ldots, n - k + 1\}$$
$$\cup \{a < N_C(r_i) < c < \overline{N_C(r_i)} < b \mid i = n - k + 2, \ldots, n\}.$$

Later, it will become clear that the $(2k-2)$-deficit of b will be used to argue that all votes in a solution of a DUAL YOUNG SCORE instance have to come from V_1.
- The set V_3 consists of $k-1$ votes to adjust the deficits of a and b so that in $V_1 \cup V_2 \cup V_3$ both a and b have a deficit of $k-1$ and all other candidates have a deficit of 0. Let $C_R := C \setminus \{a, b, c\}$. The set V_3 consists of $\lfloor k/2 \rfloor$ votes with $b < c < C_R < a$ and $\lfloor k/2 \rfloor$ votes with $b < C_R < c < a$.

Finally, the set V of votes is set to $V_1 \cup V_2 \cup V_3$ and the upper bound for the solution size of the DUAL YOUNG SCORE instance is equal to k. The key idea behind the above construction is that, to reduce the $(k-1)$-deficits of a and b by deleting at most k votes, all solutions of the DUAL YOUNG SCORE instance actually contain *exactly* k votes from V_1.

In the following, we show that the candidate c can become a Condorcet winner by deleting at most k votes iff there is a solution of size at most k for the $k/2$-RBDS-instance.

"$\Rightarrow$": We know that every solution V' of DUAL YOUNG SCORE contains exactly k votes in V_1 and, by the above construction, each vote in V_1 corresponds to a vertex in R. Denote the corresponding subset of R as D. Since V' is a solution, it must hold that every candidate $c_i \in (C \setminus \{a, b, c\})$ must be better than c in at least $\lfloor k/2 \rfloor + 1$ of the votes in V'. Therefore, choosing the corresponding red vertices to form a dominating set achieves that every blue vertex is dominated at least $\lfloor k/2 \rfloor + 1$ times.

"$\Leftarrow$": Since every dominating set $D \subseteq R$ of size at most k dominates each blue vertex at least $\lfloor k/2 \rfloor + 1$ times, we can easily extend D to a dominating set D' of size exactly k by adding $k - |D|$ arbitrary red vertices to D. Since every red vertex corresponds to a vote in V_1, we thus obtain a size-k subset V' of V corresponding to D'. According to the above construction of V_1, the removal of V' results in a new vote set where the deficits of a and b are both -1 and the deficits of all other candidates are at most -1. Therefore, c can become Condorcet winner by deleting exactly k votes.

Finally, we consider the case that k is even and give a reduction from DUAL YOUNG SCORE with an odd k to DUAL YOUNG SCORE with an even k. Given a DUAL YOUNG SCORE instance (V, C, c, k) with k being odd, we add a new vote v to V that has the form: $c < C \setminus \{c\}$ to get the new vote set V'. Then $(V', C, c, k' := k+1)$ is a DUAL YOUNG SCORE instance with k' being even. The correspondence between the solutions is easy to achieve. □

Next, we show that DUAL YOUNG SCORE is in W[2]. To this end, we can construct a parameterized reduction from DUAL YOUNG SCORE to the W[2]-complete OPTIMAL LOBBYING problem [4] also arising in the context of election systems, which is defined as follows: Given an $n \times m$ 0/1-matrix M, a length-m 0/1-vector x, and an integer $k \geq 0$, the question is whether there is a choice of at most k rows from the rows of M such that the selected rows can be *edited* such that, if x has a 0 in its ith entry, then there are more 0's than 1's in the ith column of the matrix resulting after editing the selected rows; otherwise, there are more 1's than 0's in the ith column. By *editing* a row, we mean to change some 1's in the row to 0's and/or to change some 0's to 1's. We call x the *target* vector.

Lemma 5. Dual Young Score *is in W[2].*

Combining Lemmas 4 and 5, we arrive at the main result of this section.

Theorem 4. Dual Young Score *is W[2]-complete.*

Using a similar reduction as the one in the proof of Lemma 5 (containment in W[2]) and a slightly modified version of the reduction from the W[2]-hard Set Packing problem presented by Rothe et al. [18, Theorem 2.3] (W[2]-hardness), we can also show the following theorem.

Theorem 5. Young Score *is W[2]-complete.*

5 Conclusion and Outlook

Probably the most important general observation deriving from our work is that Dodgson and Young elections behave differently with respect to the parameter "number of editing operations". Whereas for Dodgson elections we achieve fixed-parameter tractability, we experience parameterized intractability in case of Young elections. This stands in sharp contrast to traditional complexity analysis, where both election systems appear as equally hard [1,11,18] and complements the results on polynomial-time approximability of Procaccia et al. [17]. Furthermore, we found that the complexities of Dodgson elections allowing ties between the candidates are much different (fixed-parameter tractability vs W[2]-hardness) depending on the cost model for switching ties. Again, in the standard complexity framework these two cases cannot be differentiated because both lead to NP-completeness.

We conclude with few specific open questions. Bartholdi et al. [1] gave an integer linear program which implies the fixed-parameter tractability of Dodgson Score with respect to the number of candidates (also see [13] for further results in this direction). Unfortunately, the corresponding running times are extremely high and an efficient combinatorial algorithm would be desirable. The parameterized complexity with respect to the parameter "number of votes" remains open. By way of contrast, there is a trivial $O(2^n \cdot \text{poly}(m,n))$-time algorithm for Young elections with $n := |V|$ and $m := |C|$. For the parameter "number of candidates", however, only an integer linear program implying (presumably impractical) fixed-parameter tractability is known [19].

Acknowledgement. We thank Jörg Vogel for various valuable pointers to the literature and inspiring discussions.

References

1. Bartholdi III, J., Tovey, C.A., Trick, M.A.: Voting schemes for which it can be difficult to tell who won the election. Social Choice and Welfare 6, 157–165 (1989)
2. Betzler, N., Fellows, M.R., Guo, J., Niedermeier, R., Rosamond, F.A.: Fixed-parameter algorithms for Kemeny scores. In: Proc. of 4th AAIM. LNCS, vol. 5034, pp. 60–71. Springer, Heidelberg (2008)
3. Chevaleyre, Y., Endriss, U., Lang, J., Maudet, N.: A short introduction to computational social choice (invited paper). In: van Leeuwen, J., Italiano, G.F., van der Hoek, W., Meinel, C., Sack, H., Plášil, F. (eds.) SOFSEM 2007. LNCS, vol. 4362, pp. 51–69. Springer, Heidelberg (2007)
4. Christian, R., Fellows, M.R., Rosamond, F.A., Slinko, A.M.: On complexity of lobbying in multiple referenda. Review of Economic Design 11(3), 217–224 (2007)
5. Conitzer, V., Sandholm, T., Lang, J.: When are elections with few candidates hard to manipulate? Journal of the ACM 54(3), 1–33 (2007)
6. Downey, R.G., Fellows, M.R.: Parameterized Complexity. Springer, Heidelberg (1999)
7. Faliszewski, P., Hemaspaandra, E., Hemaspaandra, L.A., Rothe, J.: A richer understanding of the complexity of election systems. In: Ravi, S., Shukla, S. (eds.) Fundamental Problems in Computing: Essays in Honor of Professor Daniel J. Rosenkrantz. Springer, Heidelberg (2008)
8. Fellows, M.R.: Personal communication (October 2007)
9. Flum, J., Grohe, M.: Parameterized Complexity Theory. Springer, Heidelberg (2006)
10. Hemaspaandra, E., Hemaspaandra, L.A.: Dichotomy for voting systems. Journal of Computer and System Sciences 73(1), 73–83 (2007)
11. Hemaspaandra, E., Hemaspaandra, L.A., Rothe, J.: Exact analysis of Dodgson elections: Lewis Caroll's 1876 voting system is complete for parallel access to NP. Journal of the ACM 44(6), 806–825 (1997)
12. Homan, C.M., Hemaspaandra, L.A.: Guarantees for the success frequency of an algorithm for finding Dodgson-election winners. Journal of Heuristics (2007)
13. McCabe-Dansted, J.C.: Approximability and computational feasibility of Dodgson's rule. Master's thesis, University of Auckland (2006)
14. McCabe-Dansted, J.C., Pritchard, G., Slinko, A.: Approximability of Dodgson's rule. Social Choice and Welfare (2007)
15. McLean, I., Urken, A.: Classics of Social Choice. University of Michigan Press, Ann Arbor, Michigan (1995)
16. Niedermeier, R.: Invitation to Fixed-Parameter Algorithms. Oxford University Press, Oxford (2006)
17. Procaccia, A.D., Feldman, M., Rosenschein, J.S.: Approximability and inapproximability of Dodgson and Young elections. Technical Report Discussion paper 466, Center for the Study of Rationality, Hebrew University (October 2007)
18. Rothe, J., Spakowski, H., Vogel, J.: Exact complexity of the winner problem for Young elections. Theory of Computing Systems 36, 375–386 (2003)
19. Young, H.P.: Extending Condorcet's rule. Journal of Economic Theory 16, 335–353 (1977)

Online Compression Caching

C. Greg Plaxton*, Yu Sun**, Mitul Tiwari**, and Harrick Vin**

***Department of Computer Science
University of Texas at Austin
plaxton@cs.utexas.edu, asun@vmware.com, mitul@kosmix.com,
vin@cs.utexas.edu

Abstract. Motivated by the possibility of storing a file in a compressed format, we formulate the following class of compression caching problems. We are given a cache with a specified capacity, a certain number of compression/uncompression algorithms, and a set of files, each of which can be cached as it is or by applying one of the compression algorithms. Each compressed format of a file is specified by three parameters: encode cost, decode cost, and size. The miss penalty of a file is the cost of accessing the file if the file or any compressed format of the file is not present in the cache. The goal of a compression caching algorithm is to minimize the total cost of executing a given sequence of requests for files. We say an online algorithm is resource competitive if the algorithm is constant competitive with a constant factor resource advantage. A well-known result in the framework of competitive analysis states that the least-recently used (LRU) algorithm is resource competitive for the traditional paging problem. Since compression caching generalizes the traditional paging problem, it is natural to ask whether a resource competitive online algorithm exists or not for compression caching. In this work, we address three problems in the class of compression caching. The first problem assumes that the encode cost and decode cost associated with any format of a file are equal. For this problem we present a resource competitive online algorithm. To explore the existence of resource competitive online algorithms for compression caching with arbitrary encode costs and decode costs, we address two other natural problems in the aforementioned class, and for each of these problems, we show that there exists a non-constant lower bound on the competitive ratio of any online algorithm, even if the algorithm is given an arbitrary factor capacity blowup. Thus, we establish that there is no resource competitive algorithm for compression caching in its full generality.

1 Introduction

Recently we have seen an explosion in the amount of data distributed over hand-held devices, personal computers, and local and wide area networks. There is a

* Supported by NSF Grants CNS–0326001 and CCF–0635203.

** Supported by NSF Grant CNS–0326001 and Texas Advanced Technology Program 003658-0608-2003.

*** Most of the work was done while Yu Sun and Mitul Tiwari were Ph.D. students at UT Austin.

J. Gudmundsson (Ed.): SWAT 2008, LNCS 5124, pp. 414–425, 2008.

growing need for self-tuning data management techniques that can operate under a wide range of conditions, and optimize various resources such as storage space, processing, and network bandwidth. There is a large body of work addressing different aspects of this domain of self-tuning data management.

An important aspect of this domain that merits further attention is that data can be stored in different formats. For example, one can compress a text file using different traditional compression techniques such as gzip and bzip. Various studies [1,2,6] have experimentally demonstrated the advantages of compression in caching. A compressed file takes up less space, effectively increasing the size of the memory. However, this increase in size comes at the cost of extra processing needed for compression and uncompression. Consequently, it may be desirable to keep frequently accessed files uncompressed in the memory.

As another example, consider the option of storing only a TEX file or the corresponding pdf file along with the TEX file. One can save space by storing only the TEX file, but one has to run a utility (such as pdflatex) to generate the pdf file when needed. On the other hand, storing the pdf file may require an order of magnitude more storage space than the TEX file, but the pdf file is readily accessible when needed. In general, many files are automatically generated using some utility such as a compiler or other translator. If the utility generates a large output compared to the input, then by storing only the input one achieves a form of "compression", not in the traditional sense, but with analogous consequences. In this paper, when we refer to compression, we have in mind this broader notion of compression where one can have a wide separation between storage space and processing costs associated with different formats of a file.

In this work, we address the notion of compression and uncompression of files, while contemplating the possibility of a richer variety of separation between the sizes and processing costs associated with the different formats of a file. We focus primarily on the single machine setting, however one of our upper bound results (see Section 3.2) is applicable to a simple, but well-motivated, special case of a distributed storage problem.

Problem Formulation. We define a class of compression caching problems in which a file can be cached in multiple compressed formats with varying sizes, and costs for compression and uncompression (see Section 2 for a formal description). We are given a cache with a specified capacity. Also assume that for each file, there are multiple associated formats. Each format is specified by three parameters: encode cost, decode cost, and size. The encode cost of a particular format is defined as the cost of creating that format from the uncompressed format of the file. The decode cost of a format is defined as the cost of creating the uncompressed format. The miss penalty of a file is defined as the cost of accessing the file if no format of the file is present in the cache. To execute a request for a file, the file is required to be loaded into the cache in the uncompressed format. The goal of a compression caching algorithm is to minimize the total cost of executing a given request sequence.

The main challenge is to design algorithms that determine — in an online manner — which files to keep in the fast memory, and of these, which to keep in compressed form. The problem is further complicated by the multiple compression formats for a file, with varying sizes and encode/decode costs. Since compression caching has the potential to be useful in many different scenarios, a desirable property of an online algorithm is to provide a good competitive ratio, which is defined as the maximum ratio of the cost of the online algorithm and that of the offline algorithm over any request sequence (see [4] for more details). We refer to an online algorithm that achieves a constant competitive ratio when given a constant factor resource advantage as a resource competitive algorithm.

In a seminal work, Sleator and Tarjan [7] show that the competitive ratio of any deterministic online paging algorithm without any capacity blowup is the size of the cache, and they also show that LRU is resource competitive for the disk paging problem. Since compression caching generalizes the disk paging problem, it is natural to ask whether similar resource competitive results can be obtained for compression caching.

Contributions. In this paper, we address three problems in the class of the compression caching. Our contributions for each of these problems are as follows.

- The first problem assumes that the encode cost and decode cost associated with any format of a file are equal. For this problem we generalize the Landlord algorithm [9] to obtain an online algorithm that is resource competitive. We find that this problem also corresponds to a special case of the distributed storage problem, and hence, our algorithm is applicable to this special case.
- The second problem assumes that the decode costs associated with different formats of a file are the same. For this problem, we show that any deterministic online algorithm (even with an arbitrary factor capacity blowup) is $\Omega(m)$-competitive, where m is the number of possible formats of a file. The proof of this lower bound result is the most technically challenging part of the paper. Further, we give an online algorithm for this problem that is $O(m)$-competitive with $O(m)$ factor capacity blowup. Thus, we tightly characterize the competitive ratio achievable for this problem.
- The third problem assumes that the encode costs associated with different formats of a file are the same. For this problem we show that any deterministic online algorithm (even with an arbitrary factor capacity blowup) has competitive ratio $\Omega(\log m)$. We also present an online algorithm for this problem that is $O(m)$-competitive with $O(m)$ factor capacity blowup.

Related Work. The competitive analysis framework was pioneered by Sleator and Tarjan [7]. For the disk paging problem, it has been shown that LRU is $\frac{k}{k-h+1}$-competitive, where k is the cache capacity of LRU and h is the cache capacity of any offline algorithm [7]. In the same paper, it has been shown that $\frac{k}{k-h+1}$ is the best possible competitive ratio for any deterministic online paging algorithm. For the variable size file caching problem, which is useful in the context of web-caching, Young [9] proposes the Landlord algorithm, and shows that Landlord is $\frac{k}{k-h+1}$-competitive. For the variable size file caching problem, Cao and Irani [5] independently propose the greedy-dual-size algorithm and show that it is k-competitive against any offline algorithm, where k is cache capacity of both greedy-dual-size and the offline algorithm. For the distributed paging problem, Awerbuch et al. [3] give an algorithm that is polylog(n, Δ)-competitive with polylog(n, Δ) factor capacity blowup, where n is the number of nodes and Δ is the diameter of the network.

Various studies [1,2,6] have shown experimentally that compression effectively increases on-chip and off-chip chip cache capacity, as well as off-chip bandwidth, since the compressed data is smaller in size. Further, these studies show that compression in caching increases the overall performance of the system.

Outline. The rest of this paper is organized as follows. In Section 2 we provide some definitions. In Section 3 we present our results for the compression caching problem with equal encode and decode costs. In Section 4 we describe our results for the compression caching problem with varying encode costs and uniform decode costs. In Section 5 we discuss our results for the compression caching problem with uniform encode costs and varying decode costs.

2 Preliminaries

Assume that we are given a cache with a specified capacity and m different functions for encoding and decoding any file, denoted h_i and h_i^{-1}, where $0 \leq i < m$. Without loss of generality, we assume that h_0 and h_0^{-1} are the identity functions. We define index i as an integer i such that $0 \leq i < m$. For any index i, we obtain the i-encoding of any file f by evaluating $h_i(f)$, and we obtain the file f from the i-encoding μ of f by evaluating $h_i^{-1}(\mu)$. For any file f, we refer to the 0-encoding of f as the *trivial* encoding, and for $i > 0$, we refer to the i-encoding of f as a *nontrivial* encoding. For any file f and index i, the i-encoding of f is also referred to as an encoding of f, and we say f is present in the cache if any encoding of f is present in the cache. For any file f and index i, the i-encoding of f is characterized by three parameters: encode cost, denoted $\mathit{encode}(i, f)$; decode cost, denoted $\mathit{decode}(i, f)$; and size, denoted $\mathit{size}(i, f)$. The encode cost $\mathit{encode}(i, f)$ is defined as the cost of evaluating $h_i(f)$, and the decode cost $\mathit{decode}(i, f)$ is defined as the cost of evaluating $h_i^{-1}(\mu)$, where μ is the i-encoding of f. For any file f, $\mathit{encode}(0, f)$ and $\mathit{decode}(0, f)$ are 0.

For any file f, the *access cost* of f is defined as follows: if for some index i, the i-encoding of f is present in the cache (break ties by picking minimum such i), then the access cost is $\mathit{decode}(i, f)$; if none of the encodings of f is present in the cache, then the access cost is defined as the miss penalty $p(f)$. Without loss of generality, we assume that the miss penalty for any file f is at least the decode cost of any of the encodings of f. The cost of deleting any encoding of any file from the cache is 0. For any file f and index i, the i-encoding of f can be added to the cache if there is enough free space to store the i-encoding of f. For any file f and index i, the cost of adding the i-encoding of f to the cache is the sum of the access cost of f and $\mathit{encode}(i, f)$.

To execute a request for a file f, an algorithm A is allowed to modify its cache content by adding/deleting encodings of files, and then incurs the access cost for f. The goal of the compression caching problem is to minimize the total cost of executing a given request sequence. An online compression caching algorithm A is c-competitive if for all request sequences τ and compression caching algorithms B, the cost of executing τ by A is at most c times that of executing τ by B.

Any instance I of the compression caching problem is represented by a triple (σ, m, k), where σ is the sequence of request for the instance I, m is the number of possible encodings for files in σ, and k is the cache capacity. For any instance $I = (\sigma, m, k)$, we define $reqseq(I) = \sigma$, $numindex(I) = m$, and $space(I) = k$.

We define a *configuration* as a set of encodings of files. For any configuration S, we define the size of S as the sum, over all encodings μ in S, of size of μ. We define a *trace* as a sequence of pairs, where the first element of the pair is a request for a file and the second element of the pair is a configuration. For any configuration S and any integer k, S is *k-feasible* if the size of S is at most k. For any trace T and integer k, T is k-feasible if and only if any configuration in T is k-feasible. For any two sequences τ and τ', we define $\tau \circ \tau'$ as the sequence obtained by appending τ' to τ. For any trace T, we define $requests(T)$ as the sequence of requests present in T, in the same order as in T.

For any file f, any trace T, and any configuration S, we define $cost_f(T, S)$ inductively as follows. If T is empty, then $cost_f(T, S)$ is zero. If T is equal to $\langle (f', S') \rangle \circ T'$, then $cost_f(T, S)$ is $cost_f(T', S')$ plus the sum, over all i-encodings μ of f such that μ is present in S' and μ is not present in S, of $encode(i, f)$, plus the access cost of f in S if $f = f'$. For any file f and any trace T, we define $cost_f(T)$ as $cost_f(T, \emptyset)$. For any trace T and any configuration S, we define $cost(T, S)$ as the sum, over all files f, of $cost_f(T, S)$. For any trace T, we define $cost(T)$ as $cost(T, \emptyset)$.

3 Equal Encode and Decode Costs

In this section, we consider a symmetric instance of the compression caching problem which assumes that the encode cost and decode cost associated with any encoding of a file are equal. We present an online algorithm for this problem, and show that the algorithm is resource competitive. Interestingly, this problem also corresponds to a multilevel storage scenario, as discussed in Section 3.2.

For the restricted version of the compression caching problem considered in this section, we have $encode(i, f) = decode(i, f)$ for any file f and index i. At the expense of a small constant factor in the competitive ratio, we can assume that, for any file f, the miss penalty $p(f)$ is at least $q \cdot encode(m - 1, f)$, where $q > 1$; and by preprocessing, we can arrange encodings of files in geometrically decreasing sizes and geometrically increasing encode-decode costs. The basic idea behind the preprocessing phase is as follows. First, consider any two encodings with sizes (resp., similar encode-decode costs) within a constant factor. Second, from these two encodings, select the one with smaller encode-decode cost (resp., smaller size), and eliminate the other. While an encoding can be eliminated by one of the above preprocessing steps, we do so. After the above preprocessing phase, we can arrange the encodings of files in geometrically decreasing sizes and geometrically increasing encode-decode costs.

For ease of presentation, we assume that m encodings are selected for each file in the preprocessing phase. More precisely, after the preprocessing phase, for any file f and index $i < m - 1$, we have $\mathit{size}(i + 1, f) \leq \frac{1}{r} \cdot \mathit{size}(i, f)$ and $\mathit{encode}(i+1, f) \geq q \cdot \mathit{encode}(i, f)$, where $r > 1$. Also, we assume that the capacity of the cache given to an online algorithm is b times that given to an offline algorithm.

3.1 Algorithm

In Figure 1, we present our online algorithm ON. At a high level, ON is a credit-rental algorithm. Algorithm ON maintains a containment property on the encodings in the cache, defined as follows: If ON has the i-encoding of some file f in the cache, then ON also has all the j-encodings of f for any index $j \geq i$ in the cache. A credit is associated with each encoding present in the cache. For any file f and index i, the i-encoding of f is created with an initial credit $\mathit{decode}(i + 1, f)$, for $i < m - 1$, and credit $p(f)$, for $i = m - 1$. On a request for a file f, if the 0-encoding of f is not present in the cache, then ON creates space for the 0-encoding of f, and for other i-encodings of f that are necessary to maintain the containment property. Then, ON creates the 0-encoding of f, and any other i-encodings of f that are necessary to maintain the containment property, with an initial credit as described above. In order to create space, for each file present in the cache, ON charges rent from the credit of the largest encoding of the file, where rent charged is proportional to the size of the encoding, and deletes any encoding with 0 credit. The credit-rental algorithm described here can be viewed as a generalization of Young's Landlord algorithm [9].

We use a potential function argument similar to that of Young to show that ON is resource competitive. See [8, Section 3.3.2] for the complete proof of the following theorem.

Theorem 1. *Algorithm* ON *is resource competitive for any symmetric instance of the compression caching problem.*

3.2 Multilevel Storage

Consider an outsourced storage service scenario (for simplicity, here we describe the problem for a single user) where we have multiple levels of storage. Each storage space is specified by two parameters: storage cost and access latency to the user. The user specifies a fixed overall budget to buy storage space at the various levels, and generates requests for files. The goal is to manage the user budget and minimize the total latency incurred in processing a given request sequence. Our credit-rental algorithm for the compression caching problem with equal encode and decode costs can be easily generalized to this scenario, and we can show (using a similar analysis as above) that the generalized algorithm

{Initially, for any encoding μ of any file, $credit(\mu) = 0$}
On a request for a file f
if f is not present in the cache **then**
  $createspace(f, m-1)$
  for all indices i, add the i-encoding μ of f, with $credit(\mu) := decode(i+1, f)$, if $i < m-1$, and with $credit(\mu) := p(f)$, if $i = m-1$
else if the i-encoding μ of f is present in the cache (break ties by picking the minimum such i) **then**
  evaluate $h_i^{-1}(\mu)$
  $credit(\mu) := decode(i+1, f)$
  if $i > 0$ **then**
    $createspace(f, i-1)$
    for all indices $j < i$, add the j-encoding ν of f, with $credit(\nu) := decode(j+1, f)$
  fi
fi

$createspace(f, i)$
  $sz := \sum_{j=0}^{i} size(f, j)$
  while free space in the cache $< sz$ **do**
    $\delta := \min_{\mu \in X} \frac{credit(\mu)}{size(j, f')}$, where μ is the j-encoding of f'
    for each file f' such that there is an encoding of f' in the cache **do**
      let μ be the largest (in size) encoding of f' in the cache
      let j be the index of μ
      $credit(\mu) := credit(\mu) - \delta \cdot size(j, f')$
      if $credit(\mu) = 0$ **then**
        delete μ
      fi
    od
  od

Fig. 1. The online algorithm ON for any symmetric instance of the compression caching problem. Here, X is the cache content of ON.

is constant competitive with a constant factor advantage in the budget for the aforementioned multilevel storage problem.

4 Varying Encode Costs and Uniform Decode Costs

We say that an instance $I = (\sigma, m, k)$ of the compression caching problem is a uniform-decode instance if any file in σ satisfies the following properties. First, we consider that the decode cost associated with different encodings of any file in σ are the same; for any file f and any index $i > 0$, we abbreviate $decode(i, f)$ to $decode$. Second, we consider that for any index i, any file f and f' in σ, $size(i, f) = size(i, f')$, $p(f) = p(f')$, and $encode(i, f) = encode(i, f')$. For the sake of brevity, we write $encode(i, f)$ as $encode(i)$.

We formulate this problem to explore the existence of resource competitive algorithms for the problems in the class of compression caching. This problem is also motivated by the existence of multiple formats of a multimedia file with varying sizes and encode costs, and with roughly similar decode costs.

One might hope to generalize existing algorithms like Landlord for this problem, and to achieve resource competitiveness. However, in this section we show that any deterministic online algorithm (even with an arbitrary factor capacity blowup) for this problem is $\Omega(m)$-competitive, where m is the number of possible encodings of each file. We also give an online algorithm for this problem that is $O(m)$-competitive with $O(m)$ factor capacity blowup.

4.1 The Lower Bound

In this section we show that any deterministic online algorithm (even with an arbitrary factor capacity blowup) for any uniform-decode instance of the compression caching problem is $\Omega(m)$-competitive.

For any algorithm A, any request sequence σ, and any real number k, we define $config(A, \sigma, k)$ as the configuration of A after executing σ with a cache of size k, starting with an empty configuration.

Informal overview. At a high level, the adversarial request generating algorithm *Adversary* works recursively as follows. For a given online algorithm ON, a given number of encodings for a file m, a given cache capacity of the offline algorithm k, and a blowup b, the algorithm $Adversary(\mathrm{ON}, m, k, b)$ picks a set of files X such that any file in X is not in ON's cache, and invokes a recursive request generating procedure $AdversaryHelper(X, i, \sigma, \mathrm{ON}, k, b)$, where initially $|X|$ is the number of $(m-1)$-encodings that can fit in a cache of size k, $i = m-1$, and σ is empty. This procedure returns a trace of the offline algorithm OFF. (See Section 4.1 for formal definitions and a description of the algorithm.)

Consider an invocation of procedure $AdversaryHelper(X, i, \sigma, \mathrm{ON}, k, b)$. The adversary picks a subset of the files Y from X such that any file f in Y satisfies certain conditions. For $i > 1$, if Y contains sufficiently many files, then the adversary invokes $AdversaryHelper(Y, i-1, \sigma', \mathrm{ON}, k, b)$, where σ' is the request sequence generated; otherwise, $AdversaryHelper(X, i, \sigma, \mathrm{ON}, k, b)$ is terminated. For $i = 1$, the adversary picks a file f in Y, and repeatedly generates requests for f until either ON adds an encoding of f to its cache, or a certain number of requests for f are generated. Finally, $AdversaryHelper(X, 1, \sigma, \mathrm{ON}, k, b)$ terminates when Y is empty.

At a high level, the offline algorithm OFF works as follows. Algorithm OFF decides the encodings for the files in X when $AdversaryHelper(X, i, \sigma, \mathrm{ON}, k, b)$ terminates. For any index $j \geq i$, if ON adds the j-encodings of less than a certain fraction of files in X any time during the execution of the request sequence generated by $AdversaryHelper(X, i, \sigma, \mathrm{ON}, k, b)$, then OFF adds the i-encodings of all the files in X, and incurs no miss penalties in executing the request sequence generated by $AdversaryHelper(X, i, \sigma, \mathrm{ON}, k, b)$. Otherwise, OFF returns the concatenation of the traces generated during the execution of $AdversaryHelper(X, i, \sigma, \mathrm{ON}, k, b)$. By adding the j-encodings of a certain fraction of files in X, ON incurs much higher cost than OFF in executing the request sequence generated by $AdversaryHelper(X, i, \sigma, \mathrm{ON}, k, b)$.

Using an inductive argument, we show that ON is $\Omega(m)$-competitive for the compression caching problem with varying encode and uniform decode costs.

Adversarial request generating algorithm. Some key notations used in the adversarial request generating algorithm are as follows.

For any file f and any real number b, $eligible(f, m, b)$ holds if the following conditions hold: (1) for any index i, $size(i, f) = r^{m-i-1}$, where $r = 8b$; (2) $p(f) = p$;

(3) for any index i, $encode(i, f) = p \cdot q^i$, where $q = \frac{m}{20}$; and (4) $decode = 0$. The number of i-encodings of files that can fit in a cache of size k is denoted $num(k, i)$. Note that, for eligible files, $num(k, i)$ is equal to $r \cdot num(k, i-1)$.

For any algorithm A, any request sequence σ, any real number k, any file f, and any index i, we define a predicate $aggressive(A, \sigma, k, f, i)$ as follows. If σ is empty, then $aggressive(A, \sigma, k, f, i)$ does not hold. If σ is equal to $\sigma' \circ \langle f' \rangle$, then $aggressive(A, \sigma, k, f, i)$ holds if either $aggressive(A, \sigma', k, f, i)$ holds or, for some index $j \geq i$, the j-encoding of f is present in $config(\mathrm{ON}, \sigma, k)$.

For any request sequence σ, and any index i, any set of files X, we define $trace(\sigma, i, X)$ as follows. If σ is empty, then $trace(\sigma, i, X)$ is empty. If σ is equal to $\sigma' \circ \langle f \rangle$, then $trace(\sigma, i, X)$ is $trace(\sigma', i, X) \circ \langle (f, Y) \rangle$, where Y is the set of the i-encodings of files in X.

In Figure 2 we describe the adversarial request generating algorithm *Adversary*. The caching decisions of the offline algorithm OFF are given by the trace T generated during the execution of *Adversary* (Figure 2). See [8, Section 3.4.1.3] for the proof of the following theorem.

```
Adversary(ON, m, k, b)
T := ∅
while |T| < N do
  X := set of num(k, m − 1) files f such that (1) eligible(f, m, b) holds; and
      (2) f is not present in config(ON, requests(T), bk)
  T := T ∘ AdversaryHelper(X, m − 1, requests(T), ON, k, b)
od
return T

AdversaryHelper(X, i, σ, ON, k, b)
T, σ' := ∅, ∅
Y := X
repeat
  if i = 1 then
     Let f be an arbitrary file in Y
     count := 0
     repeat
        σ' := σ' ∘ ⟨f⟩
        S := config(ON, σ ∘ requests(T) ∘ σ', bk)
        count := count + 1
     until f is not present in S or count ≥ 8
     T := T ∘ trace(σ', 0, {f})
     σ' := ∅
  else
     X' := arbitrary subset of num(k, i − 1) files in Y
     T' := AdversaryHelper(X', i − 1, σ ∘ requests(T), ON, k, b)
     T := T ∘ T'
  fi
  reassign Y as follows: for any file f, f is in Y if and only if (1) f is in X
     (2) f is not present in config(ON, σ ∘ requests(T), bk);
     (3) cost_f(T) < (8 · e_i − 8 · e_{i−1}); and
     (4) aggressive(ON, requests(T), bk, f, i) does not hold
until (i = 1 and |Y| = ∅) or (|Y| < num(k, i − 1))
if |{f ∈ X | aggressive(ON, requests(T), bk, f, i)}| < 2b · num(k, i − 1) then
   T := trace(requests(T), i, X)
fi
return T
```

Fig. 2. The adversarial request generating algorithm for the compression caching problem with varying encode and uniform decode costs. Here, N is the number of requests to be generated.

Theorem 2. *Any deterministic online algorithm with an arbitrary factor capacity blowup is $\Omega(m)$-competitive for any uniform-decode instance I of the compression caching problem, where $m = numindex(I)$.*

4.2 An Upper Bound

In this section we present an online algorithm that is $O(m)$-competitive with $O(m)$ factor capacity blowup for any uniform-decode instance I of the compression caching problem, where $m = numindex(I)$. As in Section 3, by preprocessing, we can arrange the encodings of files in decreasing sizes and increasing encode costs; that is, after preprocessing, for any file f and any index $i < m-1$, $size(i+1, f) \leq \frac{1}{r} size(i, f)$, and $encode(i+1, f) \geq q \cdot encode(i, f)$, where $r = 1+\epsilon$, $q = 1 + \epsilon'$, $\epsilon > 0$, and $\epsilon' > 0$.

Algorithm ON divides its cache into m blocks. For any index i, block i keeps only the i-encodings of files. For any integer k and index i, let $num(k, i)$ be the maximum number of i-encodings of files that can fit in any block of size k.

Roughly speaking, ON works as follows. For any index i, ON adds the i-encoding of a file f after the miss penalties incurred by ON on f sum to at least $encode(i, f)$. We use a standard marking algorithm as an eviction procedure for each block. The complete description of ON is presented in [8, Figure 3.3].

See [8, Section 3.4.2.2] for the proof of the following theorem.

Theorem 3. *For any uniform-decode instance I of the compression caching problem, there exists an online algorithm that is is $O(m)$-competitive with $O(m)$ factor capacity blowup, where $m = numindex(I)$.*

5 Uniform Encode Costs and Varying Decode Costs

We say that an instance $I(\sigma, m, k)$ of the compression caching problem is a uniform-encode instance if any file in σ satisfies the following properties. First, we consider that the encode costs of all the nontrivial encodings of any file f in σ are the same; for any index $i > 0$, we abbreviate $encode(i, f)$ to $encode$. Second, we consider that for any index i, any file f and f' in σ, $size(i, f) = size(i, f')$, $p(f) = p(f')$, and $decode(i, f) = decode(i, f')$. For the sake of brevity, for any file f in σ, we write $decode(i, f)$ as $decode(i)$.

In this section we show that any deterministic online algorithm (even with an arbitrary factor capacity blowup) for any uniform-encode instance of the compression caching problem is $\Omega(\log m)$-competitive, where m is the number of possible encodings for each file. Further, we present an online algorithm for this problem that is $O(m)$-competitive with $O(m)$ factor capacity blowup.

5.1 The Lower Bound

In this section, we show that any deterministic online algorithm (even with an arbitrary capacity blowup) for any uniform-encode instance of the compression caching problem is $\Omega(\log m)$-competitive.

For any given online algorithm ON with a capacity blowup b, we construct a uniform-encode instance of the compression caching problem. For any file f and index $i < m - 1$, we consider that $size(i+1, f) \leq \frac{1}{r} \cdot size(i, f)$, where $r > b$. For any file f and index i such that $0 < i < m - 1$, we consider that $decode(i+1, f) \geq decode(i, f) \cdot \log m$. We also set the miss penalty $p(f)$ to be *encode*, and $encode \geq decode(m-1, f) \cdot \log m$.

Adversarial request generating algorithm. Our adversarial request generating algorithm ADV takes ON as input, and generates a request sequence σ and an offline algorithm OFF such that ON incurs at least $\log m$ times the cost incurred by OFF in executing σ. For any file f, ADV maintains two indices denoted $w_u(f)$ and $w_\ell(f)$; initially, $w_u(f) = m$ and $w_\ell(f) = 0$. The complete description of ADV is presented in [8, Figure 3.5].

Roughly, ADV operates as follows. Algorithm ADV forces ON to do a search over the encodings of a file to find the encoding that OFF has chosen for that file. Before any request is generated, ADV ensures that for any f, there is no i-encoding of f in ON's cache such that $w_\ell(f) \leq i < w_u(f)$. On a request for any file f, if ON adds the i-encoding of f such that $w_\ell(f) \leq i < w_u(f)$, then ADV readjusts $w_\ell(f)$ and $w_u(f)$ to ensure that the above condition is satisfied. If ON does not keep the i-encoding of f such that $i < w_u(f)$, then ADV continues to generate requests for f. Finally, when $w_u(f) = w_\ell(f)$, OFF claims that OFF has kept the i-encoding of f, where $i = w_\ell(f)$, throughout this process, and has executed the requests for f. Then, ADV resets the variables $w_u(f)$ and $w_\ell(f)$ to m and 0, respectively, and OFF deletes the encoding of f from its cache. In this process, OFF incurs encoding cost of adding only one encoding of file f. On the other hand ON incurs much higher cost than OFF because of adding multiple encodings of f. See [8, Section 3.5.1.2] for the proof of the following theorem.

Theorem 4. *Any deterministic online algorithm with an arbitrary factor capacity blowup is $\Omega(\log m)$-competitive for any uniform-encode instance of the compression caching problem.*

5.2 An Upper Bound

In this section we present an $O(m)$-competitive online algorithm ON with $O(m)$ factor capacity blowup for any uniform-encode instance I of the compression caching problem, where $m = numindex(I)$.

As in Section 3.1, by preprocessing, we can arrange the encodings of the files in such a way that sizes are decreasing and decode costs are increasing. In other words, after preprocessing, for any file f and index $i < m - 1$, $size(i+1, f) < size(i, f)$, and $decode(i+1, f) > decode(i, f)$. Recall that for any file f, $decode(m-1, f) < p(f)$.

For any uniform-encode instance $I = (\sigma, m, k)$, the online algorithm ON is given a $2bm$ factor capacity blowup, where b is at least $1 + \epsilon$ for some constant $\epsilon > 0$. We divide ON's cache into $2m$ blocks, denoted *i-left* and *i-right*, $0 \leq i < m$, such that the capacity of each block is bk. For any index i, *i-left* keeps only the i-encodings of files, and i-*right* keeps only the 0-encodings of files. For any file f and index i, we maintain an associated value $charge(f, i)$. Roughly, whenever the cost incurred in miss penalties or decode costs on a file f exceeds $encode$, then ON adds an encoding of the file that is cheaper in terms of the access cost than the current encoding (if any) of f. The complete description of algorithm ON is given in [8, Figure 3.6].

See [8, Section 3.5.2.2] for the proof of the following theorem.

Theorem 5. *For any uniform-encode instance I of the compression caching problem, there exists an online algorithm that is $O(m)$-competitive with $O(m)$ factor capacity blowup, where $m = numindex(I)$.*

References

1. Abali, B., Banikazemi, M., Shen, X., Franke, H., Poff, D.E., Smith, T.B.: Hardware compressed main memory: Operating system support and performance evaluation. IEEE Transactions on Computers 50, 1219–1233 (2001)
2. Alameldeen, A.R., Wood, D.A.: Adaptive cache compression for high-performance processors. In: Proceedings of the 31st Annual International Symposium on Computer Architecture, June 2004, pp. 212–223 (2004)
3. Awerbuch, B., Bartal, Y., Fiat, A.: Distributed paging for general networks. Journal of Algorithms 28, 67–104 (1998)
4. Borodin, A., El-Yaniv, R.: Online Computation and Competitive Analysis. Cambridge University Press, Cambridge (1998)
5. Cao, P., Irani, S.: Cost-aware WWW proxy caching algorithms. In: Proceedings of the 1st Usenix Symposium on Internet Technologies and Systems, December 1997, pp. 193–206 (1997)
6. Hallnor, E.G., Reinhardt, S.K.: A unified compressed memory hierarchy. In: Proceedings of the 11th International Symposium on High-Performance Computer Architecture, February 2005, pp. 201–212 (2005)
7. Sleator, D.D., Tarjan, R.E.: Amortized efficiency of list update and paging rules. Communications of the ACM 28, 202–208 (1985)
8. Tiwari, M.: Algorithms for distributed caching and aggregation (2007), `http://www.cs.utexas.edu/users/plaxton/pubs/dissertations/mitul.pdf`
9. Young, N.E.: On-line file caching. Algorithmica 33, 371–383 (2002)

On Trade-Offs in External-Memory Diameter-Approximation

Ulrich Meyer*

Institute for Computer Science,
J.W. Goethe University,
60325 Frankfurt/Main, Germany
umeyer@cs.uni-frankfurt.de
http://www.uli-meyer.de

Abstract. Computing diameters of huge graphs is a key challenge in complex network analysis. However, since exact diameter computation is computationally too costly, one typically relies on approximations. In fact, already a single BFS run rooted at an arbitrary vertex yields a factor two approximation. Unfortunately, in external-memory, even a simple graph traversal like BFS may cause an unacceptable amount of I/O-operations. Therefore, we investigate alternative approaches with worst-case guarantees on both I/O-complexity and approximation factor.

1 Introduction

We concentrate on connected undirected unweighted sparse graphs $G = (V, E)$ of n nodes and $m = O(n)$ edges. Let $d(u, v)$ be the distance between vertices u and v in G, i.e., the number of edges on a shortest path between u and v. The *diameter* of G is given by $\mathcal{D} = \max_{u,v} d(u, v)$.

Breadth first search (BFS) is a fundamental graph traversal strategy. It decomposes the input graph into at most n levels where level i comprises all nodes that can be reached from a designated source s via a path of i edges, but cannot be reached using less than i edges.

In principle, the diameter $\mathcal{D}$ can be easily computed using n BFS runs (one for each graph vertex as a root) and keeping track of the furthest level found in this process. Unfortunately, for huge sparse graphs with millions or even billions of vertices, the resulting computational complexity of $\Theta(n \cdot m) = \Theta(n^2)$ is prohibitive. Hence, it is common practice to restrict to approximations. Already the maximum depth obtained in a single BFS-run rooted at an arbitrary vertex $v \in V$ yields a lower bound $\mathcal{D}_{\mathrm{BFS}(v)}$ satisfying $\mathcal{D}_{\mathrm{BFS}(v)} \leq \mathcal{D} \leq 2 \cdot \mathcal{D}_{\mathrm{BFS}(v)}$. The approximation can be improved by performing k BFS explorations from k carefully chosen starting points [6]: in that case the additive error drops to $O(n/k)$ at the cost of increased running time.

* Partially supported by the DFG grant ME 3250/1-1, and by MADALGO - Center for Massive Data Algorithmics, a Center of the Danish National Research Foundation.

J. Gudmundsson (Ed.): SWAT 2008, LNCS 5124, pp. 426–436, 2008.

All these approaches assume that the input graphs fit into main memory. If this assumption is violated then already one single BFS exploration may take too long[1].

1.1 Computation Model

We consider the commonly accepted external-memory (EM) model of Aggarwal and Vitter [2]. It assumes a two level memory hierarchy with faster internal memory having a capacity to store M vertices/edges. In an I/O operation, one block of data, which can store B vertices/edges is transferred between disk and internal memory. The measure of performance of an algorithm is the number of I/Os it performs. The number of I/Os needed to read N contiguous items from disk is $\text{scan}(N) = \Theta(N/B)$. The number of I/Os required to sort N items is $\text{sort}(N) = \Theta((N/B)\log_{M/B}(N/B))$. For all realistic values of N, B, and M, $\text{scan}(N) < \text{sort}(N) \ll N$.

1.2 Previous Results

There has been a significant number of publications on external-memory graph algorithms; see [13,16] for recent overviews. Exact computation of the diameter on unweighted undirected graphs (via All-Pairs Shortest-Paths, APSP) has been addressed in [4,9]: both approaches require $\Theta(n \cdot \text{sort}(n))$ I/Os for sparse graphs. Taking into account that current machines easily feature several gigabytes of RAM, in the external-memory setting where $n > M \gg B$, an algorithm spending $\Theta(n \cdot n/B) = \Omega(n^2/B)$ I/Os is practically useless.

Chowdhury and Ramachandran [9] also gave an algorithm for computing approximate all-pairs shortest-paths with additive error. However, their approach only takes less I/O than exact EM APSP when $m \geq n \log n$, which is not the sparse graph case we are interested in. Of course, a constant multiplicative factor approximation of the diameter can be obtained using EM BFS. Still, this takes $\Omega(n/\sqrt{B})$ I/Os in the worst-case [12]. We are not aware of any faster EM diameter approximation algorithm for general sparse graphs, even if we are willing to accept larger worst-case approximation errors, for example a multiplicative error of $O(B)$.

1.3 Results and Ideas in a Nutshell

We provide the first non-trivial results on approximate diameter computation for sparse graphs in external-memory with I/O complexity better than that of BFS.

While EM graph algorithms are usually hard on general sparse graphs they tend to be easy on trees. Therefore, it would be tempting to extract some kind of spanning tree T from the connected input graph G using just $O(\text{sort}(n))$ I/Os [1]

[1] Even if k BFS explorations could be afforded then the approach in [6] would still have to be modified in order to find the k starting points more efficiently, e.g., by selecting each $2 \cdot n/k$-th vertex on an Euler tour around a spanning tree of the graph.

and then derive the diameter $\mathcal{D}_G$ of G from $\mathcal{D}_T$, the diameter of T, spending another $O(\text{sort}(n))$ I/Os. In fact, a recent experimental paper for internal-memory diameter estimation [11] proposes a heuristic along these lines. Unfortunately, since T is not necessarily a BFS tree the ratio $\mathcal{D}_T/\mathcal{D}_G$ may be as high as $\Omega(n)$ in the worst case. Hence, drawing conclusions from $\mathcal{D}_T$ is potentially dangerous.

Therefore, our first idea (Section 3) is as follows: instead of trying to guess $\mathcal{D}_G$ directly from $\mathcal{D}_T$ we only use T to contract G by Euler-Tour techniques in some controlled way resulting in a graph G' with n/B vertices and $O(n)$ edges. Then a subsequent BFS run on an arbitrary vertex of G' only takes $O(\text{sort}(n))$ I/Os. If it identifies $\mathcal{L} \geq 1$ BFS levels then $\mathcal{L} \leq \mathcal{D}_{G'} \leq 2 \cdot \mathcal{L}$ and by our controlled reduction we can conclude that $\mathcal{L} \leq \mathcal{D}_G \leq 2 \cdot B \cdot \mathcal{L}$.

If we choose to contract G to only n/k vertices with $1 < k \leq B$ then the approximation error is reduced to a multiplicative factor of $O(k)$. However, the respective I/O bound becomes $O(n/\sqrt{k \cdot B} + \text{sort}(n))$ I/Os. For $k \ll B$ the first term usually dominates. For example, if only about $O(n/B^{2/3})$ I/Os can be tolerated, multiplicative errors of $O(B^{1/3})$ may occur.

Using a different randomized contraction scheme (Section 4) with subsequent EM Single-Source Shortest-Paths (SSSP) computation (instead of BFS) we achieve another interesting trade-off: $O(n/\sqrt{k \cdot B/\log k} + k \cdot \text{sort}(n))$ I/Os and expected multiplicative errors of only $O(\sqrt{k})$ instead of $O(k)$. This time, if we strive for about $O(n/B^{2/3})$ I/Os, we expect multiplicative errors of $O(B^{1/6} \cdot \sqrt{\log B})$. With block sizes in EM implementations steadily increasing over the last years (currently $B \simeq 10^6$ for fast hard disks), $B^{1/6} \cdot \sqrt{\log_2 B}$ has already dropped below $B^{1/3}$, thus making the second approach not only theoretically interesting.

1.4 Organization of the Paper

In Section 2 we will first review known BFS algorithms for undirected graphs. Then in Sections 3 and 4 we appropriately modify external-memory graph clustering approaches to support efficient graph contraction and investigate the effect on the respective diameters. Finally, in Section 5 we give some concluding remarks.

2 Review of BFS Algorithms

Internal-Memory. BFS is well-understood in the RAM model. There exists a simple linear time algorithm [10] (hereafter referred as IM_BFS) for the BFS traversal in a graph. IM_BFS keeps a set of appropriate candidate nodes for the next vertex to be visited in a FIFO queue Q. Furthermore, in order to identify the unvisited neighbors of a node from its adjacency list, it marks the nodes as either visited or unvisited in a global array. When run on external-memory IM_BFS requires $\Theta(n+m)$ I/Os: (1) remembering visited nodes needs $\Theta(m)$ I/Os in the worst case and (2) unstructured indexed accesses to adjacency lists may result in $\Theta(n + m/B)$ I/Os.

EM-BFS for dense undirected graphs. The algorithm by Munagala and Ranade [15] (referred as MR_BFS) ignores the second problem but addresses the first one by exploiting the fact that the neighbors of a node in BFS level $t-1$ are all in BFS levels $t-2$, $t-1$ or t. Let $L(t)$ denote the set of nodes in BFS level t, and let $A(t)$ be the multi-set of neighbors of nodes in $L(t-1)$. Given $L(t-1)$ and $L(t-2)$, MR_BFS builds $L(t)$ as follows: Firstly, $A(t)$ is created by $|L(t-1)|$ random accesses to get hold of the adjacency lists of all nodes in $L(t-1)$. Thereafter, duplicates are removed from $A(t)$ to get a sorted set $A'(t)$. This is done by sorting $A(t)$ according to node indices, followed by a scan and compaction phase. The set $L(t) := A'(t) \setminus \{L(t-1) \cup L(t-2)\}$ is computed by scanning "in parallel" the sorted sets of $A'(t), L(t-1)$, and $L(t-2)$ to filter out the nodes already present in $L(t-1)$ or $L(t-2)$. The resulting worst-case I/O-bound is $O\left(\sum_t L(t) + \sum_t \text{sort}(A(t))\right) = O\left(n + \text{sort}(n+m)\right)$.

Somewhat better bounds can be shown for graphs with very small diameter $\mathcal{D}$: due to the intermediate sorting steps accessing the adjacency-lists for all the nodes in a BFS level actually takes at most $O(\text{scan}(n+m))$ I/Os. Thus, MR_BFS requires at most $O(\min\{n, \mathcal{D} \cdot \text{scan}(n+m)\} + \text{sort}(n+m))$ I/Os.

The algorithm outputs a BFS-level decomposition of the vertices, which can be easily transformed into a BFS tree using $O(\text{sort}(n+m))$ I/Os [7].

EM-BFS for sparse undirected graphs. Mehlhorn and Meyer suggested another approach [12] (MM_BFS) which involves a preprocessing phase to restructure the adjacency lists of the graph representation. It groups the vertices of the input graph into disjoint clusters of small diameter in G and stores the adjacency lists of the nodes in a cluster contiguously on the disk. Thereafter, an appropriately modified version of MR_BFS is run. MM_BFS exploits the fact that whenever the first node of a cluster is visited then the remaining nodes of this cluster will be reached soon after. By spending only one random access (and possibly, some sequential accesses depending on cluster size) in order to load the whole cluster and then keeping the cluster data in some efficiently accessible data structure (pool) until it is all processed, on sparse graphs the total amount of I/O can be reduced by a factor of up to $\sqrt{B}$: the neighboring nodes of a BFS level can be computed simply by scanning the pool and not the whole graph. Though some edges may be scanned more often in the pool, unstructured I/O in order to fetch adjacency lists is considerably reduced, thereby reducing the total number of I/Os.

The concrete I/O bounds for MM_BFS depend on the kind of preprocessing. Mehlhorn and Meyer [12] proposed two variants (MM_BFS_R based on parallel cluster growing and MM_BFS_D based on Euler-tours), which we will review and modify toward our goal to compute approximate diameters in the subsequent sections.

3 Simple Euler Tour Approach

3.1 Traditional Preprocessing within MM_BFS_D

The MM_BFS_D variant first builds a spanning tree T_s for the connected component of G that contains the source node. Arge et al. [3] show an upper bound of $O((1 + \log\log(B \cdot n/m)) \cdot \mathrm{sort}(n+m))$ I/Os for computing such a spanning tree deterministically. Using randomization the spanning tree I/O bound even drops to $O(\mathrm{sort}(n+m))$ with high probability [1].

Each undirected edge of T_s is then replaced by two oppositely directed edges. Note that a bi-directed tree always has at least one Euler tour. In order to construct the Euler tour around this bi-directed tree, each node chooses a cyclic order [5] of its neighbors. The successor of an incoming edge is defined to be the outgoing edge to the next node in the cyclic order. The tour is then broken at the source node and the elements of the resulting list are then stored in consecutive order using an external memory list-ranking algorithm; Chiang et al. [8] showed how to do this in sorting complexity.

Thereafter, we chop the Euler tour into *chunks* of $\mu = \max\{1, \sqrt{\frac{n \cdot B}{n+m}}\}$ nodes (*chunk size*) and remove duplicates such that each node only remains in the first chunk it originally occurs; again this requires a couple of sorting steps. The adjacency lists are then re-ordered based on the position of their corresponding nodes in the chopped duplicate-free Euler tour: all adjacency lists for nodes in the same chunks form a cluster and the distance in G between any two vertices whose adjacency-lists belong to the same cluster is bounded by μ. On the other hand, there are at most $O(n/\mu)$ non-empty clusters, which MM_BFS_D has to access using random I/Os. With $\mu = \max\{1, \sqrt{\frac{n \cdot B}{n+m}}\}$ the resulting I/O bound for MM_BFS_D is $O\left(\sqrt{\frac{n\cdot(n+m)}{B}} + \mathrm{sort}(n+m) + \mathrm{ST}(n,m)\right)$ where $\mathrm{ST}(n,m)$ denotes the I/O bound for the spanning tree computation. Thus, for sparse graphs, BFS can be solved using $O(n/\sqrt{B} + \mathrm{sort}(n))$ I/Os when applying the randomized spanning tree subroutine.

3.2 Extracting a Compacted Graph

In principle the preprocessing method of MM_BFS_D described above can be run for any chunk size $1 < k < 2 \cdot n$. Therefore, we will use it to produce a compressed graph G'_k of G in $O(\mathrm{sort}(n+m))$ I/Os: The vertices of G'_k are the non-empty clusters for G using chunk size k. As for the edges of G'_k, let $C(u)$ denote the cluster to which a vertex $u \in G$ has been mapped in the preprocessing; if $\{u, v\}$ with $u \neq v$ is an edge in G and $C(u) \neq C(v)$, then $\{C(u), C(v)\}$ will be an edge in G'_k. This may create parallel edges between $C(u)$ and $C(v)$, which can easily be eliminated in sorting complexity.

Obviously, G'_k has $n' = O(n/k)$ nodes and $m' = O(m)$ edges. Thus, a subsequent BFS run on G'_k using MM_BFS_D with randomized spanning tree computation requires only $O(\sqrt{n' \cdot (n'+m')/B} + \mathrm{sort}(n'+m'))$ I/Os, which is $O(\sqrt{n/k \cdot n/B} + \mathrm{sort}(n)) = O(n/\sqrt{k \cdot B} + \mathrm{sort}(n))$ I/Os for sparse graphs G. Note that for $k = B$, both producing G'_B and running BFS on it can be done in $O(\mathrm{sort}(n))$ I/Os if G is sparse.

3.3 Approximation Bound

Now we have to convince ourselves that a BFS run on G'_k yields a reasonable bound on the diameter of G.

Lemma 1. *Let $d_G(u,v)$ be the length of a shortest path between two nodes u and v in an unweighted and undirected graph G. Then it holds that $\lfloor d_G(u,v)/k \rfloor \leq d_{G'_k}(C(u),C(v)) \leq d_G(u,v)$.*

Proof. Let $P = \langle u = w_1, \ldots, w_l = v \rangle$ be a shortest path in G. Hence, $w_i \neq w_j$ for all $1 \leq i < j \leq l$. Since by construction at most k different vertices of G have been mapped to each cluster $C(\cdot)$, each node on a shortest path $P'_k = \langle C(u), \ldots, C(v) \rangle$ in G'_k represents at most k original vertices of P. Furthermore, any two vertices v_i and v_j of G mapped to the same cluster satisfy $d(v_i, v_j) \leq k - 1$, making it impossible to hurdle original distances larger or equal to k within any node $C(\cdot)$. Thus, $\lfloor d_G(u,v)/k \rfloor \leq d_{G'_k}(C(u),C(v))$.

On the other hand a shortest path between $C(u)$ and $C(v)$ in the unweighted graph G'_k cannot have more vertices than P: in the worst-case each vertex $w_i \in P$ is mapped to a different cluster. However, then $C(w_i)$ and $C(w_{i+1})$ are connected by an edge in G'_k, implying $d_{G'_k}(C(u),C(v)) \leq d_G(u,v)$. If $C(w_i) = C(w_j)$ for some $i < j$ then the sub-path $\langle C(i), \ldots, C(j) \rangle$ represents a detour on the shortest path between $C(u)$ and $C(v)$. Thus, $d_{G'_k}(C(u),C(v)) \leq d_G(u,v)$ still holds. □

Hence, if $\mathcal{D}_G$ denotes the diameter of G then, by Lemma 1, G'_k features a shortest path of length l where $\lfloor \mathcal{D}_G/k \rfloor \leq l \leq \mathcal{D}_G$ and no shortest path in G'_k is longer than $\mathcal{D}_G$. Therefore, the maximum depth $\mathcal{D}_{\mathrm{BFS}(v)}$ obtained in a single BFS-run rooted at an arbitrary vertex $v \in G'_k$ is bounded by $\lfloor \mathcal{D}_G/(2 \cdot k) \rfloor \leq \mathcal{D}_{\mathrm{BFS}(v)} \leq \mathcal{D}_G$ allowing us to infer $\mathcal{D}_G$ up to a multiplicative factor of $O(k)$.

As already mentioned earlier producing G'_k out of G takes $O(\mathrm{sort}(n+m))$ I/Os and running BFS on G'_k using MM_BFS_D requires

$$O\left(\sqrt{\frac{n/k \cdot (n/k + m)}{B}} + \mathrm{sort}(n/k + m) + \mathrm{ST}(n/k, m)\right) \text{ I/Os.}$$

Combining the bounds and using $m = O(n)$ we obtain:

Theorem 1. *For sparse unweighted and undirected graphs a multiplicative $O(k)$-approximation of the diameter can be computed using $O(n/\sqrt{k \cdot B} + \mathrm{sort}(n))$ I/Os.*

4 Parallel Cluster Growing Approach

In this chapter we provide another trade-off between approximation guarantee and I/O bound. While our first approach transformed the input graph G into a smaller *unweighted* graph G'_k, now we shall produce a *weighted* graph G''_k of about $O(n/k)$ vertices. Unfortunately, the larger the parameter k the more I/Os it takes to generate G''_k. Furthermore, subsequently, we need a single-source shortest-paths computation on G''_k instead of an easier and somewhat faster BFS computation on G'_k. In return, for sufficiently small k, we obtain better approximation bounds.

4.1 Modified Preprocessing

The weighted graph G''_k will be obtained by a *parallel cluster growing* method partially based on the randomized clustering in [12]. The main idea inherited from there is to choose *master nodes* independently and uniformly at random with probability $1/k$ and running a local BFS from all master nodes "in parallel": in each round, each master node c_i tries to capture all unvisited neighbors of its current cluster C_i; this is done by first sorting the nodes of the active fringes of all C_i (the nodes that have been captured in the previous round) and then scanning the adjacency-lists representation of the yet unexplored graph. If several master nodes want to include a certain node v into their cluster then an arbitrary master node among them succeeds. The selection can be done by sorting and scanning the created set of neighbor nodes.

However, this kind of preprocessing will not suffice for our purposes. The new ingredients in the clustering phase are following ones:

1. After choosing master nodes at random, $O(n/k)$ non-chosen vertices may additionally become masters using a deterministic selection procedure based on an Euler-tour for an arbitrary spanning tree of G. By turning each k-th vertex on the tour into a master (in case it has not yet been selected) we make sure that eventually each non-selected vertex in G is at distance at most $k-1$ to a least one master.
2. While growing the clusters we keep track of the distances $d_C(u,v)$ in G between captured vertices u and their respective masters v.

The graph G''_k is created as follows: its vertices are the clusters of G, that is an expected number of $O(n/k)$ clusters for the randomly chosen master vertices plus at most $O(n/k)$ further clusters for the Euler-tour based extra master nodes. An edge $\{u,v\} \in G$ will result in an edge $\{C(u), C(v)\}$ for G''_k if u and v belong to different clusters $C(u) = C_{u'} \neq C(v) = C_{v'}$ with master nodes u' and v', respectively. The weight of the created edge $\{C_{u'}, C_{v'}\}$ will be $d_C(u,u') + 1 + d_C(v,v')$. Note that this approach may create parallel edges between $C_{u'}$ and $C_{v'}$, potentially having different weights. However, using standard sorting routines only a lightest edge between $C_{u'}$ and $C_{v'}$ will be kept.

Lemma 2. *Creating graph G''_k out of $G(n,m)$ takes $O(k \cdot \mathrm{scan}(n+m) + \mathrm{sort}(n+m) + \mathrm{ST}(n,m))$ I/Os.*

Proof. By construction, each non-master vertex in G is at distance $k-1$ or less to at least one of the $O(n/k)$ master vertices on the Euler-Tour around a spanning tree. The randomly chosen master vertices not on the tour do not spoil this property. Thus, each non-master vertex is captured after at most $k-1$ rounds. Consequently, the resulting edge weights for G''_k range in $\{1, \ldots, 2 \cdot k - 1\}$.

The total amount of data being scanned from the adjacency-lists representation during the "parallel cluster growing" is bounded by

$$X := \mathcal{O}(\sum_{v \in V} k \cdot (1 + \text{degree}(v))) = \mathcal{O}(k \cdot (n+m)).$$

The total number of fringe nodes and neighbor nodes sorted and scanned during the partitioning is at most $Y := \mathcal{O}(n+m)$. Therefore, the cluster growing requires

$$\mathcal{O}(\text{scan}(X) + \text{sort}(Y)) = \mathcal{O}(k \cdot \text{scan}(n+m) + \text{sort}(n+m)) \text{ I/Os}.$$

After that each node knows the (index of the) cluster it belongs to and the distance to its master. With a constant number of sort and scan operations G''_k can be written in the adjacency lists graph format required for the subsequent SSSP computation.

4.2 Improved Approximation Bound

We aim to prove an expected multiplicative error of at most $O(\sqrt{k})$ when inferring the diameter of G from a SSSP run on G''_k. We will distinguish two cases: (a) when the diameter of G, $\mathcal{D}_G$, is at most $2 \cdot \sqrt{k}$, and (b) when $\mathcal{D}_G > 2 \cdot \sqrt{k}$. In the following we will assume that $\sqrt{k}$ is an integer in order to simplify notation.

The case $\mathcal{D}_G \leq 2 \cdot \sqrt{k}$. During the parallel cluster growing, each master node performs a limited BFS traversal in G. However, since each shortest path in the unweighted graph G has at most $2 \cdot \sqrt{k}$ edges, the resulting weights for G''_k range in $\{1, \ldots, \min\{4 \cdot \sqrt{k} - 1, 2 \cdot k - 1\}\}$. We have to consider how the total weight of a shortest path $P = \langle u = w_0, \ldots, w_{\mathcal{D}_G} = v \rangle$ in G compares to the weight of a shortest path in G''_k between $C(u)$ and $C(v)$. Using similar arguments as in the proof of Lemma 1 it turns out that in the worst-case each vertex on P in G is mapped to a different cluster while producing G''_k and furthermore the respective master nodes of these clusters are as far away from P as possible (while still capturing one node of P). Hence, the resulting weight for a shortest path P'' between $C(u)$ and $C(v)$ in G''_k is bounded from above by $4 \cdot \sqrt{k} \cdot \mathcal{D}_G$.

On the other hand, unless $C(u) = C(v)$, the weight of P'' will also not be smaller than that of P. This is because the weights of sub-paths in G running within clusters contribute to the weights of the edges connecting the nodes corresponding to these clusters in G''_k. Thus, in our SSSP computation on G''_k in order to estimate $\mathcal{D}_G$ the only problematic scenario would be if we performed SSSP on an isolated vertex. But since G is connected and we produce $\Omega(n/k) \geq 2$ clusters this cannot happen.

The case $\mathcal{D}_G > 2 \cdot \sqrt{k}$. Like in the previous case we consider a shortest path P in G defining $\mathcal{D}_G$. However, this time we split P into sub-paths P'_i, each of which comprises between $\sqrt{k}$ and $2 \cdot \sqrt{k}$ edges and consider them separately. So, let us fix a concrete sub-path $P' = \langle u = w_0, \ldots, w_{p'} = v \rangle$ of P with $\sqrt{k} \leq p' \leq 2 \cdot \sqrt{k}$ edges. Furthermore, let x be a master node with shortest distance in G to any node on P' and let $w_x \in P'$ be a node that is captured earliest among all nodes on P' during the parallel cluster growing – by either x or another master being equally close. W.l.o.g. let us assume that x is the successful master node. Note that this capturing happens in the $t := d_G(x, w_x)$-th round of the parallel cluster growing and $t \leq k$ because of the Euler-Tour based extra master nodes, which make sure that each non-master node can be captured fast. Also note that then there is a path $P_x = \langle x, \ldots, w_x \rangle$ of length t where all vertices of P_x have been captured by x.

In the absence of other nearby master nodes, after w_x has been captured by x all other nodes of P' would be captured within at most another $p' \leq 2 \cdot \sqrt{k}$ rounds by x as well.

Thus, if another master node $y \neq x$ is to capture a vertex $w_y \neq w_x$ on P', too, then a second path $P_y = \langle y, \ldots, w_y \rangle$ of length between $t = d_G(x, w_x)$ and $t + p'$ is needed and P_y must be node-disjoint from P_x. The number of different master nodes that manage to capture nodes on P' is limited by (a) the number of non-master nodes on P' (at most $p' + 1$) and (b) the number of non-master nodes in the neighborhood of P' to accommodate the paths P_x, P_y, ... (these are all graph nodes at distance less than t from P', let us call this quantity A_t).

Combining (a) and (b) we see that the nodes on P' will be captured by at most $\min\{A_t/t, p'+1\}$ different masters within rounds $t, \ldots, t+p'$. Thus, in the worst case P' in G is replaced by a path in G''_k consisting of $\min\{A_t/t-1, p'\}$ edges each of which having weight at most $2 \cdot (t + p') + 1$. Hence, when switching from G to G''_k, the detour for P' is bounded from above by $O(\min\{A_t/t, p'\} \cdot (t+p'))$. It is an easy exercise to verify the inequality $\min\{A_t/t, p'\} \cdot (t+p') \leq A_t + (p')^2 \leq A_t + 4 \cdot k$.

Keeping in mind that each vertex of G has been chosen to be a master independently with uniform probability $1/k$, we can conclude that $\mathrm{E}[A_t] \leq k$: we consider larger and larger neighborhoods around P' until we find the first level with at least one master: vertex x at distance $t = d_G(x, w_x)$. The expected number of vertices we have to check till then is given by $1/(1/k) = k$ but A_t may contain even less vertices since it only accounts for distance levels $1, \ldots, t-1$ around P' in G. Hence, the expected detour for sub-path P' is bounded from above by $O(k)$.

Now the final step is to use linearity of expectation in order to combine the partial results for the sub-paths of the diameter-defining path P in G and thus eventually obtain an upper bound on the expected approximation ratio for the case $\mathcal{D}_G \geq 2 \cdot \sqrt{k}$: we have to consider $\Theta(\mathcal{D}_G/\sqrt{k})$ sub-paths of length $\Theta(\sqrt{k})$ in G each. By our discussion above each such sub-path accounts for an expected detour of $O(k)$ in G''_k, resulting in a total detour of $O(\sqrt{k} \cdot \mathcal{D}_G)$ for the diameter in G''_k.

I/O complexity. In Lemma 2 we have already investigated the I/O complexity to produce G''_k out of G. Now we still have to consider the I/O costs to run SSSP on G''_k: we use the EM SSSP algorithm of Meyer and Zeh [14]. On undirected graphs with n nodes and m edges it takes $O(\sqrt{(n \cdot m/B) \log_2(c_{\max}/c_{\min})} + \mathrm{sort}(n+m) + \mathrm{MST}(n,m))$ I/Os where $c_{\max}$ and $c_{\min}$ are the minimal and maximal edge weights in the graph, respectively. $\mathrm{MST}(n,m)$ is the I/O-complexity to solve Minimum Spanning Trees, the currently fastest approach [1] is randomized and requires $O(\text{sort}(n+m))$ I/Os. As for G''_k we expect $n'' = O(n/k)$, $m'' = O(m)$, $c_{\min} = 1$, and $c_{\max} = O(k)$. Additionally, since we concentrate on sparse graphs G, $m = O(n)$ and $m'' = O(n)$. Thus, we can solve SSSP on G''_k using $O(n \cdot \sqrt{\log(k)/(k \cdot B)} + \text{sort}(n))$ I/Os. Together with the $O(k \cdot \text{scan}(n) + \text{sort}(n)))$ I/Os to generate G''_k, the total I/O requirements are $O(n \cdot \sqrt{\log(k)/(k \cdot B)} + k \cdot \text{scan}(n) + \text{sort}(n))$ I/Os.

Theorem 2. *An expected $O(\sqrt{k})$-approximation for the diameter of a sparse undirected and unweighted graph with n nodes and $m = O(n)$ edges can be obtained using $O(n \cdot \sqrt{\log(k)/(k \cdot B)} + k \cdot \text{scan}(n) + \text{sort}(n))$ I/Os.*

5 Conclusions

We have considered two approaches for fast external-memory approximation of diameters in sparse unweighted and undirected graphs. Both are parameterized and allow nice trade-offs between approximation guarantees and I/O requirements. Still, it is an open problem whether these trade-offs can be improved. For example, it would be tempting to try a multi-stage parallel cluster growing approach. Unfortunately, already after the first stage the shrunken graph becomes weighted and from then on the shrinking procedure will have to deal with weights, too, thus making efficient and distance-aware node reduction much more complicated.

References

1. Abello, J., Buchsbaum, A., Westbrook, J.: A functional approach to external graph algorithms. Algorithmica 32(3), 437–458 (2002)
2. Aggarwal, A., Vitter, J.S.: The input/output complexity of sorting and related problems. Communications of the ACM 31(9), 1116–1127 (1988)
3. Arge, L., Brodal, G., Toma, L.: On external-memory MST, SSSP and multi-way planar graph separation. In: Halldórsson, M.M. (ed.) SWAT 2000. LNCS, vol. 1851, pp. 433–447. Springer, Heidelberg (2000)
4. Arge, L., Meyer, U., Toma, L.: External memory algorithms for diameter and all-pairs shortest-paths on sparse graphs. In: Díaz, J., Karhumäki, J., Lepistö, A., Sannella, D. (eds.) ICALP 2004. LNCS, vol. 3142, pp. 146–157. Springer, Heidelberg (2004)
5. Atallah, M., Vishkin, U.: Finding Euler tours in parallel. Journal of Computer and System Sciences 29(30), 330–337 (1984)
6. Boitmanis, K., Freivalds, K., Ledins, P., Opmanis, R.: Fast and simple approximation of the diameter and radius of a graph. In: Àlvarez, C., Serna, M.J. (eds.) WEA 2006. LNCS, vol. 4007, pp. 98–108. Springer, Heidelberg (2006)
7. Buchsbaum, A., Goldwasser, M., Venkatasubramanian, S., Westbrook, J.: On external memory graph traversal. In: Proc. 11th Ann. Symposium on Discrete Algorithms (SODA), pp. 859–860. ACM-SIAM (2000)
8. Chiang, Y.J., Goodrich, M.T., Grove, E.F., Tamasia, R., Vengroff, D.E., Vitter, J.S.: External memory graph algorithms. In: Proc. 6th Ann.Symposium on Discrete Algorithms (SODA), pp. 139–149. ACM-SIAM (1995)

9. Chowdury, R., Ramachandran, V.: External-memory exact and approximate all-pairs shortest-paths in undirected graphs. In: Proc. 16th Ann. Symposium on Discrete Algorithms (SODA), pp. 735–744. ACM-SIAM (2005)
10. Cormen, T.H., Leiserson, C., Rivest, R.: Introduction to Algorithms. McGraw-Hill, New York (1990)
11. Magnien, C., Latapy, M., Habib, M.: Fast computation of empirically tight bounds for the diameter of massive graphs (2007), `http://www-rp.lip6.fr/~latapy/Diameter/`
12. Mehlhorn, K., Meyer, U.: External-memory breadth-first search with sublinear I/O. In: Möhring, R.H., Raman, R. (eds.) ESA 2002. LNCS, vol. 2461, pp. 723–735. Springer, Heidelberg (2002)
13. Meyer, U., Sanders, P., Sibeyn, J. (eds.): Algorithms for Memory Hierarchies. LNCS, vol. 2625. Springer, Heidelberg (2003)
14. Meyer, U., Zeh, N.: I/O-efficient undirected shortest paths. In: Di Battista, G., Zwick, U. (eds.) ESA 2003. LNCS, vol. 2832, pp. 434–445. Springer, Heidelberg (2003)
15. Munagala, K., Ranade, A.: I/O-complexity of graph algorithms. In: Proc. 10th Ann. Symposium on Discrete Algorithms (SODA), pp. 687–694. ACM-SIAM (1999)
16. Vitter, J.S.: External memory algorithms and data structures: Dealing with massive data. ACM computing Surveys 33, 209–271 (2001), `http://www.cs.purdue.edu/homes/~jsv/Papers/Vit.IO_survey.pdf`

Author Index

Printing: Mercedes-Druck, Berlin
Binding: Stein+Lehmann, Berlin

Lecture Notes in Computer Science

Sublibrary 1: Theoretical Computer Science and General Issues

For information about Vols. 1–4771
please contact your bookseller or Springer

Vol. 5130: J.v.z. Gathen, J.L. Imaña, Ç.K. Koç (Eds.), Arithmetic of Finite Fields. X, 209 pages. 2008.

Vol. 5124: J. Gudmundsson (Ed.), Algorithm Theory – SWAT 2008. XIII, 438 pages. 2008.

Vol. 5103: M. Bubak, G.D. van Albada, J. Dongarra, P.M.A. Sloot (Eds.), Computational Science – ICCS 2008, Part III. XXVIII, 758 pages. 2008.

Vol. 5102: M. Bubak, G.D. van Albada, J. Dongarra, P.M.A. Sloot (Eds.), Computational Science – ICCS 2008, Part II. XXVIII, 752 pages. 2008.

Vol. 5101: M. Bubak, G.D. van Albada, J. Dongarra, P.M.A. Sloot (Eds.), Computational Science – ICCS 2008, Part I. XLVI, 1058 pages. 2008.

Vol. 5092: X. Hu, J. Wang (Eds.), Computing and Combinatorics. XIV, 680 pages. 2008.

Vol. 5090: R.T. Mittermeir, M.M. Syslo (Eds.), Informatics Education - Supporting Computational Thinking. XV, 357 pages. 2008.

Vol. 5073: O. Gervasi, B. Murgante, A. Laganà, D. Taniar, Y. Mun, M.L. Gavrilova (Eds.), Computational Science and Its Applications – ICCSA 2008, Part II. XXIX, 1280 pages. 2008.

Vol. 5072: O. Gervasi, B. Murgante, A. Laganà, D. Taniar, Y. Mun, M.L. Gavrilova (Eds.), Computational Science and Its Applications – ICCSA 2008, Part I. XXIX, 1266 pages. 2008.

Vol. 5065: P. Degano, R. De Nicola, J. Meseguer (Eds.), Concurrency, Graphs and Models. XV, 810 pages. 2008.

Vol. 5062: K.M. van Hee, R. Valk (Eds.), Applications and Theory of Petri Nets. XIII, 429 pages. 2008.

Vol. 5059: F.P. Preparata, X. Wu, J. Yin (Eds.), Frontiers in Algorithmics. XI, 350 pages. 2008.

Vol. 5058: A.A. Shvartsman, P. Felber (Eds.), Structural Information and Communication Complexity. X, 307 pages. 2008.

Vol. 5050: J.M. Zurada, G.G. Yen, J. Wang (Eds.), Computational Intelligence: Research Frontiers. XVI, 389 pages. 2008.

Vol. 5038: C.C. McGeoch (Ed.), Experimental Algorithms. X, 363 pages. 2008.

Vol. 5036: S. Wu, L.T. Yang, T.L. Xu (Eds.), Advances in Grid and Pervasive Computing. XV, 518 pages. 2008.

Vol. 5035: A. Lodi, A. Panconesi, G. Rinaldi (Eds.), Integer Programming and Combinatorial Optimization. XI, 477 pages. 2008.

Vol. 5029: P. Ferragina, G.M. Landau (Eds.), Combinatorial Pattern Matching. XIII, 317 pages. 2008.

Vol. 5028: A. Beckmann, C. Dimitracopoulos, B. Löwe (Eds.), Logic and Theory of Algorithms. XIX, 596 pages. 2008.

Vol. 5022: A.G. Bourgeois, S.Q. Zheng (Eds.), Algorithms and Architectures for Parallel Processing. XIII, 336 pages. 2008.

Vol. 5018: M. Grohe, R. Niedermeier (Eds.), Parameterized and Exact Computation. X, 227 pages. 2008.

Vol. 5015: L. Perron, M.A. Trick (Eds.), Integration of AI and OR Techniques in Constraint Programming for Combinatorial Optimization Problems. XII, 394 pages. 2008.

Vol. 5011: A.J. van der Poorten, A. Stein (Eds.), Algorithmic Number Theory. IX, 455 pages. 2008.

Vol. 5010: E.A. Hirsch, A.A. Razborov, A. Semenov, A. Slissenko (Eds.), Computer Science – Theory and Applications. XIII, 411 pages. 2008.

Vol. 5008: A. Gasteratos, M. Vincze, J.K. Tsotsos (Eds.), Computer Vision Systems. XV, 560 pages. 2008.

Vol. 5004: R. Eigenmann, B.R. de Supinski (Eds.), OpenMP in a New Era of Parallelism. X, 191 pages. 2008.

Vol. 5000: O. Grumberg, H. Veith (Eds.), 25 Years of Model Checking. VII, 231 pages. 2008.

Vol. 4996: H. Kleine Büning, X. Zhao (Eds.), Theory and Applications of Satisfiability Testing – SAT 2008. X, 305 pages. 2008.

Vol. 4988: R. Berghammer, B. Möller, G. Struth (Eds.), Relations and Kleene Algebra in Computer Science. X, 397 pages. 2008.

Vol. 4985: M. Ishikawa, K. Doya, H. Miyamoto, T. Yamakawa (Eds.), Neural Information Processing, Part II. XXX, 1091 pages. 2008.

Vol. 4984: M. Ishikawa, K. Doya, H. Miyamoto, T. Yamakawa (Eds.), Neural Information Processing, Part I. XXX, 1147 pages. 2008.

Vol. 4981: M. Egerstedt, B. Mishra (Eds.), Hybrid Systems: Computation and Control. XV, 680 pages. 2008.

Vol. 4978: M. Agrawal, D. Du, Z. Duan, A. Li (Eds.), Theory and Applications of Models of Computation. XII, 598 pages. 2008.

Vol. 4975: F. Chen, B. Jüttler (Eds.), Advances in Geometric Modeling and Processing. XV, 606 pages. 2008.

Vol. 4974: M. Giacobini, A. Brabazon, S. Cagnoni, G.A. Di Caro, R. Drechsler, A. Ekárt, A.I. Esparcia-Alcázar, M. Farooq, A. Fink, J. McCormack, M. O'Neill, J. Romero, F. Rothlauf, G. Squillero, A.Ş. Uyar, S. Yang (Eds.), Applications of Evolutionary Computing. XXV, 701 pages. 2008.

Vol. 4973: E. Marchiori, J.H. Moore (Eds.), Evolutionary Computation, Machine Learning and Data Mining in Bioinformatics. X, 213 pages. 2008.

Vol. 4972: J. van Hemert, C. Cotta (Eds.), Evolutionary Computation in Combinatorial Optimization. XII, 289 pages. 2008.

Vol. 4971: M. O'Neill, L. Vanneschi, S. Gustafson, A.I. Esparcia Alcázar, I. De Falco, A. Della Cioppa, E. Tarantino (Eds.), Genetic Programming. XI, 375 pages. 2008.

Vol. 4967: R. Wyrzykowski, J. Dongarra, K. Karczewski, J. Wasniewski (Eds.), Parallel Processing and Applied Mathematics. XXIII, 1414 pages. 2008.

Vol. 4963: C.R. Ramakrishnan, J. Rehof (Eds.), Tools and Algorithms for the Construction and Analysis of Systems. XVI, 518 pages. 2008.

Vol. 4962: R. Amadio (Ed.), Foundations of Software Science and Computational Structures. XV, 505 pages. 2008.

Vol. 4961: J.L. Fiadeiro, P. Inverardi (Eds.), Fundamental Approaches to Software Engineering. XIII, 430 pages. 2008.

Vol. 4960: S. Drossopoulou (Ed.), Programming Languages and Systems. XIII, 399 pages. 2008.

Vol. 4959: L. Hendren (Ed.), Compiler Construction. XII, 307 pages. 2008.

Vol. 4957: E.S. Laber, C. Bornstein, L.T. Nogueira, L. Faria (Eds.), LATIN 2008: Theoretical Informatics. XVII, 794 pages. 2008.

Vol. 4943: R. Woods, K. Compton, C. Bouganis, P.C. Diniz (Eds.), Reconfigurable Computing: Architectures, Tools and Applications. XIV, 344 pages. 2008.

Vol. 4942: E. Frachtenberg, U. Schwiegelshohn (Eds.), Job Scheduling Strategies for Parallel Processing. VII, 189 pages. 2008.

Vol. 4941: M. Miculan, I. Scagnetto, F. Honsell (Eds.), Types for Proofs and Programs. VII, 203 pages. 2008.

Vol. 4935: B. Chapman, W. Zheng, G.R. Gao, M. Sato, E. Ayguadé, D. Wang (Eds.), A Practical Programming Model for the Multi-Core Era. VI, 208 pages. 2008.

Vol. 4934: U. Brinkschulte, T. Ungerer, C. Hochberger, R.G. Spallek (Eds.), Architecture of Computing Systems – ARCS 2008. XI, 287 pages. 2008.

Vol. 4927: C. Kaklamanis, M. Skutella (Eds.), Approximation and Online Algorithms. X, 289 pages. 2008.

Vol. 4926: N. Monmarché, E.-G. Talbi, P. Collet, M. Schoenauer, E. Lutton (Eds.), Artificial Evolution. XIII, 327 pages. 2008.

Vol. 4921: S.-i. Nakano, M.. S. Rahman (Eds.), WALCOM: Algorithms and Computation. XII, 241 pages. 2008.

Vol. 4919: A. Gelbukh (Ed.), Computational Linguistics and Intelligent Text Processing. XVIII, 666 pages. 2008.

Vol. 4917: P. Stenström, M. Dubois, M. Katevenis, R. Gupta, T. Ungerer (Eds.), High Performance Embedded Architectures and Compilers. XIII, 400 pages. 2008.

Vol. 4915: A. King (Ed.), Logic-Based Program Synthesis and Transformation. X, 219 pages. 2008.

Vol. 4912: G. Barthe, C. Fournet (Eds.), Trustworthy Global Computing. XI, 401 pages. 2008.

Vol. 4910: V. Geffert, J. Karhumäki, A. Bertoni, B. Preneel, P. Návrat, M. Bieliková (Eds.), SOFSEM 2008: Theory and Practice of Computer Science. XV, 792 pages. 2008.

Vol. 4905: F. Logozzo, D.A. Peled, L.D. Zuck (Eds.), Verification, Model Checking, and Abstract Interpretation. X, 325 pages. 2008.

Vol. 4904: S. Rao, M. Chatterjee, P. Jayanti, C.S.R. Murthy, S.K. Saha (Eds.), Distributed Computing and Networking. XVIII, 588 pages. 2007.

Vol. 4878: E. Tovar, P. Tsigas, H. Fouchal (Eds.), Principles of Distributed Systems. XIII, 457 pages. 2007.

Vol. 4875: S.-H. Hong, T. Nishizeki, W. Quan (Eds.), Graph Drawing. XIII, 402 pages. 2008.

Vol. 4873: S. Aluru, M. Parashar, R. Badrinath, V.K. Prasanna (Eds.), High Performance Computing – HiPC 2007. XXIV, 663 pages. 2007.

Vol. 4863: A. Bonato, F.R.K. Chung (Eds.), Algorithms and Models for the Web-Graph. X, 217 pages. 2007.

Vol. 4860: G. Eleftherakis, P. Kefalas, G. Păun, G. Rozenberg, A. Salomaa (Eds.), Membrane Computing. IX, 453 pages. 2007.

Vol. 4855: V. Arvind, S. Prasad (Eds.), FSTTCS 2007: Foundations of Software Technology and Theoretical Computer Science. XIV, 558 pages. 2007.

Vol. 4854: L. Bougé, M. Forsell, J.L. Träff, A. Streit, W. Ziegler, M. Alexander, S. Childs (Eds.), Euro-Par 2007 Workshops: Parallel Processing. XVII, 236 pages. 2008.

Vol. 4851: S. Boztaş, H.-F.(F.) Lu (Eds.), Applied Algebra, Algebraic Algorithms and Error-Correcting Codes. XII, 368 pages. 2007.

Vol. 4848: M.H. Garzon, H. Yan (Eds.), DNA Computing. XI, 292 pages. 2008.

Vol. 4847: M. Xu, Y. Zhan, J. Cao, Y. Liu (Eds.), Advanced Parallel Processing Technologies. XIX, 767 pages. 2007.

Vol. 4846: I. Cervesato (Ed.), Advances in Computer Science – ASIAN 2007. XI, 313 pages. 2007.

Vol. 4838: T. Masuzawa, S. Tixeuil (Eds.), Stabilization, Safety, and Security of Distributed Systems. XIII, 409 pages. 2007.

Vol. 4835: T. Tokuyama (Ed.), Algorithms and Computation. XVII, 929 pages. 2007.

Vol. 4818: I. Lirkov, S. Margenov, J. Waśniewski (Eds.), Large-Scale Scientific Computing. XIV, 755 pages. 2008.

Vol. 4800: A. Avron, N. Dershowitz, A. Rabinovich (Eds.), Pillars of Computer Science. XXI, 683 pages. 2008.

Vol. 4783: J. Holub, J. Žďárek (Eds.), Implementation and Application of Automata. XIII, 324 pages. 2007.

Vol. 4782: R. Perrott, B.M. Chapman, J. Subhlok, R.F. de Mello, L.T. Yang (Eds.), High Performance Computing and Communications. XIX, 823 pages. 2007.